T0342064

METAL ADDITIVE
MANUFACTURING

METAL ADDITIVE MANUFACTURING

**Ehsan Toyserkani, Dyuti Sarker, Osezua Obehi Ibhadode,
Farzad Liravi, Paola Russo, Katayoon Taherkhani**
University of Waterloo
Waterloo, Ontario
Canada

Registered Office
John Wiley & Sons, Inc., 111 River Street, Hoboken, NJ 07030, USA

Editorial Office
111 River Street, Hoboken, NJ 07030, USA

For details of our global editorial offices, customer services, and more information about Wiley products, visit us at www.wiley.com.

Wiley also publishes its books in a variety of electronic formats and by print-on-demand. Some content that appears in standard print versions of this book may not be available in other formats.

Library of Congress Cataloging-in-Publication Data

Names: Toyserkani, Ehsan, author. | Sarker, Dyuti, 1983- author. |
  Ibhadode, Osezua Obehi, 1989- author. | Liravi, Farzad, 1990- author. |
  Russo, Paola, 1986- author. | Taherkhani, Katayoon, 1989- author. | John
  Wiley & Sons, publisher.
Title: Metal additive manufacturing / Ehsan Toyserkani, Dyuti Sarker,
  Osezua Obehi Ibhadode, Farzad Liravi, Paola Russo, Katayoon Taherkhani.
Description: Hoboken, NJ : Wiley, 2021.
Identifiers: LCCN 2021028894 (print) | LCCN 2021028895 (ebook) | ISBN
  9781119210788 (cloth) | ISBN 9781119210849 (adobe pdf) | ISBN
  9781119210832 (epub)
Subjects: LCSH: Additive manufacturing. | Metal powder products–Design and
  construction. | Powder metallurgy.
Classification: LCC TS183.25 .T69 2021 (print) | LCC TS183.25 (ebook) |
  DDC 621.9/88–dc23
LC record available at https://lccn.loc.gov/2021028894
LC ebook record available at https://lccn.loc.gov/2021028895

Cover Design: Wiley
Cover Image: Courtesy of Ehsan Toyserkani (top); © MarinaGrigorivna/Shutterstock

Set in 10/12pt Times LT Std by Straive, Pondicherry, India
Printed and bound by CPI Group (UK) Ltd, Croydon, CR0 4YY

C9781119210788_071021

In memory of

Professor Pearl Sullivan (1961–2020)

Former Dean of Faculty of Engineering, University of Waterloo, Canada

A true leader, an exemplary advocate for engineering education, and a great friend

Contents

Preface

Additive manufacturing (AM) promises to change the entire manufacturing enterprise over the next two decades. No longer limited to prototyping and low-volume manufacturing, AM is being adopted for *economies of scale* without compromising *economies of scope*. The need for the digitization of manufacturing, on-demand personalized manufacturing, distributed production, and rapid production in the event of crises have all elevated the position of AM in the medical and engineering sectors. AM is now a major research target for industrialized countries as they seek to regain leadership in advanced manufacturing through innovation. The global economy is on the verge of the next industrial revolution and sector after sector is pulling away from traditional, conventional production methods to engage in and utilize AM. However, this promise does come with many challenges, particularly for metal AM. Research and development activities are progressing at full steam to address multiple technical challenges, such as speed and productivity, quality assurance, standards, and end-to-end workflow.

A major skill sets gap currently hinders efforts to tackle these challenges. For companies seeking to embrace AM, this gap translates into a limited availability of expertise to draw an entry strategy to the AM industry. The wider adoption of AM will require overcoming the limited foundational understanding of AM that currently exists within the workforce. A thorough understanding of AM capabilities is necessary for technical experts to accurately communicate the pros and cons of AM to decision-makers, while preventing misconceptions and misinformation about AM capabilities. Currently, the knowledge gap is significantly impacting progress in the sector, as companies have difficulties in recruiting AM experts to help them develop effective designs for AM as well as meaningful business cases for metal AM.

This book is designed to help academia and industry move toward filling this gap. Enhancing AM skills will require the development of foundational knowledge of AM starting at the undergraduate level. To our knowledge, there is currently no textbook available that links the basics of fundamental undergraduate Engineering courses with metal AM processes. There is a clear need to customize undergraduate concepts in technical courses related to design, heat transfer, fluid mechanics, solid mechanics, and control, with respect to AM applications. Additionally, business- and management-oriented courses should include AM to facilitate the consideration of AM in conjunction with life cycle assessment and business model developments among students.

The development of this book was motivated by our desire to provide foundational material for a core undergraduate course in Mechanical and Manufacturing Engineering, and we envision its use in graduate courses as well. Universities globally are revising their curriculum to incorporate AM-related courses. This textbook may provide an introductory platform to be adopted in such courses to promote an appreciation for and grasp of AM among both undergraduate and graduate students. This book may also fill a gap for engineers working outside academia who want to appreciate AM processes by identifying links between traditional core physics and engineering concepts courses and AM. The book provides a step-by-step understanding of metal AM and a solid foundation of the topic for readers, who will subsequently be well equipped to explore AM research in greater depth.

For a broad range of readers, this book sheds light on various key metal AM technologies, focusing on basic physics and modeling. This textbook is not a literature survey, nor is it intended for readers with no engineering background. In contrast, it is an introduction to basic physical concepts and phenomena of metal AM processes and their applications. Relevant foundational concepts, such as energy deposition, powder bed fusion, and binder jetting processes, are explained in-depth and illustrated by case studies throughout the book. Additionally, two emerging processes for metal AM: material extrusion and material jetting, are described. Basic design for AM (DfAM) and quality assurance principles are also covered.

We would like to express our sincere gratitude to several people who helped in the preparation of this book. Special thanks to Francis Dibia, Ali Keshavarzkermani, Zhidong Zhang, Yuze Huang, Mazyar Ansari, Andrew Barlow, Misha Karpinska, Donovan Kwong, and Eniife Elebute, who helped us with some materials and produced some of the figures, as attributed in the book. In addition, we acknowledge all organizations, publishers, authors, and companies that permitted use of their figures, plots, and texts; they have been cited accordingly throughout the book. Last but not least, thanks to our families, who make it all worthwhile.

Like any first edition, this textbook may contain errors and typos. We openly welcome the reader's suggestions to be considered in the second edition of this textbook in which multiple problem sets for each chapter will be introduced.

January 2021   Ehsan Toyserkani, Dyuti Sarker, Osezua Obehi Ibhadode, Farzad Liravi,
                                          Paola Russo, Katayoon Taherkhani
                                                Waterloo, Ontario, Canada

Abbreviations

2D	Two-Dimensional
3D	Three-Dimensional
3DQCN	Three-Dimensional Quasi-Continuous Network
AI	Artificial Intelligence
AE	Auto-Encoder
Al	Aluminum
AL	Absolute Limits
ALE	Arbitrary Lagrangian–Eulerian
AM	Additive Manufacturing
AMCs	Aluminum Matrix Composites
AMF	Additive Manufacturing File Format
AMGTA	Additive Manufacturer Green Trade Association
ANFIS	Adaptive Neuro-Fuzzy Inference System
ANN	Artificial Neural Network
ANOVA	Analysis of Variance
ANSI	American National Standards Institute
APG	Absorptivity Profile Group
ASCII	American Standard Code For Information Interchange
ASTM	American Society for Testing and Materials
BD	Big Data
BESO	Bidirectional Evolutionary Structural Optimization
BJ	Binder Jetting
BJP	Binder Jet Printing
BP	Backpropagation
BSE	Backscattered Electrons
CAD	Computer-Aided Design
CAE	Computer-Aided Engineering
CAGR	Compound Annual Growth Rate
CAM	Computer-Aided Manufacturing

CCD	Charged-coupled device
CCT	Continuous Cooling Transformation
CDA	Constant Drawing Area
CET	Columnar-to-Equiaxed Transition
CFD	Computational Fluid Dynamics
CL	Cathodoluminescence
CMOS	Complementary Metal-Oxide Semiconductor
CNC	Computer Numerical Control
CNN	Convolutional Neural Network
COLIN	Convex Linearization
CS	Crack Susceptibility
CT	Computed Tomography
μCT	micro Computed Tomography
CVD	Chemical Vapor Deposition
CW	Continuous Wave
DAE	Differential-Algebraic Equation
DBN	Deep Belief Network
DC	Direct Current
DDA	Decreasing Drawing Area
DED	Directed Energy Deposition
DEM	Discrete/Dynamic Element Model
DfAM	Design for AM
DfM	Design for Manufacturing
DHA	Dust Hazard Analysis
DL	Deep Learning
DMLS	Direct Metal Laser Sintering
DoD	Drop-on-Demand
DoG	Difference of Gaussian
DXF	Drawing Exchange Format
EA	Electrical Arc
EAM	Embedded-Atom Method
EB	Electron Beam
EBAM	Electron Beam Additive Manufacturing
EB-DED	Electron Beam Directed Energy Deposition
EBF3	Electron Beam Freeform Fabrication
EBF^3	Electron Beam Fusion
EBM	Electron Beam Melting
EB-PBF	Electron Beam Powder Bed Fusion
EDM	Electrical Discharge Machining
EIGA	Electrode Induction Melting Inert Gas Atomization
EKF	Extended Kafman Filter
ELT	Effective Layer Thickness
EMFs	Electric and Magnetic Fields
ESO	Evolutionary Structural Optimization
FBG	Fiber Bragg Gratings
FCC	Face Centered Cubic
FCM	Finite Cell Method
FDM	Fused Deposition Modeling
FE	Finite Element
FEA	Finite Element Analysis

FEG	Field-Emission Gun
FEM	Finite Element Method
FFT	Fast Fourier Transformation
FGM	Functionally Graded Material
FGSs	Functionally Graded Structures
FIS	Fuzzy Inference System
FMC	Ford Motor Company
FN	False Negative
FP	False Positive
FS	Free Surface
GD	Gradient Descent
GM	General Motors
GMG	Geometrically Modified Group
GP	Gaussian Process
HA	Hydroxyapatite
HAZ	Heat-Affected Zone
HDR	Heating Depth Ratio
HF	Highly Filled
HIP	Hot Isostatic Pressing
HPM	Heaviside Projection Method
ICI	Inline Coherent Imaging
IDAM	Industrialization and Digitization of Additive Manufacturing
IDT	Interdigitated Transducers
IN	Inconel
IoT	Internet of Things
ISO	International Standards Organization
ISO	International Standards Organization
KF	Kafman Filter
KNN	K-nearest neighbors
LaB$_6$	Lanthanum Hexaboride
LBM	Lattice–Boltzmann Method
LCA	Life Cycle Assessment
LCF	Low Cycle Fatigue
LDED	Laser Directed Energy Deposition
LENS	Laser Engineered Net Shaping
LGA	Lattice Gas Automata
LM	Levenberg–Marquardt
LN	Large Negative
LoF	Lack of Fusion
LP	Large Positive
LPBF	Laser Powder Bed Fusion
LPM	Laser Power Monitoring
LSF	Level Set Functions
LSM	Level Set Method
LWIR	Long Wave Infrared
MAPE	Mean Absolute Prediction Error
MC	Metal Carbide
MD	Molecular Dynamics
ME	Material Extrusion
MG	Metallic Glass

MJ	Material Jetting
MMA	Method of Moving Asymptotes
MMCs	Metal Matrix Composites
MME	Metal Material Extrusion
MMP	Micro-Machining Process
MMV	Moving Morphable Voids
MOV	Main Oxidizer Valve
MPC	Metal–Polymer Composite
MPE	Maximum Permissible Exposure
MPM	Melt Pool Monitoring
MS	Multi-Speed
MSDS	Material Safety Data Sheet
MSE	Mean Squared Error
MTPS	Multifunctional Thermal Protection System
Nd	Neodymium
NDT	Non-Destructive Testing
NFPA	National Fire Protection Association
nHA	Nano-Hydroxyapatite
NHZ	Nominal Hazard Zone
Ni	Nickle
NIR	Near-Infrared
NIST	National Institute of Standards and Technology
NN	Neural Network
NS	Navier–Stokes
OCM	Optimality Criterial Method
OCT	Optical Coherence Tomography
OEM	Original Equipment Manufacturers
OPD	Optical Penetration Depth
OTLs	Orthogonal Translational Lattices
PBF	Powder Bed Fusion
PCA	Principal Component Analysis
PDF	Point Distribution Function
PF	Powder-Fed
PI	Proportional–Integral
PID	Proportional–Integral–Derivative
PMC	Polymer Matrix Composite
PMZ	Partially Melted Zone
PPE	Personal Protective Equipment
PPHT	Post-Processing Heat Treatment
PREP	Plasma Rotate Electrode Process
PSD	Particle Size Distribution
PTA-DED	Plasma Transferred Arc Directed Energy Deposition
PVD	Physical Vapor Deposition
PW	Pulsed Wave
PZT	Piezoelectric
R&D	Research and development
RAMP	Rational Approximation of Material Properties
RDM	Relative Density Mapping
REP	Rotating Electrode Process
RF	Radio Frequency

RGB	Red-Green-Blue
RLS	Recursive Least Square
RMSE	Root Mean Square Error
RNN	Recurrent Neural Networks
ROS	Reactive Oxygen Species
RTE	Radiation Transfer Equation
SAW	Surface Acoustic Wave
SD	Signal Dynamics
SDAS	Secondary Dendritic Arm Spacing
SE	Secondary Electrons
SIMP	Solid Isotropic Material with Penalization
SINH	Sine Hyperbolic Function
SL	Sheet Lamination
SLD	Super-Luminescent Diode
SLD-OCT	Super-Luminescent Diode—Optical Coherence Tomography
SLM	Selective Laser Melting
SLP	Sequential Linear Programming
SLR	Single-Lens Reflex
SLS	Selective Laser Sintering
SN	Small Negative
SOM	Self-Organizing Map
SP	Small Positive
SQP	Sequential Quadratic Programming
SRAS	Spatially Resolved Acoustic Spectroscopy
STF	Short-Term Fluctuations
STL	Standard Tessellation Language or StereoLithography
STP	Standard for the Product Data
ST-PCA	Spatially Weighted Principal Component Analysis
SVD	Singular Value Decomposition
SVM	Support Vector Machine
TCP	Topological Close-Packed
TEM	Transverse Electromagnetic Modes
TGM	Temperature Gradient Mechanism
Ti	Titanium
TiC	Titanium Carbide
Ti-HA	Titanium-Hydroxyapatite
TMCs	Titanium-Matrix Composites
TN	True Negative
TP	True Positive
TPMS	Triply Periodic Minimal Surface
TRL	Technology Readiness Level
TTT	Transformation Time Temperature
VC	Vanadium Carbides
VED	Volumetric Energy Density
VoF	Volume-of-Fluid
VTM	Virtual Temperature Method
WF	Wire-Fed
WF-EDED	Wire-Fed Electron Beam Directed Energy Deposition
XRD	X-Ray Diffraction
XRF	X-Ray Fluorescence

YAG	Yttrium Aluminum Garnet
YLF	Yttrium Lithium Fluoride
YVO4	Yttrium Orthovanadate

Nomenclature

Unless otherwise stated in the text, these symbols have the following meanings

a	Characteristic length
a	Energy bilinear function for internal energy (Chapter 10)
A	Spot area – heat source interaction area
A	Filament or nozzle cross-section area (Chapter 7)
A_{at}	Attenuated area
A_c	Cross-section area
A_{jet}^{liq}	Intersection of melt pool area on substrate and powder stream
A_{jet}	Cross-section of powder stream on substrate
A_G	Property of filament material
A_{ij}, B_{ij}	Einstein coefficients
A_S	Surface area
b	Melt pool depth
b	Bias (Chapter 11)
B	Size of gap (Chapter 7)
B	Magnetic field
$\boldsymbol{B}$	Differential shape function matrix (Chapter 10)
c	Speed of light
c_p	Heat capacity
cyl	Function based on Bessel functions
c_s	Speed of sound in the fluid
C	Duty cycle
C	Compliance (Chapter 10)
Ca	Capillary number
C_s	Solid composition
C_L	Liquid composition
C_0	Nominal alloy composition or solute concentration
d	Spot size
d	Euclidean distance (Chapter 11)
d_0	Droplet diameter
$d_{3,2}$	Surface mean particle diameter
d_{con}	Semispherical droplet
D	Laser beam diameter
$\boldsymbol{D}$	Material matrix
D_f	Diffusion constant
D_{ijmn}	Tensor of elastic coefficients
D_L	Solute diffusion coefficient

$D_\emptyset$	Diffusion coefficient
e	When subscript or superscript, signifies a variable in its elemental form
e_i	Vector pointing
E	Laser beam energy
E	Electric field (Chapter 5)
$\boldsymbol{E}$	Young's modulus matrix (Chapter 10)
E_a	Energy of activation
E_b	E-beam energy
E_i	Input laser energy
E_{kin}	Kinetic energy
E_r	Reflected energy
E_t	Transmitted energy
E_i	Energy levels (Chapter 3)
$E_{specific}$	The energy enters the substrate from the surface in DED
f	Frequency
f	Volume fraction (Chapter 10)
f_s	Fraction of solid
f_L	Fraction of liquid
$f(R)$	Function of surface roughness
$f_i(x, t)$	Density of particles moving in the e_i direction
$f_i^{eq}(x,t)$	Equilibrium distribution
F	Force
F_0	Fourier number
$\mathcal{F}_0$	Zero-order Bessel function of its first kind
$\mathcal{F}_1$	The first-order Bessel function of the first kind
F^{Cap}	Capillary force
F_{st}	Surface tension force
F_{th}	Thermal stress load
F^{Wet}	Wetting force
g	Gravity
g_i	Effect of external forces
G	Temperature gradient
G_S	Gibbs free energy for solid
G_L	Gibbs free energy for liquid
G_z	Graetz number
ΔG	Total Gibbs free energy change
ΔG^*	Critical free energy change
ΔG_V	Free energy change per unit volume
h	Plank's constant (Chapter 3)
h	Height of a hexahedral element (Chapter 10)
h_c	Heat convection coefficient
h_i	Convective heat loss (cooling) coefficients
h_a	Average heat transfer coefficient of convection
h_w	Distance between the nozzle and surface
h_r	Radiative heat transfer coefficient

h_{min}	Minimum radius of the liquid column
H	Barrel length or height of track
H^*	Height of the melt polymer
ΔH	Enthalpy difference
HDR	Heating depth ratio
I	Intensity: energy per area
I	Current in electron beam (Chapter 5)
$I(x)$	Indicator function (Chapter 11)
I_0	Intensity scale factor or initial intensity
I_b	Beam current in the electron-beam process
J	Free electron current
k	Propagation factor
k	Equilibrium distribution coefficient (Chapter 8)
k	Solute partition coefficient (Chapter 9)
K	Thermal conductivity
$\mathbf{K}$	Global stiffness matrix (Chapter 10)
$\mathbf{K}_c$	Conductivity matrix (Chapter 10)
$\mathbf{K}_h$	Connective matrix (Chapter 10)
K^*	Modified thermal conductivity
K_0	Bessel function of the second kind and zero order
l	Layer thickness
l	Load linear function for external work (Chapter 10)
$\mathbf{L}$	Transformation matrix
L	Link intensity
L_c	Characteristic length based on domain size
L_f	Latent heat of fusion
L_l^p	Laguerre polynomial of order p and index l
m	Atomic mass
m_b	Mass of the ball (Chapter 9)
m_b	Deposited binder (Chapter 6)
m_p	Particle mass
m_p	Mass of the bound powder (Chapter 6)
m_f	Fluid mass
$\dot{m}$	Mass flow rate
M^2	Beam quality factor
Ma	Marangoni number
n	Reflection's index
N	Matrix of shape function for mesh element
N_i	Number of atoms or electrons per unit volume in the energy levels
N_L	Amount of atoms per unit volume of liquid
N_{lx}	Neumann function
N_S	Amount of atoms per unit volume of solid
Nu	Nusselt number
N_{th}	Shape vector for thermal expansion
Oh	Ohnesorge number

p	Constant characteristic of the material (Chapter 4)
p	Pressure
p	Penalty value (Chapter 10)
p_c	Capillary force
P_f	Packing fraction
P	Power
P_{at}	Attenuated laser power by particles
P_{el}	Power of electrical motor in FDM system
P_l	Net/average laser power
P_{peak}	Peak power per pulse
P_{on}	On the state of the laser power
P_{off}	Off state of the laser power
P_{tot}	Total delivered beam power
P_e	Peclet number
Pr	Prandtl number
PR	Packing density of the powder
PW	Pulsed wave laser
$P(\infty)$	Value of the extrudate property after an infinite healing time
q	Heat flux
Q	Power generated per unit volume (in all chapters except Chapter 3)
Q	Beam propagation factor (Chapter 3)
Q_c	Total energy absorbed by the substrate
Q_{ext}	Extinct coefficient
Q_l	Laser energy
Q_{rs}	Reflected energy from the substrate
Q_L	Latent energy of fusion
Q_v	Volumetric flow rate
r	Radius of nucleus (Chapter 8)
r^*	Critical nucleus radius
r_b	Ball mill radius
r_f	Filament radius
r_{0l}	Beam radius of the waist
r_{jet}	Radius of powder spray jet
r_l	Beam spot radius on the substrate
r_s	Powder stream diameter
r_p	Powder particle radius
R	Reflectivity
R	Solidification rate where it is referred to
R_c	Clad surface curvature
R_{cur}	Radius of curvature
Re	Reynolds number
Re^*	Property-based Reynolds number
R_h	Heat load by convection
$R_h(t)$	Intrinsic healing function
R_L	Local growth rate

R_N	Nominal growth rate
R_r	Radius of the actuating motor
R_{pore}	Effective pore diameter in the bed
R_q	Heat load by surface conduction
R_Q	Heat load by volume conduction
s	Hatch spacing
s_0	Specific surface area
S	Binder saturation
S_{ij}	Lateral distance between neuron i and j (Chapter 11)
S_{max}	Spreading ratio
S_{meas}	Amount of signal (Chapter 11)
$S_\emptyset$	Source term corresponding to Ø
$S(\phi)$	Shape factor
SS	Scan speed
S	Strain rate deformation tensor
ΔS	Expansion of surface area
t	Time and/or laser interaction time
t^*	Dimensionless time
$t_c = t_\mu$	Viscous time
t_{CDA}	Corresponding penetration time
t_f	Solidification time
t_I	Inertial-capillary time
t_V	Viscous-capillary time
T	Temperature
T_α	Reference temperature
T_{ave}	Average temperature
T_{in}	Filament temperature
T_g	Glass transition temperature
T_{out}	Outlet temperature
$T_{\dot{m}}$	Temperature of the liquefier wall
T_d	Drying time
T_e	Equilibrium temperature
T_l	Liquidus temperature
T_0	Ambient temperature
T_m	Materials melting point
T_p	Maximum temperature
ΔT	Undercooling temperature
ΔT_{tot}	Total undercooling temperature
ΔT_C	Undercooling temperature: solute diffusion
ΔT_T	Undercooling temperature: thermal diffusion
ΔT_K	Undercooling temperature: attachment kinetics
ΔT_R	Undercooling temperature: solid–liquid boundary curvature
TEM_{pl}	Gaussian–Laguerre transverse electromagnetic modes
U	Beam velocity
U	Travel velocity vector

U	Global displacement vector (Chapter 10)
U_p	Particle velocity vector
U_s	Rate of solidification
v	Scanning speed
v_c	Collision velocity
v_j	Jet velocity
v_p	Velocity of the particle
v_{print}	Velocity of the print head
V	Volume of melt pool
V	Design volume (Chapter 10)
V_a	Acceleration voltage
V_S	Volume of nucleus
VED	The energy enters the substrate from the surface in LPBF
w	Track or melt pool width
w	Neuron weight (Chapter 11)
w_i	Weight factor
W	Laser pulse width
We	Weber number
X_{s+c}	Weight percent of element X in the total surface of the clad region
X_c	Weight percent of element X in the powder alloy
X_s	Weight percent of element X in the substrate
y	Dendrite arm spacing (Chapter 8)
z	Distance from the surface
z_0	Waist location with respect to an arbitrary coordinate along the propagation axis
Z	Printability of a liquid
Z_h	Heat penetration depth

Greek Symbols

α	Thermal diffusivity
α_t	Coefficient of thermal expansion
β	Absorption factor
β	Absorption factor
β_p	Powder particles' absorbed coefficient
β_w	Substrate laser power absorptivity
γ	Surface tension
γ	Net electron beam energy (Chapter 5)
γ_E	Specific surface energy
γ_{SL}	Solid–liquid interfacial free energy
γ_{SV}	Solid–vapor interfacial energy
γ_{LV}	Liquid–vapor interfacial energy
$\dot{\gamma}$	Shear rate
Γ	Torque of electrical motors in FDM
Γ	Surface function (Chapter 10)
δ	Solid/liquid interface thickness
δ	Dirac delta function (Chapter 6)

ε	Total strain
ε_c	Cooling rate
ε_t	Emissivity
ε^M	Mechanical strains
ε^T	Thermal strains
ε_p	Equivalent plastic stress
ε_0	Vacuum permittivity
$\boldsymbol{\varepsilon_m}$	Mechanical strain
$\boldsymbol{\varepsilon_{th}}$	Thermal strain
Σ	Covariance matrix
η	Dynamic viscosity
η	Powder catchment efficiency (wherever it refers to throughout chapters)
η	Numerical damping coefficient for OCM (Chapter 10)
η	Learning rate (Chapter 11)
η_e	Absorption efficiency for electron beam
η_d	Dynamic viscosity
η_p	Powder catchment efficiency
θ	Representing different angles based on figures
θ	Wetting angle (Chapter 2)
θ	Far-field divergence angle (Chapter 3)
θ	Dimensionless temperature in numerical models (Chapter 7)
θ_{jet}	Angle between powder jet and substrate
θ_d	Dynamic wetting angle
θ_{eq}	Steady-state angle
Θ	Dimensionless temp in analytical models
λ	Wavelength
λ	Lagrange multiplier (Chapter 10)
λ_n	Roots of zero-order Bessel function of its first kind
μ	Viscosity
μ	Membership function (Chapter 11)
υ	Frequency (Chapter 3)
υ	Kinematic viscosity
ρ	Density
ρ_b	Density of binder
ρ_{pb}	Powder bed density
ρ_c	Density of melted powder alloy
ρ_s	Density of substrate material
ρ_s	Packing density of the pores (Chapter 6)
σ	Stefan–Boltzmann constant
σ	Covariance (Chapter 11)
σ_c	Charge density
σ_{ij}	Elastic stress
τ	Thermal time constant
τ_c	Dimensionless capillary time
ϕ	Different label for angles as indicated in the associated figures

ϕ	Powder bed porosity
ϕ	Shape factor
Φ	Interpolation function (Chapter 10)
$\phi(x, y)$	Level-set equation
ϕ_{tap}	Tapped porosity
$\phi(t)$	Rate of heat liberation in a continuous point source
$\varnothing$	Angle of incidence
Ψ	Strain energy
ω	Spinning speed
ω	Relaxation factor (Chapter 11)
ω_i	Strength of anisotropy
ω_r	Angular velocity of the actuating motor
Ω	Substrate surfaces or melt pool boundary and material domain

1

Additive Manufacturing Process Classification, Applications, Trends, Opportunities, and Challenges

---

Learning Objectives

At the end of this chapter, you will be able to:

- Understand the standard definition of additive manufacturing (AM) and seven standard classes of AM processes
- Gain basic knowledge on AM market size
- Gain basic knowledge of opportunities, threats, and trends in the AM industry
- Gain insight into applications of metal AM

1.1 Additive Manufacturing: A Long-Term Game Changer

Additive manufacturing (AM), also known as 3D printing, is a layer-by-layer fabrication technology "poised to be one of the most valued forms of manufacturing in history" [1]. AM is becoming a major research target for industrialized countries as they seek to regain leadership in manufacturing through innovation. The global economy is on the verge of the next industrial revolution, known as "Industry 4.0", i.e. the fourth industrial revolution. Sector after sector is pulling away from traditional/conventional methods of production in order to engage in and utilize AM. This fresh manufacturing method has garnered a great deal of public curiosity

---

Metal Additive Manufacturing, First Edition. Ehsan Toyserkani, Dyuti Sarker, Osezua Obehi Ibhadode, Farzad Liravi, Paola Russo, and Katayoon Taherkhani.
© 2022 John Wiley & Sons Ltd. Published 2022 by John Wiley & Sons Ltd.

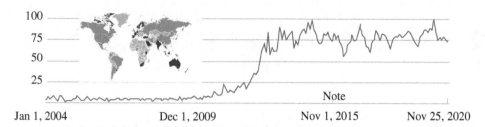

Figure 1.1 Global public interest trends for "3D Printing". Source: Extracted from "Google Trends" on 25 November 2020.

and international publicity. Every week brings news of novel and astounding AM innovations. This technology, which builds up objects by layers, has piqued the industrialized world's curiosity and imagination. This public interest was triggered around 2013–2014, as shown in Figure 1.1, in which the Google Trend suggests that a public interest in "3D Printing" has been increased more than 50-fold from 2012 to 2016. The trend stays very much the same from 2017 to 2020, but it is the highest in Google Trend compared with counterpart advanced manufacturing technologies. This figure also sheds some light on the geographical distribution of the public interest. North America, Europe, and Australia are among the regions with the highest number of news and interests related to "3D printing" technologies.

In November 2015, the United Nations encouraged countries to invest in AM technology, forecasting *major expansions in business and explosive economic growth* and comparing 3D printing to the most influential technologies of the past, such as airplanes, antibiotics, and semiconductors [2]. AM technology has been recommended for substantial research investment [3]. Major initiatives have been announced in Singapore, Australia, the United States, Canada, and Europe [4]. McKinsey estimates that by 2030, the global economic impact approaches $550 billion per year [5].

As worldwide interest in AM escalates, numerous industries are taking steps to integrate AM technologies into their applications and offerings. Many industries, including aerospace, medical, automotive, tooling, energy, natural resources, consumer, defense, etc. have started to embrace the benefits of AM processes.

For years, AM has been identified as a technology addressing "economies of scope" through customization, prototyping, and low-volume manufacturing. However, in recent years, AM has been deployed for "economies of scale," i.e. mass production, without compromising the economies of scope. This advancement from prototyping to serial production has created many research and development opportunities, especially for quality management and certification. Richard D'Aveni writes: *Today, additive manufacturing is achieving economies of scale in a variety of ways– and doing so without sacrificing economies of scope … No longer limited to product prototypes, customized one-offs, or specialized items made in small quantities, AM is now beginning to take over the kinds of mass manufacturing that have long dominated the industrial economy* [6].

AM has been considered a platform to convert digital models to physical parts in a short chain of processes, a platform facilitating a rapid move from "Art" to "Part" in a fancy analogy. The process starts with a digital model that reflects the desired design. Preprocessing is needed on the file depending on materials, applications, and AM processes. A proper AM process must be chosen that fulfills the material and application of interest. After the layered manufacturing is

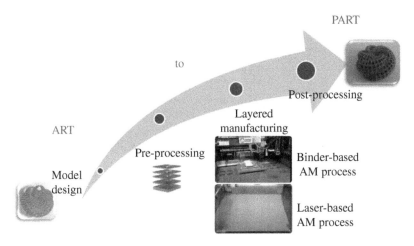

Figure 1.2   AM chain, enabling physical parts from digital design.

completed, post-processing may be needed to arrive at the physical part eventually. Figure 1.2 shows the AM process chain schematically.

The history of AM is very well connected to human civilization. The idea of "layered manufacturing" has been around as the activities of human kind have been recorded. More than five million blocks of limestone were put together "stone-by-stone" by ancient Egyptians to build the pyramids using human-made stones. Layered cakes were formally introduced in Maria Parloa's Appledore Cook Book, published in Boston in 1872, which contained one of the first layer-by-layer made cake recipes. Francois Willeme developed a method called "photographic sculpture" in 1859. In this method, 3D models of human subjects could be extracted using 24 cameras placed at different angles. Joseph E. Blanther patented an apparatus in 1892, where the apparatus uses a layering idea to create 3D topographical maps. The film industry also fancied the idea of AM processes through science fiction that speculated on "replicating" technology. The term "replicator" was used in the series "Star Trek: The Next Generation." In the animated series in 1974 and in the episode entitled "The Practical Joker"!, there were scenes where food could be requested to be replicated.

However, modern AM history starts in 1980s when computer features provided unique opportunities for people to think out of the box and introduce the first AM processes. AM first emerged in 1987 with the idea of Charles Hull, who successfully acquired a patent for his Stereolithography Apparatus, a process that solidifies thin layers of photopolymer using a laser beam. This patent opened venues for a few more 3D printing technologies in the late 1980s. Selective Laser Sintering (SLS), a method that uses a laser beam to sinter metal powder particle to form a solid object, was developed by Carl Deckard at the University of Texas in Austin.

S. Scott Crump and Lisa Crump developed another 3D printing technology based on material extrusion in the 1980s in which a material is heated and extruded through a nozzle to create an object layer by layer. The technology was called Fused Deposition Modeling (FDM), and it is now known as material extrusion. FDM is the most commonly used AM technology as of 2020.

In the 1990s, many new technologies, including Direct Metal Laser Sintering or so-called Selective Laser Melting (SLM) and Binder Jetting (BJ), were developed. The binder jetting was developed in 1993 in the Massachusetts Institute of Technology (MIT) based on an inkjet process to create 3D objects by gluing metal or ceramic powder particles.

Cost of AM machines was starting to decrease in 2000s, helping the technology to be more accessible and adopted. New AM technologies such as material jetting have been emerging and innovation in this field is at high. The internet has continued to increase accessibility to AM when open-source online libraries for AM digital models are growing rapidly. The 2020 pandemic has proved that such accessibility to digital files and the availability of inexpensive 3D printers can help the community to fill the gap in the medical supplies. There have been many stories over the internet indicating that regular people printed parts of face shields in the period that the supply chain was interrupted due to the pandemic. This feature is now encouraging industry to pay more attention to localized manufacturing to be able to address on-demand manufacturing with minimum dependencies to foreign countries.

When the skill gap is one of the major issues in the adoption of AM to industry, AM-based courses have started to be incorporated in schools, colleges, and universities curriculums to meaningfully teach AM to youth. Knowledge of AM concepts, technology, and software is a crucial element for this paradigm shift, and efforts are underway to fully integrate them in educational platforms.

1.2   AM Standard Definition and Classification

The American Society for Testing and Materials (ASTM) and International Standards Organization (ISO) have jointly established two committees to develop standards for the AM industry. The joint committees have proposed the definition of AM based on an active standard of ASTM ISO/ASTM52900, developed by Subcommittee F42.91, as

"Additive manufacturing (AM) is process of joining materials to make parts from 3D model data, usually layer upon layer, as opposed to subtractive manufacturing and formative manufacturing methodologies."

This standard also elaborates on the functionality of manufactured parts as

The functionality of a manufactured object is derived from the combination of the object's geometry and properties. In order to achieve this combination, a manufacturing process is made up of a series of operations and sub-processes that bring the shape of the intended geometry to a material capable of possessing the desired properties. The shaping of materials into objects within a manufacturing process can be achieved by one, or combinations of three basic principles: Formative shaping, subtractive shaping and additive shaping.

The same standard categorizes AM into seven processes as:

- Binder Jetting
- Directed Energy Deposition
- Material Extrusion
- Material Jetting
- Powder Bed Fusion
- Sheet Lamination
- VAT Photopolymerization

The same standard briefly defines each of these AM processes:

Binder jetting, an additive manufacturing process in which a liquid bonding agent is selectively deposited to join powder materials.

Directed energy deposition, an additive manufacturing process in which focused thermal energy is used to fuse materials by melting as they are being deposited. Note, "Focused thermal energy" means that an energy source (e.g. laser, electron beam, or plasma arc) is focused to melt the materials being deposited.

Material extrusion, an additive manufacturing process in which material is selectively dispensed through a nozzle or orifice.

Material jetting, an additive manufacturing process in which droplets of build material are selectively deposited. Note: Example materials include photopolymer and wax.

Powder bed fusion, an additive manufacturing process in which thermal energy selectively fuses regions of a powder bed.

Sheet lamination, an additive manufacturing process in which sheets of material are bonded to form a part.

Vat photopolymerization, an additive manufacturing process in which liquid photopolymer in a vat is selectively cured by light-activated polymerization.

From these processes, almost all have been used for metal AM. However, industry has embraced the following processes for metal manufacturing more widely (note that the sequence shows the domination of each process in the market): (i) Powder Bed Fusion, (ii) Directed Energy Deposition, (iii) Binder Jetting, (iv) Material Extrusion, (v) Material Jetting, and (vi) Sheet Lamination. The first three are the most popular technologies for metal AM. In Chapter 2, these processes are explained in detail.

In all these processes, when it comes to metal AM, several main concepts are shared: (i) motion systems, (ii) energy or binding sources, and (iii) material delivery mechanisms. Chapter 3 will explain the main components used in the three major metal AM processes.

1.3 Why Metal Additive Manufacturing?

From the early days of AM, the technology has been evolved substantially. Advancement after advancement in AM is announced almost daily. While AM has been substantially changed from 30 years ago, it will be unrecognizable form the current status in 2030. But why such enthusiasm exists in industry and academia to try to understand metal AM and work hard to address its challenges and adopt it to their products? There are several main factors for this motivation:

On-demand low-cost rapid prototyping: One of the major applications of AM is the manufacture of functional prototypes. Such prototyping usually carries at a fraction of the cost compared with other conventional processes and at usually non-disputable speeds. This rapid turnaround usually accelerates the design cycle (design, test, revision, and redesign). Products such as molds that would require more than 4–6 months to develop can be ready for operation in 2–3 months if being made by AM.

Simpler supply chain for effective low-volume production: Low-volume niche production usually requires more investment. Due to this issue, conventional manufacturers usually do not embrace low-volume production; however, AM companies can level this niche. Many time-consuming and expensive manufacturing techniques can be superseded by rapid and efficient metal AM for low-volume manufacturing. However, for mass production, AM is still lagging behind conventional techniques such as casting and forging. One of the reasons of this feature is that AM usually needs a simpler supply chain with fewer players involved. Although AM's supply chain is still under development by industry, it is expected to see more and more low-volume manufacturing by AM as the supply chain is reliably in place. Lowering the AM material costs will be another factor to foster AM adoption for low-volume manufacturing when the technology moves toward series manufacturing eventually. Initial costs are usually lower for AM than conventional methods because of the minimum need for tools and jig/fixtures needed for assembly costs. In conventional manufacturing (e.g. casting), each part needs a unique mold. To compensate for the cost of tools for each identical part, the number of products should be very high. AM does not usually need any specialized tooling; therefore, there are essentially no initial costs (called fixed costs too). Due to this saving, it is possible to get to the breakeven point sooner and make profits even with lower volumes.

Geometric complexity may be free: AM enables the fabrication of complex shapes that cannot be produced by any other conventional manufacturing methods (Figure 1.3). The additive nature of AM offers an opportunity where geometric complexity may not come at a higher price. Unlike conventional methods, AM offers a platform for "design for use" rather than "design for manufacture." Parts with complex or organic geometry optimized for performance may cost lower; however, attention must be given to the fact that not all complex parts and geometrical features are manufacturable by AM. Process constraints in metal AM (e.g., overhanging features) may cause issues in terms of residual stresses and defects, thus complexity may not come with full freedom!

Lightweighting: Manufacturers have been trying to fabricate both greener and more economical products. Lightweight components provide two goals: (i) the parts with reduced weight

Figure 1.3   Complex parts made by AM. The spherical nest has three spheres inside.

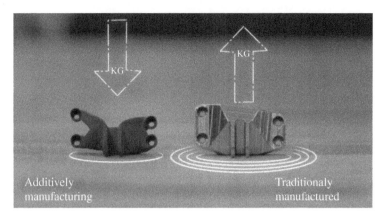

Figure 1.4 Lightweight structure made by AM. In this typical bracket, the weight has been reduced by 60% when the mechanical strength and stiffness remain the same.

take less energy to move; thus, energy consumption drops, and (ii) less raw materials are used. Both reasons indicate that the production of lightweight components has a positive impact on costs, resources, and the environment. Resource prices are virtually going up worldwide; thus, reducing material consumption is vitally important for product development. AM is nicely linked with topology optimization, making it possible to design and manufacture high-strength but lightweight structures, where conventional manufacturing processes fail to do so. Chapter 10 highlights how topology optimization and lattice structure design handshake with AM to make the fabrication of lightweight structures possible. Many lightweight but high-strength components are widely used in the aerospace industry. Any weight reduction is translated into a considerable amount of money saved in terms of the part price itself as well as fuel consumption (Figure 1.4).

Parts consolidation: Mechanical assemblies are common in industrial products. In complex mechanical machines, there are more than tens, hundreds, or even thousands of components that are either welded, or bolted, or press-fit to each other. Parts consolidation offers many advantages due to the reduction of the number of individual parts needed to be designed, manufactured, and assembled to form the final system. Part consolidations offer multiple benefits: (i) design simplification; (ii) reduction of overall project costs; (iii) reduction of material loss; (iv) reduction of weight; (v) reduction of overall risk where the number of risks associated with too many suppliers of individual parts drops; (vi) better overall performance, as it enables geometries that are desirable but cannot be made with conventional manufacturing.

AM allows for parts consolidation, even removing the need for assembly in some cases. Several applications of AM have obvious benefits for fostering product performance through lightweighting/consolidation without compromising high strength are: optimizing heat sinks to dissipate heat flux better, optimizing fluid flow to minimize drag forces, and optimizing energy absorption to minimize energy consumption. Figure 1.5 shows an example conducted by GE Additive. Almost 300 parts were consolidated in one part for the A-CT7 engine frame. This consolidation also reduced the seven assemblies to one where more than 10-pound weight was chopped off.

The A-CT7 engine mid frame

7 assemblies to 1
~300 parts to 1
>10 lbm weight reduction

Figure 1.5 Consolidation of around 300 parts to one part printed by AM. Source: Courtesy of GE Additive, open access [7], reproduced under the Creative Commons License.

Functionally graded materials (FGMs) and structures (FGSs): The integration of multiple advanced materials into one component is one of the most rapidly developing areas of AM technology. The capability to create multiphase materials with gradual variations in compositions is one of the important features of AM. During the layer-by-layer step of AM processes, the material composition can gradually be altered to obtain the desired functionality. AM also enables the development of FGSs with a single-phase material, where the density is gradually changed through the addition of cellular/lattice structures; and embedding objects (e.g. sensors) within structures. Among AM processes, DED is the most promising technology to develop such structures, where different powders can be switched insitu to develop desired composition and alloys. Figure 1.6 shows different FGMs that can effectively be developed by DED. Figure 1.7 shows a cutting tool with an embedded fiber optic, as an FGS, developed by an AM-based process.

Parts with conformal cooling channels for increased productivity: Cooling systems play a vital role in the productivity and performance of many parts. For example, in an injection molding process, the cooling period of a production cycle counts for more than 40% of cycle time. If this period drops by means of taking the heat out of the mold, the productivity increases dramatically. In an active antenna, developing conformal channels will be very important as the generated heat can be dissipated from the zone much effectively, not to affect the antenna performance. With AM, designers can have much more freedom to incorporate conformal cooling channels into their designs that facilitates uniform cooling over the entire surface. Sub-conformal channels can be included in the optimization process. Figure 1.8 shows a design of an insert used in molds. The design includes a conformal cooling channel wherein the support cells are used to enhance the heat transfer.

Parts repair and refurbishment: Machining errors or last-minute engineering changes can affect on-time delivery of tooling and potentially impact the introduction date of a new product. AM, especially DED processes, can be applied as a safe technology to repair tooling, especially on critical contacting surfaces. AM increases tool life and, in many cases, can save a high-value tool that would otherwise need to be replaced. Figure 1.9 shows an LDED process used in the in-situ repair of turbine blades.

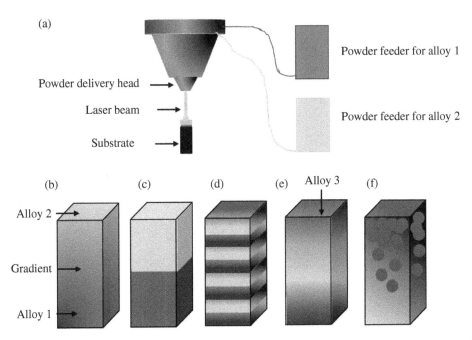

Figure 1.6 Functionally graded materials (FGMs); (a) Laser DED with multiple powder feeders is widely used for FGMs; (b) FGM with two alloys with gradual interface (c) FGM with two alloys with one sharp interface, (d) FGM with multiple interfaces, (e) FGM with three alloys, (f) FGM with selective deposition of secondary alloy. Source: Redrawn and adapted from [8].

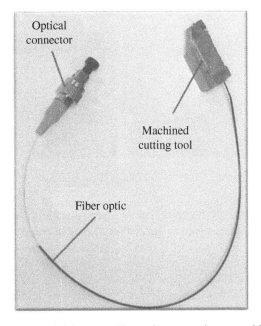

Figure 1.7 A fiber optic embedded in a metallic cutting part using a combined AM-based process. Source: Republished with permission from Elsevier [9].

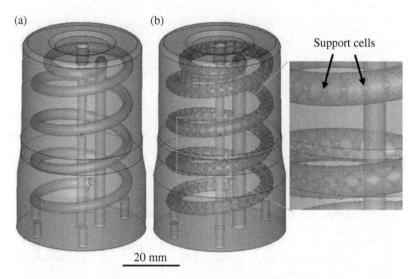

(a)   (b)   Support cells

20 mm

Figure 1.8 A mold insert with (a) conformal cooling channels, (b) conformal and lattice structures to improve heat dissipation. Source: Republished with permission from Elsevier [10].

Figure 1.9 LDED used to rebuild turbine blades. Source: Courtesy of Rolls Royce [11].

A solution to supply shortages in critical crises: Interruption to the global supply chain during crises can be catastrophic to the health and well-doing of society. The 2020 pandemic is evidence of how the supply chain of medical supplies could have been affected. AM processes can provide remedies during such crises. In March 2020, the 3D printing community got together to help in making medical devices during a very hectic period when medical centers were suffering from a lack of personal protective equipment (PPE). For example, the 3D printing community of the Waterloo region of Canada responded to the call from a local company called InkSmith to locally produce parts of PPEs. The call was received very well, and in a short period of time, more than 200 000 face shields were made and donated to hospitals until the global supply chain started to provide supplies seamlessly. The same model can be used in any crisis. The governments should proactively develop a workflow for critical crises when the AM community can be of tremendous help.

An effective solution to localized manufacturing: The 2020 pandemic has changed the world forever. The globalization idea has been hammered, and governments are now incentivizing local manufacturing to boost not only local communities but also be ready for future crises. AM will play an important role in the realization of local manufacturing. Besides, as reported in [12], more innovators and user entrepreneurs are turning into on-demand manufacturers, utilizing the opportunities of access to flexible local production. Further advancements in AM will shift the production of innovative products from a centralized to a local production platform. A shift toward local manufacturing has been started.

Health and humanitarian benefits: The medicine has been benefiting from AM for almost 20 years. Prosthetics and implants customized and tailored for specific patients are already being manufactured by AM. Many developments on the fabrication of soft tissues, for the realization of the fabrication of organs as well as a host of other personalized medical items and sensors, are underway. It has been proved that the use of precise AM replicas would reduce surgery time significantly for many patients.

Alongside the obvious benefits to industry and medicine, AM is explored as a potential aid to humanitarian issues. Intensive research is already taking place in 3D printed food and 3D printed houses to assist in the provision of food and homes/shelters in areas of humanitarian need.

Developing countries can benefit significantly from AM. In general, AM narrows the path for less developed economies to industrialize [13].

1.4 Market Size: Current and Future Estimation

AM market was around US$11 billion in 2019. The worldwide market for AM hardware, software, materials, and services is anticipated to exceed $40 billion by 2027. The industry is expected to grow at a compound annual growth rate (CAGR) of 26.4% between 2020 and 2024 [14]. Figure 1.10 shows the total AM market size under each category from 2014 to 2027. It should be noted that the aforementioned market analysis does not count for the potential adverse effect of 2020 pandemic.

Metal AM is one the fastest growing segment in the world. As shown in Figure 1.11, metal AM is one of the fastest growing sectors of the AM industry. The market size for metal AM was around US$2.4 billion in 2019 including system, material, and service sales. A 27.9% growth in the global revenue from 2019 to 2024 is estimated by the supplier side [15]. Customers, however, expect a weaker yearly growth that results in a total market size of US$7.0 billion in 2024. In fact, the annual growth in the revenue of metal AM materials has been higher than that of photopolymers, polymer powders, and filaments between 2013 and 2018 [4]. The systems that were dominant in the market of metal AM include powder bed fusions (mainly with the laser heat source) and powder-fed laser directed energy deposition as well as new technologies such as binder jetting and cold spray used for AM. Most material sales include metal powders and wire feedstock.

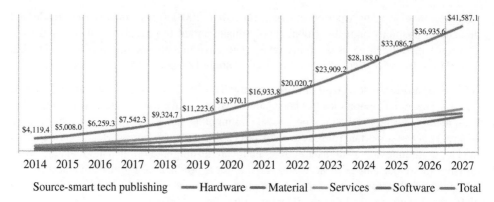

Figure 1.10 Total AM market size by segment that includes all technologies (metals and plastics) from 2014 to 2027 as forecasted by SmarTech Publishing. Source: Open access and Reproduced from [14].

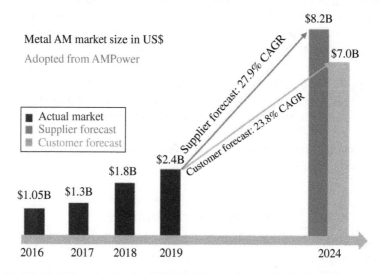

Figure 1.11 Metal AM market size in AMPower Report. Source: Redrawn and adapted from [15].

In terms of market share, the aerospace industry covers the largest share, followed by the medical sector [15]. The aerospace industry profits from the turbine, helicopters, and jet-engine components fabrication as well as new space applications such as rocket engines, attracting large venture capital worldwide, especially in the United States.

1.5   Applications of Metal AM

AM is on the path to offer disruptive solutions for mass customization, digital fabrication, and decentralized manufacturing as required by the current industrial revolution as a technology for rapid production of prototypes and evolving to reliable techniques for fabrication of

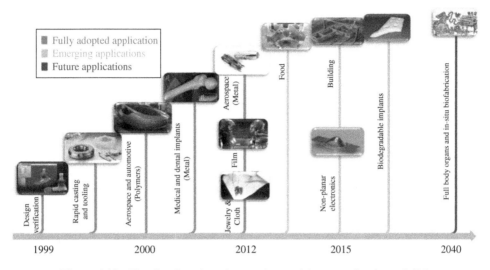

Figure 1.12 Timeline for adopted, emerging, and future applications of AM.

customized, low-volume parts we are in known as Industry 4.0. Machine developers are trying to achieve this goal by improving process throughput, build volume, process control, and available raw materials. As more and more original equipment manufacturers (OEMs) and service enterprises start to adopt AM, this transition speeds up. Due to this active progress in the AM industry, the following statement is well received as it reads, "AM is a disruptive technology that is disrupting itself regularly."

In general, AM processes started to adapt to different sectors as early as 1990. Figure 1.12 shows this timeline and applications that have embraced AM technology.

As mentioned at the beginning of this chapter, among AM processes, three classes of PBF, DED, and BJ, are integrated into mainstream metal manufacturing widely. PBF can include laser- and electron-beam-based processes where the heat source for DED can be laser, electron, arc, and plasma. The material delivery system in DED can be either powder-fed or wire-fed. As per the applications, these AM classes can be used in various applications depending on the size, complexity, and resolution of components. Figure 1.13 schematically shows these processes' applications based on the component size and accuracy/complexity associated with the component. As seen, for large-size components, the powder-fed and wire-fed DED processes are the most applicable processes, where the printed part may not require high resolution with complex features. In contrast, PBF and BJ can be used for smaller metal parts with higher resolution and complexity. In contracts, it has to be noted that the density of parts produced by DED is almost perfect when in BJ, the density cannot be high. PBF is a middle process that can produce relatively large parts up to 50 cm with high resolution and high complexity using the current state of PBF technology in 2020.

Metal AM has a track record of providing innovative solutions leading to reduced lead time, faster design-to-market cycles, or production of previously impossible parts in many industrial sectors. Chief among them are medical, dental, aerospace, defense, energy, resources, and automotive industries. In the following, we will touch based upon the status of metal AM

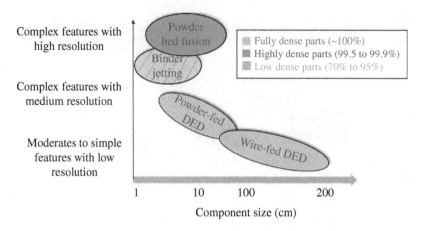

Figure 1.13   Most important metal AM processes versus part size, complexity, and resolution needed.

applications in different industrial sectors and review some of the state-of-the-art applications of metal AM in each industry.

1.5.1   Medical and Dental

The medical industry was one of the early adopters of AM for the fabrication of not only metal parts, but also ceramics, polymers, and FGMs. Metal AM has been used to produce medical devices and tools, surgery guides and prototypes, implants, prosthetics, orthotics, dental implants, crowns, and bridges from biocompatible metals such as various titanium, tantalum, and nickel alloys. These are among the main families of metal AM materials with a somewhat well-established process-property record that can be leveraged by companies, clinics, and hospitals that will use AM in the future. The design freedom in the production of complex parts with internal pores and cavities facilitating the growth of cells and the production of patient-specific parts based on the imaging of patients' anatomy are the main reasons that the medical and dental industry has shown such a high interest in AM. With personalized healthcare on the horizon, it is only expected that the scope of using AM in these sectors would increase. Due to the high precision required to produce medical parts, PBF processes are the dominant AM techniques in this sector. In addition, porosity and selective stiffness are of major importance to medical devices. Thus, BJ is playing an important role as it can produce implants with controlled porosity. Next-generation customized porous implants aim to better integrate with the surrounding bone, as they improve body fluid/cell-laden permeability. Functionally gradient porous implants/scaffolds are being designed based on interconnected triply periodic minimal surfaces (TPMS); see Chapter 10.

Figure 1.14 depicts multiple dental and orthopedic devices developed by multiple companies and centers as attributed in the figure caption. Behind all these medical devices, there is an incredible story that the authors suggest the reader look at the references provided in the caption too. In summary, metal AM (especially LPBF and E-PBF) has been an instrumental tool for the realization of these patient-tailored metal implants mainly made from titanium alloys.

(a)   (b)

(c)   (d)

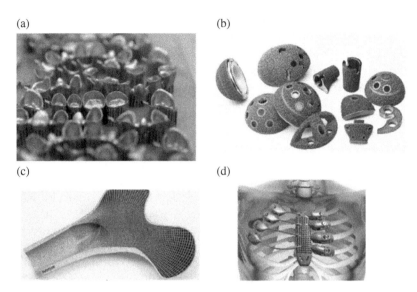

Figure 1.14 (a) Dental crowns printed by LPBF (Source: Courtesy of EOS [16]), (b) joint implants printed by E-LPF (Source: Courtesy of Orthostream [17]), (c) functionally gradient porous titanium load-bearing hip implant printed by Renishaw's LPBF (Source: Courtesy of Betatype [18]), (d) customized ribs and sternum printed by E-PBF (Source: Courtesy of Anatomics and Lab22 [19]).

1.5.2 Aerospace and Defense

The industrial adoption of metal AM was ramped up when large aviation, aerospace, and defense organizations/agencies such as GE Aviation, Lockheed Martin, SpaceX, the U.S. Department of Defense, and U.S. Air Force joined the race and started to heavily invest in R&D, machine development, advanced materials, and government-backed AM programs in mid 2010s. AM is uniquely attractive to this sector because of the lower material waste, lightweighting, reduction of the need for assembly through components consolidation, and the capability of production of highly intricate and complex parts that ultimately contribute to less fuel consumption and cost-saving due to lower level of certification as the number of parts decreases [20].

Led by safety requirements, this industry is known for having rigorous testing and certification procedures to evaluate the performance of the parts. As such, further improvements in the repeatability, reliability, and control of the metal AM systems are necessary before we can see airplanes or spacecraft with the majority of their components 3D printed. Nevertheless, it is reported by GE Additive that 28 fuel nozzles, 228 stages 5 and 6 blades and, 1 heat exchanger and 16 particles separators of GE9X engine (a new generation of high-bypass turbofan jet engine developed by GE Aviation exclusively for the Boeing 777X) are additively manufactured [7].

In space applications, the race is even wider. In 2015, the first-ever communications satellite with a design life of 16 years and weighed 4.7 metric tons satellite (named TurkmenAlem52E) with an aluminum 3D printed component was launched by SpaceX. This component was an antenna horn mounting strut. Thales developed several antenna supports with an envelope dimension of $45 \times 40 \times 21$ cm, made from aluminum. The AM-made supports have saved 22% weight and a 30% cost and an estimated decrease in the production time of 1–2 months. These brackets are subjected to a wide range of thermal stresses as during normal operation, they see temperatures of −180 to +150 °C. Designing this part for 3D printing allowed a much more efficient part to be developed because fasteners were no longer needed.

On the rocket engine applications, major activities are underway by SpaceX, NASA, and Aerojet Rocketdyn to adopt AM for rocket engine components because the qualification testing and heritage could be transferrable in many situations. SpaceX launched its Falcon 9 rocket with the first time ever AM-made Main Oxidizer Valve (MOV) in one of its nine Merlin 1D engines. The MOV operated successfully where high-pressure liquid oxygen under cryogenic temperatures and harmonic and subharmonic vibration levels were used [21]. The printed version of this traditionally casted part actually has superior strength, ductility, and fracture resistance. The 3D printed part has lower variability in material properties vs. the cast version due to uneven cooling of the part during casting. The MOV body was printed from nickel-based alloys using LPBF in less than 2 days versus the several months it takes to manufacture the cast version [21]. SpaceX also additively manufactured a SuperDraco Engine chamber. The Super-Draco engines are responsible for powering the Dragon Version 2 spacecraft's launch escape system. This system activates in an emergency to carry the astronauts to safety. These engines are also able to enable the spacecraft to land propulsively with the accuracy of a helicopter, making the spacecraft reusable, lowering the cost of space travel. This engine is being qualified. The qualification program includes testing multiple starts, extended firing periods, and extreme propellant flow and temperatures. Figure 1.15 shows AM-made parts for SpaceX engines.

Currently, the propulsion system is the primary focus of Lockheed Martin's AM efforts, with a goal to reduce the lead time on the fuel tanks from 18 months to only a few weeks.

One other major activities are related to the mission to "3D Printing in Space." While the plastic 3D printing has been tested in the international space station, there are many challenges associated with metal AM in space. The issue of "gravity" must be resolved before AM can be reliably replaced with expensive supply runs to the space stations.

Figure 1.15 LPBF-made combustion chamber (left) and the engine in finished configuration (right). Source: Courtesy of SpaceX [22].

Another development is associated with Boeing, where the company patented a 3D printing process that prints objects while they are floating in midair. This process uses magnets or acoustic waves to levitate the part while multiple AM systems apply the layers of material.

DED technology is also used in the aerospace and defense industry for repairing and refurbishment of the parts. It is a particularly important application given the long life cycle of aviatic systems and the high cost and long lead time associated with the replacement of the parts. As a result of a 2020 survey conducted by Optomec, one of the leading manufacturers of DED systems, from over 100 of their customers in the aviation market, it is claimed that over 10 million turbine blades have been repaired using DED systems. Repairing parts using DED has a lower thermal impact on the part in comparison with traditional methods such as welding. As a result, the parts will have a more favorable microstructure and mechanical performance after the repair using DED [23].

1.5.3 Communication

Developing advanced AM-made antennas is an area of growth in the communication industry because telecommunication devices on earth continue to require more and more bandwidth. Higher wave frequencies need to be used to meet these demands; however, these higher frequencies are more difficult to control. To broadcast these complex frequencies, intricately shaped antennas are required. AM processes can enable the manufacture of complex-shape metal and plastic antennas from different alloys and dielectrics, opening tremendous opportunities to the communication industry. Advanced AM-made RF antenna structures have the potential to revolutionize the design, supply, and sustainment of such devices. An AM design process can be fully integrated in the antenna design platforms to support not only customization but also antenna's performance enhancement in the field. It is reported by Optisys that the company has been able to reduce the number of parts through parts consolidation, the antenna weight through topology optimization, lead times, and production costs [24]. Figure 1.16 shows a small-size, complex, and lightweight RF antenna made by LPBF. The surface

Figure 1.16 Small-size, lightweight, one-piece, AM-made antenna. Source: Courtesy of Optisys [25].

roughness of metal parts printed by LPBF is a challenge for some frequencies; thus, efforts are underway to improve the surface roughness such that the printed antennas will be useable without any need to post-processing.

1.5.4 Energy and Resources

AM-made parts have been utilized by the energy industry to harness the natural resources of our planet for many years; however, they have not been in the media radars as widely as their counterparts in the aerospace industry. Some of these applications in this sector are revolutionary when AM is used in deep underground and oceans. As mentioned before, AM has been advanced to be more efficient through the introduction of lightweight components, cost-efficient services, and environmentally friendly materials. Several companies such as Chevron, Shell Global, BP Global, and GE Oil & Gas (Baker Hughes) have published stories about the AM adoption for prototyping and production in the energy industry. With pressure to make innovative solutions rapidly, engineers and designers in this industry use the rapid prototyping feature of AM as a key step in design verification. AM has also become an increasingly mainstream operation in the energy industry to fabricate end-use functional parts at a low-volume level. When AM-made parts need to tooling, it can offer to make lightweight structures with complex internal features. Thus, next generation of energy, oil, and gas components are being benefited from the AM features substantially, especially parts that need to exhibit performance and environmental standards. Dense, corrosive-resistant, and high-strength components can be mainly developed by DED for demands in this industry. One crucial application of AM in this industry is seen in the development of spare parts. As mentioned before, DED-based AM processes provide solutions through rapid, on-demand printing and repairing of legacy components.

So far, there are reports that AM has been used for either prototyping, low-volume production, or repairs of these parts: gas turbine nozzles, sand control screens, hydraulic components, nozzles for downhole cleanout tool, sealing accessories, liner hanger spikes, drill bits, and many more. Figure 1.17 shows multiple parts made for hydraulic devices used in the oil and gas industry.

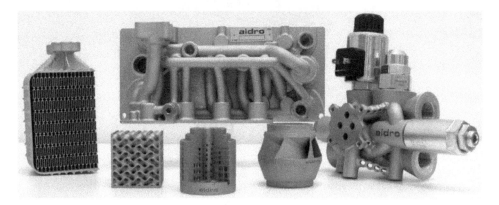

Figure 1.17 Hydraulic parts made for the oil and gas industry. Source: Courtesy of aidro [26].

1.5.5 Automotive

The automotive industry's interaction with AM goes back to 1980s. Today, some customized plastic parts used in the interior of cars, specifically luxury and highly personalized models, are 3D printed. In addition, plastic AM processes have been widely used to develop jig and fixtures and prototypes for design verification. The application of metal AM in this industry, however, has been mainly in the area of producing prototypes, heritage parts for obsolete models, and spare parts and tools. AM-made tools are an important area that has been used for mold production and tooling. In the luxury or race car sectors, however, many examples of production of end-use metal parts exist. For example, Ford Motor Company (FMC) (Detroit, MI, USA) has used the LPBF process from EOS to fabricate anti-theft wheel locks by converting the owner's recorded voice into a circular pattern, which will form the indentation needed to design a locking mechanism and a custom key. DS Automobile designers have also used EOS systems for the production of custom car accessories such as titanium door handle frames. Ford wheel locks and DS Automobile door handle frames are shown in Figure 1.18.

Even though the automotive industry has not reached the level of using AM directly for the production of final metal parts in serial production vehicles yet, a new trend for reaching that goal has already started. Many automotive companies such as Volkswagen (Berlin, Germany), BMW Group (Munich, Germany), Porsche (Stuttgart, Germany), General Motors (GM) (Detroit, Michigan), Toyota (Toyota City, Japan), etc. have entered the AM market either through investing in the improvement of their in-house AM capabilities or through making alliances with machine developers, 3D software companies, AM material producers, or research centers to expedite the adoption of metal AM. A project called "Industrialization and Digitization of Additive Manufacturing for Automotive Series Processes (IDAM)" was kicked off in 2019 in Germany for this purpose [30]. The automotive industry would be able to take advantage of the various commercial PBF and BJ systems producing final parts with properties similar to those of wrought or injection molding, respectively. A lower level of certification required by the automotive industry than aerospace and aviation should make such a transition easier.

In general, AM will be a key for supply chain transformation in the automotive industry. AM has the tremendous ability to reduce overall lead time, thus fostering market responsiveness. AM features resulting in less material usage, lightweight components, on-demand and

(a)　　　　　　　　　(b)　　　　　　　　　(c)

Figure 1.18 (a) Ford's custom anti-theft wheel lock being printed in EOS PBF system (Source: Courtesy of EOS [27]), (b) Ford's custom anti-theft wheel lock (Source: Courtesy of Ford Motor Company [28]), (c) custom titanium door handle frame in DS3 Dark Side edition from DS Automobile (Source: Courtesy of PSA [29]).

on-location production, and decentralized manufacturing at low to medium volumes will be driven force to significantly change the supply chain in terms of cost reductions, improved ability to locally manufacture parts, reduce complexity, and promoting consumer segments and markets satisfaction without any extensive capital deployment/investment.

1.5.6   Industrial Tooling and Other Applications

One of the obvious applications of metal AM is in tooling and mold production for other industries such as medical, aerospace, and automotive (see Figure 1.8). Furthermore, AM can be used to fabricate other industrial parts such as machinery components, heat exchangers, engineered structures, etc. either for parts redesigned for AM or for low volume producing legacy parts of current designs. In the consumer products sector, the promise of mass customization drives the gradually increasing usage of metal AM in various consumer products such as decorative objects, jewelry, custom sports gear, and structures such as bicycle frames. The design freeform, material graded structures, lightweighting, and fast design-to-market cycle offered by AM are predicted to have revolutionary effects on the market of industrial and personal products.

Highly scalable wire arc DED systems have also shown the potential to make larger structures in one run. MX3D (Amsterdam, The Netherlands) has been working on an award-winning smart bridge made from stainless steel, which will be placed on a canal in Amsterdam (Figure 1.19). Embedded with a network of sensors in collaboration with Alan Turing Institute (London, UK), this pedestrian bridge will collect structural data to help engineers measure the health of the structure under various environmental conditions in real time. This is a window to the future of urban planning and structural engineering world, as we are moving toward Internet of Things (IoT) smart cities with AM as one of the principal pillars of this movement realizing the physical aspect of the IoT [31].

1.6   Economic/Environmental Benefits and Societal Impact

Beyond the undisputable process flexibility of AM technologies, it also offers several major environmental benefits if the process chain and designs are properly developed. According to Diaz et al., *manufacturing activities are responsible for 19% of the world's greenhouse*

(a)   (b)

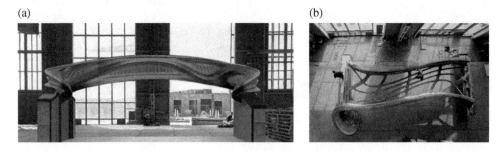

Figure 1.19   MX3D smart bridge (a) main structure (b) side wall. Source: Courtesy of MX3D [32].

gas emissions and 31% of the US total energy usage [33]. AM technologies are usually greener than conventional methods because there is less material wasted, and there may be up to 50% energy savings during part production [34]. However, it is only true if a full cycle assessment is conducted on the AM-made part to ensure its design is fully optimized for AM. Moreover, metal-based AM technologies virtually eliminate machining and the subsequent need for toxic cutting fluids; costly pollutants that presently are challenging to dispose of and have a negative environmental impact [33]. Besides, AM parts can bring about weight reductions in the range of 50–100 kg per aircraft. This will result in a significant reduction in fuel costs, based on the fact that for every kilogram removed from every aircraft in a fleet of 400 commercial jet-liners (Ex: Air Canada), annually reduces consumption by 60 000 liters.

Across the globe, manufacturing sustainability has become a crucial factor for the supply chain on all fronts. Companies are seriously working on the waste reduction, efficiency improvement of operations through the integration of AM and digitization to comply with their environmental and social obligations.

AM companies are commuting to be a major part of this adventure. Metal AM manufacturers are announcing their commitments to promote sustainability and safety through the minimization of waste, enhancement of material chemistries for safety, utilization of less energy, usage of less labor, and reduction of consumables.

AM is widely used in the repair sector to improve environmental footprint due to time and material-intensive traditional methods. Local manufacturing that is possible by AM is promoting firms' efforts to come closer to the circular/localized economy concept, an idea that fosters the optimum use of resources to avoid waste while minimizing gas emissions and energy demands.

Many companies are developing blueprints highlighting their commitments for the goals mentioned earlier. For example, the Global transit manufacturer Wabtec recently released its 2020 Sustainability Report [35]. When the report highlights several activities being undertaken by the company to improve its global environmental performance, it pinpoints the integration of AM processes such as binder jetting for sustainability by producing lightweight transportation components.

In all front of these commitments, a thorough analysis must be done to ensure metal AM is sustainable for making specific parts. Metal AM is not usually sustainable for making conventional parts that are not optimized for AM. It is recently reported by the Additive Manufacturer Green Trade Association (AMGTA) that metal AM has a higher carbon footprint per kilogram of the material used in the process compared with its counterparts processed by conventional methods while considering direct manufacturing processes exclusively [36]. This report is based on nonoptimized geometry when conventionally designed parts are directly printed without any geometrical and process optimization. However, as known, the design features have a major impact on the CO_2 footprint. When machining a block of material for simple geometries would be preferable, AM will be much appealing for making parts with hollow shells, lattices, features with complex curvatures and internal conformal channels. Thus, the sustainability of the use of AM has a direct correlation with the complexity of geometry. A simple but effective study was carried in [37], shedding light on the sustainability of AM and its CO_2 footprint when three different geometries made from Ti-6Al-4V by electron beam AM and machining were analyzed as shown in Figure 1.20. The surface finish was also a part of this analysis when the analysis considered the CO_2 footprint for gas atomization of powders required for AM as well as ingots required for processes. As seen in the figure, the energy demand and CO_2

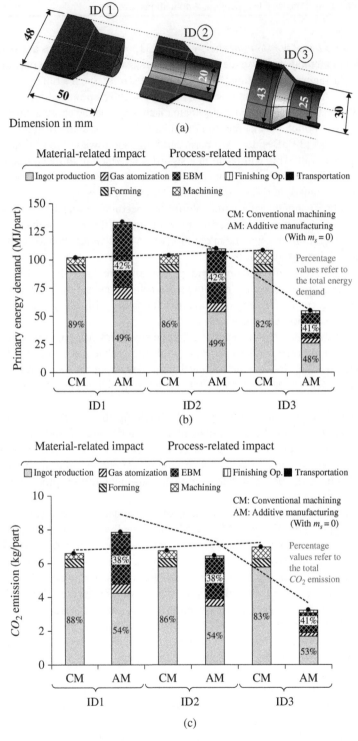

Figure 1.20 (a) Three different geometries made of Ti-6Al-4V by different processes, (b) primary energy demand, (c) CO₂ emissions. Transportation cost is neglected. Source: Republished based on CC BY 4.0 open access [37].

emission for AM are higher than of the machining if the geometry tagged as ID1 is produced. It is evidence that metal AM at its current form is not a sustainable solution for simple geometries without any internal features.

In contrast, for the geometry tagged as ID3 that features some complexity in the internal channel, metal AM offers a substantial improvement in terms of CO_2 emission and also energy demand compared with its counterpart produced by machining. The geometry tagged as ID2 that features simple internal channels, metal AM and machining both have more and less identical energy demand and CO_2 footprint.

In summary, it is extremely important to do a thorough analysis of the sustainability of metal AM upfront to understand the benefit of the process for specific geometry. In addition, life cycle assessment studies are needed to quantify the impact more precisely. The full product life cycle from production to overhaul must be included in the analysis.

1.7   AM Trends, Challenges, and Opportunities

With all the advantages that AM offers, even newcomers to the AM community appreciate the enormous potential and promise of technology that will transform the entire manufacturing enterprise in the next 10–15 years. Free complexity in designs, when industrial AM systems offer a platform to make them, is unfettered by the compromises of conventional manufacturing methods. With AM, designers do not design for manufacturing anymore; they design for end users. This is a paradigm shift.

It is reported that AM would be able to lessen the capital needed to reach minimum low-volume manufacturing [38]. This feature may lower hurdles for local manufacturing. In addition, the AM flexibility would facilitate the opportunity to produce a variety of products per unit of capital, reducing the changeovers and customization costs. To this extent, manufacturers are merging toward the understanding of different efficiencies AM can introduce throughout the supply chain, where there are four strategic paths to adopt AM for their businesses [38]:

1. Value-added proposition for existing products within traditional supply chains. In this case, companies would not radically change their products; however, they may explore AM technologies to help improve quality and reduce costs.
2. Value-added proposition by taking advantage of "economies of scale" offered by AM. In this case, companies would take the risk to transform the supply chain for their products.
3. Value-added proposition by taking advantage of "economies of scope" offered by AM. In this case, companies would be able to step into innovative personalized products and new levels of performance in their products.
4. Value-added proposition by pursuing new business models. In this case, companies will enjoy a new and effective supply chain for new business models with innovative workflow.

To follow any of the aforementioned strategies, companies need to put aside the traditional cost models and rubrics and adopt a holistic approach that will determine the impact of AM on their business. To critically assess the adoption of AM, Life Cycle Assessment (LCA) should be used. Such LCA sheds some light on the environmental and economic impact of a product or service to be offered to the market. The LCA evaluates all stages of the life cycle, from the extraction and development of raw materials, followed by AM processing, post-processing, transportation, use, and end-of-life disposal. This is a major trend in the AM industry these days.

In addition to new business models being developed by industry, there are several challenging factors that the AM community is addressing to overcome. In the following, these factors are discussed when emerging opportunities are elaborated.

Qualified materials: One of the major challenges in the field of metals and metal alloys is the number of powders that have been qualified for use with metal AM systems, including laser, electron beam, and binder-based AM processes. For example, there are currently more than 1000 steel alloys commercially available for conventional casting, but just a handful number has been verified for AM production by OEMs. In the case of aluminum alloys, the ratio is about 600:12. The shortage limits the number of parts that can be made and companies that can benefit from the technology. In addition, the relatively few qualified metal AM powders cost 5–10 times more than raw materials for casting, machining, and other traditional forms of manufacturing. Part of that problem is a lack of competition among suppliers. Another is low volume, with worldwide sales of metal AM materials totaling less than $400 million a year, a small fraction of the overall raw materials market. As the adoption of AM picks up steam, prices are expected to fall dramatically. As with most challenges, this one creates opportunities to improve powder production methods and, quite possibly, formulate entirely new powders to get the most out metal AM. For any material development, a holistic approach from material extraction to the end use and disposal must be considered.

Speed and productivity: One of the challenges of AM processes is speed. In general, production throughput speed is low for mass production. Although AM makes it possible to consolidate parts, small working volume and post-processing related to the surface enhancement add extra steps to the production time. Further process development is needed to enhance surface quality during AM processes to improve process productivity.

To address these challenges and improve AM productivity, modular flexibility is being integrated into AM processes. The scalability and modularity supported by the proper selection of processes can help to achieve the quality and speed required. Companies are working on the development of a larger working envelope into which multiple heat sources (e.g. laser beams) are incorporated. Automation and intelligent software are being developed to coordinate all subsystems in harmony with a goal of productivity enhancement.

Opportunities also exist in the field of computer modeling of AM processes to improve production via reliable and validated simulation rather than costly experimentation. Very few models have been developed to date, adding research and development costs to high material costs as deterrents for companies that might otherwise move into AM.

Repeatability and quality assurance: Although the technology has already produced impressive results, it is also true that reliability and repeatability are still significant AM problems, particularly for mass production. Failure rates for many applications remain in a range where using the technology simply is not economically justifiable due to the number of failed parts and the need to post quality checking by an expensive setup such as CT. The underlying problem is that AM is so sensitive to both environmental and process disturbances, from fluctuating temperature and humidity levels to nonuniform powder sizes. Full control of the process and surrounding environment is virtually difficult, so the focus is on solutions that employ innovative sensors to monitor conditions and quality control algorithms to automatically adjust process parameters, such as laser power or process speed, to compensate for disturbances instead. Closed-loop control is being incorporated into DED, and efforts are underway to add intermittent controllers to PBF processes due to their high speed. For PBF processes (e.g. LPBF), the major bottleneck to developing a closed-loop control system is hardware speed and accuracy. The amount of data collected by sensors are high but still not at a high frequency of 200K Hz or more to be able to effectively tune process parameters. In summary, the more advanced hardware (sensors and computational systems), the closer to the closed-loop control of PBF processes.

Industry-wide standards: Regardless of major advancements in AM over the last few years, the more nettlesome challenge is the lack of a comprehensive set of technical AM standards acceptable by industry. The absence of such standards may hinder the continued adoption of AM for industrial applications. Several main stakeholders have recognized the challenge and have started to take action. The American Society for Testing and Materials (ASTM), the American National Standards Institute (ANSI), International Standards Organization (ISO), and other standard organizations are working hard to develop standards platforms and procedures for AM. A road-map assessment of the state of standards and standards gaps in AM has recently been published [39]. ASTM through its committee F42 is currently developing standards for metal AM processes, especially for LPBF. The developed standards have the potential to help industry to effectively assess the performance of AM systems as well as the quality of printed parts. Nevertheless, with all these efforts in place, many new and reliable standards are still required. This should also be noted that concrete standards should be published rather a partially developed standard that may include flaws. If the standards are retracted and revised due to flaws regularly, such retractions will undermine the industry thrust.

End-to-end workflow, integration, and automation: In industry, anytime a new material/process/design or technology is used, a lengthy qualification process must take place. This testing is intended to prove (with lots of margin) that this new material/process/design or technology can meet all of the performance requirements. Many customers are reluctant to accept a new material/process/design or technology that does not have "heritage" in their applications. To minimize industry hesitation on the AM adoption, an effective end-to-end workflow must be developed that is simple but yet integrated and automated. All major industrial and nonindustrial AM systems providers are proposing ways to integrate their systems into complete end-to-end workflows. While AM is the core part of digital production, integration and automation of the end-to-end workflow are a different entity and beyond AM. However, any integration/automation must be well thought in harmony with AM limitations and features. The lack of digital infrastructure at the moment is a major obstacle to move with effective automated workflows for the AM industry. Automation basically begins with an effective streamline of part design and optimization through Design for AM (DfAM) tools, driven by digital warehouses and digital twins. Automated material supply lines for feedstock (e.g. powders) through manufacturing execution software should be developed to coordinate AM systems and workflow stations. Machine learning and artificial intelligence, as well as simulation, inline process monitoring software, and nondestructive testing (NDT), should work hand by hand to oversee the AM process to correct errors via when robots will remove and depowder parts from AM machines, followed by powder recycling and reuse. Automated post-processing heat treatment, polishing, etc. should be fully integrated into the aforementioned workflow. Automated AM is a part of the factory of tomorrow, a forefront of the ongoing industrial revolution within the industry 4.0 approach.

Software limitations: The commercially available software packages for the design of AM parts, support structure development, and interfacing with AM machines have limitations in assessing the feasibility of prints and identifying process constraints. In many cases, the ideas that are conceived and created in 3D modeling software exhibit major challenges to be printed mainly due to the issues with process constraints that are not included in the design. In addition, the current workflow software has limitations when it comes to AM to track individual items through each stage of the process to manage resources and delivery timelines. Another issue is associated with inter- and intra-communication and collaboration that depends on the quality of information and transmission methods. The current software and hardware still need more improvements to facilitate timely communication in AM.

Initial financial investments: One of the major barriers for metal AM adoption is a substantial investment needed to AM capital and ecosystem to deploy it to production. The AM eco-system covers software, materials, experts, post-processing equipment, certification as well as training for employees. This investment could be enormous hindering companies from embracing this technology effectively. One solution to this challenge is to rely mainly on AM service companies to integrate into the supply chain to derisk the AM adoption in the early stage. University and R&D centers can play an important role in providing funda-mental R&D required as well as training platforms for companies that are trying to adopt AM but cannot invest in it significantly.

Security: AM is fully integrated into the digital world; thus, its cyber–physical nature has raised major concerns. When AM has promoted globally distributed manufacturing, the existence of hackers is a reality. They can tweak the AM designs to create intentional defects that are not simply detectable but have catastrophic consequences while being used in actual systems. The vulnerabilities and large-scale jobs conducted by commercial AM services may inherently be difficult to substantiate the quality of printed parts. The process and supply chains must firewall to address these security concerns. These measures could be the same as other manufacturing industries such as electronic printing; however, due to typical appli-cations of AM-made parts in critical applications such as jet engines, special validation pro-cedures need to be developed to give assurance that the parts are not affected by the malicious attack in terms of intentional undetectable design alterations.

Skillsets gap: AM experts' shortage is one of the main challenges in industry. Lack of under-standing and expertise in AM is a critical factor in the AM adoption. There are a very limited workforce and highly qualified personnel to work and develop an entry strategy for new companies that want to embrace AM. A thorough understanding of AM capabilities prevents major misconceptions while introducing challenges correctly to decision-makers. The knowledge gap at the moment is significant as companies have difficulties in developing effective and meaningful business cases for metal AM. Since AM needs a new design plat-form, mechanical engineers who have been trained to design for traditional manufacturing cannot effectively undertake design for AM. They will require a steep learning curve to mas-ter it. Their mindset must be changed through training and education. Overall, learning about the capabilities and limitations of metal AM will aid companies in developing meaningful and successful applications for the technology.

A few measures to address the current gap are to develop a degree level program at different universities and colleges. The program should include a holistic curriculum that covers a vast range of areas. Designing such curriculums is of great challenge. In addition, industry certification programs should be developed when each module of the programs is taught by experts from around the world. Promoting AM consultancies is another great way to foster knowledge transfer. In addition, AM conferences and webinars are playing critical roles to fill the skill-sets gap.

1.8 Looking Ahead

This textbook includes 12 chapters. In Chapter 1, an overview of metal AM, its applications, opportunities, and challenges are provided at a high level.

Chapter 2 provides further detail on metal AM processes, their advantages and disadvantages. The chapter also lists critical process parameters in three main metal AM processes (DED, PBF, and BJ), while pinpointing the importance of the use of combined parameters in process development.

Chapter 3 highlights the major subsystems used in metal AM processes. This chapter particularly explains how a laser system works while highlighting different classes of lasers and also laser beam properties. The chapter also explains about electron beam as a major source of energy for AM processes. This chapter sheds some light on different material delivery systems, including powder feeders. In addition, some basic information on the digital files needed for AM processes is provided.

Chapter 4 starts with the fundamental of laser and e-beam material processing. It then covers major physics associated with DED processes with a particular focus on LDED. The chapter entails several case studies for modeling and analysis of DED processes such as powder-fed LDED and wire-fed DED.

Chapter 5 customizes the physics discussed for PBF processes such as LPBF and EPBF. It also sheds some light on heat source models used in the modeling of PBF processes as well as a few case studies.

Chapter 6 explains the modeling and physics of binder jetting and material jetting AM processes.

Chapter 7 covers the physics of material extrusion briefly. It sheds some light on the interaction of multiple physical phenomena.

Chapter 8 starts with a review of relevant historical background on material science. It then focuses on the comparison of conventional versus AM processes within the context of material properties. It discusses the fundamentals of solidification as well as factors affecting solidification in AM and phase transformation for different alloys used in metal AM.

Chapter 9 explains the concept of Metal Matrix Composites (MMCs). It also covers applications of AM technology in the fabrication of MMCs and possible challenges. It also sheds some light on the differentiation of categories of MMCs in AM, e.g. ferrous matrix composites, titanium matrix composites, aluminum matrix composites, nickel matrix composites.

Chapter 10 sheds some light on new design frameworks tailored to AM. It helps the reader to understand the importance of design rules and guidelines and how they differ between AM processes. It covers the theoretical framework supporting topology optimization, efforts toward including AM constraints in topology optimization models and a typical workflow of basic topology optimization in AM. It elaborates on the important terminologies that define lattice structures and helps the reader understand some practical lattice design methodologies. It also explains the significance and design strategies of support structures for the success of the printing process and product quality while improving the reader's understanding of design workflows for AM through case studies.

Chapter 11 explains different classes of in-situ sensing devices used for AM processes. It covers theories of some of the sensors used in AM, while shedding light on various applications of statistical approaches as quality assurance paradigms. It ends with fundamental and applications of machine learning techniques to AM and a case study.

Chapter 12 is mainly concerned with safety regulations in the AM industry. A basic understanding of AM process hazards, hazardous materials in AM, and associated safety matters in laser-based and electron beam AM techniques is covered. It also briefly explains human health hazards in AM and the comprehensive steps necessary for safety management.

References

1. Y. Huang, M. C. Leu, J. Mazumder, and A. Donmez, "Additive manufacturing: Current state, future potential, gaps and needs, and recommendations," *J. Manuf. Sci. Eng. Trans. ASME*, vol. 137, no. 1, p. 014001 (10pp.), 2015.
2. C. Scott, "United Nations encourages more countries to invest in 3D printing," 2015. [Online]. Available: https://3dprint.com/105046/united-nations-wipo-report.
3. Executive Office of the US President, "Capturing a domestic competitive advantage in advanced manufacturing," President's Council of Advisors on Science and Technology, 33pp., 2012. [Online]. Available: https://www.energy.gov/sites/prod/files/2013/11/f4/pcast_annex1_july2012.pdf.
4. T. Wohlers, Wohlers report 2020: 3D printing and additive manufacturing state of the industry, Wohlers Associates.
5. J. Bromberger and K. Richard, *Additive manufacturing: A long-term game changer for manufacturers*, McKinsey Co., 2017.
6. R. D'Aveni, *The pan-industrial revolution: How new manufacturing titans will transform the world*, Houghton Mifflin Harcourt, 2018.
7. M. Shaw, "Lessons learned from commercial aviation certification," 2017. [Online]. Available: https://www.nsrp.org/wp-content/uploads/2019/10/Lessons-Learned-From-Commercial-Aviation-Certification.pdf.
8. D. C. Hofmann, J. Kolodziejska, S. Roberts, R. Otis, R. P. Dillon, J.-O. Suh, Z.-K. Liu, and J.-P. Borgonia, "Compositionally graded metals: A new frontier of additive manufacturing," *J. Mater. Res.*, vol. 29, no. 17, pp. 1899–1910, 2014.
9. H. Alemohammad, E. Toyserkani, and C. P. Paul, "Fabrication of smart cutting tools with embedded optical fiber sensors using combined laser solid freeform fabrication and moulding techniques," *Opt. Lasers Eng.*, vol. 45, no. 10, 2007.
10. C. Tan, D. Wang, W. Ma, Y. Chen, S. Chen, Y. Yang, and K. Zhou, "Design and additive manufacturing of novel conformal cooling molds," *Mater. Des.*, vol 196, 109147 (10pp.), 2020.
11. T. Boon, "Rolls royce to revolutionise engine maintenance with 'snakes and beetles'," 2017. [Online]. Available: https://simpleflying.com/rolls-royce-engine-maintenance.
12. R. Kleer and F. T. Piller, "Local manufacturing and structural shifts in competition: Market dynamics of additive manufacturing," *Int. J. Prod. Econ.*, vol. 216, pp. 23–34, 2019.
13. A. Gadzala, *3D printing: Shaping Africa's future*. Atlantic Council, 2018.
14. R. Nolan, "SmarTech publishing issues 2019 additive manufacturing market outlook and summary report, estimates AM industry grew 24% percent in 2018, total market of $9.3 billion," 2018. [Online]. Available:

https://www.globenewswire.com/news-release/2018/12/13/1667021/0/en/SmarTech-Publishing-Issues-2019-Additive-Manufacturing-Market-Outlook-and-Summary-Report-Estimates-AM-Industry-Grew-24-Percent-in-2018-Total-Market-of-9-3-Billion.html.

15. AMPOIWERreport, "Metal additive manufacturing suppliers predict a market size growth of 27.9%." [Online]. Available: https://additive-manufacturing-report.com/additive-manufacturing-market.

16. EOS, "Industrial 3D printing for dental technology." [Online]. Available: https://www.eos.info/en/3d-printing-examples-applications/people-health/medical-3d-printing/dental.

17. S. Fournier, "The making of an orthopedics implant using 3D printing." [Online], 2018. Available: https://orthostreams.com/2018/12/the-making-of-a-medical-metal-implant-using-3d-printing-from-black-to-grey-metal-powder-magic-of-3d-printing.

18. Metal AM, "Renishaw showcases metal AM implants to American Academy of Orthopaedic Surgeons," 2018. [Online]. Available: https://www.metal-am.com/renishaw-showcases-metal-implants-american-academy-orthopaedic-surgeons.

19. H. R. Mendoza, "3D printing gives cancer patient new ribs and sternum in first-of-its-kind surgery," 2015. [Online]. Available: https://3dprint.com/95371/3d-printed-ribs-and-sternum.

20. J. O. Milewski, "Additive manufacturing metal, the art of the possible," in *Springer Series in Materials Science*, vol. 258, Springer Verlag, pp. 7–33, 2017.

21. A. R. Thryft, "SpaceX reveals 3D-printed rocket engine parts," DesignNews, 2014. [Online]. Available: https://www.designnews.com/design-hardware-software/spacex-reveals-3d-printed-rocket-engine-parts

22. R. Botsford End, "SpaceX's superdraco engine: Abort capability all the way to orbit," 2015. [Online]. Available: https://www.spaceflightinsider.com/organizations/space-exploration-technologies/spacexs-superdraco-engine.

23. "Optomec," 2020. [Online]. Available: https://optomec.com/optomec-customers-surpass-10-million-turbine-blade-repairs.

24. C. Clarke, "Optisys reducing antenna parts by 99% with 3D printing and simulation software," *3DPrining Industry*, 2017. Available: https://3dprintingindustry.com/news/optisys-reducing-antenna-parts-99-3d-printing-simulation-software-116509/.

25. Optisys, "Additive manufacturing transforms RF antenna design," Metal AM, 2017. [Online]. Available: https://www.metal-am.com/additive-manufacturing-transforms-rf-antenna-design/

26. B. O'Neal, "Aidro hydraulics & EOS highlighting AM processes for the oil & gas industry," SmartTech, 2019. [Online]. Available: https://3dprint.com/250954/aidro-hydraulics-eos-highlighting-am-processes-oil-gas-industry.

27. EOS, "The potential of additive manufacturing for serially produced vehicles." [Online]. Available: https://www.eos.info/en/3d-printing-examples-applications/mobility-logistics/automotive-industry-3d-printing/serially-produced-vehicles.

28. Ford, "Ford develops 3D-printed locking wheel nuts to help keep thieves at bay," 2020. [Online]. Available: https://media.ford.com/content/fordmedia/feu/en/news/2020/01/28/ford-develops-3d-printed-locking-wheel-nuts-to-help-keep-thieves.html.

29. D. Sher, "PSA's DS3 dark side edition surprises with titanium 3D printed interiors," 2018. [Online]. Available: https://www.3dprintingmedia.network/psas-ds3-dark-side-edition-surprises-titanium-3d-printed-interiors.

30. Metal AM, "IDAM project aims to integrate metal additive manufacturing in automotive series production."

31. F. Tao, M. Zhang, and A.Y.C. Nee, Digital Twin Driven Smart Manufacturing, 1st edition, Elsevier, 2019.

32. M3XD, "Smart Bridge," 2019. [Online]. Available: https://mx3d.com/projects/mx3d-bridge.

33. N. Diaz, M. Helu, S. Jayanathan, Y. Chen, A. Horvath, and D. Dornfeld, "Environmental analysis of milling machine tool use in various manufacturing environments," in Proceedings of the 2010 IEEE International Symposium on Sustainable Systems and Technology, ISSST 2010, 2010.

34. US Department of Energy, "Additive manufacturing: Pursuing the promise," 2012. [Online]. Available: https://www1.eere.energy.gov/manufacturing/pdfs/additive_manufacturing.pdf.

35. Wabtec, "2020 Sustainability report: Moving and improving teh world," 2020.

36. J. Faludi and C. Van Sice, "State of knowledge on the environmental impacts of metal additive manufacturing," 2020.

37. P. C. Priarone, G. Ingarao, R. di Lorenzo, and L. Settineri, "Influence of material-related aspects of additive and subtractive Ti-6Al-4V manufacturing on energy demand and carbon dioxide emissions," *J. Ind. Ecol.*, vol. 21, no. 1, 2017, pp. S191–S202.

38. B. McGrath, J. Hanna, R. Huang, and A. Shivdasani, "3D opportunity for life cycle assessments," *Deloitte*, 2017, 20pp. Available: https://www2.deloitte.com/content/dam/insights/us/articles/additive-manufacturing-in-lca-analysis/DUP_719-3D-opportunity-life-cycle_MASTER.pdf

39. America Makes and AMSC, *Standardization Roadmap for Additive Manufacturing – Version 2.0*, America Makes & ANSI Additive Manufacturing Standardization Collaborative, 2018.

2

Basics of Metal Additive Manufacturing

<hr>

Learning Objectives

At the end of this chapter, you should be able to:

- Understand the main additive manufacturing (AM) technologies used for metal fabrication and advantages and limitations of each technology
- Gain knowledge of various commercial AM machines and technology trends
- Understand the key process parameters for main metal AM processes and parameters interconnections
- Gain knowledge about the main alloy groups used in metal AM

<hr>

2.1 Introduction

Powder-based processes, such as powder bed fusion (PBF), directed energy deposition (DED), and binder jetting (BJ), are the most widely used AM techniques for metal processing. Furthermore, the production of metal structures using sheet lamination (SL), material extrusion (ME) where parts are made of polymer matrix composites (PMCs) reinforced with metal particles/ fibers, and material jetting (MJ) platforms is being investigated at research and development or small-scale commercial levels. This chapter introduces the main metal AM technologies: PBF, DED, and BJ. Besides, the chapter sheds some light on SL, ME, and MJ processes. Further

Metal Additive Manufacturing, First Edition. Ehsan Toyserkani, Dyuti Sarker, Osezua Obehi Ibhadode,
Farzad Liravi, Paola Russo, and Katayoon Taherkhani.
© 2022 John Wiley & Sons Ltd. Published 2022 by John Wiley & Sons Ltd.

details on the working mechanism of different modules used in more popular powder-based systems (PBF, DED, and BJ), such as lasers, electron beams, optical systems, powder delivery systems, etc., are provided in Chapter 3.

Depending on the AM technology, the quality of printed parts is affected by various complex phenomena such as powder layer formation, energy source-particle interaction, melt pool dynamics, binder–particle interaction, droplet formation, densification, etc. Each phenomenon, in turn, is controlled by multiple process parameters or material properties. In this chapter, we will touch base upon some of the most important process parameters, many of them common across multiple platforms. These factors are related to the physical phenomena taking place during the print and are investigated in depth in Chapters 4–7. Chapters 8 and 9 look at the effect of process parameters on microstructure, hardness, tensile strength, fatigue, surface roughness, density, etc., through the lens of metallurgy.

Not all of the metal powders are readily applicable to AM technologies. In fact, out of thousands of alloys available, only a handful of ferrous, titanium, aluminum, nickel, and cobalt-chromium alloys are being adopted in metal AM. The reason is that most alloys, if available in powder form, have been developed and optimized for conventional manufacturing techniques throughout decades. Metal AM, marked as a nonequilibrium process entailing rapid melting and cooling cycles, is very different from most conventional techniques in terms of thermodynamics and solidification rate. Thus, designing and optimizing alloys for this technology are necessary. Such development of material libraries is a lengthy, costly, and challenging endeavor. Nevertheless, these types of developments are one of the most active AM research fields. In addition to the common alloy families mentioned above, AM of copper and several precious metals such as gold, silver, platinum, palladium, and tantalum is being pursued. Metal AM products have applications in mainly aerospace, medical, and tooling sectors; however, new fields of application are being discovered as the industry matures. Chapter 1 covers the applications of parts made by metal AM.

2.2   Main Metal Additive Manufacturing Processes

This section mainly provides readers with the basics and principles of three classes of AM: PBF, DED, and BJ, when they are used for fabricating metal components. Besides, three emerging technologies, SL, ME, and MJ for metal parts will be briefly reviewed.

2.2.1   Powder Bed Fusion (PBF)

2.2.1.1   PBF Process Description

PBF technology is one of the seven classes of AM that accounts for the majority of metal AM market share. It is increasingly used to fabricate high end products from titanium alloys, ferrous alloys, nickel-based superalloys, etc., for various applications.

In this process, a moving heat source sinters or melts powder particles selectively as it scans a predetermined area corresponding to the cross section of the build layer to form complex parts in a layer-by-layer fashion. Powder spreading mechanisms, including hoppers and recoater, are used to distribute powder layer-by-layer. To this end, powder particles are delivered to the build compartment and leveled flat using a roller or recoater blade. After one layer is finished, the build compartment is moved down usually for a few hundred microns and a new layer of powder is deposited. Next, the particles are fused together and to the preceding layer. This process continues until the part is completed. The finished part remains in the build chamber to cool down. This approach helps avoid geometrical deformities, such as distortion. After the cooling step, the excess powder is vacuumed out and parts are cut from the substrate and/or support structures.

Different heat sources have been used, but the common sources of energy are lasers and electron beams. The process is called laser PBF (LPBF) when a laser beam is used and electron-beam PBF (EB-PBF or EPBF) when an electron beam is used in the setup. Figure 2.1 depicts both LPBF and EB-PBF, where the basics of both processes are schematically shown. A laser is a highly concentrated stream of photons that increase the temperature of metal particles when absorbed, i.e. their energy is transferred to the material through atomic-level heat transfer mechanisms via electron–phonon–photon interactions. An electron beam is a high-speed stream of free electrons concentrated using a magnetic optic system. Upon collision of the electrons with metal particles, their high kinetic energy is converted into heat through electron–nuclei interactions [2, 3]. Further descriptions about the interactions of these heat sources with materials are provided in Chapter 4. Besides, further information about the equipment used in PBF can be found in Chapter 3.

The type of thermal energy source in PBF systems leads to fundamental differences in terms of the manufacturing condition, choice of materials, process efficiency, final part quality, etc. The manufacturing process is carried out in a chamber filled with an inert gas such as argon in LPBF to avoid oxidation and powder degradation. Figure 2.2 shows the melt pool and ejected spatters from an LPBF process. However, a high vacuum chamber is required for EB-PBF to avoid the collision of electrons with molecules of atmospheric gases. EB-PBF is compatible with highly conductive powders to avoid the build-up of negative charge in the powder bed. Support structures help with the transfer of negative charge in EB-PBF and heat transfer to the substrate in both technologies. However, generally less support volume is required for EB-PBF as the manufacturing happens at a high-temperature chamber resulting

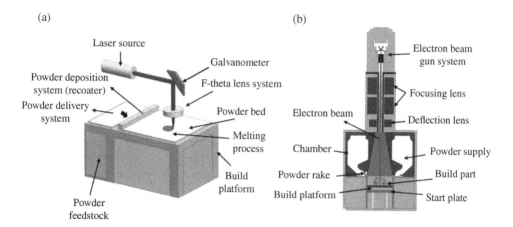

Figure 2.1 Schematic of (a) LPBF and (b) EB-PBF. Source: Redrawn and adapted from MDPI open access [1].

Figure 2.2   A view of melt pool and ejected spatters in LPBF. Source: Courtesy of Beamie Young/NIST.

in partial sintering of particles in the powder bed that restrains the parts. This higher printing temperature also leads to a lower temperature gradient and thus reduces build deformities such as warpage as a result of thermal stress. However, it can adversely affect the quality of the loose powder in the chamber through particle fusion and agglomeration, making recycling of unused powder difficult. The generation of electron beams is much more efficient than lasers because a considerable portion of the laser energy is lost through reflection and/or inherent low efficiency in lasers due to the cavities and light pumping (see Chapter 3). Furthermore, electromagneti-cally driven electron beam scanning is usually much faster than moving a laser spot using a galvanometer [4]. The focused spot size of electron beams is generally (but not always) larger than what lasers can achieve. The larger spot size combined with faster scanning of electron beams leads to the higher process throughput of EB-PBF than laser-based systems. The smaller focused beam diameter of the lasers, on the other hand, results in better surface quality and higher resolution. The key features of LPBF and EB-PBF are compared in Table 3.1 of Chapter 3.

Different nomenclatures have been used to describe different commercial/technical varia-tions of this technology, including selective laser sintering (SLS), selective laser melting (SLM), direct metal laser sintering (DMLS), etc. SLS, the first commercialized metal AM tech-nology, was developed at the University of Texas at Austin in 1980s [5]. This technology sin-ters particles at an elevated temperature at solid state or by partial melting through diffusion. SLS usually produces highly porous metal parts with inferior mechanical performance com-pared to wrought or cast counterparts. This technology was later improved in 1990s by the introduction of SLM by the Fraunhofer Institute of Laser Technology [6]. In SLM, more pow-erful ytterbium-doped (Yb) fiber lasers induce full melting and coalescence of powder particles followed by solidification in a process known as fusion. The highly concentrated energy of lasers also re-melts portions of the preceding layer resulting in the merging of successive layers and creating a strong bond between them. In this book, we concentrate on SLM and use the term LPBF to refer to this technology throughout the book.

Support structures are normally required underneath cantilever features to prevent any issues such as warping due to residual stress support. However, support-less printing with PBF is an emerging trend powered by optimum process parameters, the need for support structures would be avoided. A start-up company VELO3D has started to work on the support-less printing while offering interesting products using their optimized process recipes.

Topology optimization helps designers to develop complex shapes with high mechanical properties at a much lower weight. Chapter 10 explains the basics of topology optimization used in AM. Figure 2.3 shows samples of metallic parts with complex organic designs that are topologically optimized and printed using PBF technology.

Post-processing such as machining, polishing, or coating for improvement of the surface quality or hot isostatic pressing (HIP) for reduction of build defects might be carried out next. Figure 2.4 showcases the effect of post-processing on the quality of PBF parts.

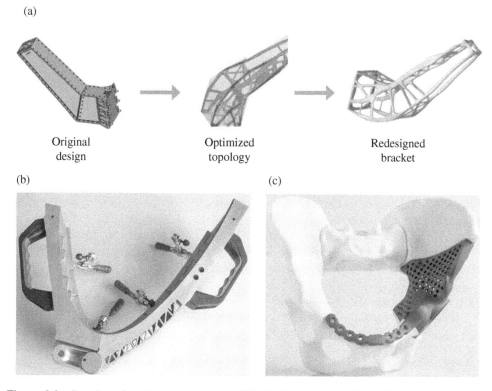

Figure 2.3 Samples of metal parts made via PBF for (a) aerospace, (b) tooling, and (c) medical industries. (a) Topology optimized antenna bracket for Sentinel satellites made from AlSi10Mg. Redesigned part demonstrated 30% increase in rigidity while being 40% lighter than the original part. (b) High precision tool with complex embossing of the inner surface made from Aluminum. (c) Lightweight titanium hip implant. (a), (b), and (c) have been manufactured by EOS GmbH in cooperation with RUAG, Any-shape, and Alphaform, respectively. Source: Courtesy of EOS [7].

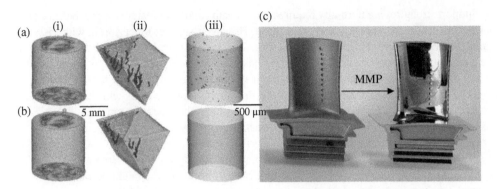

Figure 2.4 The CT scan results of (i) cylindrical, (ii) triangular prism, and (iii) magnified edge of a cuboid sample made using EB-PBF from Ti6Al4V (a) before HIP treatment and (b) after HIP treatment. The results demonstrate significant reduction in porosity after HIPing. (c) The effect of micro-machining process (MMP) on the surface properties of a PBF-made part. Source: (a, b) Republished from [8] and licensed under CC BY 4.0 and (c) courtesy of EOS [7].

2.2.1.2 PBF Advantages and Challenges

PBF is one of the most widely used AM techniques due to its high precision, the capability to produce high-quality complex structures, a relatively diverse selection of raw materials, and relatively high density of its products (nearly full dense). Similar to any other manufacturing method, various AM processes have advantages and shortcomings. Understanding the advantages and disadvantages of this process helps identify the best process based on the unique features of the target product, manufacturer capabilities, and the customer's requirements.

Process Features: PBF can produce high-resolution parts (>30 μm), making it a suitable technology for the fabrication of intricate organic designs requiring thin walls, very fine features, internal channels etc. In comparison, DED systems have a resolution usually >250 μm. The resolution of EB-PBF is slightly coarser than L-PBF (>60 μm) due to the need for larger particle sizes and larger focused beam spot size in this process. Moreover, more powder particles agglomerate on the skin surface of printed parts in EB-PBF. PBF is considered a mid-level process in terms of manufacturing speed. The laser scanner speed could be up to 5 m/s, where the EB speed, in theory, can be in the order of 10 km/s. The speed of EB-PBF process in practice is normally in the range of 10 m/s. The build rate for LBPF is 5–20 cm^3/h and for EB-PBF is about 80–100 cm^3/h. Moreover, the prolonged heat-up/cool-down cycles required before and after the print also add to the production time. These are fundamental challenges that need to be addressed.

Fusing particles at multiple locations within the build chambers simultaneously by increasing the number of lasers and improving the beam scanning systems are part of the AM industry's current solutions for the production speed issue. Moreover, modularized systems with replaceable building compartments such as M LINE FACTORY by GE Additive (Lichtenfels, Germany) have been developed to reduce the idle time of machines. In such systems, the cool-down period, powder resupplying, powder removal, etc. will be carried out offline and do not interfere with the production process.

Currently, PBF is limited to the maximum build dimensions of ~800 mm (width and length) and ~500 mm (height), making it unsuitable for producing larger functional parts in one piece with the current configurations. Manufacturers also offer smaller build chambers for specific applications such as R&D or jewelry production. It is expected that PBF systems with a larger building envelop emerge in years to come.

Challenges and remedies: More than hundred process parameters control the PBF printing process. However, many of them may not be significant. In reality, over 20 parameters normally play important roles in the quality of printed parts. Understanding a proper combination of such parameters yields a successful print with the best-finished part properties. The process optimization is currently a manual process with a high dependency on the designer and operator's judgment. Machine developers usually provide guidelines and optimized recipes for their proprietary materials; however, extending those to new alloys should be obtained through trial and error or conducting statistical experiments. This type of recipe development is usually a costly and time-consuming process. Until such time that data sets linking process conditions and part properties are developed for individual AM processes and materials, this will remain a major shortcoming for all AM classes, including PBF [9]. Once such an arrangement is fully developed, PBF can be widely adopted by industry. However, machine producers are generally moving toward the development of smarter AM machines equipped with various sensors and closed-loop quality control algorithms. With a higher level of process control and data collection possibilities at different stages of production, employing AI-based techniques might provide solutions for this issue in the early future (see Chapter 11). The current efforts for developing more intelligent systems with real-time monitoring/control capability are also important for improving the repeatability and reliability of production, an important step in the industrialization of AM, specifically for critical applications where high quality is required, and for part certification where the procedure is scrutinized substantially.

Materials: PBF has proven to be able to process various alloys as well as non-metallic materials such as ceramics and composites (see Chapter 9). It also offers a platform for fabricating highly reactive materials and high entropy alloys due to the inert printing condition under which the printing is carried out. Although many alloys have been investigated and optimized for both LPBF and EB-PBF by various research groups, there is a big gap in terms of the number of processed materials by these processes. Generally, any weldable alloy, specifically highly reactive ones, can be utilized in PBF due to similarity between the two processes [4]; however, certain criteria in terms of chemistry, morphology, and particle size distribution must be met for a powder to be fit for PBF. The number of issues with EB-PBF, when it comes to reactive materials, is less than LPBF because EB-PBF runs under full vacuum. In EB-PBF, to increase the temperature in the build plate, a defocused electron beam usually scans the surface of the build compartment several times to elevate the preheating temperature (e.g., from 300 to 1100 °C depending on the material). This preheating stage is extremely important for the EB-PBF process stability, and if the target material cannot be partially agglomerated at this stage, the process will not be successful due to a phenomenon called "smoke" [10], initiated from the charge dissipation of the electrons throughout the powder bed compartment. If this phenomenon occurs, the top powder particles may explode because particles may be pushed out of the build bed. After all, they are not agglomerated partially and there is no adequate holding force between them. Preheating would cause partial sintering among particles, holding them in place for the main process, while a sufficient electrical conductivity will be established. If material powder particles cannot create partial sinter necks during preheating, it is virtually impossible to develop parts by EB-PBF [10]. It is one of the major issues that may limit the adoption of new materials to EB-PBF.

Usually, highly spherical particles with uniform dimensions are deemed ideal for PBF. With low energy-efficient atomization techniques being the dominant methods for the production of such powder feedstock [11], the high cost of raw materials is an important obstacle that must be mitigated. The production of low-cost alloys produced by water atomization tailored for PBF is one of the more recent AM research fields. This trend results in introducing new alloys leading to the expansion of the material portfolio for PBF. Current PBF systems can only support a single material. Producing multi-material parts with PBF requires sophisticated powder delivery systems [12, 13], which are not commercially available as of 2020.

Products: If operated at optimized conditions, PBF can generate near full-density (e.g., 99.99% dense for Ti-6Al-4V) parts needed for various industrial applications. Although build flaws such as anisotropic properties, incomplete fusion, impurities, etc. are common, printed parts can be heat treated to arrive at isotropic properties. Parts made with PBF undergo a different thermal history characterized by fast heating and cooling times and smaller heat-affected zones (HAZ) compared to traditional metal processing technologies; thus, they have a different microstructure compared to their conventionally made counterparts from a similar alloy (see Chapters 5 and 8). This, in turn, causes a divergence between the mechanical properties of PBF-made and wrought parts. However, the PBF products sometimes demonstrate superior mechanical properties [4], specifically in terms of ultimate tensile strength due to finer microstructures compared to cast parts and properties very close to wrought parts. However, even the smallest level of porosity in PBF parts can reduce fatigue resistance [14]. PBF-made parts have a moderate surface finish with a roughness in the order of tens to a few hundred microns depending on the powder quality, part orientation, the process core parameters, skin parameters, etc. As a result, PBF manufacturing is usually followed by surface treatments such as polishing, sandblasting, coloring, coating, etc., which elongate the production time and increase the cost.

2.2.1.3   Commercial PBF Systems

Multiple companies produce commercial LPBF machines with different process parameters and laser types. Arcam, a Swedish company, purchased by General Electric (GE) in 2016, is the first producer of commercial EB-PBF systems in the world. Table 2.1 lists commercially available metal PBF systems and their key features, including energy source and power, beam focus diameter, scanning velocity, build area dimensions, and layer thickness.

From the data presented in Table 2.1, it can be concluded that most manufacturers of metal AM machines try to offer a wide range of key features for different applications. From low power (100 W), single energy sources matched with small cylindrical build area dimensions (80 mm ∅ × 100 mm H) and very fine layer thicknesses (10–20 μm) for special applications such as jewelry or dental to medium-sized build chambers (125 mm W × 125 mm D × 125 mm H) equipped with dual 500 W lasers for prototyping, research and development, medical applications, and small-scale productions, to the fast processing of large functional parts with high aspect ratio through navigating four or more energy sources simultaneously on large build areas (800 mm W × 400 mm D × 500 mm H), and anywhere in between. As of 2020, SLM Solutions offers the highest throughput of 1000 cm^3/h with its NXG XII 600 serial production machine in which 12 lasers can simultaneously cover a build bed of 600 mm × 600 mm × 600 mm, while alternative beam spot sizes (between 80 and 160 μm) can be selected based on the melt pool location with respect to the layer geometry.

Table 2.1 Configurations of major commercially available PBF systems.

Manufacturer	Energy source/number of lasers/power (W)	Beam focus diameter (µm)	Maximum scanning velocity (m/s)	Build area dimension (mm) (W, D, H^a) or (ø, H^a)	Layer thickness (µm)
3D Systems	Yb fiber laser/single/ Min: 100 Max: 500			Min: $100 \times 100 \times 80$ Max: $500 \times 500 \times 500$	Min: 10 Max: 100
AddUp	Yb fiber laser/single or dual/500 W		10		Min: 20 Max: 100
GE Additive (Arcam)	Single crystalline or tungsten cathode electron beam/single/ Min: 3000 Max: 6000	Min: 140 Max: 250	8	Min: $200 \times 200 \times 180$ Max: 350×380	Min: 40 Max:150
GE Additive (Concept Laser)	Fiber laser/single, dual, or quad/ Min: 100 Max: 1000	Min: 50 Max: 500	5 or 7	Min: $50 \times 50 \times 80$ Max: $800 \times 400 \times 500$	Min: 15 Max: 150
DMG Mori	Fiber laser/ Min: 200 Max: 1000	Min: 35 Max: 70		Min: $125 \times 125 \times 200$ Max: $300 \times 300 \times 300$	Min: 20 Max: 100

(continued overleaf)

Table 2.1 (continued)

Manufacturer	Energy source/number of lasers/power (W)	Beam focus diameter (μm)	Maximum scanning velocity (m/s)	Build area dimension (mm) (W, D, H^a) or (ϕ, H^a)	Layer thickness (μm)
EOS GmbH	Yb fiber laser/single, dual, triple, or quad/ Min: 100 Max: 1000	Min: 30 Max: 100	7	Min: 80 × 95 Max: 400 × 400 × 400	Min: 20 Max: 100
Renishaw	Yb fiber laser/single, dual, triple, or quad/ Min: 250 Max: 500	70		250 × 250 × 300	Min: 20 Max: 100
Sisma	Fiber laser 200	30		100 × 100	Min: 20 Max: 40
SLM Solutions	IPG fiber laser/single, dual, quad, or 12 Min: 400 Max: 1000	Min: 70 Max: 115	10	Min: 125 × 125 × 125 Max: 600 × 600 ×600	Min: 20 Max: 115
Trumpf	Fiber laser/single, triple/ Min: 30 Max: 500	Min: 30 Max: 500		Min: 98 × 100 Max: 300 × 400	Min: 10 Max: 150
Xact Metal	Fiber laser/single, dual, quad/ Min: 100 Max: 200	>50	8	Min: 127 × 127 × 127 Max: 254 × 330 × 330	Min: 20 Max: 100

[a] Including substrate height.

Another trend set out by leading companies in the field, such as EOS GmbH (Krailling, Germany), GE Additive, and 3D Systems (Rock Hill, SC, USA) is modular designs for scaling up the production. Still at an early stage, these systems are composed of multiple connected units with central control, processing, and powder management systems to simulate a production line.

Currently, most PBF system manufacturers utilize YAG fiber lasers mainly in the continuous-wave (CW) form. However, modulated lasers are used in some machines, including Renishaw (Wotton-under-Edge, UK) machines. Most of these companies offer to scale up the production rate by adding up to four laser sources with powers in the range of 100–1000 W on their larger systems. To the best knowledge of authors, the highest beam focus is <30 µm currently offered by EOS PRECIOUS M 080, suitable for making very fine structures out of precious metals in jewelry applications. Depending on the dimensions of the build area, the layer thickness and increment amount that each machine can offer vary from 10 µm offered by 3D Systems and Trumpf (Ditzingen, Germany) to 150 µm by Trumpf.

PBF companies are constantly competing to improve their systems and services by (i) integration of monitoring, sensing, and closed-loop control systems into their machines; (ii) offering faster and more efficient feedstock changing systems; (iii) testing, validating, and providing a wider envelope of metallic alloys, and (iv) providing software solutions in the form of computer-aided-manufacturing (CAM) packages customized for AM for streamlining the process from end to end.

2.2.2   Directed Energy Deposition (DED)

2.2.2.1   DED Process Description

DED is one of the main metal AM classes based upon the laser and electron beam welding, material feeding/injection, and computerized numerical control (CNC) technologies. Mainly a metal processing technique, DED melts materials as they are deposited/fed to bond them to the substrate metallurgically (Figure 2.5). In this technology, instead of fusing metal particles in a powder bed as in PBF, the raw material (powder or wire) is deposited/fed into a melt pool created at the incident point of the energy source and materials. The filler material is fully melted in the pool and solidified as the energy source passes, generating a thermal history similar to PBF. By articulating the nozzle or substrate (or both), the confocal point where the filler material meets the melt pool (or energy source) follows a programmed 3D path to deposit a layer of adjacent tracks. After completing a layer, the nozzle or substrate moves by the thickness of one layer to deposit the next layer. Consequently, a dense 3D shape is formed that normally has 100% density with minimum porosity. The fabrication is usually carried out in an inert chamber. DED thus normally provides better mechanical properties compared to PBF.

Depending on the heat source and material feeding mechanism, different technical or commercial names are used. Heat sources that are normally used in DED are:

1. Laser
2. Electron beam
3. Electrical arc (EA)
4. Plasma transferred arc (PTA)

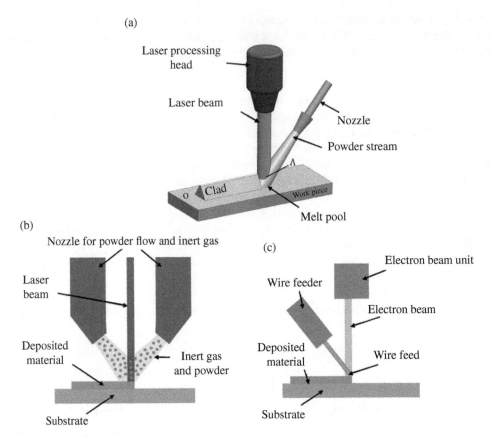

Figure 2.5   Schematics of (a) powder-fed laser DED with lateral nozzle (Source: Adopted from [15]), (b) powder-fed laser DED with a co-axial nozzle; and (c) wire-fed EB-DED. Source: Redrawn and adapted with permission from Elsevier [16].

In terms of feedstock feeding mechanism, the following systems are used:

1. Powder injection
2. Wire feeding
3. Paste feeding through extrusion

For the type of feedstock feeding mechanisms mentioned above, two types of delivery systems are used in DED technology:

1. Lateral nozzle: materials are deposited to the melt pool from the side, while the beam is located vertically above the melt pool (Figure 2.5a for powder and Figure 2.5c for wire feeding),
2. Co-axial nozzle: material flow, laser beam, and the shield gas are delivered from the same nozzle (Figure 2.5b).

For more details about the powder delivery systems (Figure 2.5), see Section 3.4.2.

Despite the type of powder delivery system used, simultaneous control of powder delivery, energy source, and substrate/nozzle movement parameters, as described in Section 2.2.2.2, makes it complicated to control and optimize a DED process. The objective is to ensure that the surface of substrate with respect to the melt pool is located within the critical beam energy density region described in [4], where adequate fusion energy for the filler material is provided. See Chapter 3 for more information.

Several names used for DED based on the heat source and feeding mechanisms are: powder-fed laser DED (PF-LDED), wire-fed electron beam DED (WF-EB-DED), and plasma transferred arc DED (PTA-DED). As a technology often associated with high throughput, large build envelopes, deposition on non-flat surfaces, multi-material printing, the high diversity of relevant raw materials, and accessible open architecture machines, DED comprises the second largest share of the metal AM industry.

DED has a coarser resolution than PBF, so it is mainly used for the following applications:

1. Production of near-net-shape structures that may need post-machining.
2. Freeform fabrication of selective features on pre-build structures
3. Repair of high-value components.

The technology can be used for coating, but we do not list it as its main application in this book. In DED, the use of support structures is avoided as much as possible. DED characteristics and its resolution limit the design freedom because extensive support structures would be needed to fabricate complex shapes with overhangs using this technology. Thus, in general, DED is used for the manufacture of parts with simpler geometries with thin walls. Figure 2.6 shows LDED used in near-net-shape production, freeform fabrication, and repair of highvalue components.

The ability of DED technology to deposit on non-flat surfaces is widely exploited for repair and remanufacturing applications. Figure 2.7 shows different stages of repairing a broken gear tooth that has caused production line stoppage using the Optomec DED system. The reported repair time is <12 hours, while a replacement would have taken 12 weeks.

In its most basic form, a DED machine includes a substrate on which the part is built, an energy source, and a material delivery nozzle mounted on a 3-axis CNC table. In complex systems, robots with a higher number of degrees of freedom, 5-axis CNC, controllable positioners for movement/rotation of the substrate, and machining tools for hybrid subtractive/AM might be available. Selecting the best configuration of DED machine parts is highly dependent on the application, size of the building part, and available upfront budget. For example, a simpler vertical material deposition 3-axis table can perform well enough for printing simple structures on flat surfaces. On the other hand, rotational movements make deposition on 3D complex surfaces easier while imposing limits on the building envelope and part weight cap (see Figure 2.6b).

As mentioned previously, DED technologies are more diverse compared to PBF since they are compatible with more forms of raw material delivery systems and heat sources.

Even though the powder morphology requirements for DED are not as strict as those of PBF, the powder is usually more expensive than wires. Still, the use of powders improves the resolution and, consequently, the ability to fabricate more complex parts. In DED, the powders are stored in one or more powder feeders and are transferred to the nozzle head and deposited into the melt pool using an inert gas to prevent contamination with oxygen (see Chapter 3 for more

(a)   (b)

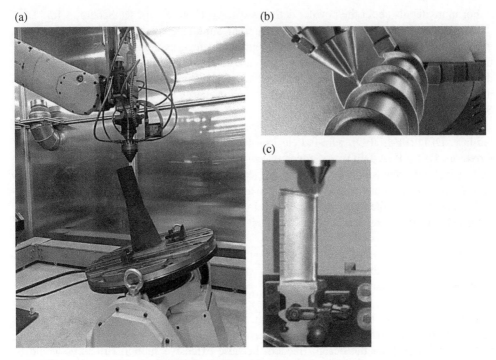

(c)

Figure 2.6 Applications of DED: (a) Near-net-shape production. Source: Courtesy of Promation Engineering. (b) Freeform fabrication of features on a shaft. Source: Courtesy of Trumpf. (c) Repairing the damaged tip of rotor blade made of precipitation-hardened CC-superalloy Inconel 738. Source: Republished with permission from Taylor & Francis Group [15]; permission conveyed through Copyright Clearance Center, Inc.

Broken gear teeth   After LENS® printed repair   Machined to spec

Figure 2.7 A sample of repaired rotating part using Optomec® LENS DED system. Source: Courtesy of Optomec [17].

information on powder feeders and inert/shield gas configuration). As there is no need to fill a powder bed, less material is used in DED and there will not be much need for powder recycling. However, a known problem with powder-fed DED systems is their low catchment efficiency.

In comparison, wires exhibit almost full catchment. Wires are normally inexpensive and available for a great number of alloys. This expands the material envelope for wire-based DED. Wire feeding is much faster than powder feeding, leading to increased printing speed and process throughput and yields full density parts. In comparison, parts made using powder-based DED might show some level of porosity [18]. The full density in wire-based DED comes at a cost as the resolution of wire fillers is less than powders. Besides, the wire must have a proper contact force with the substrate; thus, any inhomogeneity in the surface roughness would cause issues in the wire feeding system, causing major flaws and defects in the printed parts. It may even cause an interruption in the process due to wire jamming and misalignment. As a result, they are not well suited for making fine features and the products usually require to be machined to the actual dimensions. In addition, more energy is needed to melt wires compared to powders.

In terms of the thermal energy source, the advantages and disadvantages attributed to the electron beam and laser PBF in Section 2.2.1.1 apply to DED technology as well. Electron beams have a higher energy efficiency but generate a larger focused beam spot than lasers, which reduces the resolution and creates a larger melt pool. EB-DED operates in a vacuum, so only wires can be used as the raw material in this process since powder needs to be delivered using an insert gas. Arc-based techniques such as PTA or EA are similar to EB-DED in terms of the working mechanism and even have higher energy to beam conversion efficiency [18]. They have the lowest resolution among the three mentioned DED systems. Thus, their products have lower dimensional accuracy and visibly beaded surfaces. However, the main advantage of these systems is their higher deposition rate and lower price compared to EB-DED and LDED (thousands vs millions [9]). Due to the generation of high levels of heat and larger melt pools in EA-DED and PTA-DED, thermal residual stress concentration often happens, which needs to be addressed at post-processing steps. PTA-DED is best suited for repairing, remanufacturing, and overhauling.

2.2.2.2 DED Advantages and Challenges

Process features: DED processes have a coarser resolution compared to PBF. The resolution of various DED sub-categories also varies significantly and is in the order of 0.25 mm for LDED and a few millimeters for EB-DED and PTA-DED. As a result, the deposition of fine features, internal channels, and complex designs might not be achievable with this technology. Users might face limits in design complexity, specifically for arc-based systems.

The high cost of filling a powder bed specifically for expensive specialty powders and the technical requirements of controlling the articulation and temperature of larger powder beds imposes a limit on the maximum build size in PBF. Since the powder is deposited through a nozzle in DED and built in the space, the built environment is typically much larger in DED systems. This feature is specifically desirable for the tooling, automotive, and aerospace industries. The largest build areas are virtually limitless and are a function of robotic systems used in DED. For example, if a Cartesian robot is used, the building envelope could be as large as 30 m × 10 m.

In terms of process throughput, however, DED systems offer some of the highest deposition rates among different AM classes. A typical deposition rate for powder- and wire-based DED systems can be in the range of 500–4000 cm^3/h depending on the energy source, with PTA-DED offering the highest amounts of deposition rate. It is in comparison with 5–20 cm^3/h deposition rate in PBF. As noted, there is a balance between printing speed and resolution. As such, consumers or manufacturers need to be familiar with each technology's points of strength to select the most suitable system for their application. The relatively high printing speed of DED combined with its resolution can be exploited for the fabrication of parts of low complexity and repairing.

Challenges and remedies: DED is a very complex manufacturing method controlled by dozens of process parameters. On top of the thermal energy source/optics, beam/material interaction, and build environment parameters it shares with PBF, controlling the nozzle and/or baseplate motions adds another layer of complexity to this process. Fortunately, the build part is visible in DED, and open-architecture DED systems are normally available, which provides higher accessibility to incorporate metrology/control systems into the DED setup.

In configurations that the energy source is incorporated in the powder delivery nozzle, the scanning of the beam is slower and potentially less accurate than moving the beam spot using galvanometers or electromagnetic system in PBF. This is the source of some printing flaws such as over-deposition at corners (similar to overfilling seen in extrusion systems), where the speed of the nozzle is reaching zero or trying to reach the stable printing speed, but the powder delivery amount is fixed. This might not be very concerning in DED as the final product is a near-net-shape part which should be machined to the target dimensions; however, addressing such flaws during the print requires high-end motion systems, sophisticated powder delivery apparatus, and integrated closed-loop control systems.

Finally, the low catchment efficiency of the powder-based DED causes a waste of material and has proven to be a difficult flaw to address [19]. Innovative nozzles and auxiliaries systems are being developed by R&D centers to improve catchment efficiency. Wire feeding mechanisms can be complex as well, specifically at large scales. The common point among all these technologies is that many factors controlling the delivery of filler materials should be optimized for a successful print. The current method of choice for carrying out such optimizations is trial and error; however, similar to PBF, new generations of DED systems are getting smarter via various sensors and closed-loop control systems.

Materials: As mentioned previously, two types of feedstock can be used in DED: powder for laser- and arc-based systems and wire for all energy sources. Because of this, a wider range of materials is available for DED in comparison with PBF. This includes reactive metals that can be processed in the vacuum condition of EB-DED. Furthermore, a distinctive advantage of DED is the ability to produce material graded parts in-situ. It is significantly easier to introduce a second material through the addition of a secondary powder hopper or wire coil feeder compared to PBF processes.

Powder delivery using nozzles has less stringent requirements in terms of morphology, particle size distribution, etc. compared to powder bed, which leads to saving material costs in DED. However, there is a need for the development of cost-effective sources of powder for DED similar to PBF to expand the applications of this technology in industry. Based on the research conducted by Sciaky (Table 2.2), the average cost per pound of AM grade Ti-6Al-4V powder is 2.4 times higher than its wire counterpart. This ratio for tantalum, Inconel 625, and stainless steel 316 is 1.05, 2.12, and 2.04, respectively [20]. Wires are also readily

Table 2.2 The market cost of wire and AM grade powder per pound for some of the materials used in DED based on a survey of U.S. commercial market conducted by Sciaky in June 2015 [20].

Materials	Ti-6Al-4V	Tantalum	Inconel 625	Stainless steel 316
Wire (0.9 mm D)	$58	$545	$26	$5
Wire (1.1 mm D)	$54	$545	$23	$4
Wire (1.6 mm D)	$50	$524	$22	$4
Wire (2.4 mm D)	$48	$502	$21	$4
Wire (3.2 mm D)	$45	$438	$21	$4
Wire (4 mm D)	$44	$438	N/A	$4
Powder (AM grade)	$120	$522	$48	$10

available for many alloys making wire-based DED an attractive process to many applications where lower geometrical accuracy is acceptable or 100% density is required. Wires need to be of high quality and consistency in terms of diameter to produce uniform beads. For more details about material properties in DED, see Section 2.3.4.

Products: Low-dimensional accuracy and surface finish compared to PBF are common in DED systems; however, these properties might be less concerning for this technique as the parts will need to go through finishing at post-processing steps. HIP and other thermal treatments are required to address the residual stress and distortion caused by large melt pools.

Parts made using powder-based DED display minor porosity, and the wire-based DED can produce fully dense parts even before heat treatments. Like PBF, the L-DED process is characterized by fast cooling rates resulting in desirable fine microstructure leading to improved strength. The microstructure, however, is typically larger in EB- and PTA-DED due to the slower cooling rate of these processes. This is particularly significant in arc technologies where a large heat built up during the process is expected, contributing even more to a longer cool down cycle and residual stresses [4, 9].

2.2.2.3 Commercial DED Systems

DED technology has been built upon decades of experience in laser and electron beam welding. One of the first systems, called Laser Engineer Net Shaping (LENS), was developed at Sandia National Laboratories (Albuquerque, NM, USA) in 1997. The technical details of this technology at its early stage are described in a 1998 document published by the LENS project team [21]. This technology was later commercialized by Optomec (Albuquerque, NM, USA) under the same trademark name. Currently, LDED machines are offered at the commercial level with various configurations by multiple companies. Sciaky Inc. (Chicago, IL, USA) seems to be the only commercial wire EB-DED machine producer. However, NASA has similar technology known as Electron Beam Freeform Fabrication (EBF3) developed for AM in outer space [22]. Arc-based systems are usually developed by university R&D centers, either from scratch or through retrofitting CNC systems. However, a Norwegian company called Norsk Titanium (Hønefoss, Norway) sells a PTAAM system called MERKE IV^{TM}.

Table 2.3 lists some of the commercially available DED systems and their key features, including energy source and power, beam focus diameter, deposition rate, build area

Table 2.3 Configurations of commercially available DED systems.

Manufacturer	Energy source	Beam focus diameter (μm)	Deposition rate	Build area dimension (mm) (W, D, H) or (Dia, H)	Number of material feeders
Optomec	Fiber laser Min power: 400 W Max power: 3000 W	Min: 670 Max: 3000		Min: 150 × 150 × 150 Max: 900 × 1500 × 900	Up to 4 hoppers
DMG Mori	Fiber laser diode Single and dual Min power: 2500 W Max power: 3000 W	Min: 1200 Max: 4000	Min: 0.8 kg/h Max: 2 kg/h	Min: 500 × 400 Max: 650 × 560	Up to 3
Trumpf	Laser Min power: 6000 W Max power: 8000 W			Min: 800 × 600 × 400 Max: 4000 × 2000 × 750	Up to 3
BeAM (Add Up)	Laser Min power: 200 W Max power: 2000 W		Min: 15 cm^3/h Max: 130 cm^3/h	Min: 400 × 250 × 300 Max: 1200 × 800 × 800	Up to 3
DM3D	Fiber laser Min power: 1 000 W Max power: 10 000 W			Min: 300 × 300 × 300 Max: 2330 × 1670 × 1670	Up to 2 hoppers
Inss Tek	Ytterbium fiber laser Min power: 300 W Max power: 3000 W		Min: 1.5 cm^3/h Max: 42 cm^3/h	Min: 150 × 150 × 150 Max: 4000 × 1000 × 1000	3–6 hoppers
Norsk Titanium	Plasma Arc		Min: 5 kg/h Max: 10 kg/h	900 × 600 × 300	
Sciaky	Electron beam		Min: 3 kg/h Max: 9 kg/h	Min: 711 × 635 × 1600 Max: 5791 × 1219 × 1219	Dual wire feed

dimensions, layer thickness, and the number of material feeders. Although, in comparison with PBF, there are fewer DED machine manufacturers, they still try to offer a wide range of key features for different industrial and research applications.

The smallest commercially available build area for LDED is 150 mm W × 150 mm D × 150 mm H, offered by Optomec and InssTek (Daejeon, South Korea) for small-scale productions and research and development. InssTek also offers the largest working area (4000 mm W × 1000 mm D × 1000 mm H). As mentioned, EB-DED systems have higher throughputs and can produce larger parts. This capability is exhibited by one of the largest build areas for any AM system offered by Sciaky Inc. 5791 mm W × 1219 mm W × 1219 mm D in their EBAM 300 machine.

Similar to LPBF, the type of lasers used in DED systems is mainly fiber lasers. Machine manufacturers usually offer multiple laser power options for the customers to select based on their material of choice and target application. The lowest energy power in a commercial system is 400 W offered by Optomec, and the highest is 10 000 W offered by DM3D (Auburn Hills, MI, USA). Some DMG Mori (Bielefeld, Germany) systems come with a dual fiber laser diode. Although not reported by all machine manufacturers, the minimum beam focus diameter tends to be in the order of hundreds of microns (compared to tens of microns for PBF), which contributes to the high deposition rates of these systems.

The build rate is reported in various units by different manufacturers making direct comparisons difficult; however, an interesting observation is the distinctively higher possible deposition rates of PTA-DED (10 kg/h) and EB-DED (9 kg/h) compared with LDED systems (2 kg/h). The maximum layer thickness of DED systems (750–4000 μm) seems to be one to two orders of magnitude higher than those of PBF systems (40–150 μm), as reported in Table 2.1.

Alloy alteration and multi-material deposition are one of the main advantages of DED systems. Optomec, for example, provides up to four hoppers making it possible to mix various alloys on the fly and print FGMs. Siacky systems also support up to two-wire feeding systems.

2.2.3 Binder Jetting (BJ)

2.2.3.1 BJ Process Description

BJ for making metal components entails three steps: (i) green part fabrication, (ii) drying/debinding, and (iii) sintering and/or infiltration. In the first step, similar to PBF, the process starts with the deposition of a layer of powder. The binder printhead, which has at least one array of between tens and hundreds of nozzles, passes over the powder bed and dispenses small droplets of a usually polymer-based binder to bind the particles together. After completing a layer, the powder bed moves down by the height of one layer and the recoater spreads a fresh layer of powder. This process is continued until the part is completed. Figure 2.8 describes BJ technology's main components, including the powder container and delivery system, inkjet printhead, build platform, and overflow bin.

BJ machines usually provide the option to control the build chamber temperature, which might be useful in partial drying of printed parts to avoid slippage when the new layer is deposited. The parts are left in the machine for a few hours until the binder is fully cured. Since the binder has filled the pores between the powder particles, at this stage, the product is an almost fully dense composite of metal and polymer. Next, the green parts are removed from the powder bed, gently cleaned to remove the excessive loose powder, and transferred to a furnace to be heated so that the binder is first evaporated followed by partial particle sintering (debinding

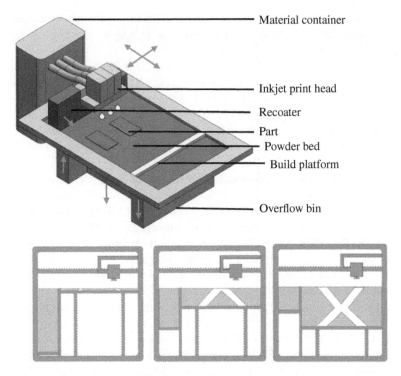

Material container

Inkjet print head

Recoater

Part

Powder bed

Build platform

Overflow bin

Figure 2.8   A schematic of binder jetting technology. Source: Redrawn and adopted from [23].

stage). At this stage, fragile green parts with a high level of porosity are obtained. Finally, either the sintering process is continued to close the pores further and improve the final part's mechanical properties or the green part is infiltrated with a different metal, normally bronze or copper, to fill the internal voids. The first option results in shrinkage of the parts that should be considered at the design stage. The second option has a significantly lower shrinkage rate and can be used to generate a two-material part with specifically engineered properties.

The final density of the BJ-made parts can be above 90%; however, the almost fully dense products of PBF and DED are difficult to obtain using this technology. HIP treatment can be used to decrease porosity. For certain applications such as implants or structural parts, porosity might be a desired property, in which case, BJ is an excellent choice of the manufacturing method. Through controlling the process parameters such as the powder bed density or binder saturation level, the degree of porosity can be controlled.

This technology shares characteristics with two other classes of AM: PBF and MJ. It is similar to PBF, where a bed of powder is printed layer by layer. However, unlike PBF, powder particles are not fused using a thermal energy source. They are rather bound together and to the previous layer using a binding liquid. This has economic benefits given the high cost of laser and electron beams and the strict safety precautions that should be considered when operating them. The binder is dispensed using a printhead similar to the MJ process with a major difference that the deposited binder in the BJ process only makes up a portion of the part, which will be largely evaporated at post-processing steps. In a sense, BJ works similar to standard

Figure 2.9   ExOne binder jetting technology in Action. Source: Courtesy of ExOne.

home inkjet printers but it glues powder particles together instead of spraying inks onto a paper. Some benefits of this technology include high printing speed, the ability to stack parts to improve efficiency, scalability, printing composites, etc. Figure 2.9 shows a BJ process in action when the binder is deposited from the printhead.

This technology was first developed at the Massachusetts Institute of Technology (MIT) in the late 1980's as three-dimensional printing (3D printing) [24]. The technology was licensed to Z Corporation (later purchased by 3D Systems in 2012), and other machine developers for commercialization. Commercial BJ machines are currently provided by 3D Systems, ExOne (North Huntingdon, PA, USA), Voxeljet (Friedberg, Germany), and HP (Palo Alto, CA, USA). The applications of BJ vary from production of visual prototypes to low-/medium-quantity functional parts to development of large parts with conformal cooling channels for casting and tooling.

Most BJ systems follow the same printing mechanism; however, HP's technique, known as multi-jet fusion, is slightly different. In this system, which was initially developed for non-metal powders and then applied to metal systems in HP's new line of machines (Metal Jet), two different binders are deposited on each layer: (i) fusing agent and (ii) detailing agent. The fusing agent is deposited where the particles need to be fused together to form the cross section of the part. The detailing agent is deposited around the fusing agent, determining the cross section's borders. At the next step, a heat source passes over the build platform. It selectively fuses the particles coated with the fusing agent while leaving sections coated with detailing agent unfused to produce finer features and sharper edges compared with regular BJ processes and control binder bleeding to the areas outside the intended part geometry. This process is shown in Figure 2.10 and explained in more detail in a white paper from HP [27].

BJ technology has been largely used with ceramic and polymer composite powders as the main materials. Its application for the fabrication of metal parts has been largely indirect through BJ of molds and models for casting. In recent years, companies such as ExOne have introduced machines dedicated to metal BJ along with the provision of proprietary alloy powders of different classes such as steel, titanium, nickel, etc. Sample metal parts made using BJ are shown in Figure 2.11.

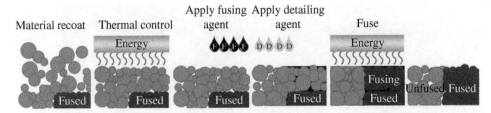

Material recoat   Thermal control   Apply fusing   Apply detailing   Fuse
                                    agent   agent

Figure 2.10   HP multi-jet fusion technique steps. Source: Courtesy of HP Development [27].

Figure 2.11   Sample part made using BJ technology. Source: Courtesy of 3dhubs [23].

2.2.3.2   BJ Advantages and Challenges

Process features: As a powder bed-based process, BJ also exhibits high vertical resolution due to fine layers of powder. The lateral resolution of this technology depends on the size, quality, and uniformity of the droplets on one side and their wetting properties in contact with the powder bed on the other. Typically, the resolution of BJ processes depends on the layer thickness when dimensional precision could be around 200 µm after sintering. However, the shrinkage due to sintering can be compensated by antistrophic correction factors applied to the original CAD model. As the green parts are not strong, printing parts with very fine features such as cellular and architectural structure might not be possible as these might break while transferring the parts to the furnace for curing. As such, some level of limitation in terms of design freedom is expected.

One major advantage of BJ is that with the parts being contained by a powder bed and in lack of high thermal gradient, there is no need for support structures to transfer the heat to the baseplate. As such, it is possible to stack parts on top of each other as far as the build dimensions and material availability permit to improve the productivity.

BJ is also considered a high-speed AM process given that its manufacturing process (unlike PBF and DED) does not involve following a raster path (in a point-wise fashion) to fuse the particles into parallel lines until the cross section of the part is formed. Instead, one or more arrays of nozzle bind the particles together in a linear fashion. Thus, the silhouette of parts on each layer is formed by one or a few passes of a printhead. Since the inkjet printheads are very inexpensive in comparison with lasers or other high-energy heat sources, BJ is the best technology for an economically feasible process scale up by simply increasing the number of arrays of nozzles or printheads.

Challenges and remedies: Like other AM systems, BJ is also a complicated process where many factors influence the final part quality. Even though no beam/optics parameters are involved, other parameters, such as printhead velocity, binder saturation, etc., should be optimized to ensure proper bonding, obtain target porosity levels, and maintain the geometrical accuracy. Even with an optimized process parameter set, it is usually required to print a few test layers until the printing condition is stable. Nozzle clogging and deterioration of the binder dispensing quality during the fabrication are common issues in the BJ process. As such, nozzle cleaning/printhead health checking stations are usually incorporated in commercial BJ systems, with periodical and automatic cleanings scheduled as part of the process.

Part shrinkage is an inherent feature of this technology; thus, it is important to consider the shrinkage rate at the design stage to make sure the final product meets the dimensional requirements. Identifying the correct sintering condition and developing master sinter curves increase the process complexity and initial setup time.

Materials: The majority of materials available for BJ are ceramics and polymers, and the choice of commercial metallic materials, although expanding, is currently limited compared to PBF and DED. A benefit of using binders instead of lasers or electron beams is that it can process metals that cannot be thermally processed. As a powder bed based process, BJ is also limited to one type of powder in the build compartment. However, the development of functionally graded structures is possible by either dispensing a binder with specific properties such as electrical conductivity, color, elasticity, etc. or infiltration of materials with lower melting points than the base powder material. For example, ExOne provides an annealed 420 stainless steel matrix infiltrated with bronze at the composition rate of 60 : 40, which yields desirable high mechanical performance (ultimate strength of 682 MPa and elastic modulus of 147 GPa), machinability, weldability, and high wear resistance [26].

Products: Generally, parts made using BJ have poorer mechanical properties than the PBF and DED products due to the higher level of porosity in BJ-AM, which leads to stress concentration, cracking, and fatigue failure. BJ has a similar surface roughness to PBF [4] but with lower dimensional accuracy. As mentioned, HP's metal jet technology is one of the recent trials to improve BJ techniques in printing sharp corners and well-defined edges. In a different approach, Desktop Metal (Burlington, MA, USA) uses printheads in their line of BJ machines that can deposit droplets with different diameters down to 1 pL to improve the dimensional accuracy and surface finish. The difference between this system and traditional BJ systems is shown in Figure 2.12. As seen in the image on the right, droplets with the largest possible volume fill the cross-sectional interior and smaller droplets are dispensed closer to the edges.

Since parts are not made by fusing particles via a heat source in BJ, the anisotropic properties imparted by such techniques do not cause property non-uniformity in x–y direction of BJ. However, weaker mechanical properties in vertical (print) direction are still an issue.

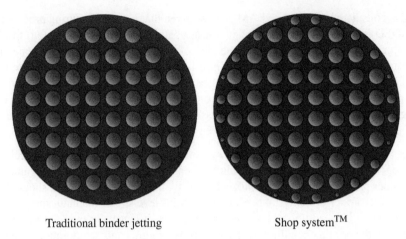

Traditional binder jetting   Shop system^{TM}

Figure 2.12 Comparison of traditional binder jetting and Desktop Metal's binder jetting with controllable droplet volume to avoid under/overprinting. Source: Redrawn; courtesy of Desktop Metal [27].

2.2.3.3 Commercial BJ Systems

There are a handful of BJ machine developers, and not all of them offer metal BJ-AM systems. For example, BJ machines of companies like HP and 3D Systems are mainly developed for non-metal powders; however, since R&D groups might use them for metal alloys, they have been included in this section. There is generally less diversity in the configuration of BJ machines each manufacturer offers. As a matter of fact, many of these manufacturers only offer one or two machine options for a few compatible metal powders. The configurations of some of the available commercial BJ systems are provided in Table 2.4.

The smallest building area belongs to ExOne's Innovent+® with dimensions of 165 mm W × 65 mm D × 65 mm H. Voxeljet, on the other hand, offers a very large platform of 4000 mm W × 2000 mm D × 1000 mm H, which is in par with some of the largest build areas in DED machines – a good indicator of the capacity of BJ for scale-up and large part manufacturing.

The minimum layer thickness in BJ systems seems to be between 30 and 50 μm, higher than the typical minimum layer thickness of 10 μm in most PBF systems. However, the average layer thickness for both of these powder bed-based technologies is similar. Considering the differences in the build platform size and layer thickness, ExOne and HP offer the highest deposition rates of 10 000 and ~5 000 cm^3/h, respectively. Desktop Metal machines with 1600 × 1600 dpi resolution and HP machines with 1200 × 1200 dpi resolution should produce the richest features.

The BJ industry is following the same trends as those of PBF and DED, albeit at a slower pace, such as the development of smarter machines equipped with sensors, cameras, and metrology systems for real-time control of the production steps, expanding materials portfolio by development of custom low-cost powders and recipes, and scaling up the process by introduction of larger build areas, faster binder deposition systems, and interconnected machines in an automated industrial platform.

Table 2.4 Configurations of commercially available BJ systems.

Manufacturer	Build rate	Print resolution	Build area dimension (mm) (W, D, H)	Layer thickness (μm)
ExOne	Min: 164 cm³/h Max: 10 000 cm³/h	Min: >50 μm voxels Max: 30 μm voxels	Min: 165 × 65 × 65 Max: 800 × 500 × 400	Min: 30 Max: 200
3D Systems	Min: 20 mm/h Max: 28 mm/h (in vertical speed)	Min: 300 × 450 dpi Max: 600 × 540 dpi	Min: 236 × 185 × 127 Max: 508 × 381 × 229	100
HP	Min: 1817 cm³/h Max: 5058 cm³/h	1200 × 1200 dpi	Min: 332 × 190 × 248 Max: 430 × 320 × 200	Min: 50 Max: 100
Voxeljet		Up to 300 dpi	Min: 300 × 200 × 150 Max: 4000 × 2000 × 1000	
Desktop Metal	700 cm³/h	Min: 1200 × 1200 dpi Max: 1600 × 1600 dpi		Min: 40 Max: 100
Digital Metal	100 cm³/h		203 × 180 × 69	

2.2.4 Emerging Metal AM Processes

The technologies reviewed throughout Sections 2.2.1–2.2.3 (PBF, DED, and BJ) are the most well-known AM techniques for producing metal parts. Considering their successful track record in high-end engineering parts production, they have received considerable attention and investment by new companies trying to gradually introduce AM into their business model, machine developers, and R&D centers. The metal AM, however, is not limited to these three categories. For example, metal SL is a technology joining layers of metal sheet using, e.g., ultrasonic consolidation, to make high-quality metal parts for special applications. Moreover, various start-up companies have been trying to bring the ME technology, a well-established technique for the production of thermoplastic parts, to the domain of metal AM. These systems are usually designed to be low-cost and easy to operate in an office environment with fewer safety requirements. Finally, researchers and start-ups have investigated the possibility of using MJ, another dominantly polymer printing technology, to produce small-sized metal parts. Despite considerable success in the fabrication of full-color, heterogeneous polymer parts, MJ currently seems to have the lowest technology readiness level (TRL) among all seven classes of AM. However, MJ is considered one of the most promising AM technologies in reaching the digital matter fabrication goal where each voxel (print volume unit) can be pre-programmed, offering limitless opportunities for digital printing of materials.

2.2.4.1 Material Extrusion

ME is a class of AM technologies that dispenses a continuous flow of a fluidic or semi-fluidic material through a nozzle and creates each layer of the part by following a predetermined tool-path. There are various ME techniques ranging from a single syringe attached to an XYZ gantry system dispensing materials using compressed gas back pressure to bioprinters made up of an array of sophisticated dispensing units in an environmentally controlled chamber to the systems that are fed with filaments of thermoplastic polymers. The latter, known as fused deposition modeling (FDM), is probably the most widely adopted type of AM used by hobbyists for prototype fabrications and industries for high-end products alike. Since FDM is the ME technique of choice for metal parts production, we will focus on this technology in this book.

Figure 2.13 shows the different components of an FDM system. As seen, at least two coils of filaments (main structure material and support) are directed to an extrusion nozzle. The filament heats up at the nozzle to become semi-liquid and ejects the nozzle in a continuous form. As the printhead moves, the extruded material bonds with the previous layer (or substrate). By following the toolpath representing the cross section of the part, a layer is completed. Support structures may also be needed to fabricate overhangs, internal voids, and other challenging geometrical features. At the next step, the printhead moves up, or the substrate moves down by the height of a layer, and the next layer is fabricated. This process continues until the part is completed. Finally, the part is removed from the substrate, and the support structures are removed, usually through chemical dissolving.

The FDM process is relatively simple, and over the years, many open-source toolpath generation, STL file correction, and process control software have been developed in support of this technology. Although the large industrial grade FDMs can be costly, this technology does not usually require a considerable upfront investment. Also, since the fine powders, compressed gases, and high energy beams do not apply to this technology, its safety regulations are considerably less stringent. As a result, they can be set up in non-laboratory environments such as offices and even homes. Increasing the number of nozzles makes it easy to increase the processing speed or print multi-material parts using FDM.

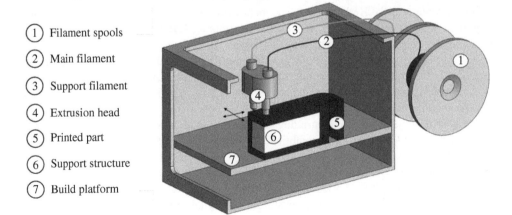

① Filament spools
② Main filament
③ Support filament
④ Extrusion head
⑤ Printed part
⑥ Support structure
⑦ Build platform

Figure 2.13 Schematic of material extrusion system. Source: Adopted and redrawn from [23].

Figure 2.14 A sample metal filament from Desktop Metal. In contrast to polymer filaments, metal filaments are not formed into a coil. Source: Courtesy of Desktop Metal [27].

Although FDM is historically a polymer manufacturing technique, some companies have developed FDM systems capable of processing metals to benefit this process. The filaments used in metal FDM are a composite of metal particles bound together using a proprietary polymer. As there is no loose powder involved, the handling of materials in this form is much safer. A sample of metal filaments is shown in Figure 2.14. The printing mechanism is the same as what described above; however, after the part is printed, it will go through debinding and sintering stages similar to BJ technology to remove the binding polymer and end up completely made of metal.

Desktop Metal is one of the companies that feature one "office-friendly" metal FDM machine with a production rate of $16\,cm^3/h$. This company currently offers FDM metal filaments in stainless steel (1704 and 316L), steel (H13 and 4140), Inconel 625, and copper. The lateral resolution, vertical resolution, and final part density, as reported by the manufacture, are 250–400 μm, 50 μm, and 96–99.8%, respectively [27]. Markforged (Watertown, MA, USA) is another company offering desktop ME systems for metals. Their metal FDM called Metal X System is similar to the Desktop Metal's Studio System in terms of the printing mechanism; however, Markforged currently offers a wider variety of materials, including Ti-6Al-4V [28]. Other varieties of metal FDM include a large build area machine AMCELL® from TRI-DITIVE (Meres, Spain) with the build rate of $400\,cm^3/h$ featuring steel and aluminum raw materials [29], and PAM series MC printers from Pollen AM (Ivry-sur-Seine, France) using low-cost pallets as the raw materials. The pallets are transferred to the nozzle using a screw feed mechanism and fluidized by shear pressure and heat. As reported by the manufacturer, this system's lateral and vertical resolutions are 5 and 40 μm, respectively [30]. The manufacturer claims that the users can use pellets from any supplier. Sample parts made using metal ME machines are displayed in Figure 2.15.

Although ME is a low-cost alternative with fewer safety concerns compared with traditional metal AM techniques for producing visual and lightweight parts, its shortcomings must be considered upfront. These shortcomings are lower final part mechanical performance, anisotropic properties in the build direction, low process speed, surface flaws (e.g., visible staircase effect, seam, overfilling, etc.), and high dependency on the proprietary metal filaments currently exclusively provided by machine developers.

(a)   (b)

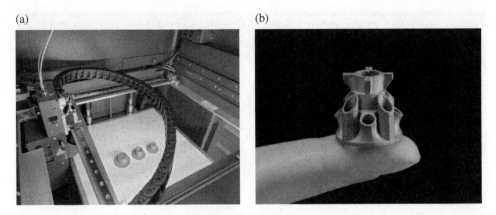

Figure 2.15   Visual prototypes made using (a) 17-4 PH stainless steel on Markforged ME system by Shukia Medical [31]; (b) Desktop Metal's Studio System. Source: Courtesy of Desktop Metal [27].

2.2.4.2   Material Jetting

In MJ technology, droplets of fluids are deposited onto a substrate where needed, followed by solidification to form a part. From an AM perspective, MJ's working mechanism is very similar to BJ. The main difference is the fact that the entire material required to fabricate a part is deposited from a printhead, and the part is made in empty space rather than a powder bed. This technology is low cost compared to AM systems using thermal energy sources such as laser and electron beams. In contrast to ME, where the nozzle needs to be near the substrate for a successful extrusion, MJ is a non-contact-based system firing droplets to the substrate from a distance. The technology is highly scalable at a reasonable cost by addition of printheads. This technology's unique advantage is the capability to fabricate true digital materials with programmable properties at the voxel level.

Figure 2.16 shows the main components of the MJ technology. The print material is stored in a container and transferred to the printhead for deposition. Only materials in liquid form can be used, so if the target material is not already a liquid, it needs to be melted or dissolved/suspended in a fluidic medium. At the next step, the backpressure creates enough energy to eject the material from the nozzle orifice and form a droplet. There are various mechanisms for creating the droplet, including (i) thermal jetting: a heater attached to the fluid reservoir evaporates the nearby material to form a bubble. The propagation of the bubble forces a droplet out of the orifice; (ii) piezoelectric jetting: the expansion/contraction of a piezoelectric ceramic actuated by electrical current causes deformation of the fluid reservoir and consequently droplet ejection. MJ printheads have hundreds to thousands of nozzles to deposit a considerable amount of material at a fast pace. Each nozzle releases a droplet where needed, known as drop-on-demand (DOD) printing. The droplets coalesce on the substrate to form a layer and are solidified either through cooling down, evaporation of the solvent, chemical reaction, or photopolymerization. The process is continued until a part is fully built. A standalone printhead is usually dedicated to the printing of sacrificial support materials.

Many academic research groups have developed in-house-built MJ systems at a low cost. The easy assembly of this system's basic components is a contributing factor to its widespread use for R&D. Various materials, including polymers, ceramics, composites, functional inks,

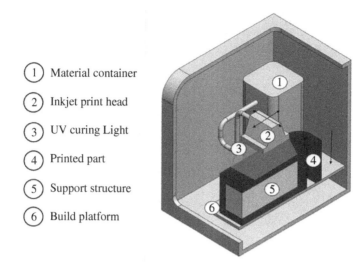

① Material container

② Inkjet print head

③ UV curing Light

④ Printed part

⑤ Support structure

⑥ Build platform

Figure 2.16 Schematic of material jetting technology. Source: Redrawn and adopted from [23].

and metals, have been tested, optimized, and characterized for MJ by these groups. Companies such as Stratasys and 3D Systems offer high-end MJ systems capable of printing heterogeneous and fully colored parts. These systems can mainly be operated where polymers offered by the machine developer are used, and the utilization of other materials might damage the printing systems or invalidate their warranty, which can be very limiting. XJET (Rehovot, Israel) is the first company that has developed a commercial MJ system for metals. In their NanoParticle Jetting™ process, described in this patent application [32], inks comprised of suspended metal nanoparticles in a liquid medium are jetted onto a substrate along with support material. The part is formed by each pass of the printhead and then goes through the support removal and sintering processes to obtain the desired mechanical properties. The system is reported to dispense very thin layers and detailed features with densities up to 99.9% after sintering. Figure 2.17 shows XJET's NanoParticle Jetting technique, where the main metallic nanoparticles and sacrificial materials are dispensed from the printhead. Some parts fabricated by this technology are shown in Figure 2.18.

Despite the large potential of MJ technology in the manufacture of true multi-material 3D printing, there are various drawbacks related to technical challenges often associated with this manufacturing method, such as nozzle clogging, effective droplet fly path control, low material content loading in suspensions, and most importantly material jettability. For a droplet to be formed, a material must meet certain rheological criteria governed by its viscosity, density, and surface tension. Even the smallest changes to any of these parameters might affect the quality and uniformity of the droplets or prevent the formation of a droplet.

2.2.4.3 Sheet Lamination

SL is an AM process during which raw materials in the form of thin sheets are bound together. These sheets can be trimmed down to represent the cross section of a layer before solid-state

Figure 2.17   XJET's NanoParticle Jetting technique is one of the emerging metal AM technologies. Source: Courtesy of XJET [33].

Figure 2.18   Sample parts made using XJET system. Source: Courtesy of XJET [33].

joining, or the excess materials may be removed after the sheets are bound together to create a part with a certain height.

The binding procedure in metal SL is ultrasonic consolidation or brazing. The brazing procedure that was introduced in 1980's has been abandoned, however, ultrasonic consolidation, which is a solid-state metal joining process, is now emerging. In this technology, two layers of metal sheets are placed on top of each other when an ultrasonic tool passes over them to remove the surface oxide due to high friction caused by the ultrasonic motion. Once the surfaces are oxide-free, two sheets form a metallurgical bond. This binding mechanism is illustrated in Figure 2.19. This process is repeated until the near-net-shape part is formed. A machining tool then removes the excess material and creates the final geometry.

This process is carried out at temperatures far below the melting point of the build material; unlike other AM technologies, it does not significantly alter the microstructure and, consequently, the mechanical properties of the raw material. Another advantage is that this technology can join dissimilar metals together in a much more efficient fashion than fusion, which

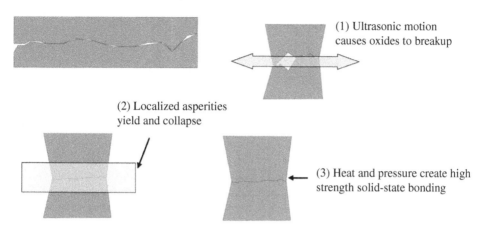

Figure 2.19 Ultrasonic consolidation mechanism. Source: Courtesy of Fabrisonic [34].

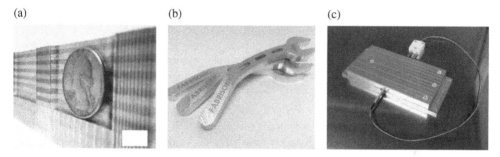

Figure 2.20 FGM parts made using Fabrisonic's ultrasonic consolidation system (a) and (b) parts composed of two materials, (c) smart structures with an embedded thermocouple. Source: Courtesy of Fabrisonic [34].

usually involves a transition area between the two alloys with poorer properties, highly prone to mechanical failure. Thus, it is suitable for the fabrication of FGMs as well as parts with embedded sensors. The final parts are highly dense without further post-processing, such as HIP treatment, as is the case with PBF and BJ products. A list of material pairs proven to be compatible with ultrasonic consolidation can be found in [34]. Among several suitable materials for this process, Al alloys can join to a wide range of materials/alloys such as Be, Cu, Au, Mg, Mo, Ni, Ti, Si, Ta, etc.

This technology has several limitations in terms of compatible materials, design freedom, and process speed. Nonetheless, a niche market is growing for ultrasonic SL applications such as effective thermal dissipation systems for casting, parts with embedded sensors and electronics to facilitate the Internet of Things (IoT), etc. Fabrisonic (Columbus, OH, USA) is the first company to commercialize this technology for AM applications showcasing FGM parts made from aluminum and copper, as shown in Figure 2.20. The right photo shows a smart structure, where a thermocouple sensor has been embedded within metallic structures. Making such smart structures with SL is the most appealing feature of this process.

2.3   Main Process Parameters for Metal DED, PBF, and BJ

As mentioned earlier, AM processes are very complex, driven by tens to hundreds of process parameters, which all can affect the quality of the final product. These parameters, their acceptable value ranges, their level of influence on various measurable output parameters, and their interdependencies are to a great degree poorly known for different AM technologies. Unlike conventional manufacturing techniques for which process parameter selection guidelines have been developed over decades, the common method for process selection for AM has been trial and error and operators' experience. The historical data for AM technologies are very limited but growing. Even the data available in the literature may not be very reliable as they might have been obtained through in-house developed systems, or non-standard procedures might have been used for data collection; thus, it is hard to duplicate and substantiate them. Furthermore, AM software developers are trying to develop software packages that assist in the development of optimized process sets based on the input material, print conditions, and part geometry; we are yet to see the wide availability of such software.

All AM process parameters can be divided into three categories, as shown in Figure 2.21:

1. Input parameters: A combination of material properties (powder rheology, chemical composition, etc.), process parameters (beam power, layer thickness, etc.), ambient properties (build chamber temperature, shield gas type, etc.), and machine features (motion system accuracy, maximum velocity, etc.). Some of these parameters, such as laser power, are controllable, and others, such as the type of shield gas, are normally constant throughout the manufacturing process. In this section, these parameters are reviewed in detail and their effect on output parameters is explained. Some of these parameters are specific to only one AM technique, while others are common across various technologies. We have identified the applicable technologies for each parameter. Any potential difference in parameter conditions from one platform to another has also been identified.

2. Process physics: Parameters, such as various thermodynamics phenomena, solidification, binder imbibition, etc., happening during the printing process. These phenomena control the transition of raw materials (powder, wire, paste) to a solid geometry due to the selection of appropriate values/conditions for input parameters. Various measurement and control systems are being developed and incorporated into commercial AM machines to monitor the process physics in situ and adjust the input parameters according to their behavior [36]. A description of these physical processes and the numerical and analytical models that govern them are provided in Chapters 4–7.

3. Output parameters: The bulk properties of the printed part are usually measured after the manufacturing process has been finished using various destructive and non-destructive tests. These include geometrical properties (tolerances, distortion, etc.), mechanical properties (hardness, strength, toughness, fatigue, etc.), metallurgical properties, and many more. Standardization bodies such as ASTM are developing new testing standards for AM. Obtaining repeatable outputs specifically in terms of density, hardness, surface roughness, and geometrical fidelity is one of the main challenges that the AM industry is currently facing [9]. These parameters are explained in detail in Chapter 8.

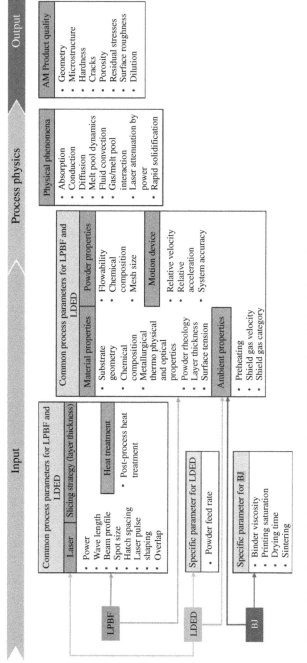

Figure 2.21 Process design parameters for LDED, LPBF, and BJ techniques. Source: Republished with permission from Elsevier [35].

2.3.1   Main Output Parameters

Many output parameters (e.g., microstructure, hardness, and cracking behaviour) are common for most metal manufacturing. In the following, a few parameters of metal AM are customized.

2.3.1.1   Melt Pool Dimensional Characteristics

Several parameters associated with the deposited/solidified geometry are shown in Figure 2.22. In this figure, h is the track height, w is the track or melt pool width, θ is the wetting angle, and b is the melt pool depth (without the track height), representing the thickness of substrate melted during the DED or LPBF and added to the track region. In PBF, the melt pool depth is considered b, whereas, in DED, $h + b$ is considered the melt pool height.

2.3.1.2   Dilution

One of the properties of the melt pool or clad, dilution has two definitions: geometrical and metallurgical [15]. The geometrical definition of dilution is illustrated in Figure 2.22. According to the specified parameters in the figure, the dilution is

$$dilution = \frac{b}{b + h} \tag{2.1}$$

Alternatively, dilution may be defined as the percentage of the surface layer's total volume contributed by the substrate's melting [37]. According to the composition, dilution is defined as

$$dilution = \frac{\rho_c(X_{s+c} - X_c)}{\rho_s(X_s - X_{c+s}) + \rho_c(X_{s+c} - X_c)} \tag{2.2}$$

where ρ_c is the density of melted powder alloy (kg/m^3), ρ_s is the density of substrate material (kg/m^3), X_{s+c} is the weight percentage of element X in the total surface of the clad region (%),

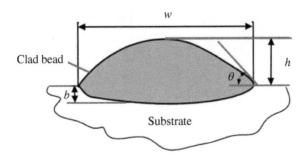

Figure 2.22   Melt pool and clad/track bead geometrical parameters. Source: Republished with permission from Taylor & Francis Group [15]; permission conveyed through Copyright Clearance Center, Inc.

X_c is the weight percentage of element X in the powder alloy (%), and X_s is the weight percentage of element X in the substrate (%).

2.3.2 Combined Thermal Energy Source Parameters for PBF and DED

Parameters such as beam power, operation mode, focused spot diameter (D), beam profile, and wavelength (λ) are critical factors in the selection and processing of materials in PBF and DED systems. Equations for calculating the average laser and electron beam energy are discussed in Chapter 4. In these equations, absorptivity plays an important role. However, in the following, multiple combined parameters are introduced that are widely used to show the process signature and quality.

A successful print on metal AM system results from an optimized high-order interaction of over a hundred different material and system parameters for a certain alloy. For example, P as laser power (W), v as scanning speed (m/s), s as hatch distance (m), and l as layer thickness (m) have been identified as some of the most significant PBF process parameters by multiple researchers [14, 37, 40]. Due to the high interdependency of energy and other process parameters on the melt pool temperature and dimensions, it is common practice to use a combined process parameter to measure the volumetric energy density (VED) or surface energy density (SED) in metal AM.

VED is the amount of thermal energy received by each unit of volume. Four combined energy density measures for various metal AM techniques are introduced [35]. These parameters can be used to establish thresholds of energy input for each material (or material group) to achieve full density products, as shown in their article based on the results reported for various ferrous alloys in the literature (Figure 2.23).

The specific energy is indicative of input energy from the surface irradiated by the laser beam [15]:

$$E_{specific} = \frac{P}{vd} \tag{2.3}$$

$E_{specific}$ (J/m^2) is mainly used in DED as the energy enters the substrate from the surface. In contrary, VED (J/m^3) is more widely used for PBF techniques [35], where the energy enters the domain through a scattered beam inside the voids between particles. In general, the four most important parameters are combined and justified based on the volumetric heat adsorption in the powder bed. Depending on the chosen parameters, two VEDs are defined as [37]:

$$VED_{Hatch} = \frac{P}{vsl} \tag{2.4}$$

where s is hatch spacing and l is the layer thickness. The spot size d provides an alternative definition more similar to specific energy:

$$VED_{Spot} = \frac{P}{vdl} \tag{2.5}$$

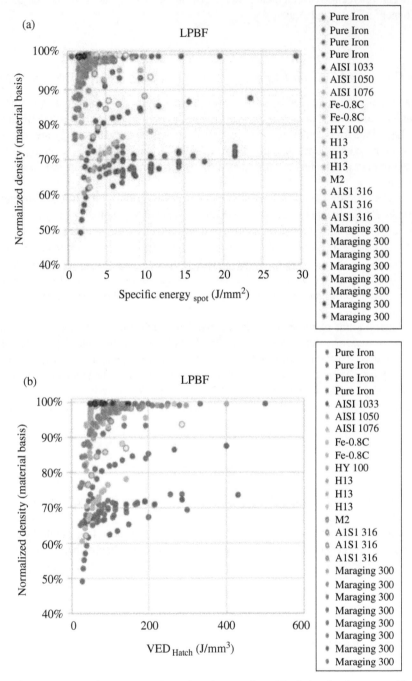

Figure 2.23 Relative density vs. energy for various ferrous alloys (a) relative density vs. specific energy, (b) relative density vs. hatch-based VED, (c) relative density vs. spot-based VED, (d) VED/H techniques. Source: Republished with permission from Elsevier [35].

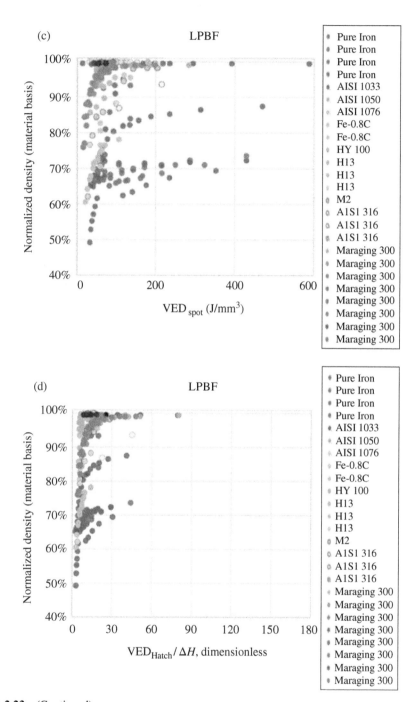

Figure 2.23 (Continued)

This is presumably related to energy requirements for fully melted materials within the proposed cube cells represented by hatch spacing, spot size, and interaction length (i.e., speed/interaction time). Therefore, to compare across materials normalized *VED*, estimated enthalpy difference is used between room temperature (T_0) and the materials melting point (T_m).

$$\frac{VED_{Spot}}{\Delta H} = \frac{P}{vdl} \cdot \frac{\beta}{C_{p,100°C}(T_m - T_0)} \tag{2.6}$$

where β is the absorption factor and C_p is the heat capacity. By using C_p at 100 °C, Figure 2.23 allows prediction that $\frac{VED_{spot}}{\Delta H}$ should be >30 for realization of full density part in steels. Based on the observed scatter, it is apparent that other factors are significant as well.

Furthermore, to incorporate material properties and the heat penetration depth that would represent the keyhole depth into the analysis, Fayazfar et al. propose a new combined number called "dimensionless depth." The one-dimensional transient model is formulated as [35]:

$$T(z,t) = \frac{P\beta}{AK} \sqrt{4\alpha t} \ iefc \left[\frac{z}{\sqrt{4\alpha t}}\right] \tag{2.7}$$

where A is the spot area, z is the distance from the surface, t is the laser interaction time, K is the thermal conductivity, and α is the thermal diffusivity. While 90% of temperature change occurs within 1 scaling parameter of the surface, the interaction time can be expressed by d/v (see Chapter 4 for more details). The heating depth can be obtained when the limit of the equation above approaches zero. This results in the following equation representing the heat penetration depth:

$$Heat\ Depth = \sqrt{4\alpha\frac{d}{v}} \tag{2.8}$$

A logical dimensionless combined process parameter may be proposed through a ratio of the above equation and layer thickness, leading to a parameter, so-called heating depth ratio (HDR).

$$HDR = \frac{\sqrt{4\alpha\frac{d}{v}}}{l} \tag{2.9}$$

The relative density versus HDR data for ferrous alloys is plotted in Figure 2.24. From the data available in the literature, it is apparent that all successful authors are operating with a heating depth greater than layer thickness. Therefore, an HDR of 1.5 appears to be a good guideline for full density parts.

2.3.3   Beam Scanning Strategies and Parameters for PBF and DED

Parameters such as beam scan velocity, hatch distance, scan pattern, pre-scanning, and re-scanning are considered beam scanning parameters and highly affect the microstructure, surface quality, internal porosity, and mechanical properties of AM-made parts. Different scan

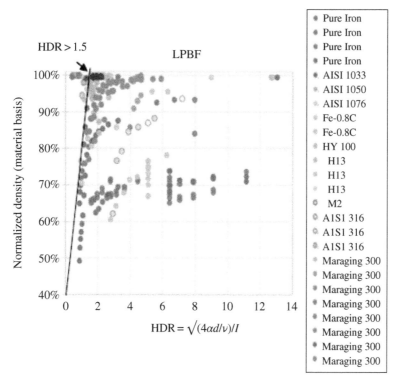

Figure 2.24 Normalized density vs. specific energy techniques. Source: Republished with permission from Elsevier [35].

patterns have been adopted to control residual stresses and part distortions. Although there are numerous scanning patterns, some of the most frequently used patterns are shown in Figure 2.25 and described below:

1. Line scanning is the most basic scanning strategy (b–f), which is achieved by following simple rectilinear scanning paths for all the layers either in the x-direction (0°) or y-direction (90°). Changing the scanning angle (e.g., 30°, 45°, 67°, …) or adopting different scanning angles for successive layers in a rotational manner (e.g., 0°, 90°, 0°, … or 45°, 135°, 45°, …) are other options commonly practiced by AM professionals. Among these angles, many machine developers use 67° between the layers.
2. Island or chessboard scanning (a) is another popular approach that divides the print layer into multiple squares or randomly shaped smaller islands with different scanning types or angles assigned to each.
3. In-out and out-in scanning strategies (g and h) that include a spiral motion.

The reasoning behind the scanning strategy affecting residual stresses is not very well understood as contradicting results have been reported by different researchers; however, the consensus is that there is more thermal stress and deformation along the scanning path.

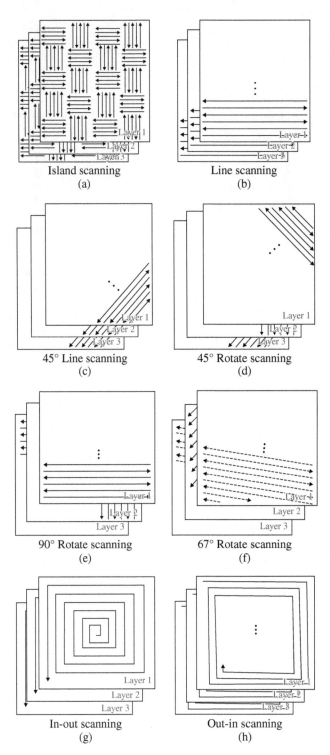

Figure 2.25 Common scanning strategies. Source: Redrawn and adapted with permission from Elsevier [39].

The island scanning strategy is expected to reduce the residual stress and geometrical deformation as a result thereof.

The effect of scan speed on the final product properties depends on other factors, specifically the beam power. However, in general, if all parameters are fixed, an increase in the scan speed results in less energy transferred to the powder bed volume, resulting in unmelted pockets of particles, known as "Lack of Fusion" (Figure 2.27b). Balling effect (due to Plateau–Rayleigh instability, see Chapter 4 for more information) generation of unstable melt pool may also occur [35], where multiple melt pool regions form on the substrate (Figure 2.27c).

On the other hand, scanning the beam on the powder bed at a very low speed could cause particle evaporation and result in keyhole-induced porosity (Figure 2.27a). Very low scanning speeds can create a deep melt pool, while the heat zone reaches the previous layers, causing remelting and resolidifying them repeatedly and changing their microstructure. Figure 2.26d depicts different manufacturing flaws as a result of the different energy and beam speed combinations.

Figure 2.27 generally shows the combined effects of laser power and scanning velocity on part density and type of potential defects in the printed parts at different conditions.

Other beam scanning properties, such as hatch distance, pattern, overlap, and scanning accuracy, control the properties of the final product as well. Hatch distance is the distance between the centers of parallel trajectory lines scanned by the beam. If this distance is too large, the gap between the lines might remain unmelted, reducing the mechanical properties of the parts. Programming a level of overlap between successive scan tracks into the trajectory design is common practice in both PBF and DED. Too much overlap, on the other hand, has the same effect as excessive energy leading to increased volumetric porosity. The laser path planning, including the hatch distance, can be determined based on the melt pool width, shown in Figure 2.28. As mentioned, hatch spacing to the beam spot size ratio is an important combined process parameter.

Figure 2.29 shows the effect of hatch distance on the porosity of aluminum alloy parts made using LPBF. The volume of pores increases as the hatch distance increases until scan tracks are separated, where there is no overlap.

2.3.4   Powder Properties for PBF, DED, and BJ

In the context of powder bed-based AM technologies, material properties are usually investigated under two separate sub-categories: (i) powder particle properties and (ii) bulk powder behavior [11, 37, 49].

At the particle level, morphology, which includes powder size distribution (PSD), sphericity, surface roughness, skewness as well as other properties like chemical composition, particle density (ρ), moisture level, etc. determine the flowability, beam/powder interaction, and the quality of final parts. Figure 2.30 shows the classification of particle level properties by [11].

The effect of PSD and particle shape on the flowability is the center of attention in AM powder characterization. In contrast, the important effect of intrinsic parameters such as the powder's chemical composition and microstructure has been studied less extensively [47]. The PSD directly dictates the lateral and vertical printing resolution and finished surface quality in PBF. The smaller the particle size, the higher the resolution of the print and the better its surface morphology. On the other hand, the electrostatic forces between the particles increase as the

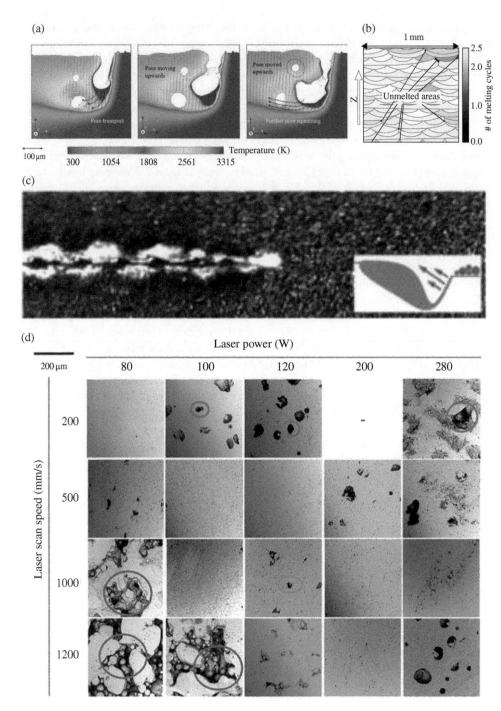

Figure 2.26 (a) Keyhole porosity and its formation mechanism. Source: Republished with permission from Elsevier [40]. (b) Lack of fusion porosity between melt pool borders. Source: Republished with permission from Elsevier [41]. (c) Balling effects. Source: Republished under CC BY 4.0 open access license [42]. (d) Effect of laser scan speed and laser power on the formation of porosity: pores marked in bottom left figures for condition 80–1000, 80–1200 and 100–1200 are lack of fusion pores, pores marked in the first row of figures are keyhole pores, and pores marked in 120–1000 and 280–1200 are close to spherical. Source: Republished under CC BY 4.0 open access license [43].

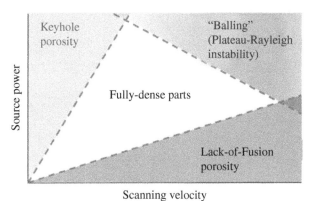

Figure 2.27 The combined effect of scanning velocity and beam power on the density of the part. Source: Republished under CC BY 4.0 open access license [44].

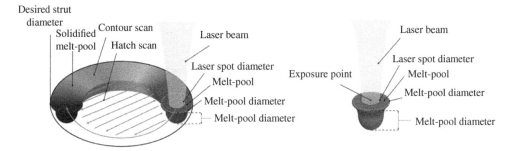

Figure 2.28 The relationship between melt-pool geometry and hatching distance. Source: Redrawn and adapted under CC BY 4.0 open access license [45].

average particle diameter decreases, which causes agglomeration and is detrimental to the powder's flow properties. Rough surface, wide PSD, and low sphericity also have similar effects on the flowability due to increased interparticle frictions. Thus, as a rule of thumb, a powder system with larger spherical particles uniform in size and a smoother surface has better flowability.

On the other hand, coarser particles with uniformly distributed diameters generate powder beds with low packing efficiency leading to increased porosity in the final part. Thus, it is not simple to come up with a straightforward recipe for particle-level powder properties. The ideal feedstock should be spherical with particles small enough to create high-density powder beds but not so small that they will not flow properly or become airborne and damage laser-optical systems or create safety issues. The particle dimension range should also be large enough so that smaller satellite particles fill the empty gaps between the coarser particles and improve the layer density, but not so wide that the flowability is reduced. It is understood that an average particle size <100 μm is desirable for powder bed-based systems [11].

Moreover, larger particle sizes are used for EB-PBF compared to LPBF to avoid the build-up of electrostatic charges. The thermal properties of alloys, such as heat capacity, thermal

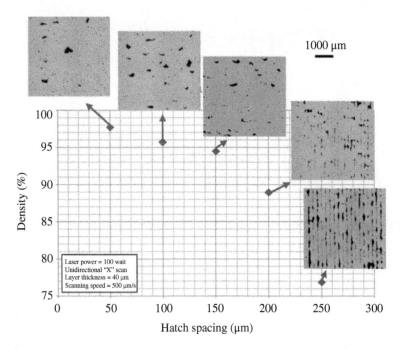

Figure 2.29 Porosity of LPBF-made parts from AlSi10Mg as a function of hatch distance. Source: Republished under CC BY 3.0 open access license [46].

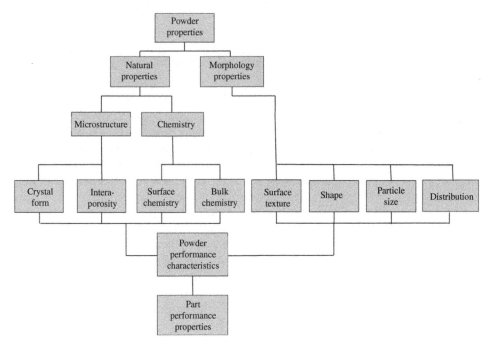

Figure 2.30 Classification of powder particle properties. Source: Redrawn and adapted with permission from Taylor & Francis [11].

diffusivity and conductivity, latent melting point, reflectance, etc., are also important factors in the processing and fusing of particles [48]. It is also extremely important to consider the difference between the thermal properties of metallic alloys in powder and solid forms.

At the bulk level, characteristics such as flowability, compressibility, cohesive strength, and bulk density control the quality and density of the deposited layer. The importance of a uniform, dense layer in manufacturing of high-density parts justifies the considerable efforts to study and adjust the behavior of bulk powder. Layer thickness is a significant material-related factor in PBF and BJ affected by PSD and affects the final part porosity and surface roughness. A large layer thickness may result in the insufficient transfer of energy to the particles. It causes manufacturing defects such as balling effect, incomplete melting and porosity, weak interlayer bonding and delamination, etc. Larger layer thicknesses also lead to a lower vertical resolution and surface flaws like staircase effect, roughness, layer visibility, and geometrical inaccuracy.

Various methods have been suggested to characterize the powder flow properties, which are usually dependent on the testing device and experimental conditions. Thus, the powder flowability properties obtained from different systems are not readily comparable to each other. One methodology that has been frequently used for characterization of powder flow properties for AM measures the powder's resistance to the movement of a blade that moves rotationally and vertically inside the powder under different conditions [49, 50]. The higher energy needed for the movement of blade inside the powder, the more cohesive the powder. Cohesiveness, in turn, adversely affects the flowability of the powder. Compressibility, which is the change in the volume of bulk powder exposed to normal stress, influences the packing efficiency of the powder bed. More compressible powders are more cohesive and result in a less packed powder bed. The relationship between powder flowability and flow and particle properties is illustrated in Figure 2.31.

The feedstock should have high flowability and low cohesiveness so that it fluidizes upon exposure to shear stress as it is being spread to the build chamber by the powder delivery device. This is crucial in obtaining a uniform and homogeneous layer, which is the key to

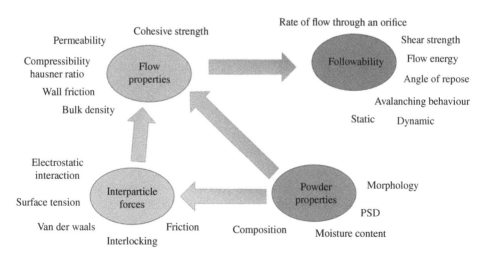

Figure 2.31 The relationship between particle and flow properties and flowability. Source: Republished under CC BY 4.0 open access license [47].

the production of a high-density part. Moreover, a highly packed powder bed transfers the heat more efficiently and reduces the thermal stress and manufacturing flaws such as warpage [4, 11]. In BJ, the packing density of the layers will affect the binder saturation as well as the surface roughness [51].

2.3.4.1 Methods of Powder Particles Production

There are five common techniques used to produce powder particles for different industries, including AM. These techniques are:

1. **Water atomization**: In this process, water is sprayed at a very high pressure (e.g., 500–1500 bar) where the melted material is presented to produce micro-size metal particles. This process is known well for its high yield rate and large particle size ranges up to 500 μm at a much lower cost. However, the process generates irregular shaped particles and cannot normally work with highly reactive materials. The irregular shape of particles as well as the presence of satellites are two major challenges when it comes to AM adoption.
2. **Gas atomization**: In this process, a high-velocity gas such as air, nitrogen, argon, or helium strikes a stream of melted material, causing distribution and particle formation. Compared to water atomization, the gas pressure is significantly lower (e.g., 5–50 bar). Gas atomized powders normally exhibit spherical shapes with a size up to 500 μm and with small satellites. The process is suitable for reactive materials, and it provides a high throughput outcome. However, satellites around particles are present, and the particle distribution is wide.
3. **Plasma atomization**: In this process, argon plasma torches are used to melt and atomize feedstock into micro-sized droplets. The feedstock (usually in the form of wire or irregular particles) melts from >10 000 °C plasma turning to atomized droplets with a size up to 200 μm. This process produces highly spherical shaped particles with much narrower particle size distribution. The process however is expensive.
4. **Centrifugal atomization:** In this process, a spinning disk interacts with the melted material to generate micron-sized droplets. It can produce narrow size range powders. When it operates continuously in an inert gas, it uses a lower level of energy compared with gas and even water atomization. The product has a lower oxide formation. However, its yield rate is low, and the cost of the system is high when the potential of tungsten contamination is high. The plasma rotate electrode process is a class of centrifugal atomization able to produce highly pure powders.

2.3.5 Wire Properties for DED

As described in Section 2.2.2, metals in the form of wire filaments are extensively used in WF-DED processes with energy sources in the form of a laser or electron beam. Similar to powders, the quality of the wires, such as their chemical composition and dimensional consistency, can control the quality of the generated melt pool. Some of the important process properties in wire-fed technologies include the wire feed rate and wire feed orientation.

The wire feed rate is simply how fast the wire is being fed from the nozzle into the deposition area. In the WF-EB-DED process, the wire feed rate limits are determined by the welding

Table 2.5 Influence of increasing process parameters on dimensions of a single bead produced by EB-DED [18].

Parameter	Deposition area	Deposition height	Deposition width
Power, P	0	−ve	+ve
Speed, V	0	+ve	−ve
Wire feed rate, λ	+ve	−ve	0

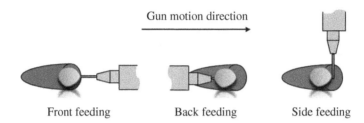

Gun motion direction

Front feeding Back feeding Side feeding

Figure 2.32 Different wire-feeding orientations. Source: Redrawn and adapted from [18].

power because for a particular power, an excessively high feed rate will translate to incompletely melted material deposition. The deposit dimensions are determined by the combined effect of feed rate, welding speed, and power. Therefore, controlling this geometry will require knowledge of the contributions of these process parameters. As summarized in Table 2.5, with other parameters kept constant, when power is increased, the width of the deposit increases, while the height decreases. Increasing the welding speed is conversely related, i.e., the width decreases and height increases [18]. The deposition area is influenced by a ratio of the wire feed rate and welding speed, and this ratio is designated by λ. When λ is increased, there will be a significant increase in the deposition area and height but no substantial increase in width.

The wire feed orientation is the position and layout of the wire feed nozzle relative to the deposition area during the process. The feed orientation, which determines drop transfer and overall deposit quality, has three types: front, back, and side feeding, as shown in Figure 2.32. As suggested in the literature, wire feed rate for Ti- and Ni-based alloy materials are limited when utilizing back and side feeding, unlike front feeding. The reason is that for the melted wire to flow smoothly into the work-piece, the wire is best fed from the front, while the next favorable position is the side. It is also discovered that back feeding provides a more stable and efficient process for aluminum. Therefore, the choice of feed orientation is material-dependent, and it should be selected such that the best deposit quality is obtained.

2.3.6 Layer Thickness for PBF, DED, and BJ

Layer thickness is the height of one layer of powder spread onto a powder bed in PBF or BJ. In DED, layer thickness is the height of deposited track. This parameter is usually considered as one of the main significant parameters for the powder bed-based processes. One of the most obvious effects of the layer thickness is on the resolution and the surface quality of the parts as well as the build time. Larger layer thickness results in a geometrical flaw known as the staircase effect on the curved or non-vertical surfaces. In contrast, the smaller layer thickness can better mimic the actual surface of the part at the cost of longer production time. Figure 2.33 shows the staircase effect and the need for smaller layer thickness as the curvature angle increases.

The minimum layer thickness achievable is controlled by (i) the precision of the moving parts of the powder bed in terms of the vertical movement and (ii) the largest particle in the feed material. The optimized layer thickness in PBF depends on the beam power, beam shape, and velocity. The maximum layer thickness should be smaller than the height of the melt pool generated as a combination of the mentioned parameters; otherwise, a portion of the particles will remain unfused, and the layers will not be completely fused together. To find out the exact maximum value, we need to consider the hatch distance. If we consider the cross section of the melt pools to be repetitive semi-ovals, the maximum layer thickness (t_{max}) is equal to the distance between the surface to the lowest point of the overlap between two adjacent melt pools (M_d^o) as shown in Figure 2.34 [44]. As seen in this figure, the larger the hatch distance, the smaller the layer thickness needs to be to avoid having unfused particles.

Similarly, for the BJ process, a layer thickness value must be selected to ensure the binders wet the entire depth of layer. This depends on the droplet nozzle spacing, printhead velocity, jetting frequency, wetting angle, binder viscosity, and binder penetration depth.

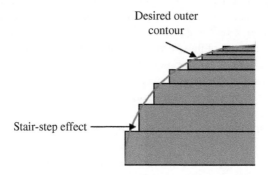

Figure 2.33   Illustration of the staircase effect. Source: Adopted from [52].

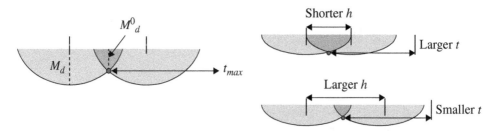

Figure 2.34   The maximum layer thickness as a measure of the overlap height of adjacent melt pools in PBF. Source: Republished under CC BY 4.0 open access license [44].

2.3.7 Ambient Parameters for PBF, DED, and BJ

The high energy required for the processing of metals in AM systems makes it necessary to consider a higher level of monitoring and control of ambient parameters in designing and developing such machines. The most important ambient parameters directly influencing the properties and quality of printed parts are related to the temperature of build chamber and the shield gas system.

The history and distribution of process temperature at various levels such as build chamber, powder bed surface, powder bed core, and local beam/powder interaction area are key parameters with direct effects on the consistency of the quality of printed parts, melt pool size and type, etc. Many studies have revealed the influence of build chamber temperature on the microstructure and mechanical properties of the metal parts [53]. Similar to other parameters discussed, the effect of temperature cannot be studied as a standalone parameter as it highly interacts with other parameters, specifically the beam source power. For example, low powder bed temperature and high power usually lead to building flaws stemming from residual stress concentration, such as distortion. Increasing the temperature of the powder bed helps with the elimination of this issue at the expense of losing the geometrical fidelity as the loose powder particles fuse together at elevated temperature, especially at the vicinity of the build part resulting in a phenomenon called part growth. This also reduces the recyclability of unused powder.

A high-temperature gradient of the powder bed is another factor that affects the quality of parts adversely by initiation and expedition of crack growth [54]. Given the importance of ambient temperature in process reliability and reproducibility, development and integration of sensors and metrology systems such as infrared cameras, thermo-couple embedded build beds, etc. into PBF machines for in-situ measurement of the temperature profile of the powder bed are essential.

Shielding gas is another significant factor in PBF and DED AM. Before the printing starts, the air in the build chamber is vacuumed and replaced with a shielding gas such as nitrogen, argon, or helium. Shielding gas mainly serves two purposes: (i) blocking the interaction of melt pool with reactive gases and (ii) removal of print by-products such as spatters [55]. Various studies show that the type of shielding gas and its flow influence the melt pool dimensions, as well as the uniformity, porosity, and mechanical performance of parts. Different shielding gases have different thermal conductivities that can change the temperature distribution around the melt pool by changing the heat transfer dynamics. For example, a lower powder bed surface temperature and higher cooling rate have been reported for nitrogen compared with argon due to its higher thermal conductivity, resulting in alteration of the microstructure and mechanical properties of parts [56]. Furthermore, a homogenous gas flow is essential for the proper removal of print by-products that might affect the laser-material interaction if present around the melt pool, leading to non-uniform mechanical properties [55].

A key characteristic of the EB-based processes is that it occurs in a high vacuum condition to avoid the electron beam dispersion. Notwithstanding, there are residual gases in the vacuum chamber and electron gun, but the pressure values are limited to 100 and 1 mPa, respectively. Also, to eliminate the risk of electrical charges formed in the powders during the melting process, an inert gas with a low pressure of about 100 mPa, typically helium, is introduced. Additionally, when the printing process is completed, helium fills the chamber again to help cool the built part.

In the BJ process, the parts are exposed to heat after the binder is deposited on each layer to get partially cured. The temperature and drying time are significant parameters that should be tuned based on the chemical composition of the binder, the thermal conductivity of the powder bed, binder saturation percentage (see Section 2.3.11), and the surface area [37, 53]:

$$T_d \propto \frac{SA}{K} \tag{2.10}$$

where T_d is the drying time, S is the binder saturation, A is the surface area, and K is the thermal conductivity of the powder.

The objective is to dry the binders enough to prevent the downward movement of freshly dispensed binders and to encourage its lateral flow to fill the voids within the target layer that is being printed while leaving enough active surface so that the successive layers are bound together properly. Long drying times might cause a lack of strong cohesion between the layers leading to structural flaws such as delamination. A certain level of shrinkage might also happen as a result of over-drying. Insufficient drying might lead to poor mechanical properties, higher porosity, and lower geometrical accuracy.

2.3.8   Geometry-Specific Parameters (PBF)

The process parameters such as beam power, velocity, and beam scanning strategy are not static throughout the entire volume of the build parts. They usually vary depending on the geometry and the location of the section of part that is being printed. In commercial systems, it is a common practice to change the parameter values at different locations of the part, i.e., contours, up-skin, down-skin, and core.

Figure 2.35 shows the top view of different areas of a layer under printing. Contour is the outer edge of each layer corresponding to the internal and external surfaces of the part for which a different power and speed can be selected. Using contours is optional, and the number of contours is usually one or two. Adjacent to the contour, a skin with a certain thickness can

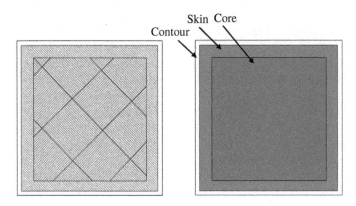

Figure 2.35   Cross section of a part as the print layer: contour/skin and core.

be considered. Similar to the contour, the selection of the skin is optional. The inner section of each layer is called the core and can be printed with a set of parameter values (e.g., hatching, power, speed, etc.) different from the skin.

Furthermore, different process parameters can be selected for printing the top most region of a part upon which there is only loose powder known as the up-skin and its lowest region below which there is only loose powder known as down-skin. Anything in between is the core of the part. Obviously, for more complex geometries, a combination of up-skin, core, and down-skin regions can be identified in each layer. Commercial machines usually calculate a percentage of overlap between these regions to ensure proper bonding.

Figure 2.36 shows up-skin vs down-skin in PBF process where upskin, inskin, and downskin hatch in the cross section of a single layer are shown.

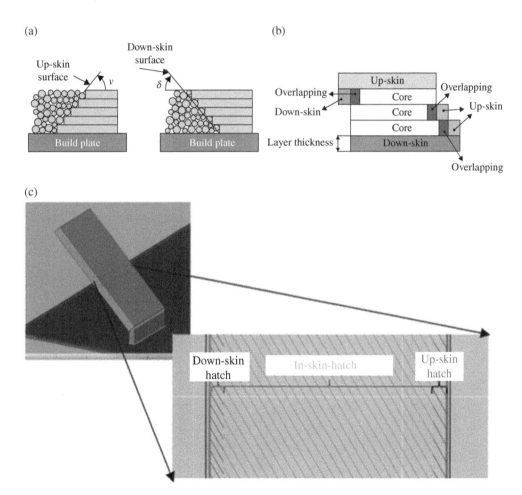

Figure 2.36 (a, b) Illustration of up-skin vs down-skin in PBF process. Source: (a) Republished with permission from Elsevier [57]; (b) republished under CC BY 3.0 open access license [58]. (c) presence of upskin, inskin, and downskin hatch in the cross-section of a single layer. Source: (c) From EOS Editor licences to Multi-Scale Additive Manufacturing Laboratory, University of Waterloo.

2.3.9   Support Structures for PBF

As mentioned before, support structures are required for PBF to make overhanging, hole, and bridge features of parts. Due to the complex features in AM-made components, support structures in different forms are necessary. They need to be removed using post-processing techniques (e.g., snapping off, EDM) to form the final component. This removal results in wasted materials and is costly. Thus, there is a trend to minimize support structures through optimum process parameters to print supportless overhanging features for some materials. As mentioned, support structures are not normally needed for BJ and DED. In DED, through a trajectory optimization when a high degree of freedom motion system is used, the support structures can be avoided. One important role of support structures is heat dissipation. In PBF, support structures facilitate the heat conduction to prevent residual stress concentration experienced during the process.

Support structures are usually added to the interface between the build plate and the main part at different densities and distributions. This interface is normally a location at which the residual stresses are at a high amount. Figure 2.37 shows the concept of support structure under an overhanging feature as well as three supports with different shapes and density.

2.3.10   Binder Properties for BJ

Before the sintering process, the BJ-made parts are composed of at least two main materials: powder and binder. The properties of powders are reviewed in Section 2.3.3. The binder properties are equally important to obtain high-quality green parts that are strong enough to be handled and transported before sintering. The design of binders with appropriate chemical,

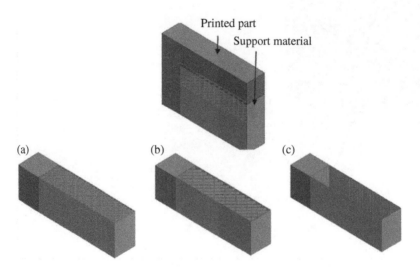

Figure 2.37   The concept of supports structures, three different support shapes and densities are shown. (a) medium density, (b) low density, and (c) high density.

rheological, and mechanical properties upon curing customized to a certain powder (or class of powders) is an essential step in BJ-AM.

A binder should be, first and foremost, jettable, a qualitative parameter controlled by the liquid binder's rheological properties such as viscosity, surface tension, and density. Viscosity is defined as the resistance of a fluid to flow and is probably the most important parameter in designing a binder for BJ and MJ. For a droplet to be formed, the energy imparted to the fluid by an internal pressure wave (created by a thermal or piezoelectric actuator) should be large enough to overcome the surface tension of meniscus at the nozzle to push the droplet out of the orifice. This is controlled by various combined parameters as described by Derby et al. [59] such as Reynolds number (Re), Weber number (We), and Ohnesorge number (Oh):

$$Re = \frac{v_j \rho a}{\eta} \tag{2.11}$$

$$We = \frac{v_j^2 \rho a}{\gamma} \tag{2.12}$$

$$Oh = \frac{\sqrt{We}}{Re} = \frac{\eta}{\sqrt{\gamma \rho a}} \tag{2.13}$$

where v_j is the jet velocity, ρ is the density, a is a characteristic of length, η is the dynamic viscosity, and γ is the surface tension. Sometimes, the parameter $Z = \frac{1}{Oh}$ is used to describe the printability of a liquid. An Oh range between 0.1 and 1 (or $1 < Z < 10$) is proposed for a stable formation of a droplet in inkjet printing, as shown in Figure 2.38. This imposes restrictions on the rheological properties of the binders used in BJ. For example, the maximum viscosity for the printable inks is usually in the range of 10–100 mPa.s for regular BJ printheads [60, 61].

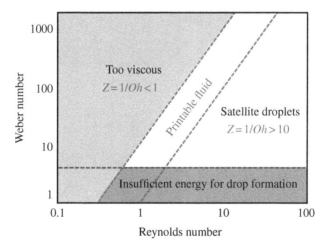

Figure 2.38 Printability of the fluids based on dimensionless Reynolds and Weber numbers. Source: Redrawn and adapted from [59].

2.3.11 Binder Saturation for BJ

Binder saturation is the fraction of the voids in a certain volume of the powder bed, which is filled with the binder. This characteristic is measured using Eq. (2.14) [35]. Binder saturation directly affects the density and surface properties of the part.

$$S = \frac{V_{binder}}{V_{air}} \tag{2.14}$$

$$V_{air} = \left(1 - \frac{PR}{100}\right) V_{solid} \tag{2.15}$$

$$PR = \frac{V_{powder}}{V_{powder} + V_{air}} \tag{2.16}$$

where S is the binder saturation level and PR is the packing density of the powder. The binder saturation level is one of the process parameters that needs to be adjusted in combination with other factors to ensure enough binding between the particles, depowdering and transferring the parts without breaking them, and the highest surface quality and dimensional accuracy possible. The effects of undersaturation and oversaturation of binder are demonstrated in Figure 2.39.

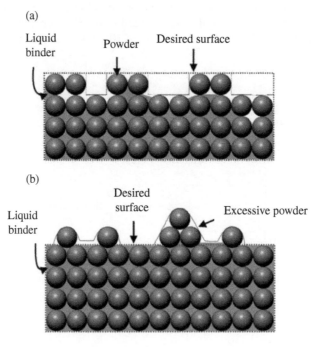

Figure 2.39 The effects of (a) undersaturation and (b) oversaturation on BJ-made parts. Source: Republished with permission from Elsevier [35].

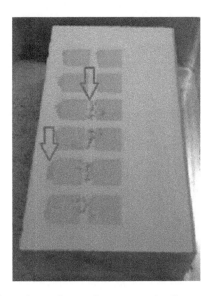

Figure 2.40 H13 tool steel powder agglomeration as a result of oversaturation. Source: Republished from DiVA open access [62].

Some of the effects of insufficient binder saturation include powder fall off from the external surface of the part and low geometrical accuracy. On the other hand, high binder saturation leads to part swelling due to bleeding of the extra binder to the areas outside the layer borders, and lowering the powder spreading quality as the overly wet particles might stick to the roller. Figure 2.40 shows the agglomeration of particles due to the shifting of powder because of oversaturation.

2.4 Materials

Chapters 8 and 9 are fully dedicated to materials used in metal AM; however, a brief review of AM metal powders/wires is provided in this section. Powders are the predominant form of material used in most AM systems. At a smaller scale, wires are used in DED, sheets are used in SL, and metallic particles suspended in a liquid formula are used in ME and MJ. Only a relatively small portion of alloys available for conventional manufacturing methods have been optimized for AM applications. The high sensitivity of AM technologies to process parameters and the unique microstructure and properties of AM-made parts require development of optimized process recipe and process-property databases for many new alloys that might be suited to AM Processes [9]. The majority of material development in metal AM has been related to ferrous and titanium alloys, followed by aluminum alloys and nickel and chrome-cobalt superalloys. AM of refractory metals (e.g., molybdenum and its alloys), precious metals (gold, silver, platinum, tantalum, etc.), tungsten, magnesium, copper, and intermetallic materials has been investigated by both academia and industry. The development of metal powders (and other forms as described above) tailored for different AM processes is an active research field. It is expected to grow as the applications of AM are increased.

2.4.1   Ferrous Alloys

Given the numerous applications of ferrous alloys such as stainless steel, tool steels, maraging steels, etc., employing them in AM processes has always been at the center of AM R&D. In a review of AM of ferrous alloys in [35], the majority of publications are attributed to PBF (68% LPBF and 1% EB-PBF), followed by LDED (28%) and BJ (3%), indicating the high popularity of PBF technology for metal manufacturing. Among various iron-based alloys, 316L stainless steel, maraging steel 300, H13 tool steel, 17-4 PH stainless steel, M2 tool steel, P20 tool steel, and 304L stainless steel are the most frequently studied alloys. The scope of work on EB-PBF of ferrous alloys has been limited to H13 tool steel and 316L stainless steel [14, 37]. Applications such as fabrication of porous biomedical implants [63], conformal cooling channels, tooling, and lattice structures have been pointed out for AM of ferrous alloys [38].

2.4.2   Titanium Alloys

Ti-6Al-4V is the most widely studied alloy for AM. Alloys of titanium with unique properties such as biocompatibility, high strength while being light, and high corrosion resistance are extensively applicable in biomedical and aerospace sectors. Being available in the form of powder and filament, all three major categories of metal AM (PBF, DED, BJ) can use titanium alloys as the raw material. Titanium is very well suited to AM as it is very difficult and costly to machine. In addition to Ti-6Al-4V and pure titanium, other alloys of interest for AM include Ti-24Nb-4Zr-8Sn, Ti-6Al-7Nb, and Ti-6.5Al-3.5Mo-1.5Zr-0.3Si [64]. Furthermore, Ti-6Al-4V, Ti–48Al–2Nb–0.7Cr–0.3Si, γ-TiAl, and Ti-24Nb-4Zr-8Sn have been used in EB-PBF [65].

2.4.3   Nickel Alloys

Nickel superalloys are another category of materials that have gained the attention of AM community due to their excellent resistance to harsh conditions and high temperature. These properties and the design freeform and optimization that AM offers provide a great combination for creative innovations in aerospace, aeronautics, and tooling sectors. Inconel alloys comprise the majority of R&D activities on AM of nickel alloys, with Inconel 718 and Inconel 625 being the most extensively studied materials of this class followed by Hastelloy X as another popular alloy of this material class for AM [38, 66].

2.4.4   Aluminum Alloys

Aluminum is a low-weight material with high electrical and thermal conductivity, high ductility, and high corrosion resistance with applications in various industries such as aerospace, construction, and automotive. As mentioned previously, the weldability of an alloy is a good indicator of its compatibility with AM processes. Aluminum alloys are not very weldable but are easy to machine. Moreover, they demonstrate low processability with lasers due to their high reflectivity and low viscosity at the melting phase [14]. As a result, aluminum has not been the focus of AM research compared with other alloy groups like titanium, nickel, and iron.

Nonetheless, given the benefits that AM of aluminum offers, such as light-weighting aviation structures, growing efforts for optimizing process parameter sets for aluminum alloys in PBF and DED have been recorded. Thus far, AlSi10Mg has been the main alloy of aluminum employed in AM; however, the possibility of printing AlSi12, AlSi9Mg, AlSi7Mg0.3, and AlSiMg0.75 has been demonstrated as well [67]. Al–Cu and Al–Sc wires are also used in EB-DED [68].

References

1. L. Dall'Ava, H. Hothi, A. Di Laura, J. Henckel, and A. Hart, "3D printed acetabular cups for total hip arthroplasty: A review article," *Metals (Basel).*, vol. 9, no. 7, p. 729, 2019.
2. L. E. Murr, "A metallographic review of 3D printing/additive manufacturing of metal and alloy products and components," *Metallogr. Microstruct. Anal.*, vol. 7, no. 2, pp. 103–132, Apr. 2018.
3. V. P. Carey, G. Chen, C. Grigoropoulos, M. Kaviany, and A. Majumdar, "A review of heat transfer physics," *Nanoscale Microscale Thermophys. Eng.*, vol. 12, no. 1, pp. 1–60, Jan. 2008.
4. I. Gibson, D. W. Rosen, B. Stucker, and M Khorasani, Additive manufacturing technologies, vol. 17, Springer, 2014.
5. C.R. Deckard – US Patent 4863538 and undefined 1989, "Method and apparatus for producing parts by selective sintering."
6. W. Meiners, K. Wissenbach, A. Gasser – US Patent 6215093 and undefined 2001, "Selective laser sintering at melting temperature."
7. "Industrial 3D Printing for Dental Technology." [Online]. Available: https://www.eos.info/en/3d-printing-examples-applications/people-health/medical-3d-printing/dental.
8. S. Tammas-Williams, P. J. Withers, I. Todd, and P. B. Prangnell, "The effectiveness of hot isostatic pressing for closing porosity in titanium parts manufactured by selective electron beam melting," *Metall. Mater. Trans. A*, vol. 47, no. 5, pp. 1939–1946, May 2016.
9. J. O. Milewski, "Additive manufacturing metal, the art of the possible," in Springer series in materials science, vol. 258, Springer Verlag, pp. 7–33, 2017.
10. M. Markl, M. Lodes, M. Franke, and C. Körner, "Additive manufacturing using selective electron beam melting," *Weld. Cut.*, vol. 16, pp. 177–184, 2017.
11. A. T. Sutton, C. S. Kriewall, M. C. Leu, and J. W. Newkirk, "Powder characterisation techniques and effects of powder characteristics on part properties in powder bed fusion processes," *Virtual Phys. Prototyp.*, vol. 12, no. 1, pp. 3–29, Jan. 2017.
12. Z. Liu, D. Zhang, S. Sing, C. Chua, and L. Loh, "Interfacial characterization of SLM parts in multi-material processing: Metallurgical diffusion between 316L stainless steel and C18400 copper alloy," *Mater. Charact.* vol. 94, pp. 116–125, 2014.
13. S. Yang and J. R. Evans, "A multi-component powder dispensing system for three dimensional functional gradients," *Mater. Sci. Eng. A*, vol. 379, no. 1–2, pp. 351–359, Aug. 2004.
14. T. Ngo, A. Kashani, G. Imbalzano, K. T.Q. Nguyen, and D. Hui, "Additive manufacturing (3D printing): A review of materials, methods, applications and challenges," *Compos. Part B: Eng.*, vol. 143, pp. 172–196, 2018.
15. E. Toyserkani, A. Khajepour, and S. Corbin, Laser cladding, CRC Press, 2004.
16. W. Y. Y. S.L. Sing, C.F. Tey, J.H.K. Tan, and S. Huang, Rapid prototyping of biomaterials, Elsevier Books, 2020.
17. T. Cobbs, L. Brewer, and J. L. Crandall, "How 3D metal printing saves time and lowers costs: DED for repair of industrial components," OPTOMEC, pp. 1–38.

18. D. Ding, Z. Pan, D. Cuiuri, and H. Li, "Wire-feed additive manufacturing of metal components: Technologies, developments and future interests," *Int. J. Adv. Manuf. Technol.*, vol. 81, no. 1–4, pp. 465–481, Oct. 2015.

19. Y. Huang, M. B. Khamesee, and E. Toyserkani, "A comprehensive analytical model for laser powder-fed additive manufacturing," *Addit. Manuf.*, vol. 12, pp. 90–99, 2016.

20. Sciaky, "Benefits of wire vs. powder metal 3D printing." [Online]. Available: https://www.sciaky.com/additive-manufacturing/wire-vs-powder.

21. C. Atwood M. Griffith, L. Harwell, E. Schlienger, M. Ensz, J. Smugeresky, T. Romero, D. Greene, and D. Reckaway, "Laser engineered net shaping (LENS): A tool for direct fabrication of metal parts," *Proc. Int. Congress Appl. Lasers Electro-Optics*, vol. 1998, no. 1, pp. E1–E7, Oct. 1998.

22. NASA, "Electron beam freeform fabrication." [Online]. Available: https://www.nasa.gov/topics/technology/features/ebf3.html.

23. 3DHubs, "Material Extrusion." [Online]. Available: 3dhubs.com.

24. E. M. Sachs, J. S. Haggerty, M. J. Cima, and P. A. Williams, Three-dimensional printing techniques, "Inventors," Massachusetts Institute of Technology, United States patent US 5, 4, 1993.

25. Hewlett Packard (HP), "HP multi jet fusion technology: A disruptive 3D printing technology for a new era of manufacturing," HP Tech. White Papers, 19 February 2014.

26. ExOne, "420 Stainless Steel Infiltrated with Bronze."

27. "Desktop Metal." [Online]. Available: https://www.desktopmetal.com/products/studio.

28. "MarkForged." [Online]. Available: https://markforged.com/.

29. "Triditive." [Online]. Available: http://www.triditive.com/en/amcell-en.

30. "Pollen." [Online]. Available: https://www.pollen.am/products.

31. "Shukla Medical Case Study." [Online]. Available: https://markforged.com/resources/case-studies/shukla-medical-case-study.

32. G. Hanan, E. Kritchman, A. Benichou, T. Shmal, G. Eytan, W. Salalha, Y. Dayagi, O. Kodinets, and L. Lavid. "Methods and systems for printing 3d object by inkjet." U.S. Patent Application 15/029,815, filed 25 August 2016.

33. "XJet." [Online]. Available: www.xjet3d.com.

34. "Fabrisonic." [Online]. Available: https://fabrisonic.com.

35. H. Fayazfar, M. Salarian, A. Rogalsky, D. Sarker, P. Russo, V. Paserin, E. Toyserkani, "A critical review of powder-based additive manufacturing of ferrous alloys: Process parameters, microstructure and mechanical properties." *Mater. Des.*, vol. 144, pp. 98–128, 2018.

36. M. Mani, S. Feng, B. Lane, A. Donmez, S. Moylan, and R. Fesperman, "Measurement science needs for real-time control of additive manufacturing powder bed fusion processes," National Institute of Standards and Technology, 2015. Available: https://nvlpubs.nist.gov/nistpubs/ir/2015/NIST.IR.8036.pdf

37. G. J. Bruck, "Fundamentals and industrial applications of high power laser beam cladding," in Laser beam surface treating and coating, vol. 957, International Society for Optics and Photonics, pp. 14–28, 1988.

38. C. Y. Yap, C. K. Chua, Z. L. Dong, Z. H. Liu, D. Q. Zhang, L. E. Loh, and S. L. Sing, "Review of selective laser melting: Materials and applications," *Appl. Phys. Rev.*, vol. 2, no. 4, p. 041101, Dec. 2015.

39. B. Cheng, S. Shrestha, and K. Chou, "Stress and deformation evaluations of scanning strategy effect in selective laser melting," *Addit. Manuf.*, vol. 12, pp. 240–251, 2016.

40. M. Bayat A. Thanki, S. Mohanty, A. Witvrouw, S. Yang, J. Thorborg, N. Skat Tiedje, J. Henri Hattel, "Keyhole-induced porosities in laser-based Powder Bed Fusion (L-PBF) of Ti6Al4V: High-fidelity modelling and experimental validation," *Addit. Manuf.*, vol. 30, p. 100835, 2019.

41. M. Tang, P. C. Pistorius, and J. L. Beuth, "Prediction of lack-of-fusion porosity for powder bed fusion," *Addit. Manuf.*, 2017.

42. P. Bidare, I. Bitharas, R. M. Ward, M. M. Attallah, and A. J. Moore, "Fluid and particle dynamics in laser powder bed fusion," *Acta Mater.*, vol. 142, pp. 107–120, 2018.

43. T. Majumdar, T. Bazin, E. M. C. Ribeiro, J. E. Frith, and N. Birbilis, "Understanding the effects of PBF process parameter interplay on Ti-6Al-4V surface properties," *PLoS One*, vol. 14, no. 8, p. e0221198, 2019.

44. J. P. Oliveira, A. D. LaLonde, and J. Ma, "Processing parameters in laser powder bed fusion metal additive manufacturing," *Mater. Des.*, vol. 193, p. 108762, 2020.

45. S. Ghouse, S. Babu, R. J. Van Arkel, K. Nai, P. A. Hooper, and J. R. T. Jeffers, "The influence of laser parameters and scanning strategies on the mechanical properties of a stochastic porous material," *Mater. Des.*, vol. 131, pp. 498–508, 2017.

46. N. Aboulkhair, N. Everitt, I. Ashcroft, Chris Tuck, "Reducing porosity in AlSi10Mg parts processed by selective laser melting," *J. Addit.Manuf.*, vol. 1, pp. 77–86, 2014.

47. S. Vock, B. Klöden, A. Kirchner, T. Weißgärber, and B. Kieback, "Powders for powder bed fusion: a review," *Prog. Addit. Manuf.*, vol. 4, pp. 383–397, Feb. 2019. Available: https://link.springer.com/content/pdf/10.1007/s40964-019-00078-6.pdf

48. M. Brandt, Laser Additive Manufacturing: Materials, Design, Technologies, and Applications, Laser Mater Process, 2016.

49. R. Freeman and X. Fu, "Characterisation of powder bulk, dynamic flow and shear properties in relation to die filling," *Powder Metall.*, vol. 51, no. 3, pp. 196–201, Sep. 2008.

50. R. Freeman, "Measuring the flow properties of consolidated, conditioned and aerated powders—A comparative study using a powder rheometer and a rotational shear cell," Powder Technol., vol. 174, no. 1–2, pp. 25–33, 2007.

51. A. Mostafaei, Elliott, A. M., Barnes, J. E., Li, F., Tan, W., Cramer, C. L., Nandwana, P. Chmielus, Markus, "Binder jet 3D printing–process parameters, materials, properties, and challenges," *Prog. Mater. Sci.*, p. 100707, 2020.

52. H. Brooks, A. Rennie, T. Abram, J. McGovern, and F. Caron, "Variable fused deposition modelling: analysis of benefits, concept design and tool path generation," in: 5th International Conference on Advanced Research in Virtual and Rapid Prototyping, pp. 511–517, 2011.

53. A. Aversa, M. Lourusso, F. Trevisan, and E.P. Ambrosio, "Effect of process and post-process conditions on the mechanical properties of an A357 alloy produced via laser powder bed fusion," *Metals – Open Access Metallurgy Journal*, vol. 7, no. 2, 2017.

54. H. Shipley, D. McDonnell, M. Culleton, R. Coull, R. Lupoi, G. O'Donnell, and D. Trimble, "Optimisation of process parameters to address fundamental challenges during selective laser melting of Ti-6Al-4V: A review," Int. J. Mach. Tools Manuf., vol. 128, pp. 1–20, 2018.

55. A. M. Philo, C. J. Sutcliffe, S. Sillars, J. Sienz, S. G. R. Brown, and N. P. Lavery, "A study into the effects of gas flow inlet design of the Renishaw AM250 laser powder bed fusion machine using computational modelling," in Solid Freeform Fabrication 2017: Proceedings of the 28th Annual International Solid Freeform Fabrication Symposium—An Additive Manufacturing Conference, SFF 2017, pp. 1203–1219, 2020.

56. P. Dastranjy Nezhadfar, M. Masoomi, S. Thompson, N. Phan, and N. Shamsaei, "Mechanical properties of 17-4 PH stainless steel additively manufactured under Ar and N 2 shielding gas." Proceedings of the 29th annual international solid freeform fabrication, Austin, TX, USA, pp. 13–15, 2018.

57. G. Nicoletto, "Directional and notch effects on the fatigue behavior of as-built DMLS Ti6Al4V," *Int. J. Fatigue*, vol. 106, pp. 124–131, 2018.

58. D. Manfredi, F. Calignano, M. Krishnan, R. Canali, E. Paola Ambrosio, S. Biamino, D. Ugues, M. Pavese, and P. Fino, "Additive manufacturing of Al alloys and aluminium matrix composites (AMCs)," Light Metal Alloys Appl., vol. 11, pp. 3–34, 2014.

59. B. Derby, "Inkjet printing of functional and structural materials: fluid property requirements, feature stability, and resolution," *Annu. Rev. Mater. Res.*, vol. 40, pp. 395–414, 2010.

60. S. V Murphy and A. Atala, "3D bioprinting of tissues and organs," *Nat. Biotechnol.*, vol. 32, no. 8, pp. 773–785, 2014.

61. E. Selbertinger, F. Achenbach, and B. Pachaly, "Method for producing silicone elastomer parts," Google Patents, Nov 2019.

62. G. Persson, "Process development for H13 tool steel powder in binder jet process," 2020.

63. R. Li, J. Liu, Y. Shi, M. Du, and Z. Xie, "316L stainless steel with gradient porosity fabricated by selective laser melting," *J. Mater. Eng. Perform.*, vol. 19, no. 5, pp. 666–671, 2010.

64. D. Herzog, V. Seyda, E. Wycisk, C. Emmelmann, "Additive manufacturing of metals," Acta Mater., vol. 117, pp. 371–392, 2016.

65. L. Zhang, Y. Liu, S. Li, and Y. Hao, "Additive manufacturing of titanium alloys by electron beam melting: A review," *Adv. Eng. Mater.*, vol. 20, no. 5, p. 1700842, 2018.

66. Z. Tian, C. Zhang, D. Wang, W. Liu, X. Fang, D. Wellmann, Y. Zhao, and Y. Tian, "A review on laser powder bed fusion of inconel 625 nickel-based alloy," *Appl. Sci.*, vol. 10, no. 1, p. 81, 2020.

67. J. Zhang, B. Song, Q. Wei, D. Bourell, and Y. Shi, "A review of selective laser melting of aluminum alloys: Processing, microstructure, property and developing trends," *J. Mater. Sci. Technol.*, vol. 35, no. 2, pp. 270–284, 2019.

68. A. I. Mertens, J. Delahaye, and J. Lecomte-Beckers, "Fusion-based additive manufacturing for processing aluminum alloys: State-of-the-art and challenges," *Adv. Eng. Mater.*, vol. 19, no. 8, p. 1700003, 2017.

3

Main Sub-Systems for Metal AM Machines

Learning Objectives

At the end of this chapter, you will be able to:

- Gain a clear understanding of main modules used in AM systems
- Gain knowledge on the physics behind the laser functioning
- Recognize the main differences among the different laser classes
- Identify the different laser-based AM processes
- Identify different powder feeders and other axillaries used in different metal AM processes

3.1 Introduction

In metal AM processes, the 3D object is built through the selective solidification/joining of powder materials in a layer-by-layer fashion. Lasers are the optimum candidates to arrive at the energy level needed for the powder particles' melting and/or sintering. Indeed, lasers can generate highly intense collimated energy sources when directional mirrors can rapidly control their directions. Laser-based AM processes generally require a laser, powder feeders, delivery nozzles (for DED technique) or recoaters as well as a powder bed (for powder bed fusion (PBF) processes), and a positioning device equipped with CAD/CAM software. In this chapter, the basics of lasers will be discussed, along with details about the most common type of

Metal Additive Manufacturing, First Edition. Ehsan Toyserkani, Dyuti Sarker, Osezua Obehi Ibhadode, Farzad Liravi, Paola Russo, and Katayoon Taherkhani.
© 2022 John Wiley & Sons Ltd. Published 2022 by John Wiley & Sons Ltd.

lasers used in laser-based AM. It will also discuss the electron beam as an energy source used in PBF and DED processes. This chapter includes information about wire feeding mechanisms, powder feeders, powder spreading mechanisms, positioning devices, and CAD/CAM systems employed in metal AM.

3.2 System Setup of AM Machines

3.2.1 Laser Powder Bed Fusion (LPBF)

A typical laser powder bed fusion system (LPBF) setup is shown in Figure 3.1. A layer of material with a specific thickness is spread over the platform, and the laser beam melts (fuses) the powder according to the CAD model loaded in the machine. The laser beam is scanned on the powder layer through specific positioning devices, which will be described at the end of this chapter. After the first layer is melted, the build platform is lowered and a new layer of fresh powder is spread and scanned. This process is repeated until the desired part is printed. During the post-processing of the part, the loose powder is removed.

3.2.2 Laser Directed Energy Deposition (LDED) with Blown Powder Known as Laser Powder-Fed (LPF)

A schematic of a typical directed energy deposition (DED) machine with blown powder, known as powder-fed, is displayed in Figure 3.2. In this type of machine, the material is deposited through a nozzle (lateral or coaxial) mounted on a multi-axis arm or table and a laser or an electron beam is employed to melt the deposited material. The material is added and melted layer by layer until the desired part is printed. The machine is equipped with powder feeders (discussed later in this chapter), where the powder is stored before flowing to the nozzle. Some

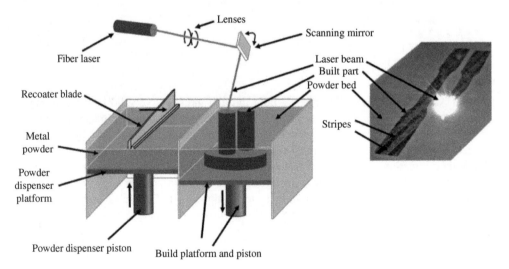

Figure 3.1 Laser powder bed fusion system (LPBF). Source: Reprinted from [1] with permission from Elsevier.

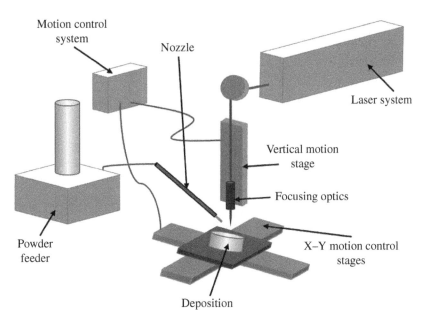

Figure 3.2 Laser Powder-Fed (LPF) system. Source: Image reproduced under the Creative Commons License CC BY-SA 3.0.

DED machines, instead of feeding powders, are equipped with metal wires, and the technique is referred to as the wire-fed deposition process. Polymers and ceramics can be used as feeding materials (powder and/or wires), but metals are the most used materials.

3.2.3 Binder Jetting (BJ)

The binder jetting (BJ) technique is used to print complex parts by injecting liquid binder into a powder bed. A schematic of the system setup is shown in Figure 3.3. The binder is in liquid form, and it is injected into the powder bed, acting as an adhesive. The binder is dispensed through an inkjet printhead, which is mounted on an x–y movable positioning system platform. Similar to the system setup described for LPBF, in BJ, a powder roller spreads the powder material which is then printed and stitched with the binder material. After the first layer is printed, the build platform is lowered and a new layer of fresh powder is spread and printed. The process is repeated until the part has been printed. Stainless steels, polymers, and glass can be 3D printed with the BJ technique.

3.3 Laser Basics: Important Parameters Needed to be Known for AM

3.3.1 Laser Theory

In 1960, T. H. Maiman invented the first ruby laser. The word "LASER" is an acronym, which stands for "Light Amplification by Stimulated Emission of Radiation," and the name itself gives information on its working principle. In particular, lasers are developed based upon

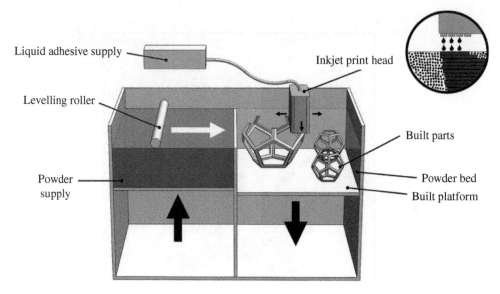

Liquid adhesive supply

Inkjet print head

Levelling roller

Built parts

Powder supply

Powder bed

Built platform

Figure 3.3 Schematic of a binder jetting system setup. Source: Redrawn and adapted from [2], Courtesy of additively.

the stimulated emission process postulated in 1917 by Einstein. The derivation of Planck's law of the spectral energy density could be achieved considering the existence of both spontaneous and stimulated emission. With the term emission, we refer to that process, in which an atom in its excited state transits to a lower energy state through emission of electromagnetic radiation, known as "photon." In the atomic model proposed by Bohr in 1915, the electrons of an atom orbit around fixed orbits. The orbital with the lowest energy is indicated as the ground state, while the orbitals with energies higher than the ground state are referred to as excited states. The electrons which revolve in these fixed orbits do not radiate energy; however, when an electron jumps to a higher or lower orbit the atom radiates energy. The term spontaneous emission indicates how an atom in the excited state spontaneously emits radiation while returning to the ground state. In Figure 3.4, the absorption, spontaneous emission, and stimulated emission processes are illustrated.

An electron in the ground state (orbit) with an energy level of E_1 is excited by "photon pumping – radiative" or "external heating – nonradiative" (energy absorption process) to an elevated level with energy of E_2. Then, after a few nanoseconds of the excitation, the electron decays to the ground state while emitting a photon (spontaneous emission), in a random spatial direction, with an energy equal to the difference between the two states ($h\upsilon = E_2 - E_1$), where h is Planck's constant (4.1×10^{-15} eV.s), $\upsilon = \frac{c}{\lambda}$, c is the speed of light (299 792 458 m/s), and λ is the wavelength. When incoming photons induce the emission, the process is called stimulated emission, as shown in Figure 3.4. In the stimulated emission, the photons emitted possess the same phase, frequency, and state of the stimulating photon's polarization. In a system with many atoms, this process can take place many times, enabling the amplification of light, which is the lasers' fundamental process. A laser beam is **coherent** (meaning the light emitted has a constant relative phase), **monocolor** (one wavelength), and **collimated** (extremely parallel rays).

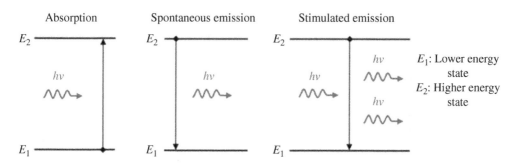

Figure 3.4 Illustration of the absorption, spontaneous emission, and stimulated emission processes.

3.3.1.1 Rate Equations and Einstein Coefficients

Let us consider an atomic system with only two energy levels E_1 and E_2, where E_1 denotes the ground state and E_2 is the excited state. N_1 and N_2 are the number of atoms or electrons per unit volume in the energy levels E_1 and E_2, respectively. Since we are considering only a two-level system, then $N_{total} = N_1 + N_2$. When the atomic or electron system interacts with electromagnetic radiation with frequency v_{12}, where $h v_{12} = E_2 - E_1$, some atoms will transit from the ground state 1 to the excited state 2. Einstein assumed that the stimulated absorption rate from level 1 to level 2 is proportional to the radiation energy density $\rho_v (v_{12})$ and the number of atoms or electrons in the ground state N_1 with a proportionality constant B_{12}.

$$\left(-\frac{dN_1}{dt}\right)_{abs} = B_{12}\rho_v(v_{12})N_1(t) \tag{3.1}$$

The absorption rate is $\left(-\frac{dN_1}{dt}\right)_{abs}$, where the negative sign indicates that $N_1(t)$ decreases with time. As aforementioned, the decay of the atoms from level E_2 to E_1 can take place through spontaneous and stimulated emissions. The spontaneous emission is proportional only to the number of atoms (N_2) in the excited state. Therefore, the spontaneous emission rate can be written as:

$$\left(\frac{dN_2}{dt}\right)_{spont} = -A_{21}N_2(t) \tag{3.2}$$

The stimulated emission rate is proportional to the radiation density $\rho_v (v_{12})$, and the number of atoms in the ground state excited N_2 with a proportionality constant B_{21}. Noteworthy, the subscripts' order for the proportionality constants A and B is 21, thus indicating that the spontaneous and stimulated emission transitions start from the excited state 2 to the ground state 1 $(2 \rightarrow 1)$. The following equation gives the stimulated emission rate:

$$\left(\frac{dN_2}{dt}\right)_{stim} = -B_{21}\rho_v(v_{12})N_2(t) \tag{3.3}$$

The proportionality constants B_{12}, A_{21}, and B_{21} are called **Einstein coefficients**. When an atomic system interacts with the light, all three processes (absorption, spontaneous emission, and stimulated emission) occur simultaneously. Therefore, the population rates in the ground state and in the excited state are equal to the sum of the absorption rate, spontaneous emission rate, and stimulated emission rate. In other words:

$$\left(-\frac{dN_1}{dt}\right) = \left(\frac{dN_2}{dt}\right) = B_{12}\rho_v(v_{12})N_1(t) - A_{21}N_2(t) - B_{21}\rho_v(v_{12})N_2(t) \qquad (3.4)$$

The three Einstein coefficients are related to each other. To obtain the relationship between them, let us consider that at thermal equilibrium, then populations 1 and 2 (N_1 and N_2) are constant,

$$\left(-\frac{dN_1}{dt}\right) = \left(\frac{dN_2}{dt}\right) = 0 \qquad (3.5)$$

Therefore, at thermal equilibrium, Eq. (3.4) becomes

$$B_{12}\rho_v(v_{12})N_1(t) - A_{21}N_2(t) - B_{21}\rho_v(v_{12})N_2(t) = 0 \qquad (3.6)$$

Or

$$\rho_v(v_{12}) = \frac{A_{21}}{(N_1/N_2)B_{12} - B_{21}} \qquad (3.7)$$

$\rho_v(v_{12})$ is the radiation energy density per unit frequency interval, and according to Planck's law, it is given by

$$\rho_v(v_{12}) = \frac{8\pi h}{c^3} \frac{v_{12}^3}{e^{hv_{12}/\sigma_B T} - 1} \qquad (3.8)$$

where $\sigma_B (=1.38 \times 10^{-23}$ J/K) is Boltzmann's constant. The number of atoms or molecules in a j state with energy E_j at thermal equilibrium is given by

$$N_j = c^* e^{-E_j/\sigma_B T} \qquad (3.9)$$

where c^* is a proportionality constant. Let us consider the ratio of the populations of levels 1 and 2 at thermal equilibrium at temperature T, which could be described by a Boltzmann distribution resulting in

$$\frac{N_2}{N_1} = e^{-(E_2 - E_1)/\sigma_B T} = e^{-hv_{12}/\sigma_B T} \qquad (3.10)$$

Therefore,

$$\rho_v(v_{12}) = \frac{A_{21}}{B_{12}e^{hv_{12}/\sigma_B T} - B_{21}} \qquad (3.11)$$

If we now compare Eqs. (3.7) and (3.11), it derives as

$$B_{12} = B_{21} \tag{3.12}$$

This result shows that the absorption and stimulated emission processes are equivalent. Moreover, we obtain that

$$A_{21} = \frac{8\pi h \upsilon_{12}^3}{c^{*3}} B_{21} \tag{3.13}$$

Asserting that absorption and spontaneous emission are proportional to each other.

3.3.1.2 The Two-Level System

To produce a laser beam or achieve light amplification through stimulated emission, it is necessary that the stimulated emission rate be higher than the spontaneous emission rate and absorption rate. This can be expressed by the following equation:

$$B_{21}\rho_{\upsilon}(\upsilon_{12})N_2 > B_{12}\rho_{\upsilon}(\upsilon_{12})N_1 \tag{3.14}$$

Since $B_{12}=B_{21}$ (Eq. 3.12), Eq. (3.14) is true only if $N_2 > N_1$, which means to have more atoms in the excited state than in the ground state. This condition is called *population inversion*. According to Eq. (3.10), at thermal equilibrium, the population density relation between levels 2 and 1 is given by the Boltzmann distribution and N_2 must be lower than N_1 ($N_2 < N_1$) because $h\upsilon_{12}/\sigma_B T$ is positive. Consequently, the population inversion in thermal equilibrium can never be reached. Let us investigate whether the laser condition of population inversion can be achieved in a two-level scheme, as shown in Figure 3.5.

The rate equation for this type of system is given by

$$\left(-\frac{dN_1}{dt}\right) = \left(\frac{dN_2}{dt}\right) = B\rho_{\upsilon}(\upsilon_{12})[N_1(t) - N_2(t)] - AN_2(t) \tag{3.15}$$

Since $B_{12}=B_{21}$, we omitted the subscripts for the Einstein coefficient B, and since the spontaneous emission takes place only from level 2 to 1, we can omit the subscripts.

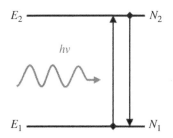

Figure 3.5 Two-level system scheme.

Let us assume that at $t = 0$, all the atoms are in ground state 1, such that $N_1 = N_{total}$ and $N_2 = 0$. Equation (3.15) can be rewritten as

$$N_2(t) = \frac{B\rho_v(v_{12})N_{Total}}{A + 2B\rho_v(v_{12})}\left\{1 - e^{-[A + 2B\rho_v(v_{12})]t}\right\} \tag{3.16}$$

If in Eq. (3.16) we put $t \rightarrow \infty$, we obtain

$$\frac{N_2(t \rightarrow \infty)}{N_{total}} = \frac{B\rho_v(v_{12})}{A + 2B\rho_v(v_{12})} \tag{3.17}$$

Since $A > 0$, from Eq. (3.17) for each given value of t, we obtain

$$\frac{N_2}{N_{total}} = \frac{N_2}{N_1 + N_2} < \frac{1}{2} \tag{3.18}$$

From Eq. (3.18), it is evident that at thermal equilibrium, the number of atoms or electrons in the excited state will not exceed the number of atoms in the ground state; therefore, in a two-level system, the population inversion cannot take place [3, 4].

3.3.1.3 The Three-Level System

Figure 3.6 shows a three-level system, where level 1 with energy E_1 represents the ground level, while levels 2 and 3 represent two excited states with energies E_2 and E_3, respectively.

We will demonstrate that in a three-level system, like the one shown in Figure 3.6, the population inversion can be obtained. In particular, a population inversion between the excited states 2 and 3 will occur such that $N_3 > N_2$, and this system will be able to lase. Since $B_{ij} =$

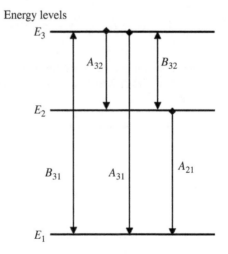

Figure 3.6   A three-level system scheme.

B_{ji}, we will employ the generic coefficient B to indicate absorption and stimulated emission between a group of two states. Before the system interacts with any radiation, all the atoms are in the ground state, which results in $N_1(0) = N_{total}$. When the system interacts with electromagnetic radiation with radiation energy density $\rho_v (v_{31})$ where $hv_{31} = E_3 - E_1$, some atoms will transit from ground state 1 to excited state 3. This type of radiation used to create atoms in the excited states is called *pump source*. The atoms in the excited state 3 can relax by stimulated emission to level 1 or by spontaneous emission to level 2 or level 1. The total number of atoms in a three-level system is equal to the sum of populations in each energy level; therefore:

$$N_{total} = N_1(t) + N_2(t) + N_3(t) \tag{3.19}$$

Similar to the two-level system, for each of the population states 1, 2, and 3, as a result of the pumping, rate equations for each of the three nondegenerate energy levels can be written and solved. At the equilibrium, the population of each level will be constant, in other words:

$$\left(\frac{dN_1}{dt}\right) = 0; \left(\frac{dN_2}{dt}\right) = 0; \left(\frac{dN_3}{dt}\right) = 0$$

Let us consider the rate equation for state 2. The population N_2 is given by, (i) the spontaneous emission from state 3 to 2 and from state 2 to 1, (ii) stimulated emission from state 3 to 2, and (iii) absorption from state 2 to 3. When the equilibrium is maintained, the rate equation for state 2 can be written as follows:

$$\left(\frac{dN_2}{dt}\right) = 0 = A_{32}N_3 - A_{21}N_2 + \rho_v(v_{32})B_{32}N_3 - \rho_v(v_{32})B_{32}N_2 \tag{3.20}$$

$$N_3[A_{32} + B_{32}\rho_v(v_{32})] = N_2[A_{21} + B_{32}\rho_v(v_{32})] \tag{3.21}$$

where it can be rewritten as

$$\frac{N_3}{N_2} = \frac{A_{21} + B_{32}\rho_v(v_{32})}{A_{32} + B_{32}\rho_v(v_{32})} \tag{3.22}$$

When $A_{21} > A_{32}$ then $N_3 > N_2$, this condition occurs when the atoms in state 2 decay much rapidly to state 1 compared to the atoms decay from state 3 to state 2. Consequently, a population inversion can occur between states 3 and 2, and this system can now lase. A three-level (or more than two-level system) system is called a *gain medium* [3, 4].

3.3.1.4   The Four-Level System

In Figure 3.7, a schematic of a four-level system is displayed. The ground level has an energy equal to E_1 and is indicated 1. Levels 2, 3, and 4 are excited states, and their energies are E_2, E_3, and E_4, respectively.

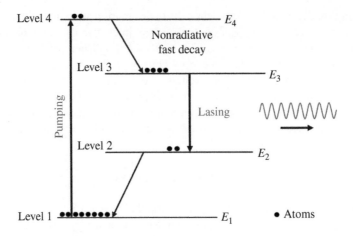

Figure 3.7   Scheme of a four-level system.

When the atoms in level 1 are excited (pumped) to level 4, they relax through a nonradiative mechanism to level 3, a metastable level with a long lifetime. When the population inversion occurs, the transition from level 3 to the lower level (2) causes the lasing within a carefully designed cavity. To maintain the lasing condition (population inversion), the lifetime of level 2 needs to be small, meaning that the atoms quickly relax from level 2 to the ground state, where they can be pumped again to the excited level 4. It is important to notice that level 2 needs to be above level 1 and unpopulated at ordinary temperature, and sometimes this is achieved by lowering the system temperature [4].

3.3.2   Laser Components

In the previous section, we have discussed the physics behind the generation of a laser beam. In this section, we will describe the main components of a typical laser system. Gain medium, pumping sources, and optical resonance systems are the three main parts [3] as discussed below. A schematic of a typical laser system is given in Figure 3.8.

3.3.2.1   Gain Medium

Gain medium (or also known as active laser medium) is materials made of atoms, ions, molecules, or electrons. Depending on the type of gain medium, different type of lasers with different wavelengths can be obtained. Gain medium can be solid, liquid, gas, dye, or semiconductor. The first laser was a ruby laser, where the gain medium was a solid ruby (Al_2O_3) in which Al^{3+} ions were replaced with Cr^{3+} ions. The latter, where the ions responsible for the lasing, since their electronic energy levels in the Al_2O_3 hosting structure, can achieve the population inversion.

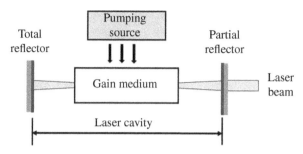

Figure 3.8 The main components of a laser are shown. The active medium or gain medium and the mirrors that reflect the emitted radiation back and forth represent the laser cavity. A pumping source is needed to excite the atoms or molecules of the gain medium.

3.3.2.2 Pumping Source

In Section 3.3.1.1, we referred to *pumping* through a radiative or nonradiative source to push atoms to the elevated states. Pumping sources are responsible for the excitation of the gain medium atoms from the ground state to upper excited states, creating the population inversion, which is a necessary condition for the occurrence of stimulated light amplification that leads to lasing. The gain medium can be pumped through optical or electrical excitation. In the first approach, the active laser medium is excited through a high-intensity light source (i.e. Radiative) such as flash lamps and lasers. Instead, the laser medium's excitation by an intense electrical discharge (i.e. nonradiative) is also known as the electrical excitation approach. This type of pumping approach is used in gas lasers.

3.3.2.3 Resonant Optical Cavity

The amplification of light intensity once arrived at the population inversion in the gain medium is achieved by forcing the stimulated photons to travel back and forth across the gain medium. In order to achieve light amplification, the laser medium is placed in a cavity also known as *resonator*. The cavity possesses two parallel mirrors placed at the ends, which direct the light back and forth, through the laser medium (see Figure 3.8). One of the mirrors is 100% reflective, while the second mirror is <100% reflective, to allow the amplified light wave to be removed from the cavity as an output beam.

3.3.3 Continuous Vs. Pulsed Laser

Lasers can be categorized as continuous wave (CW) and pulsed wave (PW) lasers. The first type of lasers produces a continuous beam of light, whose characteristics are determined by the gain medium. The first invented CW laser was a helium–neon laser with a wavelength of 1153 nm [5], and since then, other types of lasers that can be operated continuously have been developed, such as gas lasers, solid-state lasers, and dye lasers. More details on the different types of lasers will be given in Section 3.3.4. Pulsed wave lasers, in contrary, emit light in the form of pulses. Pulses can be generated by different methods, and depending on the required pulse duration, pulse energy, pulse repetition rate, and wavelength, different types of PW lasers are available. The two standard methods for the generation of a pulsed beam of light are the Q-switching and mode-locking techniques which are used for the generation of pulses in the nanosecond regime and femtosecond regime, respectively [6].

3.3.4   Laser Types

Lasers can be classified based on different parameters; however, most lasers are categorized based on the type of the active medium, as described in Section 3.3.2.1. Lasers can be grouped as follows:

- Solid-State Lasers
- Gas Lasers
- Liquid Dye Lasers
- Semiconductor Diode Lasers
- Fiber Lasers

   These lasers can produce coherent light at different wavelengths, and the output beam can be a continuous (CW) or a pulsed light beam. In the following subsections, details about these classes of lasers will be presented [3, 4, 7–9].

3.3.4.1   Solid-State Lasers

In this type of lasers, the gain medium is solid at room temperature. The ruby laser, invented in 1960, was a solid-state laser, where a ruby was used as the gain medium ($Cr3 + - Al_2O_3$). The ions responsible for the lasing in the first laser were impurities of Cr^{3+}, as mentioned in Section 3.3.2.1. Similarly, in most of the modern solid-state lasers, a dopant is embedded in a host material. Neodymium (Nb^{3+}) is the dopant used in many commercial lasers, whereas the most used host materials are yttrium orthovanadate (YVO_4), yttrium lithium fluoride (YLF), and yttrium aluminum garnet (YAG). Depending on the type of dopant and hosting materials, solid-state lasers operating at different wavelengths are produced. For instance, YAG and YLF doped with Nd^{3+} produce a laser beam with wavelengths 1064.1 and 1054.3 nm, respectively. In contrast, Ti3+ dopant ions in Al_2O_3 produce a laser with 780 nm, whereas the ruby laser's wavelength is 694.3 nm. Figure 3.9 shows the schematic of this type of lasers.

Nd:YAG Solid-State Laser
Neodymium (Nd) lasers, along with CO_2 lasers, have been the conventional and the most employed lasers in AM. However, the current trend is different, that will be discussed later. In this solid-state laser, the gain media are Nd ions doped in a host crystal. Usually, yttrium aluminum garnet (YAG, $Y_3Al_5O_{12}$) and glass is commonly employed as a host material. The emission laser wavelengths depend on the host crystal. For instance, the Nd:YAG crystal output wavelength is 1064 nm, while the Nd:glass wavelengths range between 1054 and 1062 nm. These lasers can be operated in continuous or can be pulsed, with the power of a few kW and up to 20 kW in peak power, respectively. The laser medium in the Nd:YAG lasers is pumped by a krypton or xenon flashlamps; however, some systems employ diode lasers as

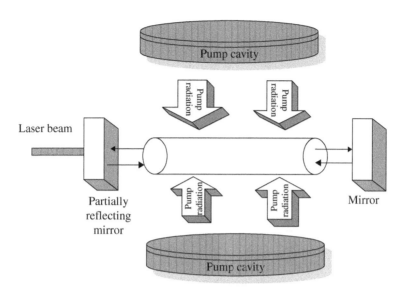

Figure 3.9 Solid-state Laser scheme. Source: Republished with permission of Taylor & Francis Group, from [7]; permission conveyed through Copyright Clearance Center, Inc.

a pumping source. The Nd laser is a four-level system, and the energy levels can be indicated as E_1, E_2, E_3, and E_4, where E_1 is the ground state and E_4 is the highest energy level. When the Nd ions are excited by the flashlamp, the electrons are excited to the highest energy level (E_4), which has a lifetime of 250 μs. The electrons through nonradiative decay will fall in the metastable energy level E_3 with a higher lifetime than the E_4 lifetime. Therefore, a population inversion occurs. The electron from E_3 will decay after some time in the lower energy level E_2 through spontaneous emission, emitting a photon of wavelength 1064 nm. The E_2 level's lifetime is very short, and the electrons will rapidly decay nonradiatively to the ground state E_1. The energy-level diagram for Nd3+ doped in YAG is shown in Figure 3.10 [7, 8, 10].

Disk Lasers

Disk lasers or active mirror is a solid-state laser characterized by a heat sink, and the laser beam is released on the opposite side of a thin layer of the active gain medium. The beam is not necessarily circular. Active medium can be an optically pumped semiconductor. The disk lasers are very high efficiency with a very small size; however, the power is limited because we have to transfer the heat out of the setup. Disk lasers have a relatively higher wall-plug efficiency, thus becoming popular in AM, especially in LDED.

3.3.4.2 Gas Lasers

In gas lasers, the gain medium is gas or a mixture of gas and depending on the type of gas used; they can be divided into atomic gas lasers and molecular gas lasers. Examples of atomic gas

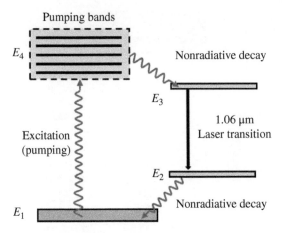

Figure 3.10 Energy-level diagram for Nd3+ doped in YAG. Source: Reproduced with permission from [8].

lasers are the helium–neon laser and the argon-ion lasers. The carbon dioxide lasers and the excimer lasers belong instead to the second class of gas lasers. Generally, in gas lasers, the low-pressure gas medium is contained in a cylinder, where at the ends, two electrodes are installed. Upon applying a voltage, an electric discharge inside the gas tube creates an electrical current, which ionizes some of the gas atoms to form free electrons, which travel from the cathode to the anode. After excitation to higher energy levels due to the collision with the free electrons, some atoms will decay back to lower energy levels. If the conditions presented in Section 3.3.1 are met, a population inversion will occur, leading to lasing.

CO₂ Laser

Carbon dioxide lasers operate in the 11 and 9 μm wavelength regions, and either CW or pulsed CO_2 lasers have been produced with powers of 100 kW and 10 kJ, respectively. The wavelength most employed in AM is 10.6 μm. The laser structures used for this molecular gas laser can be divided into longitudinally excited lasers, waveguide lasers, and transversely excited lasers. In the longitudinally excited lasers, the gaseous laser medium is enclosed in a long narrow cylindrical glass, as shown in Figure 3.11.

The electrical discharge current is applied through the electrodes, located at the opposite ends of the glass tube. In the laser systems, where the laser medium is sealed, the tube needs to be periodically changed to avoid corrosion of the electrodes due to oxygen production from the breakdown of CO_2. In other systems, the recirculation of the gas through the tube assures gas preservation. CW CO_2 lasers can be obtained by waveguide lasers, which possess a waveguide structure in the gain medium. In this type of lasers, the bore region, whose dimensions are in the order of few millimeters, is placed between radio frequency (RF) electrodes. A high-frequency alternating field across these electrodes is provided, connecting the electrodes to an RF power supply with a frequency of 80–100 MHz. The small dimensions of the bore assure an efficient cooling of the laser gas and high-pressure operation, which leads to high gain and high power output up to 100 kW. In transversely excited lasers, the gas is maintained at pressures of 1 atm or higher, and the electrodes are placed parallel to each other at a distance of a few centimeters.

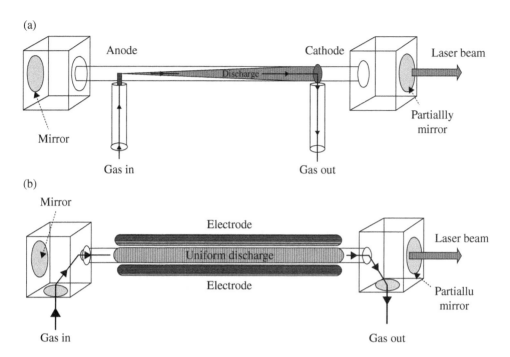

Figure 3.11 (a) Longitudinally excited and (b) transversely excited CO_2 laser. Source: Republished with permission of Taylor & Francis Group, from [7]; permission conveyed through Copyright Clearance Center, Inc.

In order to assure a uniform discharge when the high voltage is applied at the electrodes, a pre-ionization is required to ionize the gas between the electrodes. This configuration allowed the production of extremely high-energy pulsed lasers, where it has been used in some excimer lasers too. The excitation mechanism in these types of laser structures is the same. The laser gas employed in these systems contains a mixture of gases, typically made of CO_2 and N_2 in a $0.8 : 1$ ratio. In particular, the presence of N_2 increases the laser efficiency by 30%. The reason is that the excited energy level of nitrogen is very close to the vibrational level of CO_2. The N_2 excited level is metastable to radiative decay, and through collisions, the energy is transferred to the CO_2 vibrational level $(0, 0, 1)$, leading to population inversion, as shown in Figure 3.12. The possible laser transitions in CO_2 lasers are of two types: $(0, 0, 1) \rightarrow (1,0,0)$ and $(0, 0, 1) \rightarrow (0,2,0)$. These transitions result in the emission of laser beams with wavelengths of 10.6 and 9.4 μm, respectively [4, 8].

3.3.4.3 Liquid Dye Lasers

Dye lasers or organic dye lasers consist of systems where the lasing medium is liquid at room temperature. Usually, the gain media are a strong absorbing and emitting organic dye in a

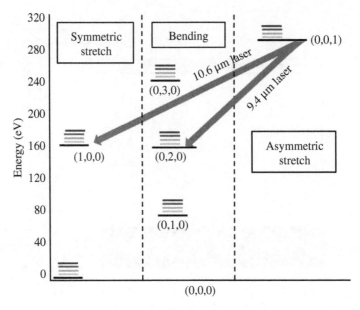

Figure 3.12 Laser transitions between vibrational levels in CO_2. Source: Reproduced with permission from [8].

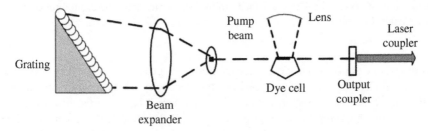

Figure 3.13 Liquid dye laser schematic. Source: Republished with permission of Taylor & Francis Group, from [7]; permission conveyed through Copyright Clearance Center, Inc.

solvent. A broad range of wavelengths can be obtained depending on the type of dye employed. A tunable laser output spanning from 320 to 1200 nm can be produced employing different dyes in sequence [8]. A high-voltage but low-current power supply and a large storage capacitor are required for such lasers. Rhodamine B is a common dye used in this type of laser. Dye lasers can be pulsed, continuous, and mode-locked, and other lasers generally pump them. Figure 3.13 shows a schematic of this type of lasers. Dye lasers have applications in medical and material texturing. No report on the use of liquid dye lasers for meaningful AM applications has been found in the literature.

3.3.4.4 Semiconductor Diode Lasers

Compared to the previously described laser systems, semiconductor diode lasers are very small, efficient and require a lower power input. Within this class of lasers, microscopic chips of gallium arsenide or semiconductors are mainly used. These lasers can be pumped electrically or optically, and the lasing medium in this type of lasers is a p–n junction made of semiconductor materials of group III–V and II–VI compounds. When a current flows through the p–n junction, the electrons are excited to a higher state in the conduction band, decaying afterward to states near the bottom of the conduction band. Simultaneously, the holes created in the valence band move at the top of the valence band. The electrons and holes recombine, emitting photons with energy near the bandgap of the material. This process can be spontaneous and stimulated, which leads to optical amplification. The beam is not necessarily circular. Large beam divergence (up to 40° half-angle) can be achievable by this laser, as shown in Figure 3.14. Non-symmetrical beam distribution is also possible. Lower energy intensity per area is one of the features of this type of lasers. Depending on the type of materials employed, different energy bandgaps can be obtained and thus different wavelengths of emission.

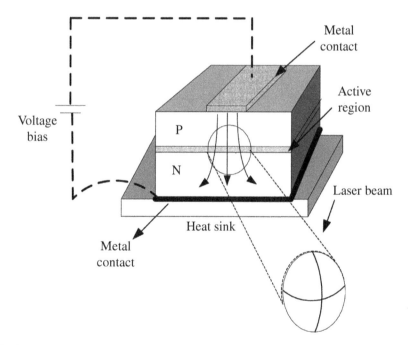

Figure 3.14   Diode laser scheme. Source: Republished with permission of Taylor & Francis Group, from [7]; permission conveyed through Copyright Clearance Center, Inc.

3.3.4.5 Fiber Optic Lasers

The gain media in this type of lasers are optical fibers where the core part of fibers is doped with rare-earth ions, such as erbium, neodymium, germanium or ytterbium, where continuous semiconductor diode lasers are used to pump light to the core section. This type of laser is also used for creating very short pulses. In a simple word, fiber lasers create a beam by using the fiber as a resonant cavity, as shown in Figure 3.15. In 2003, a 2 kW CW fiber laser was built with a spot size of 50 μm and a power density of 100 MW/cm^2. The relatively small dimensions of this laser along with the high output scalable power make this fiber laser a suitable replacement of solid-state and molecular lasers, especially for industrial processes applications [7].

Fiber strings include two sections: core and cladding. If the laser beam is pumped into the cladding around the core, the laser beam will be bounced and rebounded inside the core due to different indices of refraction for the cladding and core segments. Each time that the beam passes through the core, further pumping light is absorbed by the core. In fiber lasers, two fiber Bragg gratings (FBG) are also used to act as mirrors for special wavelengths, as shown in Figure 3.16. The efficiency of this type of lasers is relatively high due to the fact that the atom

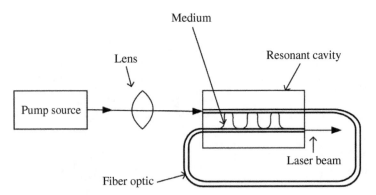

Figure 3.15 Scheme of a typical fiber laser. Source: Republished with permission of Taylor & Francis Group, from [7]; permission conveyed through Copyright Clearance Center, Inc.

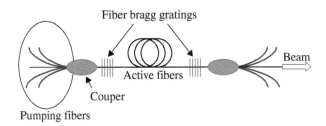

Figure 3.16 Schematic of fiber lasers that include FBGs and beam coupler Source: Redrawn and adapted from general concept available on the internet.

levels of earth elements have extremely effective energy levels, allowing the use of an inexpensive diode-laser pump source, while it will still provide high output energy. It can also use multiple pumping fibers, connected to a coupler, as shown in Figure 3.16.

Yb- and Er-Fiber Laser

Ytterbium and erbium fiber lasers are the most employed in industry. In the AM industry, single-mode Ytterbium fiber lasers are very popular. In other words, single-mode lasers provide users with perfect Gaussian energy distribution.

In the fiber lasers, the gain medium is a fiber optic, whose ends are connected to a pump source via a coupler and to an optical resonator, respectively. The pumping source is normally a CW diode laser, which continually pumps the fiber and the formed pulses passing through the system are amplified. The excitation mechanism in this type of fiber laser can be explained by taking into consideration the energy-level diagram of the erbium-doped fiber shown in Figure 3.17.

Erbium-doped fiber laser is a three-level laser, and the optical pumping will excite the electrons from the ground state (4I15/2) to the excited state 4I11/2 at a wavelength of 908 nm. The electrons' nonradiative decay to the upper laser level 4I13/2 occurs, leading to the population inversion between the upper laser level 4I13/2 and the ground state (4I15/2). In this condition, photons with a wavelength around 1550 nm are emitted and amplified as they propagate through the fiber.

3.3.4.6   Laser Deployed in Laser-Based AM Processes

Figure 3.18 summarizes the lasers employed in laser-based AM processes for specific materials [11].

3.3.5   *Laser Beam Properties*

A laser beam is coherent (meaning the light emitted has a constant relative phase), monocolor (one wavelength), and collimated (extremely parallel rays). In Section 3.3.2, it was discussed that in order to achieve the amplification of light, the laser medium is placed in a cavity or resonator, with two parallel mirrors at each end. The presence of these mirrors in Figure 3.1 leads to the creation of transverse modes superimposed upon the beam, which produce transverse characteristics of the beam known as transverse electromagnetic modes (TEM) [7, 12]. In a laser, the TEM is referred to as Gaussian–Laguerre modes and they are denoted as TEM_{pl}. The subscripts p and l are integers indicating the number of nodes of zero intensity transverse to the beam axis in the radial direction and in tangential direction, respectively. Figure 3.19 shows the TEM versus the beam distribution

For the transverse mode TEM_{00}, p and l are zero and this represents the lowest order with the same shape as a Gaussian beam. The different TEM modes have different patterns and different Gaussian energy intensities. In Figure 3.19, the different mode patterns are shown. The intensity distribution of a TEM_{pl} at a point (r, φ) (in polar coordinates) from the center of the mode is represented by Laguerre equation:

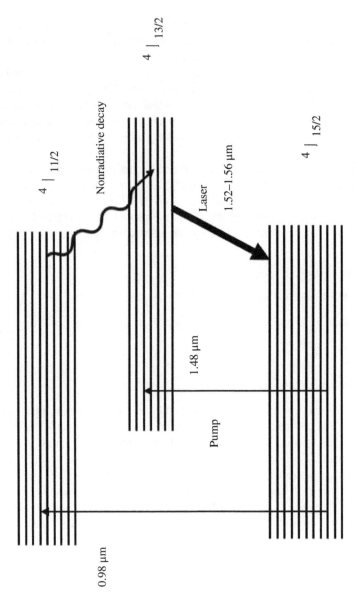

Figure 3.17 Energy-level diagram of the erbium-doped fiber. Source: Reproduced with permission from [8].

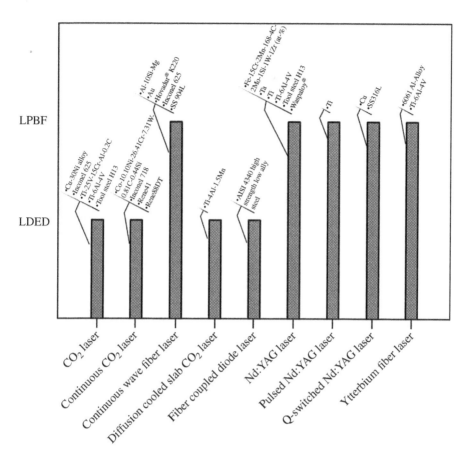

Figure 3.18 Laser employed in laser-based AM processes (i.e. laser powder bed fusion [LPBF] and laser directed energy deposition [LDED]). Source: Adapted from [11].

$$I_{pl}(r,\varphi) = I_0 \left(\frac{2r^2 M^2}{r_l^2}\right)^l \left[L_l^p\left(\frac{2r^2 M^2}{r_l^2}\right)\right]^2 \cos^2(l\varphi) \exp\left(-\frac{2r^2 M^2}{r_l^2}\right) \qquad (3.23)$$

where I_0 is the intensity scale factor (W/m²), r_l is the radius of the laser beam profile, M^2 is the beam quality factor, and L_l^p is the associated Laguerre polynomial of order p and index l [7]. In the above equation, I_0 is presented by:

$$I_0 = \begin{cases} \dfrac{2M^2}{\pi r_l^2} P_l & \text{for } l = 0 \\[4mm] \dfrac{4M^2 p!}{\pi r_l^2 (p+l)} P_l & \text{for } l = 1, 2, 3, \dots \end{cases} \qquad (3.24)$$

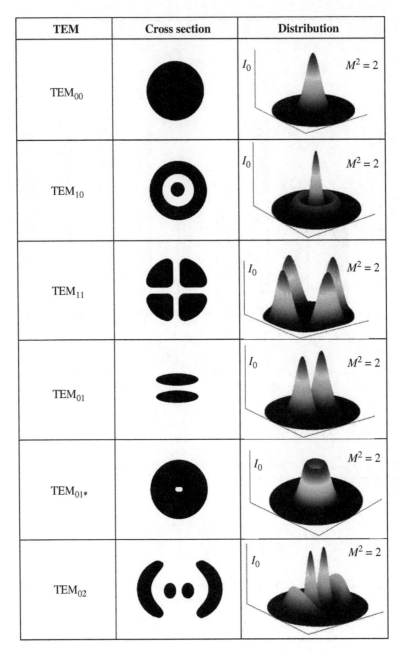

Figure 3.19 Mode patterns for different TEMs. Source: Republished with permission of Taylor & Francis Group, from [7]; permission conveyed through Copyright Clearance Center, Inc.

where P_l is the net laser power. Laguerre polynomials in general fashion can be expressed by one of the two following equations based on the indices:

$$L_l(x) = \frac{e^x}{l!} \frac{d^l}{dx^l}\left(e^{-x}x^l\right)$$

(3.25)

$$L_n^K(x) = \frac{e^x x^{-K}}{n!} \frac{d^n}{dx^n}\left(e^{-x}x^{n+k}\right)$$

(3.26)

The radius of the laser beam, r_l, depends on the propagation axis and can be expressed as

$$r_l(z)^2 = r_{0l}^2 + 4\theta^2(z-z_0)^2$$

(3.27)

where the beam radius of the waist is indicated as r_{0l}^2, the waist location with respect to an arbitrary coordinate along the propagation axis is indicated as z_0, and the far-field divergence angle is θ. A typical laser beam geometry can be represented by the following schematics (Figure 3.20):

The beam quality factor M^2, or the beam propagation factor Q, describes the propagation, and they are related as

$$M^2 = \frac{1}{Q} = \frac{n\pi r_{0l}\theta}{2\lambda}$$

(3.28)

The laser wavelength used in the medium is indicated as λ, and n is the reflection's index. We can then define the propagation factor k as

$$k = \frac{1}{M^2} = \frac{2\lambda}{n\pi r_{0l}\theta}$$

(3.29)

If $k = M^2 = 1$, the beam is Gaussian, whereas it is not Gaussian if $M^2 > 1$. The larger the M^2, the beam shape will become similar to a flat hat [7].

There is a "depth of focus" parameter in a parabolic beam profile, where it refers to a segment of the beam that can be assumed to have a cylindrical shape. The depth of focus for a 28° angle of divergence is reported to be approximately $1.05 r_{0l}$.

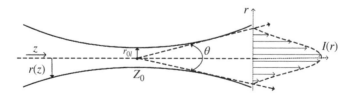

Figure 3.20 Laser beam profile. Source: Republished with permission of Taylor & Francis Group, from [7]; permission conveyed through Copyright Clearance Center, Inc.

3.4   Electron Beam Basics

In addition to the laser beam, another common source of heat for thermal-based metal powder bed additive manufacturing technologies is electron beam (EB or EBM) [13]. EB which is a collimated stream of free electrons is a very useful tool for heating and welding applications. EBs are also popular in micro- and nanoscopic imaging, machining, and spectroscopy. With a sufficiently large power, EB can be used for macroscopic cutting, machining, and welding. EB brings pros and cons to AM as explained in the following section.

3.4.1   Comparisons and Contrasts between Laser and Electron Beams

While lasers are highly coherent and have high numerical intensity sources of comparatively low energy photons, EBs have low numerical intensity sources of very high kinetic energy while remaining highly coherent. Similarities in both energy sources are observed in their capability to generate immense power; their energy sources are collimated and can be focused on a relatively small area while attaining high thermal intensity. These reasons ensure the manufacturability of parts with fine features.

The contrast between these energy sources is the heat transfer to the powder bed. Laser light provides heat through atomic photon absorption. Because of this, the amount of heat transferred depends on the laser color spectrum and the material electronic structure. Laser absorption is also highly dependent on the feed material finish and substrate finish because the reflectivity of these materials reduces the amount of energy available to be absorbed.

EB transfers heat by transforming kinetic to heat energy through inelastic collisions between the electrons and the substrate. To ensure optimum transfer of energy, the electrons are accelerated to elevated kinetic energies to collide with the material and embedding themselves while slowing the process [14]. As opposed to laser beams, material finish and electronic structure are not relevant to the heat transfer efficiency. EB can be moved with high velocity (>> 1 km/s) while being controlled by a magnetic field. The laser beam cannot be moved at velocities higher than 5–7 m/s due to the scanner mechanical constraints.

A major advantage of lasers over EBs is the ability of light to transmit through transparent materials with only tiny losses. Because of this, lasers can be used in open atmosphere work areas, with shield gases if required. Lasers can also be transmitted through optical fibers, providing greater flexibility in beam positioning. EBs require medium to high vacuum conditions to work efficiently. Because of this, EBs cannot currently be used for powder-fed DED due to the requirement of a gas or fluid to motivate the feed material to the work area. On the other hand, powder bed fusion does not require any gas or fluid for material transfer. This makes powder bed fusion an ideal additive manufacturing technique for EB heat sources.

3.4.2   Electron Beam Powder Bed Fusion Setup

The electron beam powder bed fusion (EPBF), known as electron beam melting (EBM), was developed in 2001 at Chalmers University of Technology (Sweden). It is similar to the LPBF process except for the energy source type, which is an electron beam. A schematic of an EBM apparatus is shown below (Figure 3.21).

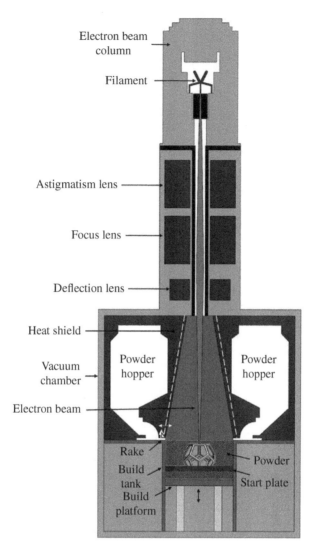

Figure 3.21   Schematic of a typical EBM apparatus. Source: Redrawn and adapted from [15].

In EBM, the powder melting is achieved using a high-energy electron beam gun, which is scanned across the material in the powder bed. Table 3.1 highlights the main differences between LPBF, EBM, and LDED with blown powder known as LPF processes. Noteworthy, numerical values for resolution, surface roughness, and layer thickness are higher in EBM than LPBF processes. This is due to the formation of a larger melt pool size resulting in creating a larger heat-affected zone. This effect is related to the type of energy source used; indeed, the kinetic energy of the beam electrons is responsible for the heating of the metal powder, which gains a negative charge as they adsorb the incoming electrons. This may lead to the creation of a powder cloud due to the expulsion of the powder particles from the powder bed because of the repulsive force between negatively charged particles and/or the creation of a more diffuse beam

Table 3.1 Comparison between AM processes adapted from [16] with several changes.

Parameters	LPBF	EBM	LDED (LPF)
Energy source	Laser	Electron beam	Laser
Layer thickness (μm)	20–90	50	100–1000
Min wall thickness (mm)	~0.2	~0.6	~0.6
Accuracy (mm)	±0.1	±0.3	±0.4
Build rate (cm³/h)	5–20	80–100	2–70
Surface roughness (μm)	5–15	15–20	15–40
Geometry limitations	Supports needed	Less supports needed	Same limitations as a typical 5 axes milling
Materials	Ferrous alloys, titanium, aluminum, nickel, ceramics, etc.	Just conductive materials (Ti alloys CrCo, Tool steel, Cu alloy, etc.)	Ferrous alloys Ti, Ni-base alloys, composites, ceramics, etc.
Productivity vs costs	Good (depends on applications)	Medium	Good
Residual stresses	Medium	Low	Medium
Part complexity	High	Medium	Low
Typical applications	Tooling, Implants, jet engines, prototypes	Near-net-shape manufacturing of Implants, turbine blades, prototypes	Repair of blades, vanes, shafts, coatings, etc.
Necessity of support structures	Medium	Low	Normally none

due to the repulsion of the incoming negatively charged electrons, which leads to a larger melt pool [17]. Electron beam guns are more efficient than lasers because, for the generation of an electron beam, most of the electrical energy is converted into a beam, while only 10–60% of the electrical energy input is used to generate a laser beam, and the remaining is dissipated as heat [17]. It should be noted that the laser wall efficiency depends on the type of lasers. More details about the working principle of EBM are given in the following paragraphs.

3.4.3   Electron Beam Mechanism

In a vacuum, the existed free electrons can be accelerated under electric and magnetic fields, leading to the formation of narrow electron beams with high kinetic energy. The working principle of an electron beam is the conversion of this kinetic energy into heat energy. In particular,

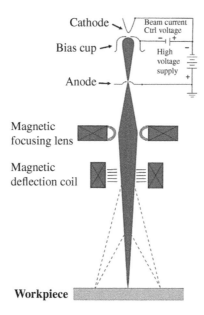

Figure 3.22 Electron beam formation schematic. Source: Redrawn and adapted from [18] under the Creative Commons License CC BY-SA 3.0.

the kinetic energy is transformed to heat upon the collisions of the electrons in a vacuum. The values of the power density of the electron beams reached upon their acceleration, collimation through strong electric fields and magnetic lenses can be as high as 10^6 W/mm^{2,} and the localized temperature could increase rapidly ~10^9 K/s. In order to generate an electron beam, the required equipment are (see Figure 3.22):

- Power supply
- Electron gun/source
- An anode
- Magnetic lenses
- Electromagnetic lenses and deflection coils

A power source is needed to supply a continuous beam of electrons. Depending on the process, low- or high-voltage equipment are available ranging from 5 to 30 to 70–150 kV. The electron gun is the cathode, which is heated up by a small filament connected to a low voltage causing the evaporation of the electrons from the cathode. Generally, tungsten is the metal of choice as a cathode. The evaporated electrons are then accelerated toward the anode, which is positively charged and placed after the electron gun. The anode possesses an aperture, thus creating an electron jet, which moves toward the magnetic lenses. These components do not allow divergent electrons to pass through and provide a very intense electron beam. The so-formed beam is then focused and deflected by electromagnetic lenses and deflection coils that it is precisely positioned with respect to the part to be printed [19, 20]. Further explanation is provided below.

3.4.3.1   Electron Beam Sources

Gun electrode types used for AM are shown in Figure 3.23. Three types are common: Tungsten (W) filament, Lanthanum Hexaboride (LaB$_6$), and field-emission gun (FEG), as shown in the figure.

3.4.3.2   Electron Beam Optics and Positioning

The electrons produced and accelerated by the beam power still lack the requisite focus and control for use. The beam must therefore be focused, shaped, and positioned to provide the control required.

Electromagnetic Lens
Similar to optical systems, EBs are focused with a lens, albeit electromagnetic, created by magnetic fields. Like a solenoid, the lens is made by winding wires around a coil. However, a solenoid creates a uniform magnetic field, whereas the electromagnetic lens, shown in Figure 3.24, is wound over a very short length, which makes the magnetic field less dense near the center of the lens. When an electron passes through the lens, it will initially experience a clockwise force (Lorentz rule, see Chapter 5) around the opening. Once the electron has some momentum in this direction, the force begins to point toward the center of the opening.

To prevent flaws in the focusing properties of the lens, the windings and other components must be constructed with high radial symmetry. The clockwise force causes the image formed by an electromagnetic lens to be rotated and inverted. Nevertheless, this is not a problem for most heating applications because the spot shape is typically circular.

Stigmators
Initially, the beam may be elliptical rather than circular because of a combination of filament geometry and the non-uniformity inside the electromagnetic lens. This phenomenon is called astigmatism, which can be corrected by stigmators. A stigmator is a device that consists of four solenoids with magnetic quadruples. An electron in this field will experience force toward the center or away from it, depending on which quadrant of the quadruple field it is in. When the

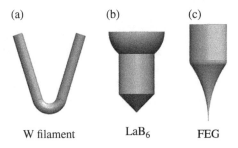

Figure 3.23   Gun electrode types: (a) Tungsten (W) filament, (b) Lanthanum Hexaboride (LaB$_6$), and (c) field-emission gun (FEG).

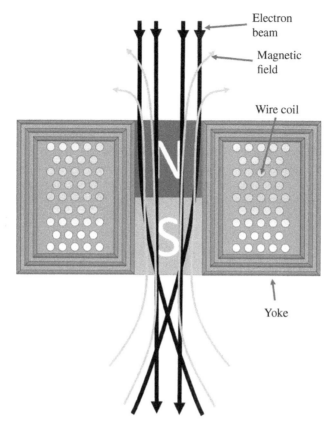

Electron
beam

Magnetic
field

Wire coil

Yoke

Figure 3.24   Electromagnetic Lens. Source: Redrawn and adapted from [21].

strength of the current in opposite pairs of solenoids is varied, the beam can be made more or less elliptic [21].

Rasterizer

The x and y position of the EB can also be controlled by a rasterizer. Similar to the stigmator, the rasterizer consists of two pairs of electromagnets, one pair for each of the x and y directions. The magnets in a pair have the same field direction. When an EB is passed through the field between these magnets, it is deflected uniformly to a direction depending on the field sign, while the field strength dictates how far the beam is deflected [21].

3.4.4   Vacuum Chambers

To successfully deploy an EB, it must be well collimated and it should be able to focus down to small spot size. When electrons are traveling in a fluid medium, they impinge on the gas molecules causing the electrons to scatter off. This scattering reduces the total beam energy available at the heating location and increases the spot size, further reducing the beam intensity. For these reasons, EB's has to be operated in vacuum conditions.

3.4.4.1   Vacuum Types and Pumping Systems for EBM

Vacuum systems are divided into ranges based on the minimum vacuum pressure that the system is capable of. Vacuum systems are often measured in Torr with one atmosphere equal to 760 Torr. When the vacuum pressure ranges from 10^3 to 10^{-3}, it is called a Soft Vacuum, which can be used by roughing pumps [22]. A vacuum pressure ranging from 10^{-3} to 10^{-8} is considered a High Vacuum used by high-performance pumps, while a pressure below 10^{-8} is considered Ultra High Vacuum and is generally too expensive to maintain except for very specific scientific and performance applications. EB heating processes typically run in the High Vacuum range.

The Soft vacuum requires roughing pumps. These can be piston-, diaphragm-, gear-, or scroll-driven pumps. In practice, a roughing pump is employed to get the system down to a low vacuum pressure before a high-performance pump can be activated. This is important to avoid stressing the high-performance system. Once the high-performance pump is on, the roughing pump stays on to serve as a "backing pump" helping to evacuate the high-performance pump's exhaust.

A turbo pump is the most common high-performance pump, which spins extremely fast to create a negative potential gradient for molecules entering into it. It does not "suck" molecules out; however, it can only act on molecules that enter into it on their own. This makes the pumping efficiency poor, and high vacuum pumps can easily be overcome by leaks and contamination inside the chamber. Consequently, the vacuum chamber and everything entering into it must be kept very clean and very dry. This may add to the production cost if pre-processing is required. Attaining a high vacuum pressure takes considerably longer than a roughing vacuum pressure; therefore, processes that can take place closer to a roughing vacuum pressure will be more cost-effective.

3.4.4.2   Chamber Types for EBM

Vacuum chambers must be sealed to function; however, if parts are to be loaded and unloaded, the seal must be lifted. Every time a new part is loaded or unloaded, the atmosphere enters the vacuum chamber and must be pumped out again, increasing cycle times and decreasing throughput and profitability.

There are two major methods of sealing a vacuum chamber, namely: open chamber and load-lock [21]. In an open chamber, the entire chamber is left open to the atmosphere and the chamber must be pumped down to the right vacuum pressure each time the door is opened. A load-lock or air-lock is a small chamber with two doors. While one door connects to the outside, the other connects to the working chamber; however, only one door is ever open at a time. When both doors are closed, the chamber can be evacuated. The working chamber is always held at high vacuum pressure and is usually much larger. The load-lock area only needs to be large enough to load and unload parts; consequently, it can be evacuated much quicker. Another advantage of the load-lock chamber is that pumping can take place while another part is being worked on. When applied in additive manufacturing, the manufacturing process takes much longer than the pumping; therefore, open chambers are permissible. In future production level systems where process times get shortened, the load-locking method will be more suitable for powder bed systems.

3.5 Powder Feeders and Delivery Nozzles Technology

Powder feeders and delivery nozzles are the most important pieces of equipment in a DED process, having a crucial role in the delivery and deposition of the powders. In a typical powder-fed DED process, the powder is conveyed using a proper powder feeder and then deposited onto the substrate through a nozzle. Moreover, it should be noted that a carrier gas is supplied to transport the powders from the feeders to the nozzles. The flowability of powders depends on powders' rheology and physical properties, on their size and shape. To have a stable powder stream, it is important to have control over the powder feed rate, which depends on the flowability of powders employed. Consequently, different types of powder feeders have been developed to accurately control the feed rate and achieve a continuous and uniform powder stream during AM processes.

3.5.1 Classification of Powder Feeders

Powder feeders can be classified into the following classes:

1. Mechanical wheel or screw powder feeder;
2. Gravity-based;
3. Fluidized bed; and,
4. Vibrating.

The combination of the above-mentioned feeders is also used. The details on these types of powder feeders are given in the next subparagraphs.

3.5.1.1 Mechanical Wheel Powder Feeder

The mechanical wheel powder feeders consist of a container containing a blade or screw driven at one end, and depending on the blade's configuration, different flow rates can be achieved. In Figure 3.25, a schematic is shown.

Powders with different sizes can be used in this type of powder feeder, and it normally operates at low or zero pressure differential between the inlet and outlet. Abrasive powders can wear the blade, which leads to an increase of maintenance costs. Backpressure using pressurized air or inert is normally applied to obtain better and consistence.

3.5.1.2 Gravity-Based Powder Feeder

In gravity feeders, powder particles with proper flowability flow through an orifice due to gravity. The amount of powder reaching the delivery nozzle can be controlled by regulating the orifice area. Moreover, to achieve a high precision feed rate, external metering components can be incorporated. For instance, a rotating disk with holes has been used to achieve an

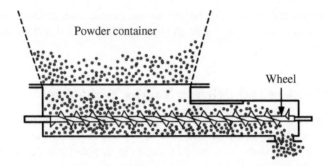

Figure 3.25 Scheme of a mechanical wheel powder feeder. Source: Republished with permission of Taylor & Francis Group, from [7]; permission conveyed through Copyright Clearance Center, Inc.

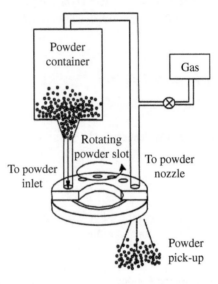

Figure 3.26 Schematic of gravity-based powder feeders with a rotating wheel for metering. Source: Republished with permission of Taylor & Francis Group, from [7]; permission conveyed through Copyright Clearance Center, Inc.

accurate volumetric feed rate by controlling the rotating disk speed. Other similar components have been produced replacing the rotating disk with a metering wheel or a lobe gear based on this concept. Schematics of the three gravity-feeder designs are shown in Figures 3.26–3.28.

3.5.1.3 Fluidized Bed Powder Feeder

The design of the fluidized bed powder feeder is based on a patent reached from the fluidics principle. The advantage of this powder feeder is the absence of mechanical parts to deliver the powder, which reduces the cost of maintenance and parts replacement. The powder stream is

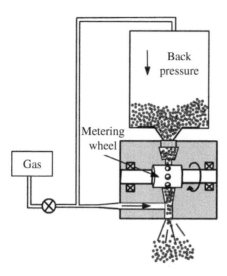

Figure 3.27 Schematic of gravity-based powder feeders with a metering wheel. Source: Republished with permission of Taylor & Francis Group, from [7]; permission conveyed through Copyright Clearance Center, Inc.

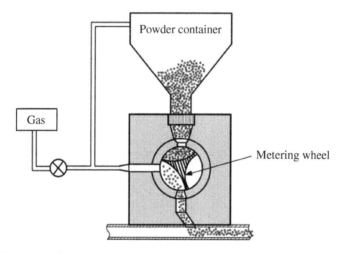

Figure 3.28 Schematic of gravity-based powder feeders with a lobe gear. Source: Republished with permission of Taylor & Francis Group, from [7]; permission conveyed through Copyright Clearance Center, Inc.

uniform and continuous, assuring a very high coating quality. The operation principle consists of conveying a specific amount of gas to a closed hopper, where the powder is contained. At the bottom of the closed hopper, a filter diffuses the gas through the powder, becoming fluidized. On top of the unit, a pickup tube allows delivering the fluidized powder under a shed on the tube and a carrier gas will propel the powder to the feed hose. Figure 3.29 shows the scheme of a generic fluidized bed powder feeder.

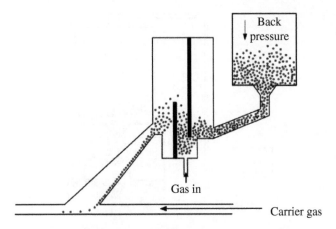

Figure 3.29 Schematic of a fluidized bed powder feeder. Source: Republished with permission of Taylor & Francis Group, from [7]; permission conveyed through Copyright Clearance Center, Inc.

3.5.1.4 Vibratory-Based Powder Feeder

A vibratory bed powder feeder, shown in Figure 3.30, consists of an oscillating tray at 50 or 60 Hz. The powder flows from the hopper to the vibrating tray, and volume flow can be infinitely adjusted to assure a steady-state powder flow. Some vibratory bed powder feeders have an oscillating tray with different plates arranged at different angles for better control of powder flow. The vibratory bed powder feeders can feed a wide range of powders from 8 to 2000 g/min with $a \pm 1\%$ precision.

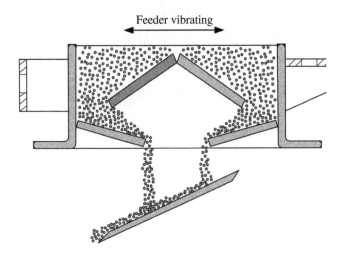

Figure 3.30 Schematic of a vibratory-based powder feeder. Source: Republished with permission of Taylor & Francis Group, from [7]; permission conveyed through Copyright Clearance Center, Inc.

3.5.2   Powder Delivery Nozzles for DED

During LDED (powder-fed AM), the interactions of powder particles with the laser beam, melt pool, and unmelted surfaces are crucial for the printed part's quality. These interactions are highly influenced by the type of nozzle, the angle between the powder delivery system and the deposition plate, and the powder stream's diameter. The nozzle choice is material-dependent and process-specific; therefore, the proper nozzle needs to be selected to achieve a good quality of the manufactured parts [23]. The delivery nozzles can have two different configurations:

1. Lateral
2. Coaxial

The coaxial nozzle's geometry configuration is based on a patent [24] that has been expired; however, over the last 20 years, different forms of nozzles have been invented based on the lateral and coaxial designs [7].

3.5.2.1   Lateral Nozzle

In the AM process equipped with a later nozzle, the powder is delivered from the side and an inert gas passing through the nozzle helps in the powder delivery stream while preventing the oxidation of the deposit. A schematic of the lateral nozzle designs is given in Figure 3.31 [7]. One issue of the lateral nozzle is its dependencies on the trajectory orientation. If the lateral nozzle is fixed to the laser beam's axis, and if the trajectory is not linear, the interaction of powder stream with the melt pool will be changed. Because of this issue, some companies like Optomec have developed a nozzle that includes four of lateral nozzles.

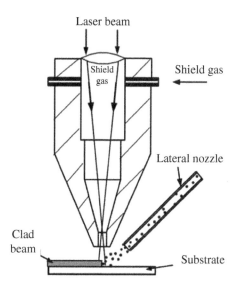

Figure 3.31   Schematic of a typical lateral nozzle. Source: Republished with permission of Taylor & Francis Group, from [7]; permission conveyed through Copyright Clearance Center, Inc.

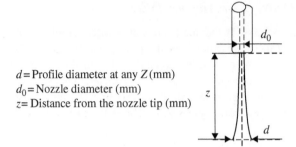

d = Profile diameter at any Z (mm)
d_0 = Nozzle diameter (mm)
z = Distance from the nozzle tip (mm)

Figure 3.32 Powder feed profile characteristics. Source: Republished with permission of Taylor & Francis Group, from [7]; permission conveyed through Copyright Clearance Center, Inc.

As mentioned in the previous section, the quality of AM-made parts strongly depends on the powder delivery system in terms of the powder stream's diameter to be deposited, the stability of the powder feed rate, speed of the particles powder, and uniform powder's profile shape. A study conducted using a different nozzle with different diameters revealed that the powder stream profile could be described by a parabolic equation (see Figure 3.32),

$$d = \lambda z^2 + d_0 \tag{3.30}$$

where z is the distance from the tip of the nozzle to the desired point (mm), d is the profile diameter at any z (mm), λ is the powder profile's quality coefficient (1/mm), and d_0 is the nozzle diameter (mm). From the study, it was able to observe that λ is valid for a specific range where the powder stream is stable [7].

3.5.2.2 Coaxial Nozzle

In this type of nozzle, the powder flow, the laser beam, and the shied gas are delivered from the same nozzle, as shown in Figure 3.33.

One of the advantages of the coaxial nozzle is the possibility to have the same deposition rates in any direction along with a better powder efficiency, which can be described as the ratio between the deposited powder on the work-piece and the powder delivered by the powder feeder in a certain time. Good quality of the manufactured parts can be achieved with a uniform powder stream at the nozzle outlet, which is obtained by maintaining a laminar flow of the powder stream parallel to the laser beam. Moreover, having the focus point at the same level of the melt pool assures both the good quality of the parts and good powder efficiency. Laminar flow, various powder stream profiles, and the powder focus point can be obtained by selecting the proper nozzle tip [7]. Figure 3.34 shows a typical powder stream out of a co-axial nozzle. The profile of powder stream can be molded by a parabolic equation like (3.23); however, the powder stream is first converged and then diverged, unlike a lateral nozzle.

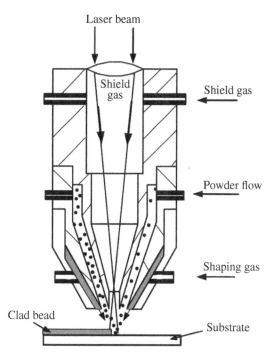

Figure 3.33 Schematic of a typical coaxial nozzle. Source: Republished with permission of Taylor & Francis Group, from [7]; permission conveyed through Copyright Clearance Center, Inc.

Figure 3.34 Powder stream at the nozzle exit to a co-axial nozzle. Source: Image available on the internet.

3.5.3   Powder Bed Delivery and Spreading Mechanisms

In powder bed processes, a layer of metal powder generally 20–100 μm thick is spread and compacted over the build platform by a counter-rotating roller, wiper/Recoater, or doctor blade. The first layer of the powder is then fully melted by a laser beam, which scans the platform according to the computer-generated path. When the first laser scan is over, the build platform descends one layer thickness and a new layer of powder is spread across the platform using the roller and then selectively melted. This process is repeated layer by layer until the desired part is manufactured. The powder delivery system needs to maximize the flowability of the powder, decrease particle clustering, and minimize the shear forces between each powder layer. The PBF is a very complex process, and to ensure good quality of the manufactured parts, it is important to have a homogenous powder thickness in each layer. A schematic of the process is shown in Figure 3.35 [23].

The built platform is contained in a metal chamber, where inert gases (argon and nitrogen) are introduced to prevent the oxidation of the melt pool and to remove the condensate produced during the process. The density of the parts manufactured by LPBF is crucial, and it strongly depends on several parameters associated with the spreading mechanism: speed of the counter-rotating roller, properties of the roller, the layer thickness of the powder, and powder particle size [25]. In order to find the optimum process parameters, trail-and-error techniques, mathematical modeling, and simulation are employed, which provides information on the compaction phenomenon [25].

In a general fashion, there are two types of recoaters: Soft recoaters and hard recoaters. Although both would spread the powder particles homogeneously, their interaction with powders is different to some extent. Such difference has a major effect on the "compaction density" of the build chamber powders that in turn affect the quality of final AM-made parts.

Soft recoaters normally composed of silicon, rubber, or soft carbon fiber (e.g. brush recoater). Due to its stretchable nature, the soft recoater behaves nicely in case of a collision with any solidified metal parts that are being built in the AM system. However, this flexibility comes

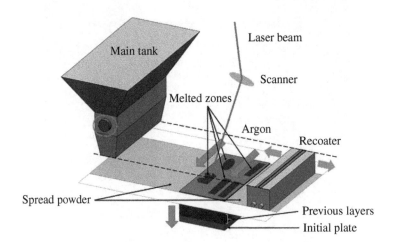

Figure 3.35   Illustration of a LPBF process system setup. Source: Reprinted from [23] with permission from Elsevier.

at some costs. They will be normally damaged very easily. However, it may imply a lower compaction density in the build chamber. This type of recoater may be ideal for concurrent additive manufacturing of different parts in the same build chamber when one part deforms differently compared to other parts.

Hard recoaters normally composed of HS steel or ceramic exerting pressure on the powder bed. This type of recoater, in comparison with the soft recoaters, prevents the part deformation to some extent. However, when the recoater collides with a substantially wrapped part, the recoater jams or some parts are separated from the support structure and dragged across the powder bed. The latter will typically damage the entire build. These types of recoaters might be useful when identical parts are printed on the same build platform. In addition, it may expose a higher compaction density to the build chamber resulting in better quality.

To avoid potential jamming, it is recommended not to put parts parallel to the recoater. They should be titled around the Z-axis for a better distribution of the recoater force.

3.5.4 Wire Feed System

This type of system is used in DED AM processes, where a solid wire feedstock can be used other than the metal powders. The feedstock capture efficiency is normally 100% [17], and the volume of the deposit is the same as the volume of the wire used as feedstock. The following Table 3.2 compares the basic features of the powder-feed/bed and wire-feed processes.

It is possible to notice that the deposition rate of the wire-feed EBF^3 process is much higher than the deposition rate of the powder-fed/bed techniques. However, the fabricated parts will have a lower geometrical accuracy [26]. In AM machines with wire feeder, both simple and complex geometries can be achieved; however, the latter requires careful control of the process parameters, and it is not possible to obtain simultaneously high geometry accuracy and low porosity of the parts. Wire-based DED machines are employed when a pore-free part is needed and when dimensional accuracy is not critical [17]. Depending on the energy source used for metal deposition, wire-feed AM can be classified into three groups: laser-based, arc welding-based, and electron beam-based, depending on the energy source employed. The electron beam-based metal deposition process (EBF^3) was developed in 1999 by Lockheed Martin

Table 3.2 Comparisons of some representative AM processes.

Type of material	Process	Layer thickness (µm)	Deposition rate (g/min)	Dimensional accuracy (mm)	Surface roughness (µm)
Powder	Laser cladding	100 to 1000	1 to 30	±50 to ±500	5 to 50
	LPBF	20 to 100	n/a	±25	9 to 10
	Selective laser sintering	50 to 75	n/a	±10	14 to 16
	DED	100 to 1000	12	±25 to ±500	~20
Wire	Arc welding	~1500	12	±50 to ±500	200
	Electron beam (EBF^3)	n/a	Up to 330	Low	High

Source: Redrawn and adapted from [26] by permission from Springer Nature.

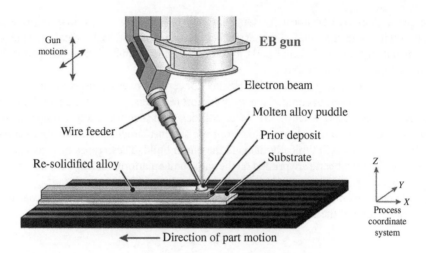

Gun motions

EB gun

Electron beam

Molten alloy puddle

Wire feeder

Prior deposit

Re-solidified alloy

Substrate

Z

Y

X

Process coordinate system

Direction of part motion

Figure 3.36 Schematic of a lateral wire-feed system equipped with EBM. Source: Redrawn and adapted from [28].

(Bethesda, MD) and then improved by the NASA Langley Research Center for the fabrication and/or repair of aerospace structures [17, 27]. The use of EBF^3 compared to the other laser-based AM techniques is very promising in aerospace applications because it can be performed in a vacuum. It is equipped with a wire feeder avoiding the use of powders difficult to be contained safely in low-gravity settings [17].

The wire feed system in this type of AM machine can be lateral (as shown in Figure 3.36), where the wire feed angle can affect the quality of the built part or coaxial (as shown in Figure 3.37). The latter setup provides an omnidirectional building performance. To achieve this, the laser beam is divided by splitting optics into three separate beams, which are then focused on a circular focal point. The optical elements' precise location allows the wire feed to be in the center of the laser beam. This configuration allows the wire to be fed into the center of the laser-generated melt pool resulting in a directional independency of the AM process [30]. Contact force is extremely important to maintain fixed; otherwise, the quality of AM-made parts will be undermined. This is why wire-fed AM may not be a good choice for cases in which a rough surface is being generated. Complicated parts may not be fabricated with this technology. Having a fix contact force and allowing the feeding mechanism to maneuver around the process zone might be challenging for making complicated shapes.

3.5.5 Positioning Devices and Scanners in Laser-Based AM

AM machines are employed for the fabrication of different parts to be used in different applications. The currently available machines can print small-scale parts since the fabrication of larger parts is hindered by the size of the machine and positioning device elements. CNC and robotic arms are used to position the material delivery systems or heat sources, and due to the workspace, there are some constraints that limit the size and shape of objects to be printed. Besides

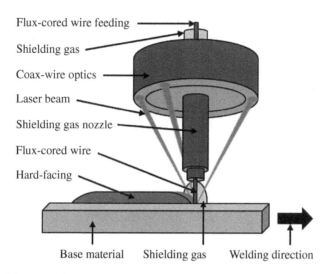

Flux-cored wire feeding

Shielding gas

Coax-wire optics

Laser beam

Shielding gas nozzle

Flux-cored wire

Hard-facing

Base material   Shielding gas   Welding direction

Figure 3.37   Schematic of a coaxial wire-feed system. Source: Redrawn and adapted from [29].

workspace, the positioning device should provide the appropriate speed and accelerations for the building of the parts. Depending on the AM machine or/and the process, the positioning devices involved are connected to a controller that determines their movements, trajectories, speed, and accelerations.

In laser powder bed fusion (LPBF) processes, the laser scanning along x and y directions is achieved by galvanometer scanners as shown in Figure 3.38. A galvo system consists of a galvo motor, a pair of mirrors controlled by the galvo motors, and the driver board "servo" that controls the system [32, 33]. Galvo motors are DC motors, which have multiple coils in a magnetic field. When the galvo motor senses an electric current, flowing through a coil, it experiences a torque proportional to the current, which is translated to a deflection of the laser beam by the galvo mirrors. The closed-loop servo control is obtained by a precision position detector mounted to the shaft, which accurately measure the mirrors' angular position. Closed-loop galvo is faster and more accurate than the open-loop galvo, which are cheaper and less accurate [32, 33].

3.5.6   Print-Head in Binder Jetting

In BJ AM processes, the parts are printed by delivering drops of binder agent onto a substrate (powder bed). Most of the common inkjet print heads can be categorized as piezo or thermal, based on the ink delivery method, as shown in Figure 3.39. In piezoelectric inkjet print heads, based on the acoustic wave theory, the piezoelectric material is the key active component. The ink head contains several walls made of the active component and a vibration plate, which form the ink channels. At the end of the ink channels, the ink nozzles are built. When electricity is applied to the piezoelectric material (channel's walls), it deforms the vibrating plate, which flexes and bends. Consequently, acoustic pressure waves are formed, which force the ink to

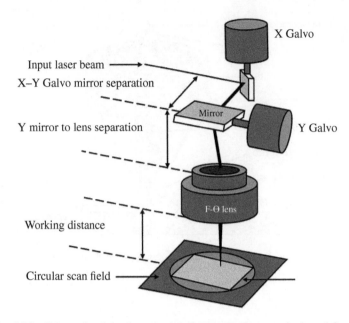

Figure 3.38   Schematic of a galvo scanner. Source: Redrawn and adapted from [31].

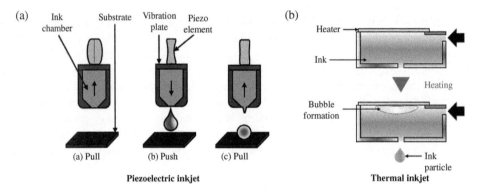

Figure 3.39   Schematics of (a) piezo and (b) thermal inkjet print heads. Source: Redrawn and adapted from [34].

be ejected from the nozzles [35]. In thermal inkjet print heads, water-based inks are used and heat is used to deliver the drops onto the substrate instead of applying an electric current. Resistors placed behind the nozzles vaporized the inks creating an air bubble of ink vapor, which quickly expand, causing a large pressure increase leading to the ink ejection through the nozzles [36]. The inkjet print heads (piezo and thermal) are moved back and forth by a stepper motor following the instructions given by a computer equipped with circuit board for image capturing and processing [37–39].

3.6   CAD File Formats

In AM processes, the part to be manufacted is designed by a computer-aided design (CAD) tool which contains all the information necessary to print the desired part. Indeed, the first important steps in all AM processes are the CAD's conceptualization and model that is converted into a file with a format readable by the AM machines. This file is then sent to the machine, and after the machine setup, the part will be printed. The CAD file can be created by an experienced designer via user interface, by 3D scanning of an existing part, or a combination of both. The CAD models used in AM are based on slicing technology, where the CAD model is made of many layers stacked together to form the desired part. The 3D objects can be designed on a computer using different CAD model formats, SAT (or ACIS), DXF, STP, and STL which are the most common ones. Among these formats, the most widely used in AM processes is STL, which stands for "stereolithography". This type of file format specifies either ASCII or binary representations describing 2D and 3D objects' surface geometry with no information on the color or the texture of the object being modeled. The STL file is composed of triangular facets, which are identified by a unit normal and by three vertices (corners) of the triangular facets, and specified using a three-dimensional Cartesian coordinate system. Information on the color of the part modeled can be included through two non-standard variations on the binary STL file using specific software packages. An image of a typical STL file is shown in Figure 3.40, and it should be noted that as the geometry changes the density of triangle facets would vary.

Some of the limitations of STL file are the impossibility to represent colors, texture, materials, and other properties of the printed object. This problem was overcome in 2009, when a new standard for additive manufacturing started to be developed by a subcommittee of the ASTM Committee F42 on AM Technologies. This standard is known as Additive Manufacturing File

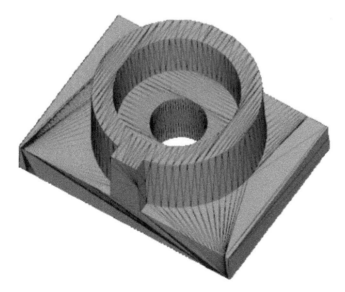

Figure 3.40   Typical STL file. Source: Republished with permission of Taylor & Francis Group, from [7]; permission conveyed through Copyright Clearance Center, Inc.

Format (AMF) (ISO/ASTM 52915 : 2016) [40], and it is an XML-based format. Any CAD/ CAM software to describe the geometry, composition, color, materials, and lattices of any object to be fabricated with AM technologies can use it.

3.6.1   CAD/CAM Software

Currently, there is a plethora of software solutions available that are used in AM machines. In particular, the software can be categorized into different categories and sub-categories based on their purpose. Different software is used for the 3D design of the parts that need to be printed, for scanning, improving the performance of the part and for process simulation studies. Examples of software that are being employed are Powermill, NX Cam, and Camufacturing/Mastercam. The PowerMiLL CAM solution provides strategies for the programming of tool paths for 3- and 5-axis CNC machines, and it is used to produce smooth surface finishes, and it provides options to control the simulating of toolpaths [41]. The software owned by Siemens, the NX Cam is an advanced high-end CAD/CAM/CAE solution employed for design, engineering analysis, and manufacturing [42]. Many software packages are also available for running robotic systems. PoweMill has recently introduced a new package called "PowerMill Additive." It is a plugin, enabling to create tool paths for additive or hybrid manufacturing that is driven by DED. As it is read from their website: "*It utilizes PowerMill's existing multi-axis CNC and robot capabilities to create multi-axis additive toolpaths with detailed process parameter control along every point of the toolpath. These process parameters enable you to finely control the operations of deposition at a toolpath point level*". Regardless of the available packages, there are still major shortcomings in the software hierarchy developed for AM. Lack of intelligence and knowledge-based process recipes is one of the major shortcomings.

3.7   Summary

In this chapter, the basics of laser have been discussed, along with the types of lasers most deployed in AM processes. The physics behind the lasing process has been reviewed. Particular attention was given to the description of three-, and four-level laser systems, which is crucial for the occurrence of population inversion, thus lasing. Information on the setup and equipment of the system required for LPB, LPF, and BJ was discussed. The types of powder feeders, wire feeders, delivery nozzles, and print heads technology used in the different AM machines were described. It has been highlighted that some AM machines, instead of a laser deploy an electron beam gun, and the process is known as electron beam melting (EBM). Working principle of an electron beam and the differences between the EBM and the other AM processes have been highlighted. The type of CAD file and software used in AM machines was presented as well.

References

1. L. E. Criales, Y. M. Arısoy, B. Lane, S. Moylan, A. Donmez, and T. Özel, "Laser powder bed fusion of nickel alloy 625: experimental investigations of effects of process parameters on melt pool size and shape with spatter analysis," *Int. J. Mach. Tools Manuf.*, vol. 121, no. March, pp. 22–36, 2017.
2. Additivley, "Binder Jet." [Online]. Available: https://www.additively.com/en/learn-about/binder-jetting.

3. J. D. Simon and D. A. McQuarrie, Physical chemistry: A molecular approach. Sausalito, CA: University Science Books, 1997.
4. K. Thyagarajan and A. Ghatak, Lasers fundamentals and applications, vol. 53, no. 9, Springer, New York, London, 650pp., 2010. DOI: https://doi.org/10.1007/978-1-4419-6442-7.
5. A. Javan, W. R. Bennett, and D. R. Herriott, "Population inversion and continuous optical maser oscillation in a gas discharge containing a he-ne mixture," *Phys. Rev. Lett.*, vol. 6, no. 3, pp. 106–110, 1961.
6. K. Thyagarajan and A. Ghatak, Lasers, Springer US, Boston, MA, 2011.
7. E. Toyserkani, A. Khajepour, and S. Corbin, Laser cladding, 1st edition, CRC Press, 2004. https://doi.org/10.1201/9781420039177.
8. W. T. Silfvast, Laser fundamentals, 2nd edition, Cambridge University Press, Cambridge, MA, 2004.
9. W. Demtröder, Laser spectroscopy 1: Basic principles, 1st edition, Springer-Verlag Berlin Heidelberg, 496pp., 2014. doi: 10.1007/978-3-642-53859-9.
10. M. Eichhorn, Laser Physics: From Principles to Practical Work in the Lab, Springer/Sci-Tech/Trade, 2014th edition (March 31, 2014), 171pp., doi: 10.1007/978-3-319-05128-4.
11. D. Gu, Laser additive manufacturing of high-performance materials, no. Lm, Springer Berlin Heidelberg, Berlin, Heidelberg, 2015.
12. S. C. Singh, H. Zeng, C. Guo, and W. Cai, "Lasers: fundamentals, types and operations." in: Singh, S.C., Zeng, H., Guo, C., Cai, W. (eds.) Nanomaterials: processing and characterization with lasers, Wiley-VCH Verlag GmbH & Co. KGaA, Weinheim, pp. 1–34, 2012.
13. M. Baumers, P. Dickens, C. Tuck, and R. Hague, "The cost of additive manufacturing: Machine productivity, economies of scale and technology-push," *Technol. Forecast. Soc. Change*, vol. 102, pp. 193–201, 2016.
14. Y. Tian, C. Wang, D. Zhu, and Y. Zhou, "Finite element modeling of electron beam welding of a large complex Al alloy structure by parallel computations," *J. Mater. Process. Technol.*, vol. 199, no. 1, pp. 41–48, 2008.
15. "Arcam." [Online]. Available: http://www.arcam.com/technology/electron-beam-melting/hardware.
16. "Sirris." [Online]. Available: www.sirris.be.
17. I. Gibson, D. W. Rosen, and B. Stucker, Additive manufacturing technologies, Springer, 2010. doi: https://doi.org/10.1007/978-1-4419-1120-9
18. "World Scientific." [Online]. Available: http://www.worldscientific.com/worldscibooks/10.1142/7745.
19. G. R. Brewer, "Electron-beam technology in microelectronic fabrication," 1st Edition, Academic Press, 376pp., 1980. ISBN: 9780121335502.
20. S.I. Molokovsky and A.D. Sushkov, "Introduction to particle-beam formation." In: Intense Electron and Ion Beams, Springer, Berlin, New York, pp. 2–17, 2005. https://doi.org/10.1007/3-540-28812-0_1.
21. V. Adam, U. Clauß, D. Dobeneck, T. Krüssel, and T. Löwer, Electron beam welding – The fundamentals of a fascinating technology. pro-beam AG & Co. KGaA, 2011.
22. R. A. Dugdale, "Soft vacuum processing of materials with electron beams," *J. Mater. Sci.*, vol. 10, no. 5, pp. 896–904, 1975.
23. M. Brandt, Laser additive manufacturing: materials, design, technologies, and applications, Woodhead Publishing, Duxford, UK, 2016.
24. I. G. E. Gore, N. Branch, T. Roman, E. Orange, K. Wayne, and R. A. Gore, "United States Patent (19)," no. 19, 1998.
25. Y. Shanjani and E. Toyserkani, "Material spreading and compaction in powder-based solid freeform fabrication methods: Mathematical modeling," In 19th Annual International Solid Freeform Fabrication Symposium, SFF 2008, 2008.
26. D. Ding, Z. Pan, D. Cuiuri, and H. Li, "Wire-feed additive manufacturing of metal components: technologies, developments and future interests," *Int. J. Adv. Manuf. Technol.*, vol. 81, no. 1–4, pp. 465–481, 2015.

27. T. S. Srivatsan and T. S. Sudarshan, Additive manufacturing innovations, advances, and applications, 1st edition, CRC Press, 2015. https://doi.org/10.1201/b19360

28. Sciaky, "Sciaky's Electron Beam Additive Manufacturing (EBAM®) Process." [Online]. Available: https://www.sciaky.com/additive-manufacturing/wire-vs-powder.

29. "Durmat." [Online]. Available: https://durmat.com/en/service/hard-facing / https://durmat.com/en/service/hard-facing.

30. S. Kaierle, A. Barroi, C. Noelke, J. Hermsdorf, L. Overmeyer, and H. Haferkamp, "Review on laser deposition welding: from micro to macro," *Phys. Procedia*, vol. 39, pp. 336–345, 2012.

31. "ii-vi." [Online]. Available: http://www.ii-vi.de/french/CO2-Laser-Optics/scanning-laser-system-optics.html.

32. H. Yeung, J. Neira, B. Lane, J. Fox, and F. Lopez, "Laser Path Planning and Power Control Strategies for Powder Bed Fusion Systems, Proceedings of the Solid Freeform Fabrication Symposium, Austin, TX, 2016. [Online]. Available: https://tsapps.nist.gov/publication/get_pdf.cfm?pub_id=921536.

33. V. C. Coffey, "What you need to know to buy a galvo-positioner," Laser Focus World, 2010. Available: https://www.laserfocusworld.com/optics/article/16567973/product-focus-galvanometer-scanners-what-you-need-to-know-to-buy-a-galvopositioner

34. "SDGMAG." [Online]. Available: https://sdgmag.com/features/digital-eye-printer-interface.

35. C. De Maria et al., "Design and validation of an open-hardware print-head for bioprinting application," *Procedia Eng.*, vol. 110, pp. 98–105, 2015.

36. N. J. Nielsen, "History of ThinkJet printhead development," *Hewlett Packard J.*, vol. 36, pp. 21–27, 1985.

37. H. Wijshoff, "The dynamics of the piezo inkjet printhead operation," *Phys. Rep.*, vol. 491, no. 4–5, pp. 77–177, 2010.

38. F. J. Primus, M. D. Goldenberg, and S. Hills, "United States Patent (19)," no. 19, 1991.

39. P. Enzymes, "United States Patent (19) 54," vol. 96, no. 19, pp. 62–66, 1973.

40. ISO/ASTM 52915:2020, "Specification for additive manufacturing file format (AMF) Version 1.2", 27pp., 2020. Available: https://www.iso.org/standard/74640.html.

41. Autodesk, PowerMiLL CAM solution that provides strategies for the programming of tool paths for 3- and 5-axis CNC machines. Available: https://www.autodesk.ca/en/products/powermill/overview.

42. Siemens PLM Software, "CAM software used for design, engineering analysis, and manufacturing. Available: https://www.plm.automation.siemens.com/global/en/products/manufacturing-planning/cam-software.html.

4

Directed Energy Deposition (DED)

Physics and Modeling of Laser/Electron Beam Material Processing and DED

Learning Objectives

At the end of this Chapter, you will be able to:

- Understand the basic of laser- and electron-beam material interaction physics;
- Understand the basic governing physics of directed energy deposition (DED);
- Gain insight into analytical and numerical modeling of DED;
- Learn about multiple case studies on modeling and analysis of DED based on lumped, comprehensive analytical, numerical, and experimental-based stochastic approaches.

4.1 Introduction

This chapter covers a brief description of the fundamental principles governing laser/electron beam material interaction and absorption. The chapter will then focus on physics, and governing equations of laser directed energy deposition (LDED) and electron beam directed energy deposition (EDED).

Lasers and electron beams provide many advantages when it comes to the thermal material processing. They effectively deliver controllable amounts of energy into selected regions of a material system to arrive at a desired state of the material: heating, melting, and evaporation (plasma).

Metal Additive Manufacturing, First Edition. Ehsan Toyserkani, Dyuti Sarker, Osezua Obehi Ibhadode,
Farzad Liravi, Paola Russo, and Katayoon Taherkhani.
© 2022 John Wiley & Sons Ltd. Published 2022 by John Wiley & Sons Ltd.

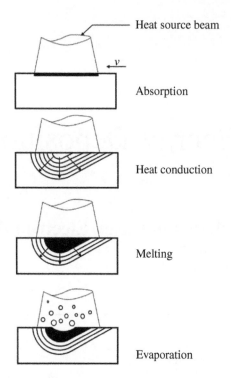

Heat source beam

Absorption

Heat conduction

Melting

Evaporation

Figure 4.1 Interaction of a moving heat source and a substrate and the associated stages.

The interaction time between the heating source and material leads to different stages, as shown in Figure 4.1. The relative velocity between the heat source and the substrate will form the shape of the melt pool. When a moving heat source passes over elastically bound charged particles that form a solid object, the particles will be set in the melting phase by absorbing energy. This energy absorption causes the particles to resonate, which eventually causes melting and evaporation as particles cannot stand together at the elevated heat flux. More energy and molecular resonance cause "evaporation" or "boiling," where the vapor can also absorb energy. Excessive energy can turn the electrons to shake freely, thus forming "Plasma."

The flow of heat is explained by Fourier's law, that is,

$$\frac{q}{A} = -K\frac{\partial T}{\partial x} \tag{4.1}$$

where q is the heat flux (W), A is the interaction area (m^2), T is the temperature (K), K is the thermal conductivity (W/m · K), and x is the longitudinal dimension (m).

Melting of material by any moving heat source of energy, including lasers and electron beams, depends on heat flow in the material. As Eq. (4.1) shows, the heat flow is a linear function of the thermal conductivity. However, in addition to the thermal conductivity, there are other factors that influence the heat flow. The rate of temperature change also depends on the specific heat of materials, c_p (J/kg · K):

$$heat\ rate \approx \frac{1}{\rho c_p} \tag{4.2}$$

where ρ is the density (kg/m^3). As a matter of fact, the most important factor for heat flow is thermal diffusivity (m^2/s), that is,

$$\alpha = \frac{K}{\rho c_p} \tag{4.3}$$

The depth of heat penetration in terms of (m), Z_h, for a moving heat source with an interaction time of Δt can approximately be calculated by:

$$Z_h = \sqrt{4\alpha\Delta t} = \sqrt{4\alpha\frac{D}{v}} \tag{4.4}$$

where v is the heat source speed (m/s) in terms of and D is the overlap area between the substrate and the heat source during the exposure time (m).

Figure 4.2 shows a schematic where a mild steel substrate is exposed to a stationary heat source for a short period of time (10^{-6} s). As the energy per area increases, different modes will occur, including absorption, melting, boiling, and ionization.

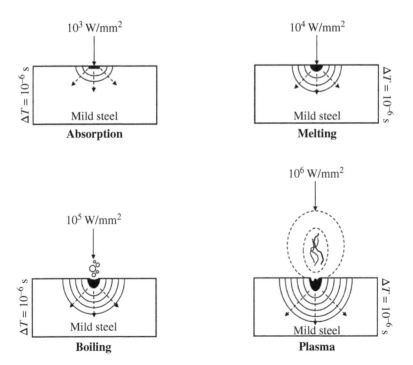

Figure 4.2 Schematic of phases formed in a mild steel substrate while being exposed to a stationary heat source for a short period of time (10^{-6} s).

4.2 Laser Material Interaction and the Associated Significant Parameters to Laser AM

A laser beam has many unique characteristics that make it a proper tool for material processing. The electromagnetic radiation of a laser beam is absorbed by the surface of opaque materials (e.g. metals). The laser beam with respect to the substrate causes the thermal cycle in the surface layer. The schematic representation of the physical phenomena, shown in Figure 4.1, is applicable to a wide range of laser material processing techniques. These processes are differentiated due to different combinations of absorption, heat conduction, melting, material addition, and rapid solidification. However, the common physical phenomenon of all laser material processing techniques is rapid solidification, which causes a superior and fine metallurgical structure [1].

Once the laser beam emits on the material, it gives electrons of fixed energy to material regardless of the laser beam energy E, given by:

$$E = hf - p \tag{4.5}$$

where h is the Planck's constant (i.e. 6.625×10^{-34}) (J.s), f is the frequency (i.e. $\frac{c}{\lambda}$, where c is the speed of light (i.e. 2.99×10^8 m/s) (1/s) and λ is the wavelength of the light (m)), and p is the constant characteristic of the material.

Figure 4.3 shows a laser beam in the form of an electromagnetic wave with a spinning speed of ω. E_i is the input laser energy, E_r is the reflected energy, and E_t is the transmitted energy decayed in the material due to heat transfer and the ejection of electrons. As shown in the figure,

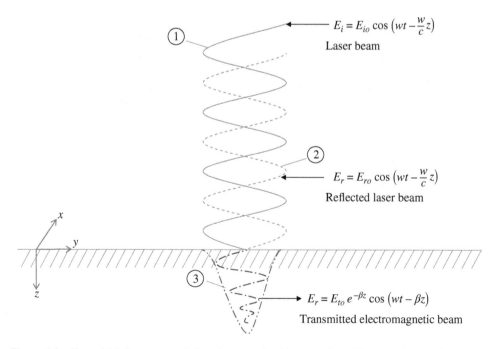

Figure 4.3 Sinusoidal electromagnetic laser beam: emitted beam, reflected beam, and transmitted beam.

as the beam travels inside the new medium, the energy will be absorbed and decayed. Beer–Lambert's law can express this:

$$I = I_0 e^{-\beta z} \tag{4.6}$$

where I is the energy/power per area or intensity (J/m^2), I_0 is the initial intensity, β is the absorption factor, and z is the vertical axis. It should be noted that β is a function of material, intensity, wavelength, and temperature.

When the electromagnetic field passes over small elastically bound charged particles, the particles will be set in the melted phase, as explained in Section 4.1. All phenomena explained in the mentioned section is valid for the case of a moving laser beam.

The laser beam's relative velocity with respect to the substrate causes the thermal cycle in the surface layer. However, this thermal cycle is dependent on too many parameters such as the laser beam properties and material diffusivity. For example, the material interaction with the beam should be approximately equal to the "thermal time constant" for metallic samples. Thermal time constant τ is defined based on the temperature–time plot represented in Figure 4.4. The thermal time constant can be found using the equation in the figure. Although T_1 and T_2 can be general, T_1 may be considered as ambient temperature and T_2 as melting temperature. This value can provide the process developer with the minimum interaction time (process speed) as well as the pulse duration if a modulated/pulsed laser is used.

In the following subsections, several key factors that the reader should pay attention to them are mentioned:

4.2.1 Continuous Versus Pulsed/Modulated Lasers

The laser beam can be in the form of a continuous-wave (CW), pulsed, or modulated wave. A pulsed laser beam refers to optical pulses rather than the CW [1]. There are many methods of creating laser pulsation; however, the outcome will be in the shape of pulses. The pulse duration can range from nanoseconds (10^{-9} s) to picoseconds (10^{-12} s) to femtosecond (10^{-15} s) or even attosecond (10^{-18} s). In the form of the pulse wave, several parameters associated with the shape of pulses are defined: laser pulse energy E, laser pulse width (laser pulse duration) W,

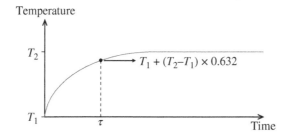

Figure 4.4 Graphical concept of the thermal time constant.

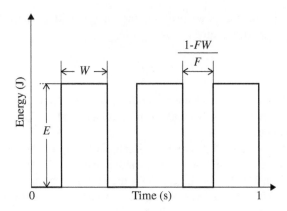

Figure 4.5 Laser pulse shaping, including pulse width W, pulse energy E, and pulse frequency F. Source: Republished with permission of Taylor & Francis Group, from [2]; permission conveyed through Copyright Clearance Center, Inc.

laser pulse frequency (laser pulse repetition rate) F, average power P_l, and duty cycle C. These parameters which are shown in Figure 4.5 can be expressed by:

$$C = FW \tag{4.7}$$

$$P_l = EF \tag{4.8}$$

Peak power per pulse P_{peak} can be derived by:

$$P_{peak} = \frac{E}{W} \tag{4.9}$$

At a given pulse energy, the shorter the pulse width, the larger the peak power. Also, If the amount of energy per pulse is very small (e.g. nano-joules to pico-joules), the peak power could be very large for ultra-short pulse lasers. For pulsed laser additive manufacturing processes, the peak power has a substantial role in the local heat-affected zone (HAZ) where the periodic fluctuation in microstructure and banding effect can be explained by the peak power frequency.

A modulated is basically a manipulated CW beam, where the output power is triggered selectively. Figure 4.6 shows a typical modulated laser beam. As seen, the modulated pulse is very similar to what is known as a pulsed laser. Thus a common misconception about the identical nature of modulated lasers and pulsed lasers exists. In the modulated laser, the beam simply turns on and off in a periodic fashion using mechanical or optical shatters, whereas the pulsed laser releases a burst of energy with the desired frequency. The modulated laser thus only turns on to set maximum output power, regardless of the frequency. This will offer a unique feature in which the modulated diode laser can be used for pulse shaping.

Although the period and pulse width for both pulsed and modulated lasers can be measured identically, the overshoot, ringing, rising time, and falling time for modulated lasers must be taken into account. The concept of all these factors can easily be understood from Figure 4.6. An instantaneous transition from ON mode to OFF mode is impossible; thus a short duration of

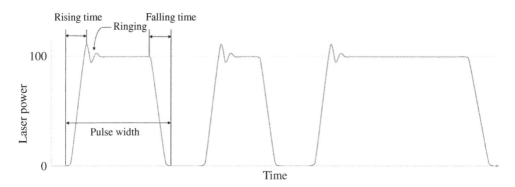

Figure 4.6 A typical modulated/pulsed laser beam with rising time, falling time, pulse width, and ringing.

time, known as rising and falling time, is required for the laser to turn ON and OFF, due to this technical limitation of electronic switching units.

Multiple parameters represent effective power given to the specific volume of the material in the process zone. The extinction ratio, known as the depth of modulation, is defined by:

$$Extinction\ Ratio = \frac{P_{on}}{P_{off}} \qquad (4.10)$$

P_{on} is the on state of the laser power, and P_{off} is the off state of the laser power. Another important parameter is the duty cycle. The same equation as Eq. (4.7) is used for the modulated laser. Average power can also be calculated using Eq. (4.8).

4.2.2 Absorption, Reflection, and Transmission Factors

In order to calculate the net power delivered to the process zone, the important parameters of laser beams, including reflectivity and polarization, must be understood. The values of absorptivity, reflectivity, and transmissivity are related by the following equation:

$$R = \begin{cases} 1-\beta\ for\ opaque\ materials \\ 1-\beta-T_R\ for\ transparent\ materials \end{cases} \qquad (4.11)$$

where R is the reflectivity, β is the absorptivity, and T_R is the transmissivity. The reflectivity R for normal angles of incidence from air to opaque materials with a perfectly flat and clean surface is derived by [3]:

$$R = \left[(1-n)^2 + k^2\right] / \left[(1+n)^2 + k^2\right] \qquad (4.12)$$

where n is the refraction coefficient and k is the extinction coefficient of the material. Table 4.1 lists the optical properties for several materials for a light wavelength of 1060 nm [2].

Table 4.1 Optical properties and Brewster angle of several materials for 1060-nm light wavelength in room temperature [2–4].

Materials	k	n	Brewster angle
Al	8.50	1.75	60.2
Cu	6.93	.015	—
Fe	4.44	3.81	75.2
Ni	5.26	2.62	~76
Pb	5.40	1.41	—
Ti	4.00	3.80	70.8
Zn	3.48	2.88	—
Glass	0.10	0.50	56.3

Photons with shorter wavelengths are easier to be absorbed by the materials than photons with longer wavelengths; thus, at shorter wavelengths, R reduces normally [3]. When the temperature rises, the photon population increases too. Therefore, the probability of interaction between the electrons and material increases, causing a decrease in the reflectivity and an increase in the absorptivity [3].

4.2.3   Dependencies of Absorption Factor to Wavelength and Temperature

The absorption factor is highly dependent on the laser wavelength. At shorter wavelengths, the reflectivity normally decreases, whereas the absorptivity of the surface increases. While this statement is mainly valid for the majority of metals, it is not the case for some metals and non-metal elements. For some elements such as copper, the reflectivity varies nonlinearly at some wavelengths as indicated in Figure 4.7. The reflectivity has to be validated for new alloys and elements that have not been used widely for laser material processing.

The reflectivity factor is a function of temperature as well as seen in Figure 4.8. As the temperature elevates, an increase in the phonon population occurs that, in turn, results in more photon–electron energy interactions. Thus, the likelihood that electrons interact with the part rather than oscillate is increased [5].

4.2.4   Angle of Incidence

The "Drude Reflectivity" theory indicates that the reflectivity changes with both the angle of incidence and the plane of polarization [6]. There are two terminologies: (i) "*p*" ray (parallel): it refers to the case when the plane of polarization is in the plane of incidence, (ii) "*s*" ray ("perpendicular" for which the term of "Senkrecht" is used): it refers to the case when the ray has its plane of polarization normal to the plane of incidence.

The reflectivity for these two rays can be obtained by:

$$R_p = \left[n - (1/\cos\emptyset) \right]^2 + k^2 \Big] / \Big[\left[n + (1/\cos\emptyset) \right]^2 + k^2 \right]$$ (4.13)

$$R_s = \left[n - \cos\emptyset \right]^2 + k^2 \Big] / \Big[\left[n + \cos\emptyset \right]^2 + k^2 \right]$$ (4.14)

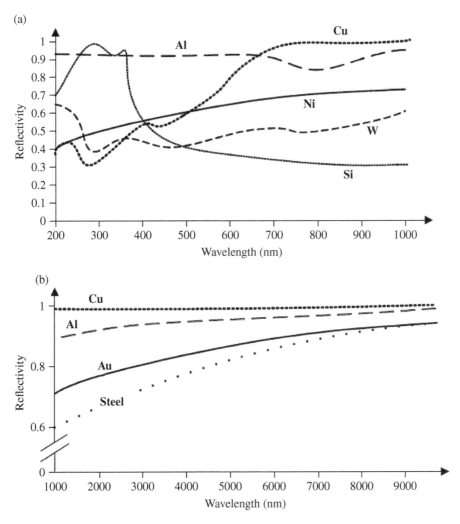

Figure 4.7 Dependencies of reflectivity to wavelengths, (a) from 200 to 1000 nm wavelength, (b) from 1 000 to 10 000 nm wavelength. Source: Republished with permission of Taylor & Francis Group, from [2]; permission conveyed through Copyright Clearance Center, Inc.

where n is the refraction coefficient, k is the extinction coefficient of material, and $\emptyset$ is the angle of incidence. The reflectivity changes with the angle of incidence at 1060 nm laser beam are shown in Figure 4.9. As seen in this figure, there is an angle for the p-ray, known as the "Brewster angle," where the angle of reflection and the refraction angle are normal to each other. At this angle, the electric vector in the plane of incidence cannot be reflected due to the absence of components at a normal degree to itself. Any beam with only one plane for the electric vector is called a "polarized" beam. The refractive index and the Brewster angles for several materials are listed in Table 4.1.

Most laser beams are polarized due to the amplification within the cavity compatible with the one plane concept (see Chapter 3).

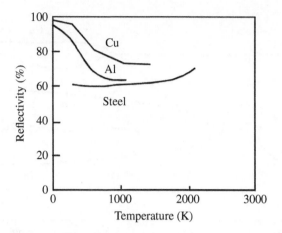

Figure 4.8 Temperature dependencies of reflectivity for Al, Cu, and steel at 1060-nm laser beam [4]. Source: Reproduced by Permission from the publisher.

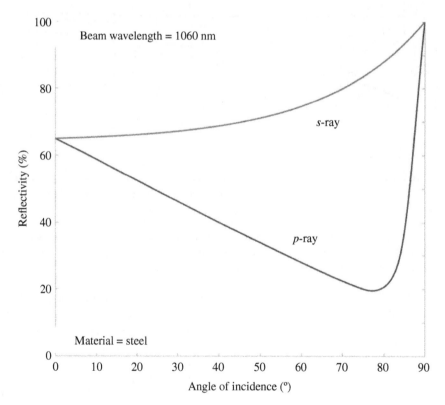

Figure 4.9 Dependencies of reflectivity to the angle of incidence for *s*-ray and *p*-ray at 1060 nm for steel. Source: adopted from [4].

Table 4.2 Reflectivity of different surface roughness for a 10.6-μm wavelength at normal angles of incidence.

Surface roughness	Reflectivity (%)-direct	Reflectivity (%)-diffuse	Total (%)
Sandpaper-roughened (1 μm)	90.0	2.7	92.7
Sandblasted (19 μm)	17.3	14.5	31.8
Sandblasted (50 μm)	1.8	20	21.8
Oxidized	1.4	9.1	10.5
Graphite	19.1	3.6	22.7

Source: Adopted from [4].

4.2.5 Surface Roughness Effects

Surface roughness has a major effect on the absorption factor because the surface waviness and ripples cause the beam to reflect multiple times. Table 4.2 lists the effect of surface roughness on the reflectivity factor. It is also hypothesized that there may exist a "stimulated absorption" due to beam interference with sideways-reflected beams [4]. It should be noted that if the surface roughness is less than the beam wavelength, the radiation, the surface roughness does not create any variation in the reflectivity, and the surface can be perceived as perfectly flat. It should be noted that "the reflected phase front from a rough surface, formed from the Huygens wavelets, will no longer be the same as the incident beam and will spread in all directions as a diffused reflection [4]."

4.2.6 Scattering Effects

The presence of a large number of inhomogeneities in the media (e.g. powder particles in laser powder bed fusion (LPBF)) will cause the scattering phenomenon. This means that the beam starts to travel in non-straight lines. The level of the scattering depends on many factors, including particle size, shape, and distribution. In general, three types of scattering exist:

1. **Rayleigh scattering**: It refers to the light elastic scattering by particles that are much smaller than the light wavelength. As a rule of thumb, if the light frequencies are significantly below the resonance frequency of the scattering particle, the scattering value is proportional to $\frac{1}{\lambda^4}$. Rayleigh scattering is mainly caused by the particles' electric polarizability. The oscillating electric field of light affect the particle charges that, in turn, oscillate the particle at the same frequency, making the particle a small radiating dipole where its radiation is seen as scattered light [7]. It should be noted that the individual atoms/molecules may be considered as particles, particularly when light travels through transparent solids, liquids, and gases.

   In laser-based AM, there are new possibilities to use the Rayleigh scattering concept to develop noncontact measurements of the laser beam within a fraction of a second and without compromising beam quality.
2. **Mie Scattering**: Mie scattering usually occurs in the media, where there may be many spherical particles with diameters approximately equal to the beam wavelength. This scattering is very relevant to laser AM, especially LPBF or at the hot zone of LDED. If the

keyhole phenomenon occurs in these processes, there will potentially be some aerosol due to boiled or ablated particles in the keyhole that will cause beam scattering, affecting the focus and processing conditions. It is shown that for particles with a radius of r, and for the condition when $2\pi r/\lambda \gg 1$, there is a strong tendency for forward scattering which is a form of beam refocusing [8].

3. **Bulk Scattering:** This type of scattering occurs when particles are greater than the beam wavelength. In this case, the scattered intensity is nearly independent of the wavelength. This type of radiation should be presented in LDED with powder injection.

4.3   E-beam Material Interaction

Several aspects of electron beam interactions with materials are the same as laser material interaction; however, there are a few phenomena that need further understanding. E-beam entails accelerated electrons that may be scattered through the part without any interaction. The scattering could be elastic or inelastic scattered [9], resulting in various signals used for imaging, quantitative/semi-quantitative analysis, and X-ray generated from the target material. These signals can be categorized as [9]:

1. **Signals for imaging:** Tagged and used in secondary electrons (SEs), backscattered electrons (BSEs), cathodoluminescence (CL), auger electrons, and characteristic X-rays.
2. **Signals for quantitative and semi-quantitative analyses of materials**: Tagged and used in characteristic X-rays, where Bremsstrahlung radiation as a continuous spectrum of X-rays, energy creates a background where characteristic X-ray is considered.
3. **Signals from X-rays generated**: Tagged and used in X-ray diffraction (XRD) and X-ray fluorescence (XRF).

Figure 4.10 shows the developed electron interaction volume when E-beam irradiates on a sample. The purpose of the book is not to explain each of these phenomena, so the reader is encouraged to study many books and papers available for this topic.

In terms of material interaction, the concept is very the same as laser interaction; however, the E-beam energy (E_b) entered the material can be expressed as a two-dimensional Gaussian as follows:

$$E_b(x,y) = \gamma \frac{V_a I_b c^2 \Delta t}{2\pi\sigma^2} e^{\left(-\frac{1}{2\sigma^2} < x - x_b, y - y_b > \right)} \tag{4.15}$$

where (x, y) is the center position in one individual cell under the beam (m), (x_b, y_b) is the beam center position (m), σ is the standard deviation associated with the Gaussian energy distribution, V_a is the acceleration voltage (V), and I_b is the beam current (A). Like a laser beam, electrons may be vanished because of reflection at the surface, so γ represents the net electron beam energy percentage that penetrates into the material.

One difference that should be considered in the E-beam modeling is initiated from the absorption length. The absorption length of E-beam is much higher than the thermal length causing the E-beam to penetrate through the material almost instantaneously [10]. Thus, the energy source should be considered as a volumetric flux like the heat source model used for LPBF (see Chapter 5).

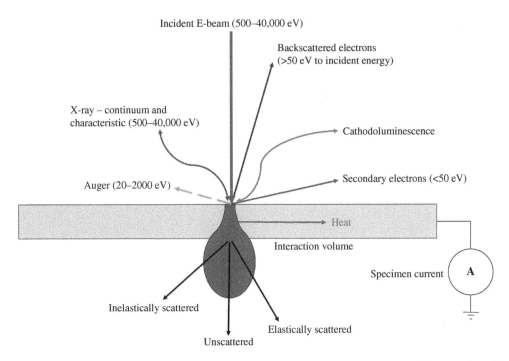

Figure 4.10 E-beam interaction with a substrate and the associated signals generated. Source: Adopted from [9].

Basically, two different absorption approaches have to be considered: (i) an exponential absorption and (ii) a constant absorption. For example, in [10, 11], an accelerated voltage of 60 kV for the first approach and a constant absorption type for a 120-kV E-beam have been considered. Figure 4.11 shows the absorption coefficient for these two scenarios at accelerated and fixed voltages at two different cell sizes underneath the E-beam versus the penetration depth.

As seen, the shape of E-beam volume varies substantially as a function of voltage and whether it is fixed or accelerated. This is an important fact that needs to be incorporated into any numerical and analytical modeling.

4.4 Power Density and Interaction Time for Various Heat Source-based Material Processing

For metals, there are specific boiling thresholds that are a function of energy intensity and interaction time. All AM processes with a heat source use a moving heat flux; thus the interaction time between the beam and material becomes an important factor.

A few general plots in the literature suggest an acceptable range of energy intensity and interaction time for each process. Figure 4.12 shows different processes and where they stand for a ballpark power intensity and interaction time. For example, the graph suggests that for directed energy deposition (DED) (either laser or E-beam), a power intensity range of 20–60 W/mm^2 is

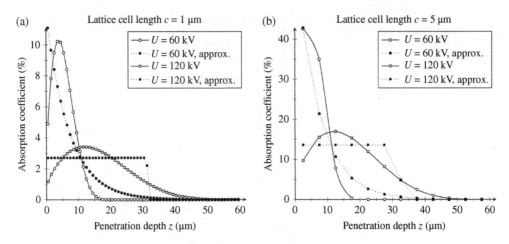

Figure 4.11 Penetration depth versus absorption coefficient for accelerated 60 kV and fixed 120 kV at two cell sizes underneath the beam, (a) 1 μm, (b) 5 μm [10, 11]. Source: Reproduced with permission from [10].

required when the interaction time could be in the range of 0.01–1 s. This plot is obviously a start point to design a process, but no means should be considered as a guideline.

4.5 Physical Phenomena and Governing Equations During DED[1]

Both LDED and EDED are complex processes that include multiple physical phenomena, such as energy absorption, heat conduction, heat losses, phase transformation, and fluid dynamics. The following section will explain each physical phenomenon, and then a few case studies mainly for LDED, either wire-fed or powder-fed, are presented.

4.5.1 Absorption

In the LDED process, absorption takes place by the injected powder particles or fed wire as well as the top surface of the substrate. Reflection from the substrate and powder particles or wire loses the remaining part of laser beam energy. Usually, the amount of net energy absorbed by the substrate is expressed by the laser energy multiplied by an overall absorption factor. In particular, as explained in the previous sections, the absorption factor is determined by the material, the laser wavelength, the temperature, and the angle of incidence. Moreover, the surface roughness has a significant effect on the absorptance of laser power in the material, as explained in the previous sections. See Figure 4.1 and also Equations (4.11) and (4.12) for further information.

---

[1] This Section 4.5 is partially taken from the previous work of one of the authors [2] with permission from the publisher. Permission conveyed through Copyright Clearance Center, Inc.

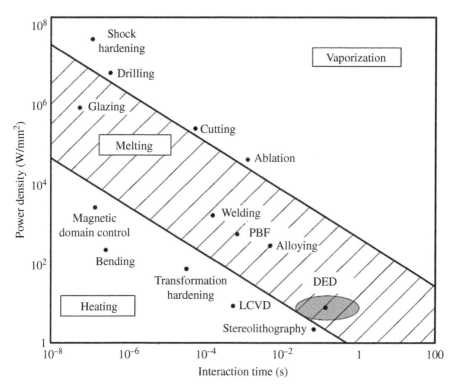

Figure 4.12 Power density and interaction time for various heat source-based material processing, mainly for ferrous, Ti, and Ni alloys [2, 4]. Source: Republished with permission of Taylor & Francis Group, from [2]; permission conveyed through Copyright Clearance Center, Inc.

4.5.2 Heat Conduction

The heat generated by absorption is penetrated into the metal substrates by thermal conduction accompanied by losses through convection and radiation. Conduction is one of the main physical phenomena that accounts for the temperature distribution in the melt pool in DED. The most general equation to clarify conduction phenomenon in the three-dimensional space of DED for numerical and analytical analysis is Fourier's law of heat conduction equation. However, this equation should be modified to incorporate the role of the moving heat source into it. This is done through the application of green functions to a moving domain. For a DED process, a moving heat source (e.g. laser beam) with a general distribution intensity irradiates on the substrate at $t = 0$ is shown in Figure 4.13. Due to additive material, the deposited track forms on the substrate, as shown in the figure. The transient temperature distribution $T(x, y, z, t)$ is obtained from the 3D heat conduction in the substrate as [12]:

$$\frac{\partial(\rho c_p T)}{\partial t} + \nabla.(\rho c_p \mathbf{U} T) - \nabla.(K \nabla T) = Q \tag{4.16}$$

where Q is the power generated per unit volume of the substrate (W/m^3), K is the thermal conductivity (W/m·K), c_p is the specific heat capacity (J/kg·K), ρ is the density (kg/m^3), t is the time (s), and $\mathbf{U}$ is the travel velocity vector of the substrate (process speed) (m/s).

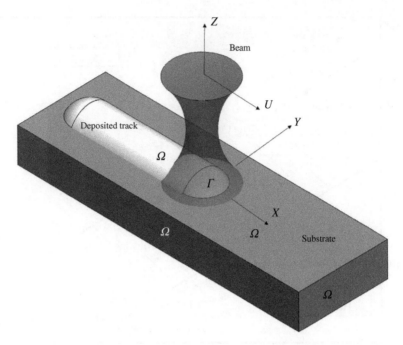

Figure 4.13 Schematic of physical domains of DED. Source: Adopted from [2].

In DED, the beam source is mainly modeled through boundary conditions when Q is assumed to be zero. If the beam energy is considered to be added to the physical domain from the top surface, the following equation should be considered as one of the boundary conditions:

$$-K(\nabla T.\mathbf{n})|_\Gamma = \beta I(x, y, z, t) \tag{4.17}$$

where n is the normal vector of the surface, $I(x, y, z, t)$ is the laser energy distribution on the substrate (W/m^2), and β is the absorption factor.

4.5.3 Surface Convection and Radiation

During DED, the heat input from the laser beam is conducted into the substrate locally. Since the substrate usually remains at room temperature, the dissipation of the heat occurs in the surrounding. Forced gas due to shield gases increases the level of heat loss from the surface that would be reflected in the convection coefficient. Due to the infrared light emitted from the melt pool, radiation plays a role in energy loss.

Surface convection to the ambient atmosphere and/or to the shielding gas flow is considered as a loss and should be considered in any analysis. In general, the effect of energy losses can be defined by the boundary condition as:

$$-K(\nabla T.\mathbf{n})|_\Omega = -h_c(T - T_0) - \varepsilon_t \sigma\left(T^4 - T_0^4\right) \tag{4.18}$$

where h_c is the heat convection coefficient (W/m^2·K), ε_r is the emissivity, σ is the Stefan–Boltzman constant (5.67×10^{-8} W/m^2·K^4), Ω is the substrate surfaces (m^2), and T_0 is the ambient temperature (K).

It should be noted that the magnitudes of surface convection and thermal radiation are relatively small compared to the heat conduction that is generated by the beam flux. Therefore, these two terms are usually neglected in analytical modeling but not in numerical modeling.

4.5.4 Fluid Dynamics

In the DED processes, a significant temperature variation can be found in the melt pool and the HAZ which results in both density and surface tension difference throughout the melt pool volume. Because of these gradients, both free convective and Marangoni flows are initiated within the melt pool. The melt pool fluid dynamics are depended on the secondary and tertiary thermocapillary-type forces, which are driven by wetting angle variations as well as capillary effects. Powder or wire feed rate and deposition angle significantly affect the free surface stability and the melt pool dynamics. All of these forces would change the melt pool free surface and its melt pool momentum, impacting its morphology/shape, solidification rate, and eventually the properties of solidified tracks [2, 4].

The peak velocity of melt pool flow increases with laser power for a given spot, and therefore the temperature difference between the irradiated top surface and its surrounding area will cause an increase in convection flow with higher peak velocities. As the laser power decreases with higher speeds, the melt pool peak temperature decreases, when the buoyancy surface tension flows would have a lower peak speed [3].

In DED, the conservation of momentum is one of the important governing laws. The equation of momentum is Newton's second law applied to fluid flow, which yields a vector equation. The momentum equation is represented as:

$$\frac{\partial(\rho \mathbf{U})}{\partial t} + \nabla.(\rho \mathbf{U} \otimes \mathbf{U}) = \rho \mathbf{g} - \nabla p + \mu \nabla.(\nabla \mathbf{U}) \tag{4.19}$$

where $\mathbf{g}$ is the gravity field (m/s^2), μ is the viscosity (kg/s·m), and p is the pressure (N/m^2).

The equation of continuity must be sustained in the DED process, represented by:

$$\nabla \mathbf{U} = \mathbf{0} \tag{4.20}$$

In a general case, on the surface of the melt pool, if $\mathbf{g}$ is vertical, the surface tension is derived by:

$$\Delta p + \rho \mathbf{g} z = \left(2\mu \frac{\partial \mathbf{U}}{\partial n}.\mathbf{n} \right) + \frac{\gamma}{R_c} \tag{4.21}$$

and

$$\mathbf{U}.\mathbf{n} = 0 \tag{4.22}$$

where z is a vertical coordinate (m), γ is the surface tension (N/m), and R_c is the track (known as clad) surface curvature (m) [13].

At the solid–liquid interface,

$$f(x, y, z, t) = \text{constant where } u_x = u_y = u_z = 0 \tag{4.23}$$

and

$$T = T_m \tag{4.24}$$

where $f(x, y, z, t)$ is a function that presents the melt pool interface with the substrate, and u_x, u_y, and u_z, are fluid velocity components in x, y, and z directions, respectively (m/s) [14]. This condition is valid for pure elements. For alloys, the freezing range (where alloys transform from all liquid to all solid) should be considered.

4.5.4.1 Wetting Angles, Interfacial Free Energies, and Capillary Effects

In DED, wetting angle and interfacial free energies are important parameters that indicate the quality and shape of the deposited track. In general, three types of cross section may be produced by DED, particularly LDED, with blown powder, as shown in Figure 4.14. These cross sections represent the amount of dilution, corresponding wetting angle θ, and interfacial free energies γ (J/m^2). Three interfacial energies for DED can be considered: solid–liquid interfacial free energy γ_{SL}, solid–vapor interfacial energy γ_{SV}, and liquid–vapor interfacial energy γ_{LV}.

A balance between the afore mentioned energies governs the shape of the deposited bead. This balance is expressed by:

$$\gamma_{SV} - \gamma_{SL} = \gamma_{LV} \cos \theta \tag{4.25}$$

The substrate becomes wet by the liquid if $\cos\theta$ goes toward one or if $\gamma_{SV} - \gamma_{SL} > \gamma_{LV}$. The spreading factor is defined as:

$$S = \gamma_{SV} - \gamma_{SL} - \gamma_{LV} \tag{4.26}$$

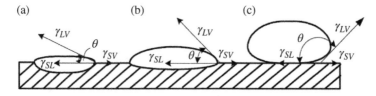

Figure 4.14 Track cross section created by DED, (a) high dilution, well wetting, (b) ideal track, and (c) no dilution, non-wetting. Source: Republished with permission of Taylor & Francis Group, from [2]; permission conveyed through Copyright Clearance Center, Inc.

A large positive S leads to major spreading, whereas a lower number causes a non-wetting system. When the laser energy is high, dilution increases and wetting angle decreases, as shown in Figure 4.14.

Capillary forces play a critical role in DED processes. The following equation can present capillary forces:

$$\mathbf{F}^{Cap} = R_{cur}\gamma. \, dA. \, \mathbf{n} \tag{4.27}$$

where R_{cur} is the radius of curvature. Wetting forces can be calculated by:

$$\mathbf{F}^{Wet} = \gamma\left(\cos\theta_d - \cos\theta_{eq}\right) \tag{4.28}$$

where θ_d is the dynamic wetting angle (being changed) and θ_{eq} is the steady-state angle. Figure 4.15 shows these parameters.

The Marangoni effect in the melting process is the mass transfer along with an interface between two fluids due to the surface tension gradient. Generally, it can be represented by the Marangoni number Ma, which is defined as:

$$Ma = \frac{\Delta\gamma L}{\mu D_f} \tag{4.29}$$

where L is a distance parallel to the surface (m), μ is the viscosity (Pa.s), and D_f is a diffusion constant.

However, the Ma due to thermal gradients is presented by:

$$Ma = -\left(\frac{\partial\gamma}{\partial T}\right). \frac{L\Delta T}{\mu\alpha} \tag{4.30}$$

where α is the thermal diffusivity. In general, Ma is the ratio of heat transfer driven by the surface tension to the heat transfer by conduction in the melt pool. For small values for Ma, thermal diffusion is more dominant when there is not much flow convection. On the contrary, for large Ma values, convection flow occurs governed by the surface tension gradients, known as Bénard–Marangoni convection.

It is shown numerically and experimentally that the mean bulk temperature of the melt pool increases with the assistance of Marangoni flow compared to heat transfer only by the conduction. Whereas the bulk mean temperature decreases with increasing Marangoni number in a long-time interaction. Overall, it was observed that Marangoni flow enhances the heat flux

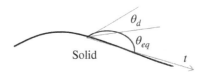

Figure 4.15 Dynamic and equilibrium wetting angles.

at the melt pool rim. It should be noted that the final shape of the melt pool is very much dependent on this region [15]. In the literature, the use of a corrected thermal conductivity factor to count for Marangoni flow is recommended. Experimental work and estimations in the literature [16] suggest that the effective thermal conductivity in the presence of thermocapillary flow is at least twice the stationary melt conductivity. This increase can be generally presented by:

$$K^* = aK \tag{4.31}$$

where a is the correction factor and K^* is modified thermal conductivity (W/m·K).

4.5.5   Phase Transformation

A series of metallurgical phase transformations or oxidation happen during DED in both directions: melting and resolidification. Phase changes either generate or consume energy according to the direction of evolution of enthalpy and Gibb's free energy [17]. Examples can be the transformation from ferrite to austenite or oxidation. In a temperature field, the exceeded generated or used energy via phase transformations can affect significantly by changing the temperature gradient as well as changing density. In general, these deviations are presented by [17]:

$$-K_s \mathbf{n}.\nabla T_s = -K_l \mathbf{n}.\nabla T_l + \rho U L_m \tag{4.32}$$

where s refers to solidus and l to liquids properties, and $\mathbf{U}$ is the normal velocity of the phase front in the direction $\mathbf{n}$. See Figure 4.16.

Usually, diffusion is not a dominant physics in DED due to rapid process and nonequilibrium process. However, for the nonequilibrium process, a phase field solidification method may be applicable.[2] In this method, two main variables are considered: a temperature field T and a phase field p both as a function of location r and time t. When $p = 0$, it refers to liquid and when $p = 1$, it refers to solid. At the liquid–solid interface, p values between 0 and 1. To track changes in phase field parameter given p, the whole system is expressed by the following free energy, which includes m parameter [18]:

$$\emptyset[p, m] = \int \frac{1}{2} \delta^2 |\nabla p|^2 + F(p, m) \, dr \tag{4.33}$$

where δ is defining the solid–liquid interface thickness with a very small value. It usually controls the mobility of the solid–liquid interface and is used to introduce the anisotropy of solidification. F is a double-well potential function in which its two local minimums (at $p = 0$, 1) may vary by changing the m parameter:

$$F(p, m) = \frac{1}{4}p^4 - \left(\frac{1}{2} - \frac{1}{3}m\right)p^3 + \left(\frac{1}{4} - \frac{1}{2}m\right)p^2 \tag{4.34}$$

---

[2] Courtesy of Dr. Ali Keshavarz for providing the authors with phase field formulas.

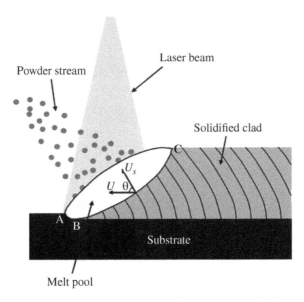

Figure 4.16 A schematic of the process zone during LDED powder-fed. Melting takes place from A to B, while solidification takes place from B to C. The lines within the solidified track deposit indicate the direction of compression. Source: Republished with permission of Taylor & Francis Group, from [2] permission conveyed through Copyright Clearance Center, Inc.

The m parameter is set to change in the range $\left[-\frac{1}{2}, \frac{1}{2}\right]$. Since the global minimum of the F function is completely dependent on the m value, it is assumed that m is the function of the temperature field:

$$m(T) = \left(\frac{A}{\pi}\right) \tan^{-1}[B(T_e - T)] \tag{4.35}$$

where T_e is the equilibrium temperature, A and γ are positive constants. To ensure the former restriction on the m parameter ($|m| < \frac{1}{2}$) is satisfied, the α parameter must be less than 1.

To introduce the anisotropy in growth of crystals, δ is defined as a function of direction:

$$\delta(\theta) = \bar{\delta}\eta(\theta) \tag{4.36}$$

where $\bar{\delta}$ is the average value of the solid–liquid interface thickness and $\eta(\theta)$ represents the anisotropic behavior of growth as:

$$\eta(\theta) = 1 + \omega_i \cos[k_\varepsilon(\theta - \theta_0)] \tag{4.37}$$

where ω_i is a constant implying the strength of anisotropy, k_ε is the mode of anisotropy that is normally 4 for a FCC material, θ_0 is the easy growth direction of the crystal, and θ is the direction perpendicular to the solid/liquid curvature and derived from the primary phase field parameter [19]:

$$\tan \theta = \frac{dp/dy}{dp/dx} \tag{4.38}$$

Considering the fact that the phase parameter p is not a conserved quantity, the evolution of the phase field parameter can be described as [18]:

$$\tau \frac{\partial p}{\partial t} = -\frac{\delta \emptyset}{\delta p} \tag{4.39}$$

To consider the enthalpy law, temperature field T should be derived from the following equation:

$$\frac{\partial T}{\partial t} = \nabla^2 T + \frac{L_f}{K} \frac{\partial p}{\partial t} \tag{4.40}$$

4.5.6 Rapid Solidification

Melt pool composition is determined by alloying between the deposited material(s) and the substrate. Furthermore, the phases being formed during solidification have a vital role in the bonding strength of the interface. The microstructure created in the bulk of deposited tracks substantially determines the final product's physical and mechanical properties [2].

LDED includes a localized heat source, interacting for a very short time (usually less than 0.5 s), leading to a high solidification rate. Figure 4.16 represents that the rate of solidification U_s or R can be related to the beam velocity U by the following equation:

$$U_s = R = U \cos \theta \tag{4.41}$$

where θ is the angle between the vector representing the direction of the substrate motion and the vector normal to the solidification rate at a particular point. Inspection of Figure 4.16 reveals that the solidification rate varies throughout the process volume from zero at point B to a maximum at the melt pool–solid interface.

Both solidification rate U_s and temperature gradient ∇T can be determined numerically. Also, both of them increase with the increase in processing speed. At the bottom of the melt pool, the thermal gradient is highest where the solidification rate is zero at that point.

4.5.7 Thermal Stresses

Material expansion or contraction depends on temperature variations; therefore, thermal strain depends on current and initial temperatures independent of stress. The total strain ε in a body can be expressed by:

$$\varepsilon_{mn} = \varepsilon_{mn}^M + \varepsilon_{mn}^T \qquad (m, n = 1, 2, 3) \tag{4.42}$$

where ε^M (m/m) and ε^T (m/m) are mechanical and thermal strains, respectively. The plasticity may not be dominant in DED as the entire part is mainly investigated for residual stresses.

Then, the constitutive equation for linear elastic materials is

$$\sigma_{ij} = D_{ijmn}\varepsilon_{mn} \quad (i,j,m,n=1,2,3) \tag{4.43}$$

where σ_{ij} (Pa) is the elastic stress and $\boldsymbol{D}_{ijmn}$ (Pa) is the tensor of elastic coefficients with 81 components. It should be noted that there are only 36 independent components because the strain and stress tensors are mainly symmetrical.

Initial conditions for stress and strain fields can be expressed by:

$$\varepsilon\,(x,y,z,0)|_{T=T_0} = 0 \text{ and } \sigma\,(x,y,z,0)|_{T=T_0} = 0 \tag{4.44}$$

4.5.8 Flow Field in DED with Injected Powder

In DED with injected powder, known as powder-fed DED, a flow field in the exit of the feeding nozzle needs to be understood. Powder-fed DED is a complex process involving interaction between the heat source, the powder particles, and the melted region of the substrate. In order to build the track with accurate dimensions and high efficiency of the powder deposition in a coaxial LDED, it is essential to analyze the powder flow structure [20]. In coaxial DED, the powder is carried by a flow stream impinging on the substrate. Some designs also include a shaping gas flow helping the powder flow stream to concentrate on the melt region of the substrate. In the coaxial DED process, three different flows are encountered. There is an air (or argon) flow at the center for protecting the lenses from the hot powder particles that may bounce off the substrate. Next is the flow with powder particles aiming at the irradiated region, and finally, the shaping gas, as shown in Figure 4.17. All these flows and their interactions affect the catchments of the depositing powder at the laser-irradiated region and, therefore, affect the efficiency and the quality of the deposited track.

The flow at the nozzle's exit can be laminar or turbulent, depending on the nozzle exit Reynolds number. It has been shown that turbulence-free jet cannot be sustained for $Re < 1000$ [19]. Typical flow parameter values for flow at the coaxial nozzle exit indicate that both laminar and turbulent jet can exist depending on the size and the exit velocity of the powder stream. Therefore, both flow patterns and the associated equations are discussed in the following subsections.

4.5.8.1 Laminar Flow for LDED

The governing equations for laminar flow are the Navier-Stokes and continuity equations as:

$$\rho\frac{\partial \mathbf{U}}{\partial t} - \eta_d\nabla^2\mathbf{U} + \rho(\mathbf{U}.\nabla)\mathbf{U} + \nabla p = \mathbf{F} \tag{4.45}$$

$$\nabla.\mathbf{U} = 0 \tag{4.46}$$

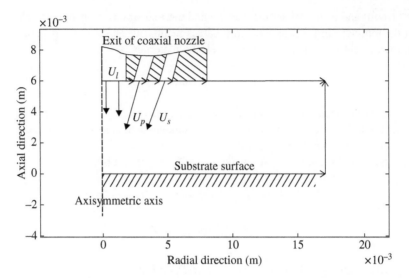

Figure 4.17 Geometry and boundary conditions for a typical coaxial nozzle exit (U_l is shield gas velocity, U_p is powder stream velocity, and U_s is shaping gas velocity). Source: Republished with permission of Taylor & Francis Group, from [2]; permission conveyed through Copyright Clearance Center, Inc.

where **U** is the velocity field (m/s), ρ is the density of the gas (kg/m³), η_d is the dynamic viscosity (m²/s), p is the pressure field (N/m²), **F** is the external force (N). The boundary conditions depend on the physical domain of interest. As a case study, a domain with the boundaries shown in Figure 4.17 can be elaborated as:

- On the solid surface (i.e. substrate and solid parts of the nozzle), the conditions are set to a no-slip boundary condition, in which $v = 0$ and $u = 0$, where v and u are the velocity components (m/s).
- On the top free surface, a neutral condition is considered, in which **n**. $(\eta \nabla \mathbf{U}) = 0$, where **n** is a normal vector on the free surfaces.
- On the side free surface, a straight outflow is considered, in which $t \cdot \mathbf{U} = 0$.
- On the axisymmetric axis, the slip condition can be considered, in which $\mathbf{n} \cdot \mathbf{U} = 0$.

On a different physics, in order to investigate the trajectory of particles in the above flow field, the following equations were employed:

$$m_p \frac{d(\mathbf{U}_p)}{dt} = (m_p - m_f)\mathbf{g} - 6\pi r_p \eta (\mathbf{U}_p - \mathbf{U}) \qquad (4.47)$$

$$\frac{d(\mathbf{x}_p)}{dt} = U_p \qquad (4.48)$$

where m_p is particle mass (kg), m_f is the fluid mass that the particle has displaced (kg), $\mathbf{U}_p$ is the particle velocity vector (m/s), **U** is the fluid velocity vector (m/s), r_p is the particle radius (m), **g** is the gravity (m/s^2), and x_p is the tracing particle position in the flow field (m). Inherent in Equation (4.47) is the assumption that the acceleration of a particle is influenced by gravity force and drag force [21].

4.5.8.2   Turbulent Flow for LDED

In a turbulent flow, in addition to Navier-Stokes and continuity equations, the kinetic energy of turbulence k and dissipation of kinetic energy of turbulence ε should be solved. This type of turbulence modeling is referred to as $k - \varepsilon$ turbulence modeling. The general form of governing equation is presented by:

$$U.\nabla\emptyset + \nabla.(D_\emptyset \nabla\emptyset) = P_\emptyset + S_\emptyset \tag{4.49}$$

where $\emptyset = (u, v, k, \varepsilon)$, $D_\emptyset$ is the diffusion coefficient and $S_\emptyset$ is the source term corresponding to each $\emptyset$ component.

In general, a comparison between the laminar flow field and turbulent patterns indicates a major change in the flow field. The vortex strength is weaker, and the flow streams exit the domain mainly from the side free boundary condition.

4.6   Modeling of DED

Modeling is an essential platform to understand the physical phenomena happening in the process and to predict the quality of the manufactured product without any experiments. To create a heat distribution in the DED substrate, three approaches have been made:

1. Analytical modeling;
2. Numerical modeling;
3. Experimental-based modeling.

In the next sections, these approaches will be discussed.

4.6.1   Analytical Modeling: Basics, Simplified Equations, and Assumptions

Most of the analytical solutions for laser-based, E-beam-based, or other DED heat sources originate from the basis of laser welding modeling.

There are several heat models that have been used for DED. Considering types of the heat source and boundary conditions, they can be categorized into:

1. Moving point heat source where the substrate has an infinite thickness;
2. Moving line heat source, where the substrate has an infinite or semi-infinite thickness;
3. Moving hypersurface line heat source where the substrate has a semi-infinite thickness;
4. Integrated moving point and line heat sources;
5. Modified point/and line moving heat sources;
6. Periodic moving point and line sources where the substrate has semi-infinite thickness.

The number of approaches abound; however, amongst the analytical solutions proposed in the literature, the moving point heat source with the medium of infinite thickness is widely used for which Rosenthal introduced an equation for DED with a Gaussian energy intensity [22]. To this extent, in the next section, Rosenthal's equation will be discussed in detail as well as a modified approach for finding maximum temperature known as Adam's equation. However, before addressing these equations, a heat transfer equation solution for a semi-infinite plate is presented in the form of differential equations.

4.6.1.1 Analytical Solution of Heat Transfer Equation for Constant Extended Surface Heat

The solution of heat transfer Equation (4.16), with the following assumptions:

1. No convective/radiative heat losses,
2. No heat generation in substrate,
3. Constant thermal properties,
4. Constant extended surface heat input,
5. Large heat source compared with the depth,
6. Not considering the beam size and substrate thickness,

is

$$T_{x,t} = \frac{2\beta P}{K} \left\{ \sqrt{\alpha t}.ierfc\left[\frac{x}{2\sqrt{\alpha t}}\right] \right\} \tag{4.50}$$

where, P is the surface power density $(x = 0)$, β is the absorption factor, α is the thermal diffusivity, and $ierfc(u) = \frac{e^{-u^2}}{\sqrt{\pi}} - u[1 - erf(u)]$ with u being an integration variable [4]. When the power is turned off, the material will cool for $t > t_1$ according to the relationship [4]:

$$T_{x,t} = \frac{2\beta P}{K} \sqrt{\alpha} \left[\sqrt{t}.ierfc\left(\frac{x}{2\sqrt{\alpha t}}\right) - \sqrt{t_0 - t_1}.ierfc\left(\frac{x}{2\sqrt{\alpha(t_0 - t_1)}}\right) \right] \tag{4.51}$$

This model applies when the motion is unidirectional and all assumptions mentioned above are considered [4].

4.6.1.2 Analytical Model for an Instantaneous Point Source of Energy

If we consider an instantaneous point source of energy $Q\rho C$, the differential equation for the conduction of heat in a stationary medium (Eq. 4.7) is satisfied by:

$$T = \frac{Q}{8(\pi\alpha t)^{2/3}} \exp\left\{ -\frac{\left[(x-x')^2 + (y-y')^2 + (z-z')^2\right]}{4\alpha t} \right\} \tag{4.52}$$

The total quantity of heat in the infinite region is

$$\iiint_{-\infty}^{+\infty} \rho C T dx dy dz = \frac{Q\rho C}{8(\pi\alpha t)^{3/2}} \int_{-\infty}^{+\infty} \exp\left[\frac{-(x-x')^2}{4\alpha t}\right] dx \int_{-\infty}^{+\infty} \exp\left[\frac{-(y-y')^2}{4\alpha t}\right] dy$$

$$\int_{-\infty}^{+\infty} \exp\left[\frac{-(z-z')^2}{4\alpha t}\right] dz$$

(4.53)

4.6.1.3 Analytical Model for Continuous Point Source

For a continuous point source, if heat is liberated at the rate of $\varphi(t)\rho C$ per unit time from $t = 0$ to $t = t'$ at the point (x', y', z'), the temperature at (x, y, z) at time t is

$$T(x, y, z, t) = \frac{1}{8(\pi\alpha)^{3/2}} \int_0^t \varphi(t') e^{\frac{-r^2}{4\alpha(t-t')}} \frac{dt}{(t-t')^{3/2}}$$

(4.54)

where $r^2 = (x-x')^2 + (y-y')^2 + (z-z')^2$. If $\varphi(t)$ is constant and equal to the net heat flux distribution (q), the above equation can be written as:

$$T = \frac{1}{4(\pi\alpha)^{3/2}} \int e^{\frac{-r^2}{4\alpha}} d\tau = \frac{q}{4\pi\alpha r} - erfc\left\{\frac{r}{\sqrt{4\alpha t}}\right\}$$

(4.55)

where $\tau = (t - t')^{-1/2}$.

For $t \to \infty$, $T = q/4\pi\alpha r$ describes a steady temperature distribution in which a continuous source of heat is applied continuously. The above equation can calculate the heating from different distribution (line sources, disc sources, or Gaussian sources).

4.6.1.4 Rosenthal's Equation

Rosenthal proposed a 2D and 3D analytical solution to analyze heat flow in the substrate, considering a moving heat source [23]. In most analyses, the 3D heat flow equation is usually used. In the proposed equations, Rosenthal used the following assumptions for simplification [22, 23]:

1. Steady-state heat flow should be considered. No transient thermal model is incorporated.
2. A point moving heat source is considered.
3. No heat of fusion is incorporated into the model.
4. All thermal and optical properties are assumed fixed and not temperature-dependent.
5. No convective and radiative heat losses are considered.
6. No Marangoni effect; however, its effect can be incorporated through a modified thermal conductivity.

Rosenthal's equation for 3D heat flow in a semi-infinite substrate, when the heat source moves along the x axis, is given by:

$$T - T_0 = \frac{\beta P}{2\pi RK} e^{-v(x+R)/(2\alpha)} \tag{4.56}$$

where P is the laser power (W), v is the process speed along x-axis (m/s), and $R = \sqrt{x^2 + y^2 + z^2}$. Other parameters are introduced before.

Cooling rate for the centerline surface spot ($x > 0$, $y = 0$, $z = 0$) will be estimated by:

$$\frac{\partial T}{\partial t} = -2\pi K \left(\frac{v}{\beta P}\right)(T - T_0)^2 \tag{4.57}$$

According to the material thermal properties and the DED process parameters, the isotherm T on a plane at a given x has a radius of R that can be used to calculate the steady-state temperature $T(x, y, z)$, considering the moving heat source at any location in the substrate (x, y, z), as shown in Figure 4.18.

Although Rosenthal's assumptions simplify the mathematical analysis involved, this model suffers from a few issues that need to be considered. At the center of the Gaussian/point heat source, the temperature is theoretically infinity, which is practically wrong. One remedy to this issue is to set the ultra-high temperature of the center to the material boiling temperature. Also, thermal properties' values vary with the temperature and neglecting the thermal fusion gives substantial errors. However, if correctly used and calibrated, this model is a straightforward solution for the analytical modeling of DED.

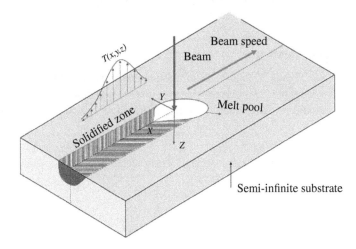

Figure 4.18   Schematic of 3D heat flow during DED used for the development of Rosenthal's equation.

4.6.1.5   Adam's Equation

Adam modified Rosenthal's equation by considering the thickness of the HAZ [24]. The model is supplementary to Rosenthal's equation to predict the temperature profile across the HAZ [24]. The equation for 2D heat flow is

$$\frac{1}{T_p-T_0} = \frac{\sqrt{2\pi e}}{\beta P}v\rho c_p hy + \frac{1}{T_m-T_0} \tag{4.58}$$

where T_p is the maximum temperature (K) at the substrate adjacent to the melt pool and far from the fusion line by y (m), where h is the substrate thickness (m). Other parameters are the same as introduced before. At $y = 0$, $T_p = T_m$.

The equation for 3D heat flow is given by:

$$\frac{1}{T_p-T_0} = \frac{1.31\sqrt{2\pi e}}{\beta P v}K\alpha\left[2 + \left(\frac{vy}{2\alpha}\right)^2\right] + \frac{1}{T_m-T_0} \tag{4.59}$$

These equations can be used to calculate the HAZ surrounding the melt pool. It should be noted that the previous equations are based on the point source when basically no melt pool effects are included in the solution.

4.6.2   Numerical Models for DED[3]

Models, which are based on physical laws, contribute to a better understanding of DED processes. A precise model can support the required experimental research to develop an optimum DED process. In general, numerical models for DED are categorized into two main classes:

1. Steady-state models; and,
2. Dynamics models.

Steady-state models are those models that are independent of time, whereas dynamic models refer to those models that take into account the transient response of the process.

In these models for DED, several important physical phenomena are addressed, such as thermal conduction, thermocapillary (Marangoni) flow, powder and shield gas forces on melt pool, mass transport, diffusion, laser–powder interaction, melt pool–powder interaction, power attenuation, and laser–substrate interaction in the process zone. The previous equations represented these phenomena.

The models are also categorized as:

1. Microscale;
2. Mesoscale;
3. Macroscale.

---

[3] This Section 4.6.2 is taken from the previous work of one of the authors [2] with permission from the publisher. Permission conveyed through Copyright Clearance Center, Inc.

Microscale models include a broad class of computational models that simulate fine-scale details of DED in scale from a few µm to 500 µm. Whereas the mesoscale models cover the range from 300 to a few millimeters. Macroscale models refer to structural modeling that could be from a few millimeters to large size components. Microscale and macroscale models can be used together to understand different aspects of the same problem.

These models can include ordinary, partial, and integral–differential equations, where each physics will be coupled to boundary conditions or complementary physics through algebraic equations. A simplified macroscale model may be combined with more detailed microscale models. Physics can be coupled to develop a multi-physics domain for DED. But in all cases, simplifications are required to enable to solve the finite element models effectively. In Section 4.7.3, a model for powder-fed LDED (PF-LDED) will be discussed. It should be noted that the same approach can be applied to other DED processes when boundary conditions are customized for such a process.

4.6.3 Experimental-based Models: Basics and Approaches[4]

This section addresses the experimental-based modeling techniques and how they can apply to DED. As we learned, DED is a complex process for which it is very difficult to develop reliable numerical and analytical models due to process complexities. In DED processing, the complexity arises from the nature of the governing equations, which are partial differential and also the interaction of thermal, fluid, and mass transfer phenomena in the process, as addressed in the previous sections.

Experimental-based models can fill the gap and provide the users with a reliable dynamic model that can be used for process prediction and also real-time control system development. Two approaches can be used:

1. Stochastic techniques, known as system identification in engineering, such as autoregressive exogenous;
2. Artificial intelligent models such as neural networks and deep learning.

All experimental-based modeling techniques such as the stochastic, artificial neural network, and neuro-fuzzy approaches are essentially based on optimizing the parameters of a given model to result in the minimum error between the measured and model prediction data. It should be noted that nonlinearity in the process should be linearized around the operating point.

There are basically three general model structures that are used for nonlinear model prediction, based on prior and physical knowledge:

1. White-box, when the model is perfectly known;
2. Gray-box, when some physical insights are available, and;
3. Black-box, when the system is completely unknown.

---

[4] This Section 4.6.3 is mainly taken from the previous work of one of the authors [2] with permission from the publisher. Permission conveyed through Copyright Clearance Center, Inc.

A black-box model is much more complex than the other two cases due to the variety of possible model structures. One of the model structures for black-box modeling is artificial intelligence. Furthermore, experimental-based modeling techniques are well developed for linear systems; however, for nonlinear systems, the techniques are very limited. They require many considerations for selecting the model structure, inputs/outputs, and optimization techniques used to find the system parameters.

Selecting a proper set of inputs and outputs and collecting data are critical in any dynamic model development. The collected data due to the excitation signals should be rich enough and allow for identifying necessary higher modes to accurately present the dynamics of the system. Independent of the chosen model architecture and structure, the data's characteristics determine a maximum accuracy that can be achieved by the model. For a nonlinear system, the minimum and maximum amplitude and length of the excitation signals are essential to the identification process.

The maximum and minimum of the amplitude reflect the range of process parameters over which the model should accurately predict the process. The magnitude of amplitude should also be changed around the desired points of operation. The length of excitation signals (duration) chosen should not be too small nor too large. If it is too small, the process will have no time to settle down, and the identified model will not be able to describe the static process behavior properly. On the other hand, if it is too long, only a very few operating points can be covered for a given signal length. The other concern about the data collection is noise within the data. The noise can arise from sensors or from the side effects of the other process parameters that are not included in the model. It is essential to generate a rich excitation signal for the data collection in terms of amplitude and duration to compensate for the effects of noisy signals.

DED is a thermal process, and for a thermal process, the response to an excitation signal is essentially slow. As a result, the minimum length for excitation signals is set to 10 s. It is experimentally tested that the process response is settled down after 10 s, which indicates the required time for obtaining a steady-state response.

Laser pulse energy, width, frequency, and table velocity are important excitation signals in a pulsed LDED process. The deposition geometry and microstructure are two geometrical and physical properties that can be selected as the output signals. See Section 4.7.5 for a case study on this type of modeling.

Generally, the model identification procedure can be itemized as follows:

1. Design an experiment and collect input–output data from the process. The richness of the data must be validated thoroughly.
2. Examine the data and select useful portions of the original data.
3. Select and define a model structure.
4. Compute the best parameters associated with the model structure according to the input–output data and a given cost function.
5. Verify the identified model using unseen data which are not used in the identification step.
6. If the model verification is acceptable, the desired model is identified; otherwise, Steps 3–5 should be repeated by another model structure or with more data.

A case study is provided in Section 4.7.5; however, several points are critically important to be considered in a general fashion regarding the experimental-based modeling of AM processes:

- The most appropriate modeling algorithm to use for prediction at each scale and physics must be carefully analyzed and chosen.
- The following questions should be answered upfront: what level of error in the predictions from these algorithms can be tolerated, relative to the process? Is there a trade-off between the accuracy and training time of different algorithms? If so, how to address this trade-off?
- To be a state of the art in the experimental-based modeling for DED, if bespoke algorithms are required, they must be accurate. This accuracy must be compared with that from existing algorithms to justify the need to design new algorithms.
- Error analysis: To what extent do the errors in these model predictions accumulate when the linking is performed across different scales and physics?
- What design of experiments are the best to utilize scarce resources and ensure only the most relevant experiments are performed to obtain rich data.

4.7   Case Studies on Common Modeling Platforms for DED

In the following section, multiple models that are more popular for DED will be discussed. These models are explained, and typical results are presented. It should be noted that the essence of this textbook is not to cover the literature but to provide the readers with essential basic information.

4.7.1   Lumped Analytical Model for Powder-Fed LDED[5]

In a lumped model, the dependency of the process (equations) and spatial variables are ignored and time becomes the only independent variable. This simplification will render ordinary differential equations as opposed to partial differential equations. In PF-LDED, a lumped model can be proposed by a balance of energy in the process. No flow dynamics, no phase transformation, etc., are considered in this model. Only energy balance is considered.

The balance of energy in the process is shown in Figure 4.19.

In the figure, the total laser energy absorbed by the substrate and powder particles as well as a different source of losses (e.g. reflection, radiation and convection) are shown. The balance of energy can be expressed by:

$$Q_c = Q_l - Q_{rs} - Q_L + (\eta - 1)Q_p - Q_{rp} - Q_{radiation} - Q_{convection} \tag{4.60}$$

where Q_c is the total energy absorbed by the substrate (J), Q_l is the laser energy (J), Q_{rs} is the reflected energy from the substrate, Q_L is the latent energy of fusion (J), η is the powder catchment efficiency, Q_p is the energy absorbed by powder particles (J), Q_{rp} is the reflected energy from powder particles (J), $Q_{radiation}$ is the energy loss due to radiation (J), and $Q_{convection}$ is the energy loss due to convection (J).

---

[5] This revised case study is mainly reproduced from the one of the authors' previous work [2] with permission from the publisher. Permission conveyed through Copyright Clearance Center, Inc.

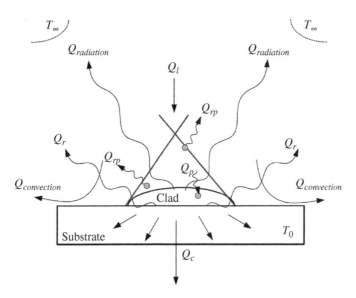

Figure 4.19 Balance of energy in PF-LDED. Source: Republished with permission of Taylor & Francis Group, from [2]; permission conveyed through Copyright Clearance Center, Inc.

The laser energy is presented by:

$$Q_l = \beta_w P_l t_i \tag{4.61}$$

where β_w is the absorption factor of the substrate, P_l is the average laser power (W), and t_i is the interaction time between the material and laser (s), which is presented by:

$$t_i = \frac{2r_l}{U} \tag{4.62}$$

where r_l is the beam spot radius on the substrate (m) and the process speed is shown by U (m/s).

The reflected energy from the substrate is

$$Q_{rs} = (1 - \beta_w)(Q_l - Q_p) \tag{4.63}$$

The latent heat energy can be expressed by:

$$Q_L = L_f \rho V \tag{4.64}$$

where L_f is the latent heat of fusion (J/kg), ρ is the average density in the deposited area, so-called clad (kg/m^3), and V is the volume of melt pool, including the clad region (m^3).

In order to find an expression for V in a lumped fashion, we can consider a portion of a cylinder laying on the substrate, as shown in Figure 4.20, where the width of the track is assumed to be equal to the laser beam diameter on the substrate.

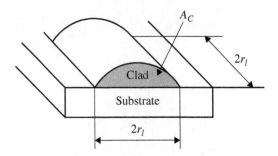

Figure 4.20 Lumped cross section of single track deposited in LDED. Source: Republished with permission of Taylor & Francis Group, from [2]; permission conveyed through Copyright Clearance Center, Inc.

Also, we assume that the length of the melt pool is equal to the laser beam diameter. The cross-section area A_c can be found by:

$$A_c = \frac{\dot{m}}{\rho_p U} \eta \tag{4.65}$$

where ρ_p is the particle density (kg/m^3) and $\dot{m}$ is the mass flow (kg/s). If dilution is ignorable, which is an appropriate assumption for PF-LDED, and the width of the melt pool is equal to laser diameter, the volume of melted area V can be expressed by:

$$V = 2r_l A_c \tag{4.66}$$

In a lumped model, the powder catchment efficiency can be assumed as the ratio between the area of the laser beam and powder stream on the substrate. Therefore,

$$\eta = \frac{r_l^2}{r_s^2} \tag{4.67}$$

where r_s is the powder stream diameter on the substrate (m).

In order to derive an equation for the energy absorbed by powder particles in a lumped model, consider a homogeneous distribution of powder particles over the laser beam volume as shown in Figure 4.21. If the powder particle radius r_p is known, the number of particles n in the laser beam volume over a time period of t_i is estimated by this equation if we assume the powder particles are distributed uniformly over the cylindrical volume with a height of $2r_s$ and diameter of $2r_l$:

$$n = \frac{3\dot{m} t_i}{4\pi \rho_p r_p^3} \tag{4.68}$$

where ρ_p is powder particle density (kg/m^3).

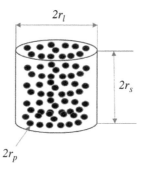

$2r_l$

$2r_s$

$2r_p$

Figure 4.21 Attenuated laser volume in PF-LDED.

The overall volume of the powder particles in the laser beam indicates the attenuated volume. In order to find the attenuated area A_{at} (m^2) by the powder particles, we can assume that particles are stacked uniformly in the height of the cylinder shown in Figure 4.21. With this assumption, the following equation can be estimated:

$$A_{at} = \frac{n\pi r_p^3}{r_s} = \frac{3\dot{m}t_i}{4\pi\rho_p r_s} \tag{4.69}$$

As a result, the absorbed energy by the particles can be obtained by:

$$Q_p = Q_l \frac{A_{at}}{A_l} \tag{4.70}$$

The reflected energy from the powder particles can be derived from

$$Q_{rp} = (1 - \beta_p)Q_p \tag{4.71}$$

where β_p is the powder particles' absorbed coefficient.

The radiative loss can be presented by:

$$Q_{radiation} = A_l \varepsilon_t \sigma (T^4 - T_0^4)t_i \tag{4.72}$$

where ε_t is the emissivity, σ is the Stefan–Boltzman constant (5.67×10^{-8} W/m^2·K^4), T is the melt pool temperature (K), and T_0 is the ambient temperature (K).

The convective loss in a lumped model, assuming a concentrated heating zone in the laser beam area, can be presented by:

$$Q_{convection} = A_l h_c (T - T_0)t_i \tag{4.73}$$

where h_c is the heat convection coefficient (W/m^2·K). Calculating h_c is difficult, and Goldak [25] and Yang [26] suggested an experimental expression, which is

$$h_c = 0.00241\varepsilon_t T^{1.61} \tag{4.74}$$

Equation (4.68) introduces a nonlinear term in the final energy balance equation. This term can, however, be ignored for the simplification of the final differential equation.

The energy Q_c can be presented in an integral form as:

$$Q_c = \rho c_p \int_{V_s} T(x, y, z, t_i) dV_s \tag{4.75}$$

where V_s is the heat-affected volume in Cartesian coordinates (x, y, z) (m^3), and ρ is the average density in the melt pool region (kg/m^3).

The substitution of the parameters in Equation (4.60) leads to an equation for Q_c. Plugging the derived equation for Q_c into Equation (4.75) leads to a lumped differential equation that presents the lumped model of the process.

This differential equation can be solved for $T(x,y,z)$ by different numerical approaches in mathematical packages such as MATLAB.

As an example, considering the following parameters:

- Substrate material: pure iron (reflectivity factor: 45%);
- Powder material: pure aluminum (reflectivity factor: 85%);
- Particle size: 50 μm spherical;
- Powder feed rate: 0.5 g/min;
- Powder stream diameter on the substrate: 1.6 mm;
- Laser spot diameter: 1.2 mm;
- Laser energy per pulse: 4 J/pulse;
- Laser pulse frequency: 200 Hz;
- Process speed: 10 mm/s;
- Ambient temperature: 25 °C;
- Substrate temperature: 25 °C;

The temperature distribution ballpark, found by Equation (4.75), is shown in Figure 4.22.

4.7.2　Comprehensive Analytical Model for Powder-Fed LDED (PF-LDED)[6]

In this section, a comprehensive analytical model for PF-LDED is developed. The following assumptions have been made:

1. The lateral nozzle is assumed where it has a perfect circular outlet.
2. The effects of gravity and drag forces are considered negligible. Thus, the powder stream has the same velocity as the gas near the nozzle outlet.
3. No convective and radiative losses in the powder stream.
4. Powder particles, impinging onto the molten pool, are considered effectively added to and mixed with the liquid flow on the melt pool surface.

---

[6] This section is a reproduced summary from one of the authors' prior papers [27] with permission from the publisher. Courtesy of Dr. Yuze Huang.

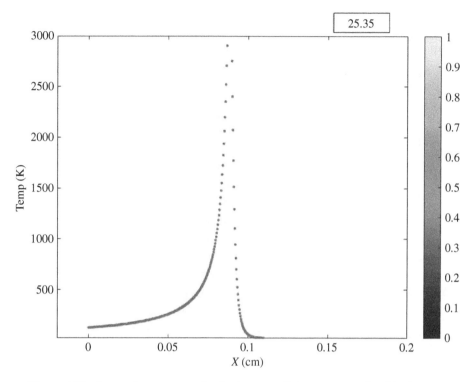

Figure 4.22   Lumped temperature distribution at $y = 0$ for parameters listed in the text.

5. The thermophysical properties for both powder and substrate are considered to be temper-ature-independent. Average values over the temperature variation were considered in the model.

4.7.2.1   Model for Powder Spatial Distribution

The schematic of PF-LDED is shown in Figure 4.23 in the positive y-direction with the process velocity v, and the origin of its coordinates is fixed at the center of the laser beam spot on the substrate. The nozzle has an inclined angle ϕ and distance H concerning the substrate plane. The laser beam and the powder stream interact with each other after point P.

The powder concentration mode in the transverse direction can be identified by a Gaussian distribution. One way to express the powder concentration distributions in the transverse plane is to show its powder stream luminance distributions. Based on Mie theory, the lumi-nance of the powder stream is proportional to particle concentration, so the luminance dis-tribution in the image expresses the particle concentration distribution within the powder stream. As an example, the Inconel 625 powder stream grayscale image is shown in Figure 4.24, where the powder concentration has Gaussian distribution in the transverse plane with varying distances.

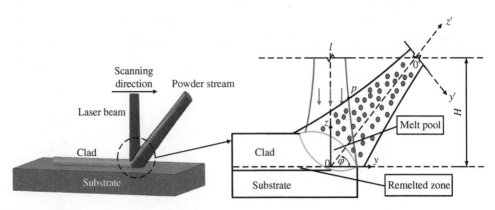

Figure 4.23 Schematic diagram for laser powder-fed laser-directed deposition (PF-LDED) [27]. Source: Reproduced with permission from the publisher.

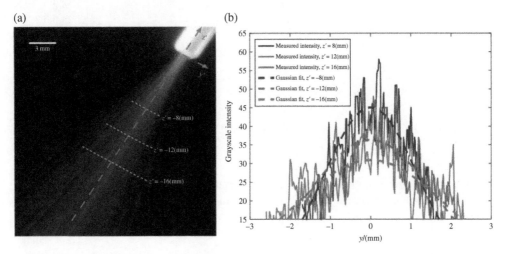

Figure 4.24 Inconel 625 powder stream grayscale intensity distribution measurement ($\dot{m} = 5(\text{g}/\min)$, $\phi = 60°$, $\dot{g} = 2.5(\text{dL}/\min)$, $r_0 = 0.7(\text{mm})$): (a) grayscale image with transversal lines and (b) measured grayscale intensity distribution with Gaussian fitting results on transversal lines [27]. Source: Reproduced with permission from Elsevier.

Powder spatial mass concentration $\rho(x', y', z')$ and the number concentration $n(x', y', z')$ can be derived as:

$$\rho(x', y', z') = \frac{2\dot{m}}{v_p \pi r^2(z')} \exp\left[-\frac{2\left(x'^2 + y'^2\right)}{r^2(z')}\right] \tag{4.76}$$

$$n(x', y', z') = \frac{2\dot{m}}{v_p m_p \pi r^2(z')} \exp\left[-\frac{2\left(x'^2 + y'^2\right)}{r^2(z')}\right] \tag{4.77}$$

where $\dot{m}$ is the powder feed rate (kg/s), m_p is the average mass for each particle (kg), $v_p = \dot{g}/\pi r_0^2$ is the average powder velocity (based on the assumption provided) (m/s), $\dot{g}$ is the gas flow rate [m³/s], r_0 is the nozzle internal radius (m), and $r(z')$ is the effective radius of powder stream reaching $1/e^2$ of the peak concentration value of the powder stream center (m). Based on the information provided in [27], the effective radius of powder stream can be expressed as:

$$r(z') = r_0 - z' \tan\theta, z' < 0 \qquad (4.78)$$

where θ is the effective divergence angle. Considering the coordinates transformation from $x'y'z'$ to xyz:

$$\begin{aligned} x' &= x \\ y' &= (y - H/\tan\phi)\sin\phi - (z - H)\cos\phi \\ z' &= (y - H/\tan\phi)\cos\phi + (z - H)\sin\phi \end{aligned} \qquad (4.79)$$

Then the powder mass concentration in xyz coordinates will be derived as:

$$\rho(x, y, z) = \frac{2\dot{m}}{v_p \pi r^2(z)} \exp\left[-\frac{2\left[x^2 + [(y - H/\tan\phi)\sin\phi - (z - H)\cos\phi]^2\right]}{r^2(z)}\right], \qquad (4.80)$$

$$r(z) = -[(y - H/\tan\phi)\cos\phi + (z - H)\sin\phi]\tan\theta + r_0$$

4.7.2.2 Laser Beam and Gas–powder Stream Interaction

The laser beam and consequently its power intensity are attenuated by the powder stream during their interaction [28, 29]. Assuming a Gaussian TEM$_{00}$ mode laser power with intensity distribution as [2, 4]:

$$I(x, y, z) = \frac{2P_l}{\pi R_L^2(z)} \exp\left[-\frac{2(x^2 + y^2)}{R_L^2(z)}\right] \qquad (4.81)$$

where P_l is the laser power (W), $R_L(z)$ is the effective radius of the laser beam (m) at a distance $\Delta z = z_0 - z$ from the beam waist position z_0 with R_{0L} radius for the laser beam and far-field divergence angle θ_L. $R_L(z)$ is expressed as:

$$R_L(z) = \sqrt{R_{0L}^2 + 4\theta_L^2(z_0 - z)^2} \qquad (4.82)$$

The laser power intensity that is attenuated by powder stream with dz distance could be calculated based on Mie' theory [28]:

$$dI = -\sigma I(x, y, z)n(x, y, z)dz \qquad (4.83)$$

where $I(x, y, z)$ is the laser intensity at point (x, y, z), σ is the extinction cross section of a sphere particle ($\sigma = Q_{ext}\pi r_p^2$), Q_{ext} is the extinct coefficient, and r_p is the mean radius of the particles. $n(x, y, z)$ can be calculated by:

$$n(x, y, z) = \frac{2\dot{m}}{v_p m_p \pi r^2(z)} \exp\left[-\frac{2\left[x^2 + \left[(y - H/\tan\phi)\sin\phi - (z - H)\cos\phi\right]^2\right]}{r^2(z)}\right] \tag{4.84}$$

In the PF-LDED process, as the particle size is much bigger than the laser wavelength, it is reasonable to assume that the extinction coefficient $Q_{ext} = 1$ [28] and most of the attenuated laser energy are absorbed by the particle. Equation (4.77) is derived from the first-order approximation of the Mie's theory, which calculates the total attenuation or extinction power for the laser beam travels through the powder stream and is valid when $2\pi r_p/\lambda_L < 300$[28]. If the average particle radius would be larger than 20–30 μm and laser wavelength is $\lambda_L = 1.06$(μm), this condition can be satisfied. We should point out that this criterion is only considered when the scattering is included for total attenuated power calculation. In PF-LDED, r_p is normally much larger than λ_L, and the power absorption is the predominant attenuation type.

By seeing Figure 4.24, the attenuated laser beam intensity $I_A(x, y, z)$ may be calculated with the integration of Equation (4.77) over the interaction length in z-axis as:

$$I_A(x, y, z) = I(x, y, z) \exp\left[-\sigma \int_z^{z_p(x, y)} n(x, y, z)dz\right] \tag{4.85}$$

where $z_p(x, y)$ is the upper surface of the powder stream, which is approximated with the powder stream top boundary line $z_p(y)$:

$$z_p(y) = K + (y - K/\tan\phi)\cdot\tan(\phi - \theta), \quad (K = H + r_0 \sin\phi/\tan\theta, \ \phi > 0) \tag{4.86}$$

The powder absorbs energy from the laser beam during their interaction, and the absorbed energy increases the temperature of the particles. Based on the assumption, the temperature increment ΔT for time interval $\Delta t = d_z/v_p \sin\phi$ follows the energy balance equation:

$$\beta I_A(x, y, z)\pi r_P^2 \Delta t = c_p \rho_p \frac{4}{3}\pi r_P^3 \Delta T \tag{4.87}$$

where β is the powder laser absorptivity, c_p is the material-specific heat capacity, and ρ_p is the average particle density (kg/m^3). Integrating Equation (4.87) over the interaction length in z-axis allows the particle temperature $T_p(x, y, z)$ to be calculated as:

$$T_p(x, y, z) = T_0 + \frac{3\beta}{4\sin\phi v_p c_p \rho_p r_P}\left|\int_z^{\bar{z}_p(y)} I_A(x, y, z)dz\right| \tag{4.88}$$

where $\bar{z}_p(y) = z_p(-y)$ is the symmetrical line of $z_p(y)$ about z-axis, and T_0 is the ambient temperature. The path of integration from z to $\bar{z}_p(y)$ was used to approximate the particle traveling distance component in the laser beam transverse direction.

As the powder impinges onto the melt pool, it draws energy to increase its enthalpy s and the heated powder energy intensity $I_p(x, y, z)$ is expressed as a negative energy source [30]:

$$I_p(x, y, z) = c_p\rho(x, y, z)\left[T_p(x, y, z) - T_m\right] \tag{4.89}$$

4.7.2.3 Thermal Conduction on the Substrate

On the substrate surface, the coming energy from the attenuated laser beam $I_A(x, y, z)$ and heated powder flux $I_p(x, y, z)$ are summed up and treated as a bulk heating source. Thus, the resultant energy source intensity $I_{net}(x, y, z)$ is expressed as:

$$I_{net}(x, y, z) = I_A(x, y, z) + I_p(x, y, z) \tag{4.90}$$

For a moving heat source on a semi-infinite workpiece surface, if we use Rosenthal's equation (4.50) and considering the resultant energy intensity distribution, the temperature field on the substrate is derived by integrating Equation (4.56) over the laser beam area based on superposition:

$$T(x, y, z) - T_0 = \frac{1}{2\pi k} \int_{\xi = -r_L(z)}^{\xi = r_L(z)} \int_{\eta = -\sqrt{r_L^2(z) - \zeta^2}}^{\eta = \sqrt{r_L^2(z) - \zeta^2}} \left[\beta_w I_A(\xi, \eta, z) + c_p\rho(\xi, \eta, z)\left(T_p(\xi, \eta, z) - T_m\right)\right]$$

$$\times \frac{\exp\left[-\nu(y - \eta + R)/2\alpha\right]}{R} d\eta d\xi, \qquad R = \sqrt{(x - \xi)^2 + (y - \eta)^2 + z^2} \tag{4.91}$$

where β_w is the substrate laser power absorptivity. In order to approach reality or a more precise platform, the latent heat of fusion, thermocapillary phenomena (Marangoni effect), and the varying laser power absorptivity (Brewster effect) are taken into consideration with the following approximations:

- The effect of latent heat fusion L_f on temperature distribution can be approximated by increasing the specific heat capacity c^* as [31]:

$$c_p^* = \frac{L_f}{T_m - T_0} + c_p \tag{4.92}$$

- The effect of Marangoni flow can be accounted using the modified thermal conductivity k^* as explained in Equation (4.31).
- When the beam is circularly polarized, the absorptivity $\beta_w(\varphi)$, due to Brewster effect (see Section 4.2.4), can be considered as the following equation:

$$\beta_w(\varphi) = \beta_w(0)(1 + \alpha_w\varphi) \tag{4.93}$$

where φ is the inclination angle, $\beta_w(0)$ is the laser power absorptivity for a flat plane, and α_w is a constant coefficient that depends on the material. For the sake of simplification, the inclination angle is approximated based on the clad/track height h and laser beam diameter D:

$$\varphi = \tan^{-1}\left(\frac{h}{D}\right) \tag{4.94}$$

4.7.2.4 Melt Pool and Deposited Track Geometry

Melt pool projection limits on the substrate surface are approximated by the solid–liquid line. Equation (4.90) is simplified as:

$$T(x, y, z) - T_0 = \Gamma(x, y, z) \tag{4.95}$$

As shown in Figure 4.25, points A, B, C, and D are located on the boundary of the melt pool. According to Equation (4.95), the coordinate values can be calculated as:

$$\begin{cases} \Gamma(y_A, 0, 0) = T_m - T_0, y_A > 0 \\ \Gamma(y_B, 0, 0) = T_m - T_0, y_B < 0 \\ \Gamma(0, x_D, 0) = T_m - T_0, x_D > 0 \\ \Gamma(0, x_C, 0) = T_m - T_0, x_C < 0 \end{cases} \tag{4.96}$$

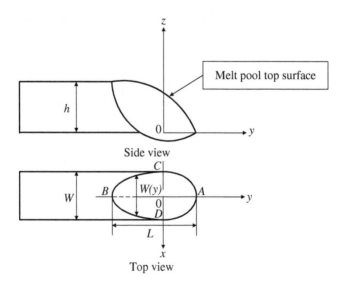

Figure 4.25 Schematic diagram for melt pool geometry and deposited track [27]. Source: Reproduced with permission from Elsevier.

The projection of the melt pool on the substrate plane has the boundary Ω:

$$\Omega = \{y_B \leq y \leq y_A, x_C \leq x \leq x_D\}, \qquad L = y_A - y_B, \qquad W = |x_C| = |x_D| \qquad (4.97)$$

where L and W are the melt pool length and width for the boundary Ω, respectively. Two half ellipses approximate the boundary of the melt pool with the same minor semi-axis of $W/2$ (the left part has a major semi-axis of $|y_B|$ and the right part has major semi-axis of $|y_A|$). With considering the melt pool top surface boundary as a parabolic curve, the melt pool top surface is expressed as [32]:

$$z(x, y) = \left[-h\frac{(y-y_B)^2}{(y_A-y_B)^2} + h \right] \cdot \left(1 - \frac{4x^2}{W^2(y)} \right), \qquad (x \in [x_C, x_D], y \in [y_B, y_A]) \qquad (4.98)$$

The width of the melt pool $W(y)$ can be expressed as:

$$W(y) = \begin{cases} W \times \sqrt{\left(1 - \dfrac{y^2}{y_A^2}\right)}, y \geq 0 \\[4mm] W \times \sqrt{\left(1 - \dfrac{y^2}{y_B^2}\right)}, y < 0 \end{cases} \qquad (4.99)$$

By dividing the substrate surface into small elemental patches dS, each patch is so small that the powder flow density over it is essentially uniform. The powder mass dm that impinged onto dS in an elemental elapsed time can be approximated as:

$$dm = \rho(x, y, 0) \cdot dS \cdot v_p \sin\phi \qquad (4.100)$$

The clad height on dS, the area can be calculated based on the sum of the powder impinging onto it during the dwelling time:

$$h_{dS} = \frac{1}{\rho_p dS} \int_0^{D/v} dmdt \qquad (4.101)$$

Then, for any point within the melt pool boundary, the clad height can be derived as:

$$h(x, y) = \lim_{\delta \to 0} h_{dS} = \frac{2 \sin\phi D\dot{m}}{v\rho_p \pi r^2(z = 0)} \exp\left[-\frac{2\left[x^2 + [(y-H/\tan\phi)\sin\phi + H\cos\phi]^2\right]}{r^2(z = 0)} \right],$$

$$(x, y) \in \Omega$$

$$(4.102)$$

where δ is the largest diameter of the patch dS. Based on the equation, the original clad height h_0 is estimated with an average value over the laser beam spot boundary and the process track height h (along the laser scanning direction) is calculated over the melt pool boundary Ω. The deposited width is approximated with the melt pool width $W(y)$ as expressed by Equation (4.99).

4.7.2.5 Catchment Efficiency

The melt pool has a curved top surface, as derived in the previous section. Particles that fall into the melt pool surface are considered to be effectively integrated with the melt pool. To calculate the melt pool effective area in the coming powder stream, the melt pool is approximated by an inclined surface S_1, which is then projected to powder transverse planes S_2 and S_3 that cuts through the highest and lowest point of the melt pool top surface, respectively. The particles number probability density for per unit time per unit area on powder stream transverse plane can be expressed as:

$$f(x',y',z') = \frac{n(x',y',z')}{\int\limits_{-\infty}^{+\infty}\int\limits_{-\infty}^{+\infty} n(x',y',z')dx'dy'} = \frac{2}{\pi r_p^{\,2}(z')} \exp\left[-\frac{2\left(x'^2 + y'^2\right)}{r_p^{\,2}(z')}\right] \qquad (4.103)$$

Then the catchment efficiency can be derived as the integration of the particle number probability over the melt pool projection area on the powder transverse plane. By integrating Equation (4.103) over the effective projection area A_{S_2} and A_{S_3}, the overall catchment efficiency is derived as:

$$\eta = \frac{\int\limits_{A_{S_2}}\int f(x',y',z')dx'dy' + \int\limits_{A_{S_3}}\int f(x',y',z')dx'dy'}{2} \qquad (4.104)$$

4.7.2.6 Cooling and Solidification Rates[7]

After solving the transient thermal field based on Equation (4.91), the cooling rate $\dot{T}$ for any interest point $X = (x, y, z)$ may be derived as:

$$\dot{T} = \partial T(X,t)/\partial t \qquad (4.105)$$

The temperature gradient G on the solidification front can also be derived from the temperature field solution as:

$$G = \sqrt{\left[\frac{\partial T(X,t)}{\partial x}\right]^2 + \left[\frac{\partial T(X,t)}{\partial y}\right]^2 + \left[\frac{\partial T(X,t)}{\partial z}\right]^2} \qquad (4.106)$$

---

[7] This section is a reproduced summary from one of the authors' previous paper [33] with permission from the publisher. Courtesy of Dr. Yuze Huang.

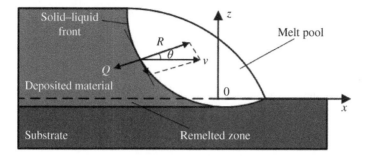

Figure 4.26 Schematic diagram of the solidification front in the longitudinal centerline cross section ($y = 0$) of the single-track deposition. Laser scanning speed v, solidification rate R, and local predominant heat flow Q [33]. Source: Reproduced with permission from Elsevier.

In the PF-LDED process, the laser beam moves continuously, and the solidification front keeps pace with the advancing laser source, as shown in Figure 4.26. As the local solidification front moves along the maximum temperature gradient, which is normal to the solid–liquid front and opposite to the local predominant heat flow direction Q, the solidification rate R may be calculated by [2]:

$$R = v \cos \theta \tag{4.107}$$

where θ is the angle between the local solid–liquid front normal and the laser scanning direction. Since θ is nearly 90° at the bottom part of the melt pool and approach 0° at the top region, the solidification rate will reach the minimum at the bottom region and approach the maximum at the top part.

The angle θ may be derived based on the thermal solution in the longitudinal centerline cross section as [34]:

$$\cos \theta = \frac{\partial T(X,t)/\partial x}{\sqrt{[\partial T(X,t)/\partial x]^2 + [\partial T(X,t)/\partial z]^2}} \tag{4.108}$$

It has to be noted that there is no temperature gradient in the transverse direction ($\partial T(X,t)/\partial y$) at the longitudinal centerline cross section. From Equations (4.107) and (4.108), the solidification rate in the longitudinal centerline cross section may be derived as:

$$R = \frac{\partial T(X,t)/\partial t}{\sqrt{[\partial T(X,t)/\partial x]^2 + [\partial T(X,t)/\partial z]^2}} \tag{4.109}$$

Similarly, for the spatial point $X = (x, y, z)$ in the Cartesian coordinate system, $\cos \theta$ can be directly calculated as:

$$\cos \theta = [\partial T(X,t)/\partial x]/\sqrt{[\partial T(X,t)/\partial x]^2 + [\partial T(X,t)/\partial z]^2 + [\partial T(X,t)/\partial z]^2} \tag{4.110}$$

and the corresponding solidification rate can be expressed as:

$$R = \dot{T}/G \tag{4.111}$$

The dendrite arm spacing (DAS) λ has been established based on the solidification parameters (G and R) by the well-tested Kurz and Fishers' model [35]:

$$\lambda = AG^{-n}R^{-m} \tag{4.112}$$

where A, n, and m are material-dependent parameters. The DAS provides a useful approach to establish the precise effect of solidification conditions on microstructure. Especially, the SDAS can be calculated by setting $n = m$ based on Equation (4.105).

4.7.2.7 Sample Results

Based on the analytical model for PF-LDED, derived from the previous section, several results as samples are provided here. The laser beam is attenuated by the powder stream before it reaches the substrate.

Figure 4.27 shows the original laser beam intensity distribution on the substrate with a maximum intensity of 660 (J/mm^2). The result has incorporated the energy taken by the powder distribution shown in Figure 4.24. Compared to the largest attenuation intensity loss with the initial maximum laser intensity, a maximum laser beam attenuation percentage can be predicted around 4%.

With coupling the attenuated laser beam and the heated powder stream as the resultant moving heat source, the temperature field on the substrate surface is shown in Figure 4.28. Melt pool projection geometry on the substrate surface is approximated to two half-ellipses (the dashed line shown in the figure), which fit well with the calculated melt pool temperature 1563(k) that identify the solid–liquid interface.

This model can be used for real-time temperature evolution prediction. Figure 4.29 shows that the model predicted results are in good agreement with the experimental measurements of the real-time melt pool peak temperature under different laser scanning speeds and various energy densities (PL/vD). Further information/discussion of the results can be found in the author's work in [33].

According to the mathematical model described in the previous section, the melt pool peak temperature map was plotted as a function of laser power and scanning speed, as shown in Figure 4.30. As seen, the maximum temperature occurs at the largest laser power and minimum scanning speed.

According to the thermal model described in this case study, the local thermal history of the SS 316L deposition at different depth locations of the melt pool bead is calculated and shown in Figure 4.31. It shows the local thermal cycles in the single-layer deposition. It is observed that both top and bottom locations will experience a thermal cycle with the motion of the laser beam along the track path. In this situation, the temperature of the top and bottom locations increases instantly from the low ambient value to a high magnitude (e.g. T_s, T_l) and then decreases over time. Further information can be found in [33].

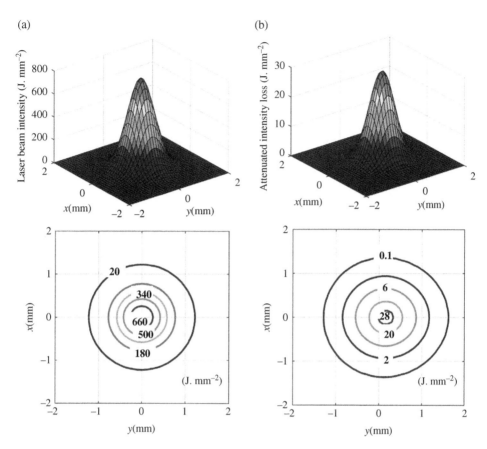

Figure 4.27 Laser beam intensity distribution on the substrate surface: (a) without attenuation and (b) attenuated laser intensity loss by Inconel 625 powder ($\dot{m} = 7(\text{g}/\min)$, $P_L = 1000(\text{W})$, $r_0 = 0.7(\text{mm})$, $\dot{g} = 2.5(\text{dL}/\min)$)[27]. Source: Reproduced with permission from Elsevier.

Temperature gradient (G) and solidification rate (R) are the two main parameters that affect the solidification microstructure. The effect may be illustrated by the solidification map, as presented in Figure 4.32. The solidification map is constructed by G and R in the combination forms with $G \times R$ (cooling rate) and G/R, where the G/R ratio governs the solidification mode. At the same time, their product ($G \times R$) controls the scale of the solidification microstructure [36]. As seen in Figure 4.32, the solidification mode may transform from planar to cellular, columnar dendritic and equiaxed dendritic as the G/R ratio decreases. Additionally, a higher value of $G \times R$ will induce a finer substructure and, consequently, may improve the mechanical properties of the fabricated parts.

The calculated results of $G \times R$ and G/R for different depth locations of SS 316L and Inconel 625 melt pool beads are shown in Figure 4.33. As seen, by keeping all parameters constant but increasing the scanning speed, the cooling rates ($G \times R$) increase significantly for all three locations, while G/R ratios decrease. Moreover, the cooling rate ($G \times R$) increases from the bottom toward the top of the melt pool bead. By contrast, the G/R value decreases from the bottom location to the top region. Further information can be found in the caption of the figure as well as in [33].

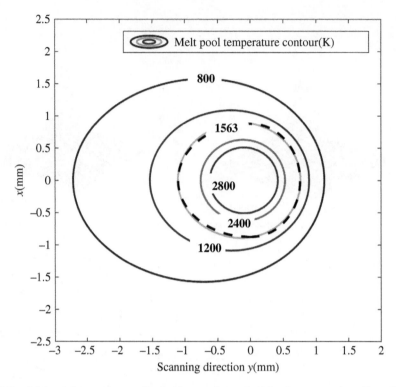

Figure 4.28 Melt pool temperature distribution on Inconel 625 substrate surface ($\dot{m} = 5(\text{g}/\text{min})$, $v = 7.5(\text{mm/s})$, $P_L = 1000(\text{W})$, $r_0 = 0.7(\text{mm})$, $\dot{g} = 2.5(\text{dL}/\text{min})$) [27]. Source: Reproduced with permission from Elsevier.

4.7.3   Numerical Modeling of LDED: Heat Transfer Model[8]

As a case study, a numerical model for the PF-LDED process is described in this section. The laser heat source is assumed to be pulsed. The main objective of developing a 3D transient finite element model of PF-LDED is to investigate the effects of laser pulse shaping, traveling speed, and powder feed rate on the deposited track so-called clad geometry as a function of time.

In order to develop a more precise model, a solution strategy is proposed. In this strategy, the interaction between the powder and the melt pool is assumed to be decoupled, and as a result, the melt pool boundary is first obtained in the absence of powder spray. Once the melt pool boundary is calculated, it is assumed that a layer of coating material based on powder feed rate and elapsed time is deposited on the intersection of the melt pool and powder stream in the absence of a laser beam. The new melt pool boundary is then calculated by thermal analysis of the deposited powder layer, substrate, and laser heat flux.

For the implementation of the proposed solution strategy, a finite element technique is used to develop a novel 3D transient model for PF-LDED. The model can be used to shed some light

---

[8] This Section 4.7.3 is a reproduced summary from one of the authors' previous work [2] with permission from the publisher. Permission conveyed through Copyright Clearance Center, Inc.

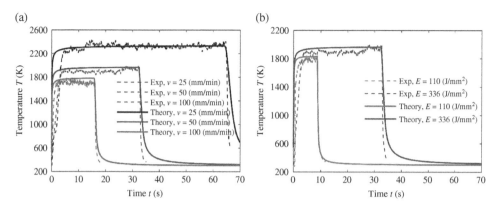

Figure 4.29 Real-time melt pool top surface peak temperature of SS 316L deposition under following conditions. (a) Different laser scanning speeds (laser power 700 W) and (b) different energy densities (E 110 J/mm² with laser power 917 W, scanning speed 200 mm/min; $E = 336$ J/mm² with laser power 700 W, scanning speed 50 mm/min). Powder feed rate 4 g/min [33]. Source: Reproduced with permission from Elsevier.

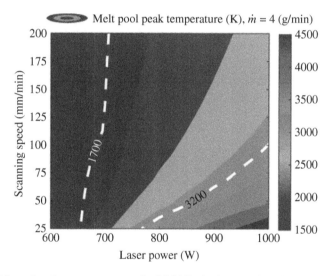

Figure 4.30 Melt pool peak temperature map for SS 316L single-track deposition. The peak temperature was calculated at the transient time moment ($t = L/2v$, track length $L = 30$ mm). The melting temperature (1700 K) and boiling temperature (3200 K) of SS 316L are marked with white dash lines [33]. Source: Reproduced with permission from Elsevier.

on the effects of laser pulse shaping parameters (laser pulse frequency and energy) on the clad geometry when other process parameters such as travel speed, laser pulse width, powder jet geometry, and powder feed rate are constant. Or, it can be used to change multiple parameters simultaneously and understand the correlation among these parameters.

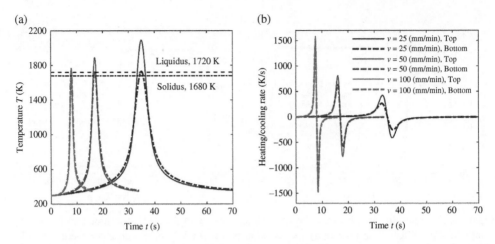

Figure 4.31 Real-time local thermal profiles at different clad height locations during single-layer SS 316L deposition. (a) Temperature cycles. (b) Heating/cooling rate. Laser power 700 W, powder feed rate 4 g/min. The liquidus and solidus temperatures of SS 316L are marked with black dash lines [33]. Source: Reproduced with permission from Elsevier.

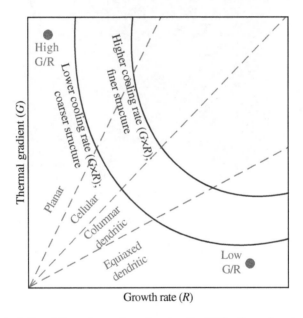

Figure 4.32 Effect of G and R on the mode and scale of solidification microstructure [33]. Source: Reproduced with permission from Elsevier.

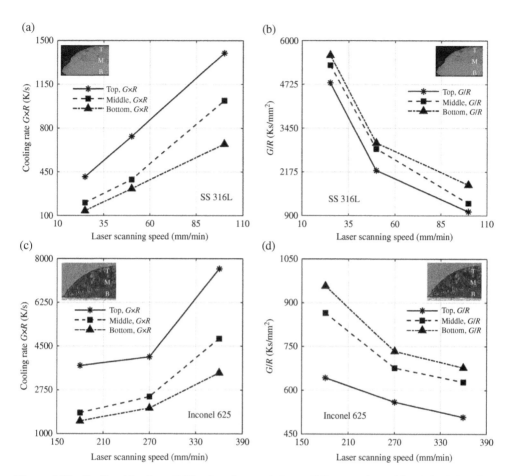

Figure 4.33 Predicted in situ solidification characteristics at different melt pool bead depth locations versus the laser scanning speed for the transient time moment $t = L/2v$. (a), (c) and (b), (d) are the cooling rate $G \times R$ and G/R ratio for SS 316L and Inconel 625, respectively. 700 W laser power and 4g/min powder feed rate were used for SS 316L simulation, and 1000 W laser power and 7 g/min powder feed rate were used for Inconel 625 simulation. Track length $L = 30$ mm. The three different height locations in the final deposit were marked with circular blocks representing the corresponding depth locations in the melt pool bead. T, M, and B represent the top, middle, and bottom positions, respectively.[33]. Source: Reproduced with permission from Elsevier.

For this process, let's assume that a moving laser beam with a Gaussian distribution intensity strikes the substrate at $t = 0$ as it is shown in Figure 4.13. Due to the material added, the clads form on the substrate, as shown in the figure.

The transient temperature distribution $T(x, y, z, t)$ is obtained from the three-dimensional heat conduction in the substrate expressed by Equation (4.16).

As discussed earlier, the boundary conditions for the heat transfer process are expressed in Equation (4.17), when the temperature at $t = 0$ and $t = \infty$, is equivalent to the ambient temperature.

Equation (4.16), along with the boundary conditions, cannot comprehensively express the physics of the process. Therefore, to incorporate the effects of the laser beam shaping, latent heat of fusion, Marangoni phenomena, geometry growing (changing the geometry), and Brewster effect, the following adjustments are considered:

- A pulsed Gaussian laser beam with a circular mode (TEM00) is considered for the beam distribution. The laser power distribution profile or intensity I (W/m^2) is

$$I(r) = \delta I_0 e^{-\left(\frac{\sqrt{2}}{r_l}\right)^2 r^2}$$
(4.113)

where $r = \sqrt{x^2 + y^2}$ and $I_0 = \frac{2}{\pi r_l^2} P_l$ where $P_l = EF$. if the laser is pulsed. r_l is the beam radius (m), I_0 is intensity scale factor (W/m^2), P_l is the average laser power (W), E is the energy per pulse (J), and F is the laser pulse frequency (Hz). When the laser beam is on $\delta = 1$ and when it is $\delta = 0$, the parameter δ is changed based on the laser pulse shaping parameters such as frequency F and width W that is the time that the laser beam is on in one period. For the CW laser, the pulsation parameters are irrelevant.

- The effect of latent heat of fusion on the temperature distribution can be approximated by increasing the specific heat capacity expressed in Equation (4.92).
- The effect of fluid motion due to the thermocapillary phenomena can be taken into account using a modified thermal conductivity for calculating the melt pool boundaries, as explained in Equation (4.31).
- Power attenuation is considered using the method developed by Picasso et al. [36] with some minor modifications. Figure 4.34 shows the proposed geometrical characteristics in the process zone which is used in the development of the following equations. Based on their work:

$$P_1 = P_l \beta_w(\theta)\left(1 - \frac{P_{at}}{P_l}\right)$$
(4.114)

$$P_2 = P_l \eta_p \beta_p \frac{P_{at}}{P_l}\left[1 + (1 - \beta_w(\theta))\left(1 - \frac{P_{at}}{P_l}\right)\right]$$
(4.115)

where P_l is the total power directly absorbed by the substrate (W), P_2 is the power that is carried into the melt pool by powder particles (W), P_{at} is the attenuated laser power by the powder particles (W), $\beta_w(\theta)$ is the substrate absorption factor, β_p is the particle absorption factor, and θ is the angle of the top surface of the melt pool with respect to the horizontal line, as shown in Figure 4.34 (deg). Consequently, the total power absorbed by the substrate P_w (W) is

$$P_w = P_1 + P_2 = \beta P_l$$
(4.116)

where β is the modified absorption factor.

The ratio between the attenuated and average laser power can be obtained by [36]:

$$\frac{P_{at}}{P_l} = \begin{cases} \dot{m}/2\rho_c r_l r_p v_p \cos\theta_{jet} \; if \; r_{jet} < r_l \\ \dot{m}/2\rho_c r_{jet} r_p v_p \cos\theta_{jet} \; if \; r_{jet} \geq r_l \end{cases}$$
(4.117)

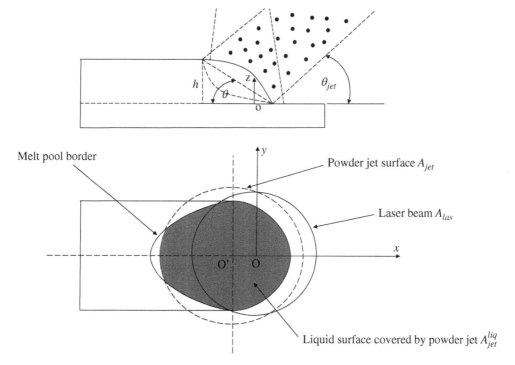

Figure 4.34 Schematic of the laser beam, powder stream and substrate interaction. Source: Adopted from [2].

where $\dot{m}$ is the powder feed rate (kg/s), ρ_c is the powder density (kg/m^3), r_l is the radius of the laser beam on the substrate (m), r_p is the radius of powder particles (m), v_p is the powder particles velocity (m/s), θ_{jet} is the angle between powder jet and substrate (deg), and r_{jet} is radius of powder spray jet (m).

The powder catchment efficiency η_p can be considered as the ratio between the melt pool surface and the area of powder stream (Figure 4.34) as:

$$\eta_p = \frac{A_{jet}^{liq}}{A_{jet}} \tag{4.118}$$

where A_{jet}^{liq} is the intersection between the melt pool area on the substrate and powder stream, and A_{jet} is the cross-section area of the powder stream on the substrate.

If we assume the absorption of a flat plane inclined to a circular laser beam depends linearly on the angle of inclination as shown in Figure 4.34, $\beta_w(\theta)$ can be calculated from Equation (4.92).

- The temperature dependency of material properties and absorption factor on the temperature is taken into account in the model.
- In order to reduce the computational time, a combined heat transfer coefficient for the radiative and convective boundary conditions is calculated by Equation (4.75).

- Marangoni effect can be considered through a correction factor to thermal conductivity as represented in Equation (4.31).
- Using Equation (4.17), the boundary condition can be simplified to:

$$-K(\nabla T.\mathbf{n})|_\Omega = \begin{cases} \dfrac{2}{\pi r_l^2} \delta \beta P_l \left[-\left(\dfrac{\sqrt{2}}{r_l}\right)^2 r^2 \right] - h_c(T-T_0) \text{ if } \Omega \in \Gamma \\[4mm] -h_c(T-T_0) \text{ if } \Omega \notin \Gamma \end{cases} \tag{4.119}$$

In order to bring the physical domain and the boundary conditions to a reliable numerical platform, a solution algorithm is required. A method can be proposed to obtain the deposition geometry in a 3D and the time-dependent LDED process. This proposed numerical solution has two steps, as follows:

1. Obtaining the melt pool boundary in the absence of the powder spray. In this step, the interaction between the powder and melt pool is assumed to be decoupled, and, as a result, the melt pool boundary can be obtained by solving the heat transfer equation described before.
2. Adding a layer of the powder to the substrate in the absence of the laser beam. In this step, once the melt pool boundary is calculated, it is assumed that a layer of coating material based on the powder feed rate, elapsed time, and intersection of melt pool/powder jet is deposited on the substrate. The new deposited layer creates a new tiny object on the previous domain which is limited to the intersection of the powder stream and the melt pool. For each increment in time, its height is given by:

$$\Delta h = \frac{\dot{m}\Delta t}{\pi r_{jet}^2 \rho_c} \tag{4.120}$$

where Δh is the thickness of the deposited layer (m) and Δt is the elapsed time (s). For numerical convergence, the temperature profile of the added layer is assumed to be the same as the temperature of the underneath layer. The new temperature profile of the combined substrate and the powder layer is then obtained by repeating Step 1. Or after deposition of these tiny stacked layers, a standard elliptical shape can be fit to represent the deposited track.

Figure 4.35 shows the sequence of the proposed numerical modeling. The numerical solution is carried out in two different time periods. The first one is the time between two deposition steps, and the second one is the time period for calculating the melt pool area.

Many numerical methods for solving the heat transfer equation have been reported since 1940. The finite element method (FEM) is one of the most reliable and efficient numerical techniques, which has been used for many years. FEM can solve different forms of partial differential equations with different boundary conditions. In this work, the governing partial differential equation (PDE) is highly nonlinear due to material properties with dependency on temperature and a moving heat source with a Gaussian distribution.

To implement the numerical solution strategy, for meshing, the domain should be partitioned into tetrahedrons (mesh elements) or regular. Due to the deposited layer and changes in the substrate geometry, an adaptive meshing strategy should be implemented. Normally, much finer meshes underneath the laser beam and coarser in the surrounding area. A "mesh independency analysis" must be conducted to ensure that the number of meshes is optimum.

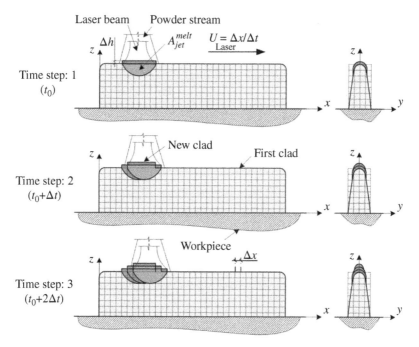

Figure 4.35 Sequence of calculation in the proposed numerical model [37]. Source: Reproduced with permission from Elsevier.

A time-dependent solver should be used to solve the nonlinear time-dependent heat transfer equation. For example, an implicit differential-algebraic equation (DAE) solver with automatic step size control can be suggested. This solver is suitable for solving equations with singular and nonlinear terms.

If the standard elliptical shape is fit to the deposited layer, it will help the convergence of the numerical solution and reduction of the computational time by eliminating the large temperature gradient between the added and underneath layers which cause instability in most of today's numerical solvers.

4.7.3.1   Sample Results

Using the above numerical approach, many results can be extracted. A few examples of a limited set of data is shown in this textbook from [37]. In this work, the material properties of AISI 4340 steel were considered for both the substrate and powder particles. Process parameters are $\dot{m} = 2$ g/min, $U = 1.5$ mm/s, $P = 300$ W, $T_0 = 300$ K, $r_{jet} = 7.5$e-4 m, and $r_l = 7.0$e-4 m. It is assumed that four layers are deposited

In this case, Figure 4.36 shows the maximum temperatures and the stresses for all four layers along with the deposited layers. For instance, for the first layer at the simulation time 2 s, the laser beam is at point P which is located at $x = 8$ mm on the substrate or $x = 3$ mm on the deposition track, as shown in this figure.

Figure 4.37 shows temperature distribution after 9 s for layer one. The maximum temperature is calculated to be around 1894 K.

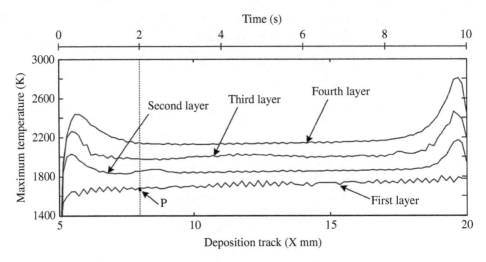

Figure 4.36 Maximum temperatures for each layer, when $\dot{m} = 2$ g/min, $U = 1.5$ mm/s, $P = 300$ W, $T_0 = 300$ K, $r_{jet} = 7.5e-4$ m, $r_l = 7.0e-4$ m [37]. Source: Reproduced with permission from Elsevier.

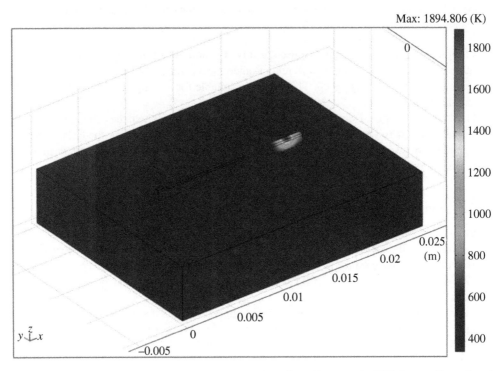

Figure 4.37 Temperature distribution for the second layer deposition at $t = 9$ s [37]. Source: Reproduced with permission from Elsevier.

4.7.4 *Modeling of Wire-Fed E-beam DED (WF-EDED)*

The process model description and development of WF-EDED are relatable to its LDED counterpart involving complex transient heat transfer processes such as conduction, convection, and radiation, as shown in Figure 4.38 and described in Section 4.3. The basis of the energy source for E-beam, in general, is a focused electron beam made up of a large number of electrons. An electron is a particle with mass $m = 9.1\text{E-}31$ kg and negative electrical charge $e = 1.6\text{E-}19$ C. Electrons are produced when a tungsten (W) electrode is heated up to temperatures close to 2500 °C. These electrons are accelerated and collimated using electric and magnetic fields within a vacuum chamber, so that electrons do not scatter due to atmospheric conditions. With kinetic energy of about 60 keV, this high-energy electron beam is made to impinge on the wire material and workpiece, producing heat energy resulting in a melted deposition area. The current density of this beam can be calculated by the Richardson–Dushman thermionic emission equation:

$$J = AT^2 e^{-\frac{W}{\sigma T}} \tag{4.121}$$

where J is the current density of emission (J/m^2), A is Richardson's constant ($\sim$1202 mA/ mm^{2}K^2), T is the temperature (K), W is the work function of the cathode material (J or eV), and σ is the Boltzmann constant (5.67×10^{-8} W/m$^2 \cdot$ K^4).

As explained in the previous section, the electron beam travels at high kinetic energy and is made to impinge on the substrate–wire combination. This kinetic energy is transferred to the material and transformed into internal energy high enough to cause localized melting within the deposition zone resulting in a melt pool. Conduction occurs between the melt pool and surrounding material areas, convection within the melt pool, and radiation from the melt pool to surrounding areas away from the substrate; these are illustrated in Figure 4.38. There also exist some evaporation and condensation within the interaction zone like laser material

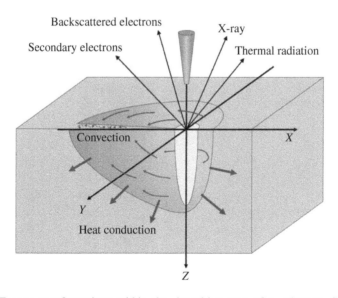

Figure 4.38 Energy transformations within the deposition area of a substrate. Source: Adapted from [38].

processing. As a result of the thermal load and heat transfers, stresses and distortions are induced. Temperature gradients cause inhomogeneous thermal expansion and cooling that have a significant influence on the magnitude of induced stresses and distortions. Like LDED, this complex thermomechanical process can be simplified when reasonable assumptions are made. These assumptions are made based on the insignificance of a physical phenomenon on the overall process model. Convection and radiation are not as significant as conduction. Therefore, they can either be neglected or considered within boundary conditions instead of a full-scale thermofluid CFD analysis. Also, evaporation and condensation can be neglected due to their insignificant contributions in most processes. Another assumption popularly made on the mechanical side of the process model is considering the substrate as an elastic–plastic body neglecting any hardening characteristic. Like LDED, the heat transfer equation can be expressed as Equation (4.16). Q in this equation is the heat source into the system, and some popular models are the rotated Gaussian volumetric model, Goldak double ellipsoidal model, uniform heat generation, and constant heat reservoir (See Chapter 5 for more information about heat source models). From investigations in research, the Gaussian volumetric heat source and Goldak double ellipsoidal model have been known for their accuracy and relatively low computational cost and are therefore the most popular. The rotated Gaussian volumetric energy Q_{GV} can be expressed as [39]:

$$Q_{GV} = -\eta_e P_B \frac{\ln(0.1)}{\pi r_b^2 l_B} e^{\left[4\ln(0.1)\frac{(x^2+y^2)}{d_b^2}\right]} \times \left\{-3\left(\frac{z}{l_b}\right)^2 - 2\left(\frac{z}{l_b}\right) + 1\right\} \tag{4.122}$$

where η_e is the absorption efficiency for electron beam, which is dependent on the material and beam used in WF-EDED, P_B is the power of the electron beam (W), r_b is the radius of the focused beam (m), and l_B is the electron beam's penetration depth (m). The beam's penetration depth has been approximated to be dependent on the beam's voltage and material density thus [40]:

$$l_B = 2.1 \times 10^{-5} \frac{U^2}{\rho} \tag{4.123}$$

U, which is the beam's voltage, is usually a value of about 60 kV.

The Goldak's double ellipsoidal heat source energy Q_{DE} can be expressed as:

$$Q_{DE} = \frac{6\sqrt{3}P_B\eta}{abc\pi\sqrt{\pi}} e^{-\left(\frac{3x^2}{a^2} + \frac{3y^2}{b^2} + \frac{3(z+v_s t)^2}{c^2}\right)} \tag{4.124}$$

where a, b, c are width, depth, and length respectively of ellipsoid; P is the source power (W), η is the absorption efficiency, v_s is the beam's speed (m/s), and t is the time (s).

Radiation occurs as heat loss from melt pool to the vacuum surroundings and the radiative heat transfer coefficient is written as:

$$h_t = \sigma\varepsilon_t(T + T_0)(T^2 + T_0^2) \tag{4.125}$$

where σ is the Stefan–Boltzmann constant (5.67×10^{-8} W/m$^2 \cdot$ K^4), ε_t is the emissivity, and T_0 is the temperature of ambient (K).

Large and uneven temperature distributions give rise to thermal stresses that result in distortions within the built part. Since the printed part is built on a substrate, the part substrate assembly can be regarded as a cantilever assembly with the substrate providing a complete fixture to one end of the part. Neglecting any hardening rule, the part can be approximated to an ideal elastic–plastic body. To obtain the mechanical response of the part, a quasi-static mechanical analysis should be carried out using the thermal history, which can be a coupled or de-coupled thermomechanical solution approach. The constitutive equation for the layer-by-layer process can be written thus [41]:

$$^n\sigma = {}^{n-1}\sigma + \Delta\sigma \tag{4.126}$$

where n is the current layer while $n-1$ is the previous layer. The stress increment can be written as:

$$\Delta\sigma = \Delta C\left({}^{n-1}\varepsilon - {}^{n-1}\varepsilon_p - {}^{n-1}\varepsilon_T\right) + C\left(\Delta\varepsilon - \Delta\varepsilon_p - \Delta\varepsilon_T\right) \tag{4.127}$$

where C is the fourth-order material stiffness tensor. The plastic and thermal strains can be obtained thus:

$$\varepsilon_p = \int_\Omega \frac{\partial f}{\partial\sigma} d\varepsilon_q \tag{4.128}$$

$$f = f_0(\varepsilon_q, T) \tag{4.129}$$

$$\varepsilon_T = \int_\Omega \alpha dT \tag{4.130}$$

where Ω is the material domain, f is the yield function which a function of the material's yield strength f_0 and in turn is dependent on the equivalent plastic stress, ε_p, and temperature T as defined in Equation (4.129). In Equation (4.130), α_t is the coefficient of thermal expansion (1/K).

In implementing these models in a commercial or custom-developed FEA program, the pattern of deposition has to be taken into account. Two major deposition patterns in WF-EDED are reciprocating and unidirectional modes shown in Figure 4.39.

This thermomechanical model can be calibrated and/or validated by several experimental techniques. Residual stresses can be experimentally determined by hole-drilling and XRD methods for a printed sample, while thermocouples or IR Camera can be used to determine *in situ* temperature evolution.

4.7.5   A Stochastic Model for Powder-Fed LDED[9]

In this case study, a stochastic model based on a Hammerstein–Wiener model structure is discussed. The Hammerstein–Wiener model is one of the structures used in nonlinear system identification. Figure 4.40 shows the proposed model structure where f and g are the Hammerstein and Wiener memoryless nonlinear elements, respectively. The nonlinear memoryless elements' addition allows us to incorporate our physical knowledge of the process into the model while keeping the overall model as simple as possible.

---

[9] This Section 4.7.5 is a reproduced summary from one of the authors' previous work [2] with permission from the publisher. Permission conveyed through Copyright Clearance Center, Inc.

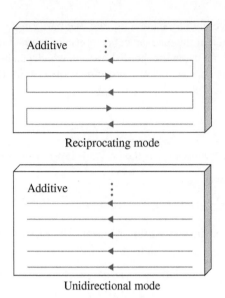

Figure 4.39 Deposition patterns in wire-fed EBAM. Source: Adapted from [39].

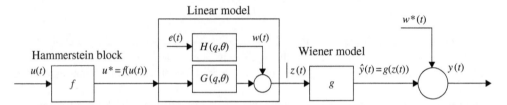

Figure 4.40 Proposed Hammerstein–Wiener nonlinear structure (Source: Republished with permission of Taylor & Francis Group, from [2]; permission conveyed through Copyright Clearance Center, Inc.)

In order to find the nonlinear elements of the model (f and g), it is shown an inverse dependency of the deposited height on the square root of the relative beam velocity. It is also shown the dependency of temperature and clad height on a sigmoid function of the beam velocity. As a result, it can be inferred that the clad height depends on at least two nonlinear functions in the form:

$$h = f\left(\frac{1}{\sqrt{v}}, \frac{1}{1 + \exp(-v)}\right) \tag{4.131}$$

This reciprocal relationship between the laser velocity v and height h is also evident from experimental results. Therefore, the Hammerstein–Wiener nonlinear parts of the model can be defined as:

$$f = \frac{1}{\sqrt{v}} \tag{4.132}$$

and

$$g = \frac{c_1}{c_2 + c_3 \exp(c_4 z)} \tag{4.133}$$

where c_1 to c_4 are constant coefficients that should be identified and z is a discrete function operator in the system identification domain (see Figure 4.40)

In the Hammerstein–Wiener model structure, disturbances are modeled as additive terms in the linear part and the output signals, as shown by $w(t)$ and $w*(t)$ in Figure 4.40, respectively [42, 43].

Based on these assumptions and the notation of Figure 4.32, the output of the linear block is

$$z(t) = G(q, \theta)u^* + H(q, \theta)e(t) \tag{4.134}$$

where $G(q, \theta)$ and $H(q, \theta)$ are the linear models (rational functions) of the shift operator q for the system and disturbances, respectively, and $e(t)$ is assumed to be white noise. The other parameters are shown in Figure 4.40.

Removing the bias problem from Equation (4.134) (see [44] for details), it can be written as:

$$A(q)z(t) = B(q)u^*(t) + e(t) \tag{4.135}$$

where

$$A(q) = 1 + \sum_{i=1}^{nf} a_i q^{-i} \tag{4.136}$$

and

$$B(q) = 1 + \sum_{i=1}^{n-(nk+nb-1)} b_i q^{-i} \tag{4.137}$$

and nk, nf, and nb are orders of the delay, denominator, and numerator, respectively.

Using Equation (4.137) and applying the operator q, $z(t)$, Equation (4.134) can be written as:

$$z(t) = -\sum_{i=1}^{nf} a_i z(t-i) + \sum_{i=1}^{n-(nk+nb-1)} b_i u^*(t-i) + e(t) \tag{4.138}$$

Assuming the Wiener nonlinear part in Figure 4.40 is invertible, the last equation can be converted to:

$$z(t) = -\sum_{i=1}^{nf} a_i g^{-1}(\hat{y}(t-i)) + \sum_{i=1}^{n-(nk+nb-1)} b_i u^*(t-i) + e(t) \tag{4.139}$$

and $y(t)$ becomes

$$y(t) = g(z(t)) + w^*(t) \tag{4.140}$$

In general, all linear and nonlinear parameters are included in the optimization procedure to minimize the output error. However, the implementation of this algorithm usually suffers from numerical divergence. In the following, an improved algorithm is proposed to predict the model

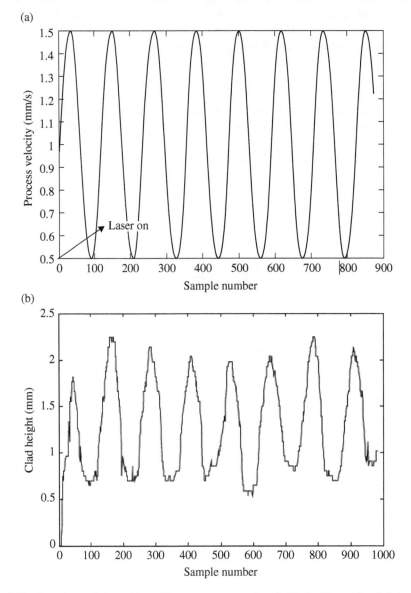

Figure 4.41 Experimental data: (a) multistep process speed and (b) clad/deposition height. Source: Republished with permission of Taylor & Francis Group, from [2]; permission conveyed through Copyright Clearance Center, Inc.

parameters. Since the Hammerstein part of the system is assumed to be known for the LDED process, the algorithm only identifies the linear and Wiener nonlinear parts.

The steps of the algorithm are

1. Remove the mean value from the output data $y(t)$.
2. Ignore the Wiener block g; guess the order of linear part $\hat{G}$.
3. Use $u*(T)$ and $y(t)$ as the input and output; find the primary linear model $\hat{G}$.
4. Repeat steps 2 and 3 by changing the order of the linear model to minimize $\sum (y(t) - z(t))^2$ where $z(t)$ is the output of the linear system.
5. Find the nonlinear parameters of $g(z)$ based on $z(t)$ using Gauss–Newton minimization method.
6. Find $z^*(t) = g^{-1}(y(t))$.
7. Re-identify the linear model based on $u*(t)$ and $z*(t)$.
8. Repeat from Step 5 until $\left| \epsilon_1^k + \epsilon_2^k - \left(\epsilon_1^{k-1} + \epsilon_2^{k-1} \right) \right| \leq \delta$ where k and δ are the iteration index and a small positive number, respectively, and: $\epsilon_1^k = \|\theta_k\|, \epsilon_2^k = \left\| y_k(t) \right\|$.

As an example, a set of data were fed to this algorithm. A typical data is shown in Figure 4.41.

The optimum value for the order of the linear subsystem was 7 for the denominator and 4 for the numerator with a delay of 1. A sample time of 0.08 s was used in the identification process. After system identification using the algorithm described above, the coefficients for the LDED at the process condition when $P = 300\,\text{W}$ and mass flow is 1 g/min were derived as

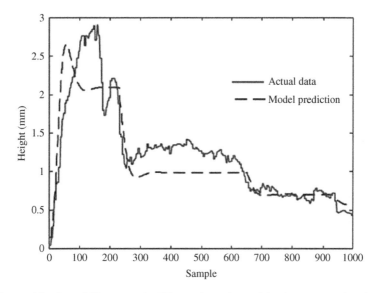

Figure 4.42 Verification of Hammerstein–Wiener dynamic model using unseen data in the training phase. Source: Republished with permission of Taylor & Francis Group, from [2]; permission conveyed through Copyright Clearance Center, Inc.

a1 = −1.0764, a2 = −0.0672, a3 = 0.0610, a4 = −0.0069, a5 = 0.0024, a6 = 0.0235, a7 = 0.0924, b1 = 0.0035, b2 = 0.0700, b3 = −0.0265, b4 = −0.0196, c1 = 0.157, c2 = 0.0020, c3 = 1.0125, and c4 = −1.6722.

Figure 4.42 shows the outcome of the dynamic model for a set of new data not used in training. As seen, the model can predict the process well. This model can be used for the development of the controller.

4.8 Summary

This chapter covered critical physics and modeling principles of DED. As mentioned before, this book's goal is not a literature survey but to consolidate major physics governing various DED processes. To this end, the basic of laser and E-beam material processing interaction physics was discussed. The basic governing physics of DED was explained. Different approaches for analytical and numerical modeling of DED were disclosed.

Furthermore, multiple case studies on modeling and analysis of DED based on lumped, comprehensive analytical, numerical, and experimental-based stochastic approaches were explained. It was explained that the role of experimental stochastic and machine learning models to DED is essential as they can rapidly predict the process. This will open up an opportunity to develop adaptive quality assurance algorithms and closed-loop controllers for DED that will be discussed in Chapter 11.

References

1. "Overview of Modulated and Pulsed Diode Laser Systems." Market Tech Inc, Scotts Valley, 2018.
2. E. Toyserkani, A. Khajepour, and S. Corbin, *Laser cladding*, CRC Press, 2004.
3. W. M. Steen, "Laser material processing – An overview," *J. Opt. A Pure Appl. Opt.*, vol. 5, no. 4, pp. S3–S7, 2003.
4. W. M. Steen and J. Mazumder, *Laser material processing*, Fourth edition, Springer, 2010.
5. T. Higgins, "Non-linear optical effects are revolutionising electro optics," *Laser FocusWorld*, Peconic Publishers, pp. 67–74, 1994.
6. N. A. Kent, "REVIEW: The Theory of Optics by Paul Drude," *Astrophys. J.*, vol. 18, p. 75, 1903.
7. L. Rayleigh, "On the transmission of light through an atmosphere containing small particles in suspension, and on the origin of the blue of the sky," *Philos. Mag. J. Sci.*, vol. 47, no. 287, pp. 375–384, 1899.
8. F. Hansen and W. W. Duley, "Attenuation of laser radiation by particles during laser materials processing," *J. Laser Appl.*, vol. 6, no. 3, p. 137, 1994.
9. Chamina, "Electron Beam – Specimen Interaction." [Online]. Available: http://academic.uprm.edu/pca-ceres/Courses/CHAMINA/HO5.pdf.
10. M. Markl, R. Ammer, U. Ljungblad, U. Rüde, and C. Körner, "Electron beam absorption algorithms for electron beam melting processes simulated by a three-dimensional thermal free surface lattice boltzmann method in a distributed and parallel environment," *Proc. Comput. Sci.*, vol. 18, pp. 2127–2136, 2013.
11. K. Kanaya and S. Okayama, "Penetration and energy-loss theory of electrons in solid targets," *J. Phys. D. Appl. Phys.*, vol. 5, pp. 43–58, 1972.
12. H. S. Carslaw, J. C. Jaeger, L. R. Ingersoll, O. J. Zobel, A. C. Ingersoll, and J. H. Van Vleck, "Conduction of heat in solids and heat conduction," *Phys. Today*, vol. 1, no. 7, p. 24, 1948.
13. A. F. A. Hoadley and M. Rappaz, "A thermal model of laser cladding by powder injection," *Metall. Trans. B*, vol. 23, pp. 631–642, 1992.

14. C. Chan, J. Mazumder, and M. M. Chen, "Two-dimensional transient model for convection in laser melted pool.," *Metall. Trans. A, Phys. Metall. Mater. Sci.*, vol. 15, pp. 2175–2184, 1983.

15. X. F. Peng, X. P. Lin, D. J. Lee, Y. Yan, and B. X. Wang, "Effects of initial molten pool and Marangoni flow on solid melting," *Int. J. Heat Mass Transf.*, vol. 44, pp. 457–470, 2001.

16. C. Lampa, A. F. H. Kaplan, J. Powell, and C. Magnusson, "An analytical thermodynamic model of laser welding," *J. Phys. D. Appl. Phys.*, vol. 30, pp. 1293–1299, 1997.

17. A. Kaplan, "Keyhole welding: the solid and liquid phases," in Springer Series in *Materials Science*, Springer, 2017.

18. R. Kobayashi, "Modeling and numerical simulations of dendritic crystal growth," *Phys. D Nonlinear Phenom.*, vol. 63, no. 1–3, pp. 410–423, 1993.

19. V. L. Streeter and J. Kestin, "Handbook of fluid dynamics," J. Appl. Mech, vol. 28, no. 4, p. 640, 1961.

20. J. Lin, "Simple model of powder catchment in coaxial laser cladding," *Opt. Laser Technol.*, vol. 31, no. 3, pp. 233–238, 1999.

21. Comsol, M ultiphysics C OMSOL, 2010. Available: https://cdn.comsol.com/doc/4.3/COMSOL_Release-Notes.pdf.

22. T. Chande and J. Mazumder, "Two-dimensional, transient model for mass transport in laser surface alloying," *J. Appl. Phys.*, 1985.

23. D. Rosenthal, "The theory of moving source of heat and its application to metal transfer," *ASME Trans.*, vol. 43, pp. 849–866, 1946.

24. C. Adams, "Cooling rates and peak temperatures in fusion welding," *Weld. J.*, vol. 37, pp. 210-s–215-s, 1958.

25. J. Goldak, M. Bibby, J. Moore, R. House, and B. Patel, "Computer modeling of heat flow in welds," *Metall. Trans. B*, vol. 17, pp. 587–600, 1986.

26. L. X. Yang, X. F. Peng, and B. X. Wang, "Numerical modeling and experimental investigation on the characteristics of molten pool during laser processing," *Int. J. Heat Mass Transf.*, vol. 44, no. 23, pp. 4465–4473, 2001.

27. Y. Huang, M. B. Khamesee, and E. Toyserkani, "A comprehensive analytical model for laser powder-fed additive manufacturing," *Addit. Manuf.*, vol. 12, 2016.

28. A. Frenk, M. Vandyoussefi, and J. Wagniere, "Analysis of the laser-cladding process for stellite on steel," *Mater. Trans. B*, vol. 28, pp. 501–550, 1997.

29. A. J. Pinkerton, "An analytical model of beam attenuation and powder heating during coaxial laser direct metal deposition," *J. Phys. D. Appl. Phys.*, vol. 40, no. 23, pp. 7323–7334, 2007.

30. A. J. Pinkerton, R. Moat, K. Shah, L. Li, M. Preuss, and P. J. Withers, "A verified model of laser direct metal deposition using an analytical enthalpy balance method," in 26th International Congress on Applications of Lasers and Electro-Optics, ICALEO 2007 – Congress Proceedings, 2007.

31. S. Brown and H. Song, "Finite element simulation of welding of large structures," *J. Manuf. Sci. Eng. Trans. ASME*, vol. 4, pp. 501–533, 1992.

32. A. Fathi, E. Toyserkani, A. Khajepour, and M. Durali, "Prediction of melt pool depth and dilution in laser powder deposition," *J. Phys. D. Appl. Phys., vol.* 39, no. 12, 2006.

33. Y. Huang et al., "Rapid prediction of real-time thermal characteristics, solidification parameters and microstructure in laser directed energy deposition (powder-fed additive manufacturing)," *J. Mater. Process. Technol.*, vol. 27, p. 12, Article Number: 116286, 2019.

34. H. L. Wei, J. Mazumder, and T. DebRoy, "Evolution of solidification texture during additive manufacturing," *Sci. Rep.*, vol. 5, Article Number: 16446, 7pp., 2015.

35. W. Kurz and D. J. Fisher, "Dendrite growth at the limit of stability: tip radius and spacing," *Acta Metall.*, vol. 29, no. 1, pp. 11–20, 1981.

36. M. Picasso, C. F. Marsden, J. D. Wagniere, A. Frenk, and M. Rappaz, "A simple but realistic model for laser cladding," *Metall. Mater. Trans. B*, vol. 25, pp. 281–291, 1994.

37. M. Alimardani, E. Toyserkani, and J. P. Huissoon, "A 3D dynamic numerical approach for temperature and thermal stress distributions in multilayer laser solid freeform fabrication process," *Opt. Lasers Eng.*, vol. 45, no. 12, 2007.

38. U. Dilthey, "Electron beam welding," in *New Developments in Advanced Welding*, Woodhead Publishing Limited, pp. 198–228, 2005.

39. Z. Chen, H. Ye, and H. Xu, "Distortion control in a wire-fed electron-beam thin-walled Ti-6Al-4V freeform," *J. Mater. Process. Technol.*, vol. 251, pp. 12–19, 2018.

40. Y. Qin et al., "Temperature-stress fields and related phenomena induced by a high current pulsed electron beam," *Nucl. Instr. Methods Phys. Res. Sect. B Beam Interact. Mater. Atoms*, vol. 225, no. 4, pp. 544–554, 2004.

41. E. Denlinger, Residual stress and distortion modeling of electron beam direct manufacturing Ti-6Al-4V, *Thermo-mechanical modeling of additive manufacturing*, M. Gouge and P. Michaleris (eds.), Butterworth-Heinemann, 2017.

42. E. W. Bai, "An optimal two stage identification algorithm for Hammerstein-Wiener nonlinear systems," in Proceedings of the American Control Conference, 1998.

43. S. A. Billings and S. Y. Fakhouri, "Identification of nonlinear systems using the wiener model," *Electron. Lett.*, vol. 13, no. 17, pp. 502–504, 1977.

44. L. Ljung, *"System identification, theory for the user"*, Prentice Hall, 1999.

5

Powder Bed Fusion Processes

Physics and Modeling

Learning Objectives

At the end of this chapter, you will be able to:

- Gain a clear understanding of the physics behind the powder bed fusion processes.
- Identify the main governing equations for heat transfer by conduction and fluid flow in the melt pool.
- Gain an understanding of the numerical models and techniques employed in the description of the heat transfer during laser powder bed fusion.

5.1 Introduction and Notes to Readers

There are substantial similarities between the physics and governing equations of directed energy deposition (DED) processes, explained in Chapter 4, and physics and governing equations of powder bed fusion (PBF) processes. Many equations highlighted in Chapter 4 will be used and customized for PBF. If required, similar equations in Chapter 4 are listed again in Chapter 5 to back the discussion and maintain the continuity. If not required, the equation number in Chapter 4 is simply provided. In this chapter, two main PBF processes will be discussed:

Metal Additive Manufacturing, First Edition. Ehsan Toyserkani, Dyuti Sarker, Osezua Obehi Ibhadode,
Farzad Liravi, Paola Russo, and Katayoon Taherkhani.
© 2022 John Wiley & Sons Ltd. Published 2022 by John Wiley & Sons Ltd.

(i) laser powder bed fusion (LPBF) and (ii) electron beam powder bed fusion (EB-PBF). For each process, a few case studies are provided. It should be noted that the goal of this textbook is not a literature survey but to provide basic vital information for engineering students to link AM processes to fundamental physics they learn in undergraduate programs.

5.2   Physics of Laser Powder bed Fusion (LPBF)

The underlying physics of LPBF needs to be taken into consideration to improve the quality of the printed parts. In particular, the quality of a part is influenced by over 130 parameters such as laser power, scan velocities, scanning strategies, core/skin parameters, layer thickness, hatching distance, laser–powder interaction, and many more [1]. While describing the physics of LPBF, simulations and modeling are used to achieve a better understanding of the process researchers have developed an integrated model for LPBF through combining four different submodels: thermal, metallurgical, chemical, and melting/sintering submodels [2]. The development of the integrated model requires identifying and modeling all the factors influencing the LPBF process. Table 5.1 lists the direct input material parameters identified for each of the four submodels. For more details, see Chapter 2.

In the LPBF process, four stages can be identified based on the laser–powder bed interaction (see Figure 4.1 in Chapter 4):

1. Absorption,
2. Heat conduction,
3. Melting and shrinkage,
4. Resolidification.

To understand how the process parameters affect the quality of the build and to evaluate distortions and residual stresses, finite element modeling is often used. Of particular interest to researchers is the thermal modeling, which can be used to perform the mechanical analysis, the characterization of the melt pool, and the microstructures of the builds and also to guide researchers in the selection of the printing parameters [4].

Table 5.1   Material input parameters for LPBF [2, 3].

Parameters	Description
Optical parameters	Reflection/absorption, optical penetration
Characteristics of powder grains	Particle size, particle size distribution, particle shape, particle roughness
Thermal properties	Thermal conductivity, specific heat, latent heat, melting temperature, thermal expansion, melting temperature, boiling temperature
Chemical properties	Reaction enthalpy, chemical solvency
Metallurgical properties	Oxidation potential, alloy composition, diffusion coefficients, liquid–solid range
Rheological properties	Viscosity, surface tension
Mechanical properties	Elastic modulus, yield point, tensile strength

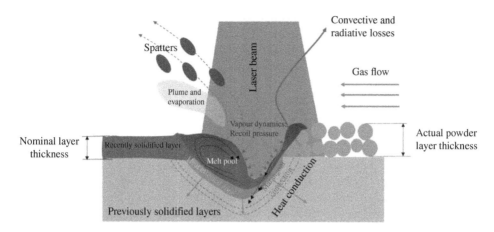

Figure 5.1   Schematic of LPBF, showing physical phenomena surrounding the melt pool.

The physics underlying LPBF is very complex but very the same as Laser Directed Energy Deposition (LDED) that includes absorption, reflection, radiation, heat transfer, phase transformations, mass transportation within melt pool, chemical reactions, etc. (see Chapter 4 for further information). Figure 5.1 shows the schematic of the process where several physical phenomena are highlighted. Generally, the powder bed absorbs a substantial extent of heat during the heating process, which causes an increase in the temperature and the melting of the powder. This leads to the formation of a melt pool where a strong fluid-flow exists, and it is driven by the surface tension gradient and capillary effects, known as Marangoni convection. This liquid metal flow is highly dynamic, and consequently, the shape of the melt pool is changes; thus in LPBF, the laser–powder interaction needs to be taken into consideration more carefully. The particle size distribution and packing density of the powder bed affect the laser beam's absorptance; moreover, the shape of the melt pool, the heat transfer, and fluid flow are also affected by the powder particle distribution [4–6]. In the next section, the equations governing the thermomechanics in LPBF are presented. These equations are the basics of the modeling approaches used to gain a better understanding of the physics of the LPBF process.

5.2.1   Heat Transfer in LPBF: Governing Equations and Assumptions

The heat transfer process involved in LPBF can be described as a result of different mechanisms:

1. Diffusion of the energy absorbed through laser radiation onto the highly conductive particles.
2. When the particles absorb adequate energy, heat is diffused to a low conducting area away from the particles.
3. There is thermal conduction between the solid particles of the powder and also between the trapped atmospheric particle such as air and protective gas.

4. Thermal radiation between the surfaces of the powder particles and between the neighboring atmospheric particles.
5. Thermal conduction over the fluid film joined to a surface.
6. Conduction, radiation and convection (thermal) from the melt pool.

Referring to Chapter 4, the laser–powder interaction can be described by the classical theory of electron structure of matter assuming the formation of an "electron gas made of the free electrons," which move between the ions of the crystal lattice. Upon the absorptance of the incident laser energy by the metal particles, the free electrons adsorb this energy and through their oscillation, they transmit the energy to the neighboring electrons, ions, and lattice defects. The heat transfer process involved during the laser–matter interaction depends on the pulse duration or irradiation time (τ_p). For irradiation time less than 1 ms, a nonthermal model may be used, but for longer than 1 ms, the process can be described by the classical heat transfer theory [3, 7, 8]. Most of the unsteady-state heat transfer models have to solve Fourier's second law (see Chapter 4).

The heat balance equation can be written as:

$$heat_{in} - heat_{loss} - heat_{conducted\ out} = heat_{accumulated} + heat_{generated} \tag{5.1}$$

The different rates of conduction and convention influence the difference between the heat in and the heat out. If we consider the heat flow by conduction in the x-direction at the absence of any losses, and when the heat source moves along the x-axis, we can rewrite Equation (5.1) as follows:

$$(heat_{in} - heat_{out}) = \left\{ -k\frac{\delta T}{\delta x} + k\left[\frac{\delta T}{\delta x} + \frac{\delta\left(\frac{\delta T}{\delta x}\right)}{\delta x}\Delta x\right] \right\}\Delta y\Delta z = K\nabla^2 T \Delta x\Delta y\Delta z$$

$$\tag{5.2}$$

Convection can be expressed by the following equation:

$$Heat_{accumulated} = -\rho c_p U_x \frac{\partial T}{\partial x}\Delta x\Delta y\Delta z \tag{5.3}$$

$$Heat_{generated} = Q\Delta x\Delta y\Delta z \tag{5.4}$$

Therefore, the total balance is

$$K\nabla^2 T - \rho c_p U_x \nabla T - \rho c_p \frac{\partial T}{\partial t} = -Q \tag{5.5}$$

$$-\frac{U_x}{\alpha}\nabla T - \nabla^2 T - \frac{1}{\alpha}\frac{\partial T}{\partial t} = -\frac{Q}{K} \tag{5.6}$$

where Q is power generation per unit volume of the substrate (W/m^3), K is thermal conductivity (W/m·K), c_p is specific heat capacity (J/kg·K), ρ is density (kg/m^3), t is time (s), U is the travel velocity (process speed) (m/s), and α is the thermal diffusivity (m^2/s) [9, 10].

In LPBF, unlike LDED, Q can be considered as a volumetric heat source added to the physical domain. Due to the gap between particles and also the formation of keyhole due to Marangoni effects (see Figure 5.1), the laser beam will be scattered and penetrated directly into the bed, thus its reflection scattered inside the hole and voids between particles (see Figure 5.1). Due to this reason, a volumetric heat source can present this phenomenon very well. There are eight popular heat source models used for LPBF, where these source models represent Q in Equation (5.6). In the following section, these eight heat source models are explained.

5.2.1.1   Volumetric Heat Source Models for LPBF

The modeling of LPBF is extremely dependent on the heat source information since they influence the precision, accuracy, sensitivity, and reliability of the models. The developed heat sources can be distinguished into two groups:

1. Geometrically modified group (GMG), in which the shape of the heat sources is modeled using different geometries (i.e. cylinder, semi-ellipsoid, semi-sphere, etc.) and,
2. Absorptivity profile group (APG), in which the heat sources have the general form of a two-dimensional Gaussian distribution placed on top of the surface, and the laser beam is adsorbed along with the powder layers.

Figure 5.2 shows eight different heat source models used for modeling. These heat source models will be discussed in the following sections.

GMG Group
1. **Cylindrical Heat Source Model**: A schematic for this type of GMG heat source model is displayed in Figure 5.2a. The model considers the optical penetration depth (OPD) of the laser beam into the powder bed [12]. The OPD is defined as the depth at which the laser intensity is 36.8% (1/e) of the laser energy absorbed at the powder bed surface. According to this model, the power per volume of the heat source or so-called volumetric intensity is described as:

$$Q(x, y, z) = \frac{\beta P}{S.\alpha_{OPD}.OPD} \tag{5.7}$$

where x, y, and z are the spatial variables (m), β is the absorptivity of the laser beam, the laser power is indicated by P (W), S is the area of the laser spot (m^2), OPD is chosen as the layer thickness (m), and α_{OPD} is a correction factor for the assumed OPD. The term $S \cdot \alpha_{OPD} \cdot OPD$ is the volume that interacts with the laser beam. In one study, an OPD of 20 μm and α of 1 were used due to a lack of data for 17-4 PH stainless steel [11, 12].
2. **Semi-spherical Heat Source Model**: The intensity of a laser beam in a two-dimensional heat source model has a Gaussian distribution; however, a more precise model is obtained considering a three-dimensional heat source. A schematic of the semi-spherical heat source model is given in Figure 5.2b. The energy density in 3D has a semi-spherical Gaussian distribution, which could be expressed as:

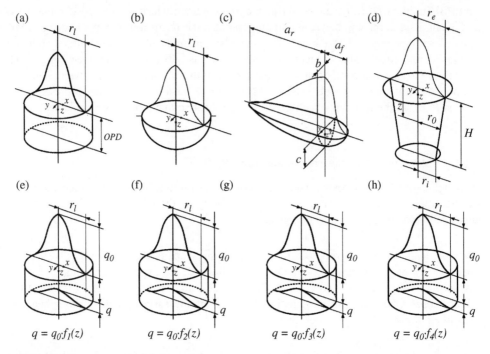

Figure 5.2 Heat source models schematics: (a) cylindrical shape; (b) semi-spherical shape; (c) semi-ellipsoidal shape; (d) conical shape, (e) radiation transfer method; (f) ray-tracing method; (g) linearly decaying method; (h) exponentially decaying method. Source: Reprinted from [11] with permission from Elsevier.

$$Q(x, y, z) = q_0 e^{\left[-2\frac{x^2 + y^2 + z^2}{r_l^2} \right]}$$

(5.8)

where r_l is the radius of the laser beam (m), x, y, and z are the coordinates of the heat source, and q_0 is a coefficient derived by the energy balance. Since the integration of power over the semi-infinite domain should be equal to the total energy input, based on energy conservation,

$$\beta P = \int_0^\infty \int_{-\infty}^\infty \int_{-\infty}^\infty q_0 e^{\left[-2\frac{x^2 + y^2 + z^2}{r_l^2} \right]} dx\, dy\, dz$$

(5.9)

where β is the laser beam absorptivity and P is the power of the laser source (W). The coefficient q_0 can be written as:

$$q_0 = \frac{2^{5/2} \beta P}{\pi^{3/2} r_l^3}$$

(5.10)

Therefore, the semi-spherical power per volume distribution is

$$Q(x,y,z) = \frac{2^{5/2}\beta P}{\pi^{3/2} r_l^3} e^{\left[-2\frac{x^2+y^2+z^2}{r_l^2}\right]}$$

(5.11)

3. **Semi-ellipsoidal Heat Source Model:** The semi-ellipsoidal heat source model proposed in [13], takes into consideration that the melt pool in LPBF is non-spherical. A more precise simulation of the melt pool dimensions is obtained using the Gaussian distribution of an ellipsoid with the center at (0,0,0) and semi-axes a, b, and c parallel to the coordinate axes x, y, and z, respectively:

$$Q(x,y,z) = \frac{2^{5/2}\beta P}{\pi^{3/2} abc} e^{\left[-2\left(\frac{x^2}{a^2}+\frac{y^2}{b^2}+\frac{z^2}{c^2}\right)\right]}$$

(5.12)

A schematic of the semi-ellipsoidal heat source model shape is displayed in Figure 5.2c. As depicted in the schematic, a double ellipsoidal energy density distribution is proposed since the front part (indicated with the ellipsoid's subscript f) could differ from the rear part (subscript r). In particular, the front power distribution can be expressed as:

$$Q_f(x,y,z) = f_f \frac{2^{5/2}\beta P}{\pi^{3/2} a_f bc} e^{\left[-2\left(\frac{x^2}{a_f^2}+\frac{y^2}{b^2}+\frac{z^2}{c^2}\right)\right]}$$

(5.13)

And the rear power distribution is

$$Q_r(x,y,z) = f_r \frac{2^{5/2}\beta P}{\pi^{3/2} a_r bc} e^{\left[-2\left(\frac{x^2}{a_r^2}+\frac{y^2}{b^2}+\frac{z^2}{c^2}\right)\right]}$$

(5.14)

where a_f and a_r are the front and rear semi-axes, as shown in Figure 5.2c.

4. **Conical Heat Source Model:** The conical shape heat model has been used for the simulation of the heat source in the LPBF process. A schematic of the shape used in this model is depicted in Figure 5.2d. The power distribution, according to this model, can be written as:

$$Q(x,y,z) = q_0 e^{\left[-2\frac{x^2+y^2}{r_0^2}\right]}$$

(5.15)

$$r_0(z) = r_e + \frac{z}{H}(r_e - r_l)$$

(5.16)

As shown in Figure 5.2d, r_e and r_l are the top and bottom radius of the cone (m), respectively, while H is the height of the cone (m). The integration of the power per volume over the conical shape domain should be equal to the total power input, therefore:

$$\int_{-H}^{\infty}\int_{-\infty}^{\infty}\int_{-\infty}^{\infty} q_0 e^{\left[-2\frac{x^2+y^2}{r_0^2}\right]} dx\,dy\,dz = \beta P \tag{5.17}$$

From this equation, it is possible to derive q_0, which can be expressed as:

$$q_0 = \frac{6\beta P}{\pi H\left(r_e^2 + r_e r_l + r_l^2\right)} \tag{5.18}$$

If we substitute the term q_0 in the expression of the power distribution, the heat source expression will be

$$Q(x, y, z) = \frac{6\beta P}{\pi H\left(r_e^2 + r_e r_l + r_l^2\right)} e^{\left[-2\frac{x^2+y^2}{r_0^2}\right]} \tag{5.19}$$

APG Group

The APG general form, as discussed previously, is that of a two-dimensional Gaussian distribution on top of the surface and the laser beam is absorbed along the depth of the powder bed. The intensity distribution can be written as:

$$Q(x, y, z) = \frac{2P}{\pi r_l^2} e^{\left[-2\frac{x^2+y^2}{r_0^2}\right]} \frac{d\beta}{dz} \tag{5.20}$$

The term $\frac{d\beta}{dz}$ is the absorptivity profile function and it can be replaced with the parameter $f(z)$. The term $\beta(z)$ is the absorptivity coefficient function.

1. **Radiation Transfer Equation Method**: We previously discussed the radiation transfer equation (RTE) [14], where the laser is assumed to penetrate through the powder bed with thickness z_{bed} and with an optical media with an extinction coefficient of η. According to this method, the volumetric heat source due to radiation transfer is expressed as follows:

$$Q(x, y, z) = \frac{2P}{\pi r_l^2} e^{\left[-2\frac{x^2+y^2}{r_0^2}\right]} f_1(z) \tag{5.21}$$

$$f_1(z) = \left(-\eta \frac{dq}{d\xi}\right) \tag{5.22}$$

In this equation, $\xi = \eta z$ where z is a dimensionless local depth coordinate and q is the dimensionless form of the net radiative energy flux density, which has been discussed previously. A schematic of the volumetric heat source due to the radiation transfer method is shown in Figure 5.2e.

2. **Ray-tracing method:** The absorptivity profile function is derived using the ray-tracing analytical method or numerical methods. This method uses Monte Carlo ray-tracing simulations to calculate the absorptivity profile of the powder bed model with randomly distributed particles [15]. According to this method, the volumetric heat source model can be expressed as:

$$Q(x, y, z) = \frac{2P}{\pi r_l^2} e^{\left[-2\frac{x^2+y^2}{r_0^2}\right]} f_2(z) \tag{5.23}$$

Note: $f_2(z) = \frac{d\beta}{dz}$ is the absorptivity function derived by the Monte Carlo ray-tracing simulations. A schematic of the volumetric heat source derived from the ray-tracing method is shown in Figure 5.2f.

3. **Linearly Decaying Equation:** The absorptivity profile can be described by a linearly decaying function [16]. A schematic of the proposed heat source model is shown in Figure 5.2g.

   According to this method, the volumetric power distribution can be written as:

$$Q(x, y, z) = \frac{2P}{\pi r_l^2} e^{\left[-2\frac{x^2+y^2}{r_0^2}\right]} f_3(z) \tag{5.24}$$

where

$$f_3(z) = \frac{2\beta}{\delta}\left(1 - \frac{z}{\delta}\right) \tag{5.25}$$

and δ represents the penetration depth (m), which is normally assumed equal to the layer thickness. However, it can be different based on experimental measurements.

4. **Exponentially decaying equation method:** The exponentially decaying equation model as a method to describe a volumetric heat source [17] and a schematic is shown in Figure 5.2h.

   The equation used for the intensity distribution can be expressed as:

$$Q(x, y, z) = \frac{2P}{\pi r_l^2} e^{\left[-2\frac{x^2+y^2}{r_0^2}\right]} f_4(z) \tag{5.26}$$

where

$$f_4(z) = \frac{\beta}{H} e^{\left(-\frac{|z|}{H}\right)} \tag{5.27}$$

5.2.1.2 Effective Thermal–Optical Material Properties in the Presence of Powder Particles

In LPBF, the interaction between the laser and the matter is very important. In particular, a major component is the description of the absorption of the laser light by the powder layer and the distribution of the absorbed energy. All thermal properties are also changed due to gaps between powder particles. For a description of absorption and reflection factors, see Chapter 4. As elaborated in Chapter 4, the absorptivity is influenced by various parameters, such as laser wavelength, temperature, and surface roughness.

The absorption of the laser energy by a metallic powder bed is influenced by the powder characteristics. In particular, particle size, size distribution, particle shape, and roughness affect reflection, absorption and scattering. Figure 5.3 shows the phenomena associated with laser light reflection, scattering, and absorption in a powder bed.

The powder parameters that have effects on thermal and optical properties are listed in Table 5.2 [18].

Many producers/guidelines/datasets provide effective thermal properties based on the above-mentioned parameters; however, given the powder characteristics, the thermal conductivity of powder material can be derived by [24]:

$$
\frac{K_e}{K_g} = \left(1 - \sqrt{1 - \phi}\right)\left(1 + \frac{\phi K_R}{K_g}\right) + \sqrt{1 - \phi}
$$

$$
\left\{(1 - \emptyset)\left[\frac{2}{1 - \frac{BK_g}{K_s}}\left(\frac{B}{\left(1 - \frac{BK_g}{K_s}\right)^2}\left(1 - \frac{K_g}{K_s}\right)Ln\frac{K_s}{BK_g} - \frac{B + 1}{2} - \frac{B - 1}{1 - \frac{BK_g}{K_s}}\right) + \frac{K_R}{K_g}\right] + \emptyset\frac{K_{Contact}}{K_g}\right\}
$$

$$(5.28)$$

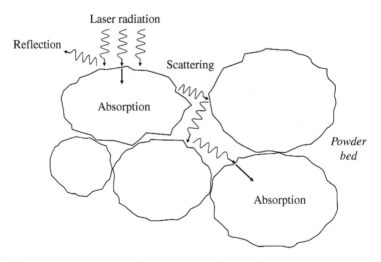

Figure 5.3 Powder–laser interaction mechanisms. Source: Redrawn and adapted from [3].

Table 5.2 Equations used for powder particles shape characterization [18].

Shape factor	Formula	References
Circularity	$\dfrac{4\pi A}{P^2}$	[3, 19]
Aspect ratio	$\dfrac{l}{L}$	[20]
Elongation	$1 - \dfrac{D_b}{D_a}$ or $\log_2\left(\dfrac{D_a}{D_b}\right)$	[3, 21]
Dispersion	$\log_2(\pi D_a D_b)$	[21]
Roundness	$W_i = \dfrac{\sum\limits_{i=1}^{N} D_i}{N D_{in}}$ or $\dfrac{r_1}{R}$	[3, 22]
Flatness	$\dfrac{r_2}{R}$	[22]
Perimeter to area ratio	$\dfrac{P^2}{A}$	[23]

A: Projected area of particle (m^2),
P: Perimeter of particle (m),
l: Minor axis length perpendicular to major axis (m),
L: Major axis length connecting two most distant points on the projection (m),
D_a and D_b: Major and minor axis length axes of the Legendre ellipse, respectively (m),
D_{in}: Sieving diameter (m),
r_1: Radius of curvature of sharpest developed edge of particle (m),
r_2: Radius of curvature of most convex direction of particle (m),
R: Mean radius of particle (m)

where K_e is the effective thermal conductivity of powder bed (W/m·K), K_g is the thermal conductivity of the gas filled between particles (W/m·K), K_s is the thermal conductivity of solid (W/m·K), φ is the experimentally measured porosity of the powder bed [25], K_R is thermal conductivity of the powder bed due to radiation (W/m·K), $\emptyset$ is flattened surface fraction between particles, B is deformation parameter of the particle where $K_{contact}$ can be derived by [26]:

$$K_{contact} = 18\emptyset K_s \qquad \text{if } \emptyset < 3 \times 10^{-4}$$

$$K_{contact} \approx K_s \qquad \text{if } \emptyset > 0.01 \qquad\qquad (5.29)$$

Figure 5.4 shows a major difference between the bulk and powder thermal conductivity for a nickel-based alloy. It also depicts the difference between the densities taking into account the density/porosity of powder bed by:

$$\rho_{powder} = (1 - \varphi)\rho_{bulk} \qquad\qquad (5.30)$$

To consider phase change from solid to liquid, apparent heat capacity should be calculated by [28]:

$$C_p = \begin{cases} C_{p,sensible} \text{ if } T < T_m - 0.5\Delta T_m \qquad \text{or} \qquad T > T_m + 0.5\Delta T_m \\[2mm] C_{p,modified} = C_{p,sensible} + \dfrac{L}{\Delta T_m} \quad \text{if} \quad T_m - 0.5\Delta T_m < T < T_m + 0.5\Delta T_m \end{cases} \qquad (5.31)$$

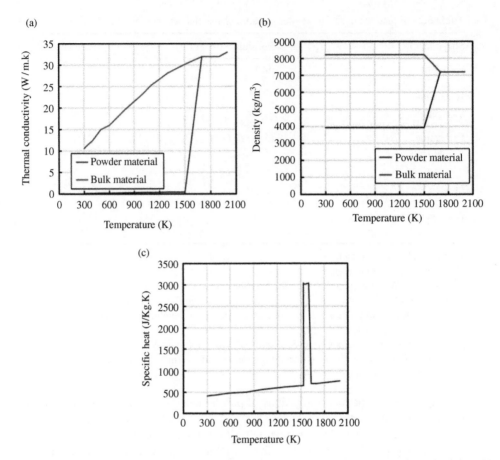

Figure 5.4 Thermophysical properties of bulk and powder material: (a) thermal conductivity, (b) density, and (c) specific heat for a nickel-based alloy. Source: Reprinted from [27] with permission from Elsevier.

where T_m is the melting temperature (K), which is considered as the center point between solidus and liquidus temperature. ΔT_m is the temperature difference between liquidus and solidus temperature, and L is the specific latent heat of fusion (kJ/kg).

5.2.1.3   Effective Layer Thickness (ELT)

In LPBF, it is reported that "*the actual thickness of powder particles that spread on solidified zones, so-called effective layer thickness (ELT), is higher than the nominal layer thickness*" [29]. As seen in Figure 5.1, the powder particles shrink substantially after melting and solidification. ELT highly depends on many factors, including process parameters and material properties. This parameter should be identified experimentally. It is reported that ELT of 17-4 PH stainless steel for a nominal build layer thickness of 20 μm is measured to be larger than 100 μm [29]. This value must be considered in any modeling because the use of nominal value rather than ELT clearly affects the outcome of any analysis.

5.2.2   Fluid Flow in the Melt Pool of LPBF: Governing Equations and Assumptions

The fluid flow in the melt pool, created by the heat transfer explained in the last section, strongly affects the properties of parts printed by LPBF. In particular, they are extremely important in determining the printability of materials, the microstructure of printed parts, and the size/shape of the melt pool because the direction and magnitude of the fluid-flow can change the local heat transfer schemes. The convection in the melt pool is driven by different forces, which can be listed as follows: 1) surface tension, 2) buoyancy, and 3) electromagnetic forces. The latter contributes to the convection only when an electric current is employed; they result from the interaction of the divergent current path in the melt pool with the magnetic field that it generates [3, 30]. The main equation governing the fluid flow in the melt pool is the Navier–Stokes equation (see Chapter 4, Equation (4.45)).

The Peclet number (Pe), a dimensionless number, can be used to predict the dominance of fluid dynamics during LPBF. Pe represents the ratio of heat transport by convection to heat transfer by thermal conduction. The value of Pe is important for the assessment of which type of heat transfer will take place in the melt pool. The Peclet number can be expressed by the following equation:

$$Pe = \frac{u\rho c_p L}{K} \tag{5.32}$$

In this equation, u is the velocity (m/s), ρ is the density (kg/m^3), c_p is the specific heat at constant pressure (J/(kg · K)), L is a characteristic dimension (m), and K is the thermal conductivity of the melt pool (W/(m · K)).

For values of $Pe \gg 1$, the heat transport is governed by convection, and the heat transfer by conduction plays a minor role. For low Pe numbers $\ll 1$ (metals with high thermal conductivities at low velocities and for small melt pool size), the heat transport governs mainly by conduction [30].

In the following, the main fluid-based phenomena are explained and customized for LPBF.

5.2.2.1   Marangoni Convection

The heat flux that comes from the heat source will cause spatial and temporal variations of the liquid–metal density, as explained in Chapter 4. This will lead to an upward flow at the center of the melt pool due to the presence of buoyancy forces. In laser metal processing, surface tension forces dominate all the other forces in the melt pool [30]. The flow is known as Marangoni convection, and the surface tension gradient that is stress is known as the Marangoni effect/stress. The origins of the spatial gradient of surface tension are due to temperature and composition variations at the melt pool surface, and it can be expressed as:

$$\tau = \frac{d\gamma}{dT}\frac{dT}{dy} \tag{5.33}$$

where τ is the shear stress due to temperature gradient (N/m^2), γ is the interfacial tension (J/m^2), T is the temperature (K), and y is the distance along the surface from the axis of the heat source (m). The shear stress τ can be expressed also as a function of the density (ρ), viscosity (μ), and local velocity (u) by the following equation:

$$\tau = \frac{0.332\rho^{1/2}\mu^{1/2}u^{1/2}}{y^{1/2}} \tag{5.34}$$

From the combination of the above equations, it is possible to calculate the maximum velocity (u_m) at about halfway between the heat source axis and the melt pool edge, i.e. at $y = W/4$, being W the width of the melt pool [30]:

$$u_m^{3/2} \approx \frac{d\gamma}{dT}\frac{dT}{dy}\frac{W^{1/2}}{0.664\rho^{1/2}\mu^{1/2}} \tag{5.35}$$

The above equations are valid when the velocity is large, and the shear stress cannot be accurately calculated for small velocities. A more appropriate equation for the calculation of the shear stress at small velocities is $\tau = 2\mu u_m/d$. Noteworthy, $d\gamma/dT$ is negative for pure metals and alloys, which means that at the periphery of the melt pool (the temperature is lower than in the center of the melt pool), the surface tension is the highest, causing an outward flow. An inward flow is instead created by the positive $d\gamma/dT$ of certain surface elements such as sulfur. A combination of inward and outward flows is originated by the spatial variation of surface tension, depending on the external and internal conditions [3, 30].

5.2.2.2 Buoyancy and Electromagnetic Forces

Buoyancy forces could be dominant when maximum velocities are small as long as the surface tension gradient is not the main driving force. In this case, maximum velocity can be approximated by:

$$u_m \approx \sqrt{g\vartheta\Delta Td} \tag{5.36}$$

where g is the acceleration due to gravity (m/s^2), ϑ is the thermal expansion coefficient/volume expansion at constant pressure (1/K), ΔT is the temperature difference (K), and d is the depth of the melt pool (m) [30].

5.2.2.3 Melt Pool Size and Shape: Keyhole Versus Conduction Modes

The size and shape of the melt pool can be affected by the change of the temperature coefficient of surface tension, $d\gamma/dT$ induced by the presence of heat flux, and certain elements in the melt pool. The value of the surface tension results in the direction of convective flow in the melt pool. The melt pool penetration depth is not only a function of laser power but also the concentration of surface-active elements such as oxygen. The interfacial surface tension has been reported that may be described by the integration of Gibbs and Langmuir adsorption isotherms through the following equation [30]:

$$\gamma = \gamma_m^0 - A(T - T_m) - RT\Gamma_s \ln\left(1 + k_1 a_1 e^{-\Delta H^0/RT}\right) \tag{5.37}$$

where γ is the surface tension as a function of composition and temperature (J/m^2), γ_m^0 is the interfacial tension of the pure metal at the melting point T_m (J/m^2), T is the absolute temperature (K), A is the temperature coefficient of surface tension for the pure metal (J/m^{2}K), R is the gas constants (J/K.mol), Γ_s is the surface excess of the solute at saturation solubility (m^2), k_1 is the entropy factor, a_1 is the activity coefficient of the solute, and ΔH^0 is the enthalpy of segregation (J).

It has been reported that the presence of oxygen at a certain concentration will lead to a change of the value of $d\gamma/dT$ from positive in a low-temperature regime to a negative value at high temperature [30].

Depending on the surface tension and Marangoni convection (see Figure 5.1), the melt pool can fall into two modes:

1. Keyhole
2. Conduction

In the conduction mode, the laser power intensity per volume is not as large as the power intensity per volume needed for the keyhole mode. In the conduction mode, the heat penetration is governed by the heat conduction into the metal from the surface. In the conduction mode also, the melt pool width is typically wider than its depth (see Figure 5.5).

In the keyhole mode, the power density is large, causing evaporation (see next section for the vaporization theory). The pushing of gas outward, so-called recoil pressure, creates a keyhole from the surface down to the melt pool's depths. As long as the process remains at high power

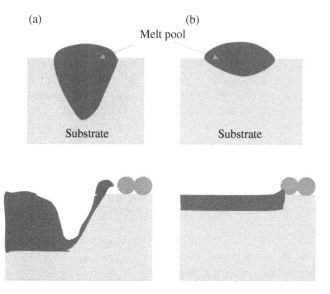

Figure 5.5 Schematic of (a) keyhole mode and (b) conduction mode. Top figures are the transverse view and down figures are the longitudinal view.

density given to the meltpool volume, the keyhole remains open, creating a narrower but deep melt pool shape (see Figure 5.5).

As described in Chapter 2, the conduction mode may cause porosity due to "lack of fusion," and the keyhole mode may cause gas porosity due to the collapse of the keyhole and gas entrapment in the melt pool region.

5.2.3   Vaporization and Material Expulsion: Governing Equations and Assumptions

5.2.3.1   Evaporation Rate

The maximum melt pool temperatures during LPBF are normally higher than the material melting temperature; thus, this situation leads to the vaporization of the metals and/or alloying elements if the peak temperature get closer to the boiling temperature. Consequently, information on the maximum temperature reached in the melt pool can be obtained from the melted metal's vapor composition. The Langmuir equation can be used for the calculation of the evaporation rate at low pressure (no condensation of the vapors) [30]:

$$J = \frac{p^0}{\sqrt{2\pi MRT}} \tag{5.38}$$

The term J indicates the vaporization flux (m^{-3}), p^0 is the vapor pressure of the vaporizing species over the liquid (N/m^2), M is the molecular weight of the vaporizing species (mol), R is the gas constant (J/K.mol), and T is the temperature (K).

Different models can be used to obtain the vaporization rates of pure metals and the loss of alloying elements during LPBF. These calculations can be based on the numerical solution of the equations of conservation of mass, momentum, and translational kinetic energy of the vapor near the melt pool surface. Noteworthyly, the pressure gradient-driven mass transfer should also be considered in the calculations. All these calculations are dependent on the maximum temperature that may be underestimated in many numerical methods.

Researchers have developed the following equation to estimate the peak temperature during melting, taking into consideration the evaporation flux (W), which is temperature-dependent, the enthalpy of vaporization (L_e), the enthalpy of mixing ($\Delta \bar{H}$), and the laser power density (P) [31]:

$$W(T)(L_e - \Delta \bar{H}) = xP \tag{5.39}$$

In the above equation, the term x corresponds to the fraction of the input power used for vaporization. The peak temperatures of melt pool under different conditions can be estimated based on the above equation and Equation (5.38), which express the relation between vaporization flux and temperature [30].

5.2.3.2   Material Expulsion and Recoil Force

The fast evaporation in the melt pool causes a difference in pressure between the area near the heat source and the surroundings leading to the formation of a shockwave. As a consequence, the vapor moves away from the surface and it exerts a large recoil force, which causes the metal to be expelled from the cavity.

The recoil force F_{recoil}, (N), is a function of the peak temperature T_{max}, and for a given peak temperature (K), the recoil force can be calculated as follows:

$$F_{recoil} = 2\pi \int_0^{r_b} r \Delta P(r) dr \tag{5.40}$$

The radial distance is indicated by r_b (m), the surface temperature is equal to the boiling point, and the difference between the local equilibrium vapor pressure and the atmospheric pressure is indicated by $\Delta P(r)$ (N/m^2), and it is a function of radial distance from the beam axis [30]. The material expulsion takes place when the recoil force overcomes the surface tension force; therefore, if we describe the critical temperature as T_{cr}, where the recoil force is equal to the surface tension force, then for value of $T_{max} > T_{cr}$, liquid expulsion occurs.

In LPBF, the laser beam will defocus after the liquid metal expulsion, causing a decrease in the heat flux. In this situation, liquid expulsion will no longer occur since the peak temperature will be lower than the critical temperature T_{cr} [30]. This may cause a cyclic expulsion/relaxation throughout the process.

5.2.4   *Thermal Residual Stresses: Governing Equations and Assumptions*

Like DED, PBF processes are characterized by rapid heating and cooling rates, which introduce residual stresses in the printed parts. All equations in Section 4.5.7 related to thermal stresses are valid for LPBF.

Residual stresses can be categorized based on the scale at which they occur, namely, type I residual stresses vary over large distances resulting in parts' deformations and are of major concerns. Types II and III are due to the materials' different phases and atomic dislocations. The mechanical properties of the materials are not largely affected by these types of residual stresses and they are harder to measure due to the insufficient instrument's resolution [32, 33]. The mechanisms responsible for residual stress are mainly two, which are [32–39]:

1. Temperature gradient mechanism (TGM), and
2. Cool-down phase of the melt pool top layers.

The TGM mechanism arises from the large temperature gradient that is formed during the rapid heating of the powder bed particles and the slow heat conduction. A schematic of the TGM mechanism is displayed in Figure 5.6.

During the heating step, the top layer material expands; however, the underlying layer at lower temperature prevents this expansion inducing a compressive strain (ε_{th}) at the heated zone (top layers) as illustrated in Figure 5.6. Noteworthyly, the compressive strain will be partially elastic (ε_{el}) and partially plastic (ε_{pl}), if the compressive stress will exceed the compressive yield stress (σ_{yield}) of the material. During the cooling step, the compressed upper layers cool down, and a shrinkage of the material will occur. The shrinkage is partially inhibited by the plastic deformation formed during the heating step, and this results in the formation of residual tensile stress (σ_{tens}) at the irradiated zone, which is surrounded by a zone of compressive stress (σ_{comp}) [32–34, 38]

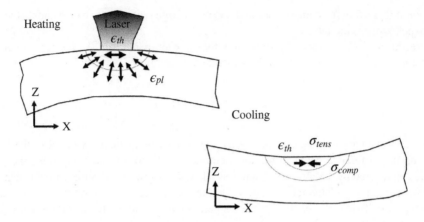

Figure 5.6 Temperature gradient mechanism inducing residual stress. Source: Redrawn and adapted from [32].

The second mechanism responsible for introducing residual stresses in PBF is the cool-down phase of the top layers, and it is due to the characteristic of the process. In particular, in PBF processes, the parts are printed layer by layer, and during these cycles, the material will experience remelting and resolidification. The resolidification is followed by a cooling down step, which leads to shrinkage of the material. However, this deformation is inhibited by the previously deposited material, inducing tensile stress in the added top layer and compressive stress in the underlying layer [32, 33, 38].

5.2.5 Numerical Modeling of LPBF

LPBF, like DED, can be simulated (see Chapter 4):

1. At micro-scale that is in the range of the powder particles' size,
2. At meso-scale that is in the range of deposited track's size, and
3. At macro-scale that in the range of printed part's size.

In the first case, we need to consider the laser interaction with the powder, powder melting, and evolution of the melt, while in the second and third cases, the laser heating and melting would be treated as a thermal source and it would provide a guide for the calculation of residual stress [1]. However, due to a major difference in terms of macro-sized printed parts and micro-scale laser interaction with powder particles, it is computationally extensive to use the same micro-scale heat source for macro-size modeling; thus, an effective heat source should be developed to apply equivalent heat flux to the part instantaneously while mimicking a micro-scale modeling to minimize computational time. While the finite element modeling methods are widely used for meso- and macro-scale modeling, there are three methods used for micro-scale modeling that are described in the next section.

5.2.5.1 Micro-Scale Numerical Modeling Methods for LPBF

Different types of modeling methods have been developed for the optimization of the LPBF process, which take into consideration simulations at the atomic, particle, and component levels. In particular, the most employed simulations are the molecular dynamics (MD), the discrete/dynamic element model (DEM), and the finite element model (FE).

Molecular Dynamics Model

The MD model is employed to simulate the laser melting and sintering kinetics during the AM process, which occurs in a very short timescale, typically in the range of nanoseconds. The sintering mechanism and atomic diffusion of nickel particles in the PBF process have been modeled at an atomistic scale using the embedded-atom method (EAM) [40]. This method is an approximation used to describe the energy between atoms, where the resulting energy is obtained by embedding an atom into the local electron density provided by the remaining atoms of the system and also considering an electrostatic interaction [41]:

$$E_i = F_\alpha \left(\sum_{j \neq i} \rho_\beta \left(r_{ij} \right) \right) + \frac{1}{2} \sum_{j \neq i} \emptyset_{\alpha\beta} \left(r_{ij} \right) \tag{5.41}$$

where F_α (eV) is the embedding energy, ρ_β (A^{-3}) is the averaged atomic electron density, and $\emptyset$ (eV) is an electrostatic, two-atom interaction. The EAM is a multi-body potential, since the electron density derives from the sum of many atoms.

Discrete/Dynamic Element Model

The discrete element model is used to describe and calculate the interaction between a larger number of particles. It is suitable to model the powder bed heating by laser, powder deposition, and the powder particles flow during the relayering step in PBF. The particle flow simulation could be achieved by calculating the particles' displacements and rotations using Newton's second law and rigid body dynamics equation, as reported in [40]. The equations used are listed below:

$$m_i \ddot{x}_i = m_i g + \sum_j F_{ij} \tag{5.42}$$

$$I_i \ddot{\theta} = \sum_j \left(r_{ij} \cdot F_{ij} \right) \tag{5.43}$$

where $\ddot{x}_i$ is the translational acceleration (m/s^2), $\ddot{\theta}$ is angular acceleration (1/s^2), the mass of the particles is indicated with m_i, (kg), g is the gravitational acceleration (m/s^2), F_{ij} (N) represents the force at contact with neighboring particles j, I_i the moment of inertia of the particle (kg · m^2), while r_{ij} is a vector (m) directed from the center of the particle i to the contact point with particle j [40].

Finite Element Model

As explained in Chapter 4, different numerical models have been developed to simulate and investigate different processes in LPBF. However, finite element method (FEM) is employed to analyze the temperature distribution from micro-scale powder particles to the substrate, to predict the evolution of transient temperature, and to simulate the thermal stresses or the evolution of distortions at the macroscale level. The rapid transient nature and small size of the melt pool in LPBF are two major challenges when it comes to FEM. Mesh independence analysis, solving engines, and convergence time are important steps to be optimized to ensure the validity of results.

5.2.6   Case Studies on Common LPBF Modeling Platforms

In the following sections, two case studies are provided. As mentioned before, the goal of this textbook is not a literature survey but to provide basic information needed to appreciate the reading of scholarly papers on the modeling of LPBF, while linking the topics to the main undergraduate engineering topics in heat transfer and solid mechanics. Within this context, the following two case studies shed some light on the procedure of modeling LPBF, but they do not enter into an in-depth discussion of results.

5.2.6.1   Melt Pool Size Estimation in LPBF[1]

Properly estimating the geometries of the melt pool in the LPBF process is critical because the melt pool forms single tracks, single tracks form single layers, and single layers form the final products. In this case study, the goal is to estimate the melt pool geometry by modeling and compare the results with experimental data.

For collecting experimental data, single-layer multiple-track experiments can be carried out on an EOS M 290 LPBF machine, in which the beam spot diameter is 100 μm. The powder used in this case study is the gas-atomized Stainless Steel 17-4PH powder. The processing parameters are listed in Table 5.3. All other process parameters are kept as default. The printed samples were cross-sectioned, mounted, polished, and etched. The melt pool dimensions (width and depth) and the single-track surface profiles were measured by a laser scanning confocal microscope.

It is necessary to mention the scanning path. An area of 1 × 20 mm was scanned by the laser as shown in Figure 5.7. There are nine single tracks and the laser scanning started from the left bottom corner.

The corresponding simulation of this process is implemented using the commercial software package, COMSOL 5.3a Multiphysics. The simulations can be performed considering nonlinear transient thermal analyses within the metal powder and the base plate, as shown in Figure 5.8. To reduce the requirements of the computational resources, only a portion of the whole scanning area, the left 1 × 1 mm area, is modeled and simulated. The dimensions

Table 5.3   Parameters for the single-layer multiple-track scanning.

Laser power (W)	Scanning speed (mm/s)	Layer thickness (μm)
195	800	20

[1] The authors would like to acknowledge the contribution of Dr. Zhidong (Brian) Zhang for preparing several photos of this section.

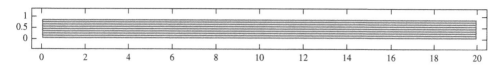

Figure 5.7 The laser scanning path used for the case study.

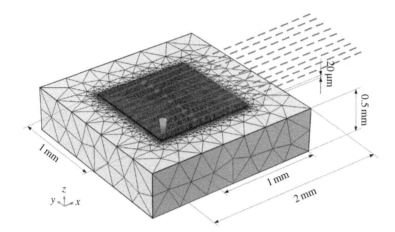

Figure 5.8 Geometry and mesh used in the finite element simulation of the case study.

of the solid substrate in the simulation were $2000 \times 2000 \times 500$ μm, and those of the powder layers are appointed to be $1000 \times 1000 \times 20$ μm. The dash lines represent the scanning path out of the modeling area. When moving out of the modeling area, the laser beam keeps moving with the scanning path as shown in Figure 5.7 but does not heat the modeling area. Several boundary conditions are considered. The radiative and convective heat losses are assigned on the top surface of the simulation domain. The other sides of the domain are set as a fixed preheating temperature. Further details of the material properties can be referred to in the previously published paper [11].

Printed Track Surfaces: Experimental and Modeling Results
In LPBF, the laser moves rapidly, making the isotherm curves elongated [11]. The elongated isotherm curves can produce ripples on the final printed track surfaces, as depicted in Figure 5.9. θ is the tail angle of a triangle ripple. From these single track results, the wavy surface just looks like the water ripples after a fast-moving boat. In Figure 5.9b, the simulation employed the heat source (APG4) model explained in Section 5.2.1.1.2.

For multiple tracks, similar features can also be observed, as shown in Figure 5.10a, but they are overlapped with each other. The 3D height profile is shown in Figure 5.10b. The laser scanning started from the bottom-left corner. The first track is higher than the other eight tracks. This may be due to the powder was sufficient for the first track, while the powder for the following scanning was less due to the denudation phenomenon [42]. The height for the last track on the top of the picture is almost as high as the substrate indicating there was no adequate material to be added. This can also be implied by the result of cross-section results, which will

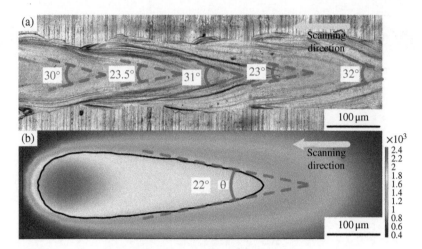

Figure 5.9 Ripples of a single track: (a) experimental results and (b) numerical results. Source: Reprinted from [11] with permission from Elsevier.

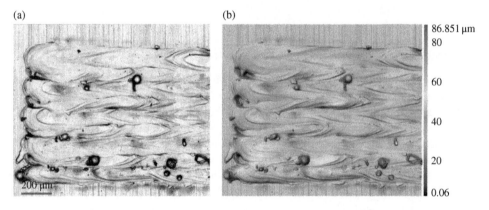

Figure 5.10 Experimental surface of the multiple-track scanning: (a) microscopy picture and (b) 3D height profiles.

be discussed in the next section. Besides, periodical deep dimples can be found at the left edge in Figure 5.10b. This may be caused by the recoil pressure that pushing melting material away to form deep holes, as discussed comprehensively in [43]. These dimples may cause porosity and increase surface roughness in the final products; therefore, contour scanning could be important in the manufacturing of real parts.

Prediction of Melt Pool Dimensions: Experimental and Modeling Results

Experiments can be carried out with the process parameters listed in Table 5.3. Firstly, single-track scanning can be implemented. The melt pool width and depth of a single-track scanning can be measured through the analysis of the microscopic images as shown in Figure 5.11. Good agreement between the experimental and simulation results can be derived.

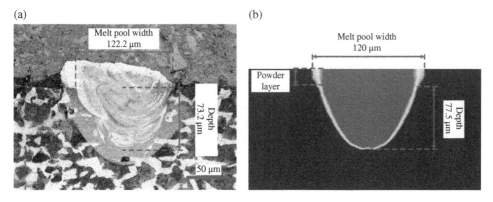

Figure 5.11 Melt pool cross section of a single track from (a) experiment (b) simulation.

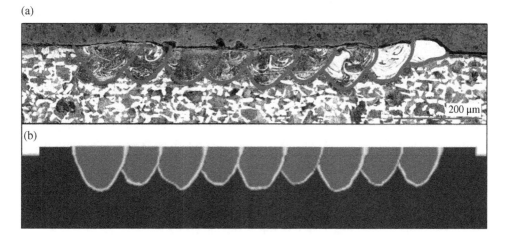

Figure 5.12 Multi-track melt pool cross sections: (a) experiment and (b) simulation.

A single-layer multiple-track experiment can also be done. The melt pool cross sections are shown in Figure 5.12a. The scanning began with the most-right melt pool, which represents the first track. It can be observed that the melt pool of the first track is more swallow than the others. This may due to the LPBF machine automatically treated the first track as a contour scan and therefore scanned it with lower laser power and higher speed.

The first track can then be excluded from the comparison between the experimental and simulation results. For the other eight tracks, the melt pool depths are comparable. Interestingly, the second melt pool is much brighter than the remaining seven tracks. This may indicate more material of stainless steel exists in it because the stainless steel cannot be etched by the etchant, Nital. In other words, during the scanning of the remaining seven tracks, there was no adequate stainless steel powder exists on the scanning path because of the denudation phenomenon discussed in [42]. So the laser had a high chance of only melting the substrate material, low carbon steel, which can be easier etched. Besides, the inherent characteristic of

Table 5.4 Melt pool depths in the multi-track scanning.

Track no.	Experiment (μm)	Simulation (μm)
2	69.6	72.5
3	83.4	85.5
4	69.7	69.5
5	85.1	82.0
6	78.5	71.5
7	75.3	82.0
8	78.9	72.0
9	76.1	84.5

rapid cooling in LPBF may produce much smaller grains [44], making the melt pools look darker after being etched.

The melt pool depths of both the experimental and simulation results have listed in Table 5.4. The average error between the experimental and simulation results is 6.0%, indicating the simulation can correctly predict the melt pool depths in single-layer multiple-track LPBF printing. Moreover, a periodic trend can be observed that the melt pool depths for adjacent tracks increase and decrease intermittently. This trend is because the laser employed a raster scanning pattern and the cross-sectioned plane is near to the left end. When the laser is coming to the left end, the substrate shall be heated up for the next scanning track when the laser is going away from the left end. Thus, the melt tracks will become deeper. However, the trend disappears after the sixth track, according to the experimental results. This may be due to the decrease in the absorptivity because of the powder denudation phenomenon. Less powder could lead to less absorptivity, as discussed in [45]. At the same time, the periodic trend is kept in the simulation results because such powder dynamics are not considered in the present numerical model.

5.2.6.2 Modeling of Residual Stress Induced During LPBF

In this case study, a simplified model to quantify the residual stress profiles in the LPBF process based on the work in [32] is presented. Let us consider a base plate of height h_b and width w_b, on top of which a part has been built of height h_p, and the thickness of the top layer is indicated as t (see Figure 5.13).

The following assumptions are taken into account [32]:

- The printed part and the substrate are at room temperature.
- The tensile stress due to thermal shrinkage ($\alpha_t \Delta T$, where α_t is expansion coefficient and ΔT is temperature gradient) induced by the top layer as a consequence of the shrinkage is equal to the yield strength of the material (σ).
- The stress in the x-direction (σ_{xx}) is independent of the y coordinate.
- No external forces are acting on the system.

Based on the last assumption, the equilibria of force and momentum equations for a unit width are

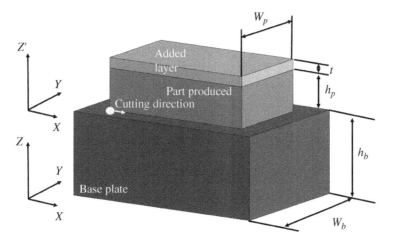

Figure 5.13 Schematic of the theoretical model of the LPF process. Source: Redrawn and adapted from [32].

$$\int \sigma_{xx}(z)\, dz = 0 \tag{5.44}$$

$$\int \sigma_{xx}(z)z\, dz = 0 \tag{5.45}$$

The strain profile upon the addition of the top layer distributes linearly as:

$$\varepsilon_{xx}(z) = az + b \tag{5.46}$$

The substrate and the part possess different stiffness, which contributes to the formation of different stress levels. The ratio between the substrate stiffness and the stiffness of the part can be indicated with $m = E_{substrate}/E_{part}$. The equilibrium conditions can be expressed according to Mercelis and Kruth as [32]:

$$\int_0^{h_b} m(az + b)dz + \int_{h_b}^{h_b + h_p} (az + b)dz + \int_{h_b + h_p}^{h_b + h_p + t} \bar{\sigma}dz = 0 \tag{5.47}$$

$$\int_0^{h_b} m(az + b)zdz + \int_{h_b}^{h_b + h_p} (az + b)zdz + \int_{h_b + h_p}^{h_b + h_p + t} \bar{\sigma}zdz = 0 \tag{5.48}$$

The coefficients a and b are

$$a = -6\bar{\sigma}t\,\frac{\left(2mh_bh_p + mh_bh_pt + h_p^2 + h_pt + mh_b^2\right)}{\left(4mh_b^3h_p + h_p^4 + m^2h_b^4 + 6mh_b^2h_s^2 + 4mh_bh_p^3\right)} \tag{5.49}$$

$$b = \bar{\sigma}t \frac{\left(2mh_b^3 + 6h_p mh_b^2 + 3mh_b^2 t + 6h_b h_p^2 + 6h_b h_p t + 2h_p^3 + 3h_p^2 t\right)}{\left(4mh_b^3 h_p + h_p^4 + m^2 h_b^4 + 6mh_b^2 h_s^2 + 4mh_b h_p^3\right)} \tag{5.50}$$

If we consider that, the part and the substrate have different widths, and then the term m can be rewritten as $m = E_{substrate} W_{base}/E_{part} W_{part}$. Another parameter to consider in the model of the stress profile of the part is the relaxation stress, which is related to the removal of the part from the substrate plate. The relaxation stress ($\sigma_{relaxation}$) can be expressed as:

$$\sigma_{relaxation}(z') = cz' + d \tag{5.51}$$

where $z' = z - h_b$ as illustrated in Figure 5.13. The constants c and d are obtained, calculating the equilibrium conditions for the part and indicating with h_c the height of the part in the X, Y, Z' new coordinate system. The relaxation stress causes the shrinkage (the constant part of the relaxation stress) and bending of the part upon its removal (the linear part of the equation) [32].

$$c = -6 \frac{-2\int_0^{h_c} z' \sigma(z')\, dz' + h_c \int_0^{h_c} \sigma(z')\, dz'}{h_c^3} \tag{5.52}$$

$$d = 2h_c \frac{\int_0^{h_c} \sigma(z')\, dz' - 3h_c \int_0^{h_c} z' \sigma(z')\, dz'}{h_c^2} \tag{5.53}$$

5.3   Physics and Modeling of Electron Beam Additive Manufacturing

The basics of electron beam interaction with materials are provided in Chapter 4. Also, the main components of electron beam melting (EBM) gun are explained in Chapter 3. In the following, several key parameters and major physical equations are explained.

5.3.1   Electron Beam Additive Manufacturing Parameters

The controlling parameters for an EB powder bed AM can be broadly subdivided into two types: EB parameters and environmental parameters.

5.3.1.1   Electron Beam Parameters

EB parameters comprise the beam current, accelerating voltage, beam focal diameter, shape, position, and velocity. The total heating power P (W) is a product of the beam current I (A) and the accelerating voltage V:

$$P = I \times V \tag{5.54}$$

Typically, beam currents are in the order of milli-amps. Free electrons, produced by an emission, have a low electrical potential of less than 10 V, insufficient to cause heating; therefore, they are accelerated by an elevated voltage potential. The electric potential of the electrons is efficiently converted to kinetic energy due to their significantly small mass. AM requires high-power densities and heating rates. Although EBs typically attain kilowatts of power, they are not able to meet this requirement. To meet this high power density need, the focal spot size is required to be small. EBs have nanometer-scale beam spots and are comparatively much smaller than laser spots, but an extremely small focal spot is not needed because powder sizes are in the order of microns. Therefore, EBs and lasers typically produce similar feature sizes in PBF processes [46]. Beam shape can also be controlled; however, circular shapes are the most common and preferable. Another possible shape is an ellipse, while more complex shapes are not normally possible. Beam spot position is easily adjusted using EB than lasers and forms one of the important advantages of EBs over lasers. With a combination of rapid repositioning and extreme heating rates possible in EB systems, it is possible to teeter between several melt pools ensuring the EB acts simultaneously on multiple locations in the powder bed [47].

Another attractive characteristic of EB in AM is the rapid movement of melt pool due to the high heating rates. This is important because the high heating rate will vaporize the powder if the beam velocity is too low, resulting in inferior product quality [47]. However, if a slow melt pool change is needed, the beam must be pulsed necessary [48].

5.3.1.2 Environmental Parameters

Examples of environmental parameters in an EB-PBF system include the chamber vacuum level, powder characteristics, powder bed shape, powder ambient temperature, and layer thickness. An EB can operate in medium to high vacuum conditions; however, the higher the vacuum level, the less the beam scattering. Also, high vacuum conditions take longer to attain; therefore, this must be factored in when considering EB for a time-conscious production cycle.

The type of powder material is important when considering electrical conductivity. For materials with very little electrical conductivity, if they are unable to discharge current, electrons will build up in the powder creating a strong field powerful enough to cause them to escape thereby leading to an explosion in the powder bed. Other than electrical conductivity, the material is neither limited to reflectivity as it not a focused light nor reactivity because it takes place in vacuum conditions. Consequently, EB-PBF is an excellent candidate for parts made by very reactive materials that would normally be unsafe to use under ambient conditions. However, caution must be taken when handling such materials outside vacuum conditions. Aluminum and titanium are not normally considered as very reactive materials, but they can form a thin oxide layer that seals in the material when processed in ambient or near ambient conditions. This thin oxide layer is inimical for additive manufacturing, and its formation will not occur when processed in vacuum conditions making EB ideal for such materials [46, 47].

Another important consideration is the powder bed shape because it exhibits poor thermal conductivity in vacuum conditions. There will exist a large thermal gradient at the boundary of the powder bed because heat is stored up in the powder bed. This is the reason why the powder bed in EB-PBF is significantly higher than that for laser. Also, oxidation that can occur in laser AM is absent due to the vacuum environment.

Any additive manufacturing system predicated on heating the base material is subject to the induction of thermal stresses due to the repeated and frequent heating cycles as each layer is added [49]. The induced temperature gradients and in turn thermal stresses can be significantly reduced when the temperature of the powder bed is kept close enough to the powder's melting temperature.

The layer thickness in a typical EB system, like other powder bed systems, is kept in the order of microns and a few particle layers. However, unlike in LPBF, larger layer thicknesses are permissible for parts with significantly high vertical dimensions because of the remarkable heat penetration depths attainable [50]. It should be noted that other common parameters (e.g., scanning strategies, preheating) are discussed in Chapter 2.

5.3.2   Emissions in Electron Beam Sources

Thermionic, field, and Schottky emissions are the three most common methods of producing free electrons, as discussed in the following sections.

5.3.2.1   Thermionic Emission

This is the emission of electrons from a heated metal material. The valence electrons in metals can readily move from one atom to another. Nevertheless, these electrons are incapable of escaping the metals totally because ambient conditions offer a substantially large energy potential acting as a barrier. However, electrons at or close to the surface of the metal can gain enough energy to overcome this barrier when the metal is adequately heated. As long as these "free" electrons are expunged from the metal, they can be collected. This energy needed to release valence electrons from the metal is referred to as the work function. Typically, most metals need to be heated to temperatures over 1500 K to give electrons enough energy to become free. Thermionic emission is the most adopted method to generate EBs for industrial applications and it can be activated in a medium vacuum. Filament materials such as tungsten and lanthanum hexaboride are commonly used because they have high melting temperatures and low reactivities [47, 51].

5.3.2.2   Field Emission

A field emission source is one usually made of tungsten crystal, where the electrode in the electron gun assembly is raised to a potential of 4 kV such that free electrons are emitted from the tip of the crystal. From electrostatics, the charge density σ_c (C/m^3) for a conducting sphere is related to the radius of the sphere r (m) and the potential around the sphere V (C/m^2) as [52]:

$$\sigma_c = \frac{\varepsilon_0 V}{r} \tag{5.55}$$

where ε_0 (C^2/N/m^2) is the vacuum permittivity. As expressed in this equation, the charge density increases when the radius decreases. Therefore, if the potential of an electrode is very large, and the radius is very small, the charge density will be extremely high. This, in turn, will increase the repulsive force between electrons, thereby decreasing the potential barrier to their escape away from the surface of the metal even at room temperature. Additionally, the smaller the radius is, the greater the coherence of the electron source is. This ensures the possibility of reducing the focal point and increasing the beam intensity. As aforementioned, the tip is made very sharp, and this increases the susceptibility to contamination; therefore, unlike thermionic sources, field emission sources must be operated at very high vacuum conditions. Field emission filaments are typically produced from single-crystal tungsten sharpened to a tip radius of about 100 nm [53]. These filaments are often used in the EB microscopy sources but may also be adopted in nano-scale additive manufacturing, as explained in Chapter 4.

5.3.2.3   Schottky Emission

These emission sources employ the working operations of both thermionic and field emission. This is done by increasing the temperature of the filament and applying a potential field to a very sharp point on the filament, ensuring that electrons are agitated at a high thermal state while simultaneously lowering the potential barrier. Schottky filaments are typically made of tungsten coated in zirconium oxide.

5.3.3   Mathematical Description of Free Electron Current

The mathematical model of the free electron current is described by the Richardson equation [54]:

$$J = A_G T^2 e^{-W/\sigma T} \tag{5.56}$$

where the free current J (A) is a function of the temperature T (K) and the work function W (J). A_G (A/cm^2/K^2) is a property of the filament material and a constant property of the filament material, while σ (J/K or eV/K] is the Boltzmann constant. While increasing the temperature creates an exponential increase in the free current, a reduction in the work function exponentially increases the free current. The work function of a material can be modified to be a function of the external field to observe the effect of the external field thus [54]:

$$W_{eff} = W - \Delta W \tag{5.57}$$

$$\Delta W = \sqrt{\frac{e^3 F}{4\pi\varepsilon_0}} \tag{5.58}$$

where F (N/C or V/m) is the external field and ε_0 is the vacuum permittivity. When the strength of the external field is increased, the effective work function is decreased, ensuring an exponential increase in the free current generated. The range in which this is possible is called the Schottky emission.

As identified in the previous sections, EBs typically have much smaller focal spot sizes than lasers. The Rayleigh limit for angular resolution, from [55], is

$$\theta = 0.610 \frac{\lambda}{R} \rightarrow \Delta l = 0.610 \frac{f\lambda}{R} \tag{5.59}$$

where θ (rad) is the angular resolution, Δr (m) is the spot radius in the image plane, λ (m) is the wavelength, and R (m) is the lens radius. For a unique lens radius, there exists a linear relationship between the wavelength and the spot radius. This is a result of the de Broglie correlation between particle energy and wavelength:

$$\lambda = \frac{h}{p} \tag{5.60}$$

λ is the particle's wavelength (m), h is Planck's constant (m^{2}kg/s), and p is the particle momentum (m $\cdot$ kg/s). When an electron is accelerated to 10 kV, it possesses a wavelength of 0.012 nm. Most lasers utilized for heating have wavelengths in the range of 400–10 000 nm; therefore, EBs can resolve very small feature sizes. However, in practice, the feature size is constrained by the particle size of the powder.

Electromagnetic-based forces govern the dynamics of EB, as explained below.

5.3.3.1 Lorentz Forces

The Lorentz equation can be used to describe the behavior of EBs, thus:

$$F_{Lorentz} = q(E + v \times B) \tag{5.61}$$

The equation shows that a charge q(C), which is subjected to an electric field E (N/C) will be accelerated by a force which is in the same direction of the field B (T). Based on this principle, free electrons are accelerated to give them sufficient energy for heating purposes. The right-hand side of the positive sign shows that when charge q is moved with a velocity v (m/s) in a magnetic field B, it will be accelerated in a direction that is perpendicular to both the direction of motion and magnetic field. These principles form the foundation of EB optics.

5.3.3.2 Electron Acceleration

Free electrons have very low energies in the order of a few eVs. Typical filament currents are in the range of hundreds of millivolts and can be calculated by Equation (5.54). Additionally, the electrons are emitted from the filament in all directions, so only a small fraction will propagate through the electron optics system toward the work piece, further lowering the beam power. To maximize the fraction of useful electrons and to generate significantly more power from the EB, a method is required to "funnel" the electrons into one direction and accelerate them in that direction, as shown in Figure 5.14. Using the first part of the Lorentz equation, we can apply a very large voltage with the cathode near the filament and the anode placed in the direction we wish the electrons to travel. The filament and the bias are both held at a strong negative voltage forming the cathode and repelling the free electrons. The bias cathode is very carefully designed in a bell shape, so that the electric field forms a "funnel" guiding all of the electrons toward the much smaller ring-shaped anode.

The system is designed in such a way to have a "crossover point" similar to a focal point in between the cathode and anode. The diameter of this crossover point is a critical parameter concerning the minimum spot size of imaging systems. Still, it is not as critical in welding

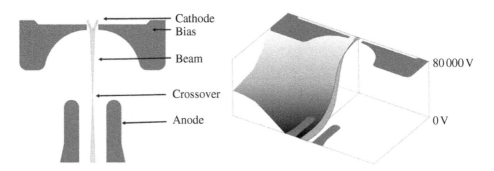

Figure 5.14 Electron acceleration between anode and cathode and electric potential shape. Source: Redrawn and adapted from [47].

and cutting systems where the EB spot size is usually much larger. Once the electrons reach the anode, the electric field is slightly positive so that the electrons are guided through the center of the anode, and once through the anode, the electric field becomes zero so that the electrons are free to propagate. With a typical voltage difference of tens of kilovolts, the beam power is now in the order of kilowatts, which is sufficient for most welding and cutting processes. As mentioned earlier, the filament produces electrons with a few eVs of energy, meaning that they have an initial potential of a few volts. Since some of these electrons are emitted while traveling away from the anode rather than toward it, there will be a difference in the distribution of the final electron voltage. This distribution is dependent on the spherical θ coordinate of the free electrons' initial momentum. For a filament producing electrons with an initial potential of 1 V, the maximum difference is ±1 V and for imaging systems, this small distribution can affect the focusing ability of the system. For welding and cutting applications, this value can be ignored since the practical working spot size is much larger than the limiting value. Also, the final beam power will have a distribution in the order of a few watts in a beam with kilowatts of total power, so this distribution can be neglected.

Using a single large electric field to accelerate the beam, the process is very efficient, with up to 60–70% of the wall energy being converted to beam kinetic energy [47]. This is a major reason why EB systems can be highly efficient. However, the cost of operation, calibration dilemmas, sensitivity of beam quality to surrounding environment even far from the beam source and other as discussed in Chapter 2 are among the challenges of adoption of electron beam to AM.

5.3.4 Modeling of Electron Beam Powder Bed Fusion (EB-PBF)

5.3.4.1 Motivations for Modeling

With the understanding of the more fundamental aspects of the EB-PBF process, the knowledge can be used to model the process. Accurately modeling a process can be difficult, especially for one as complex as EB-PBF. However, there are some distinct advantages that modeling can provide. The entire EB-PBF process is relatively slow (though the EB scanning is quick), which makes testing through experimental methods quite long and expensive. Although running a model can also be time-consuming, it is usually faster than actually making the part. Once a model is developed, it only requires intensive computing power to run it. However, producing actual parts requires the machine to be run, using materials, energy, and time, in addition to an operator's time and skill (unlike running a simulation). A model, once developed and validated, can be used to more quickly and inexpensively determine the outcome of making a part with unique processing parameters.

5.3.4.2   Energy Transfer and Transformation

EBs transfer energy to the substrate by converting kinetic energy to heat through inelastic collisions. The electrons embed beneath the surface of the substrate. This is very efficient because up to 70% of the kinetic energy is turned into heat [47, 56]. The efficiency is mainly governed by material properties such as density and hardness and the angle of incidence on the substrate [57].

When the beam energy is deposited in the powder bed, the transformation and transference of energy increase the temperature of the powder bed. When this temperature increase exceeds the solidus temperature of the powder material, a solid–fluid phase conversion takes place, which consumes latent heat L. After a short while, the liquid phase exceeds a limit, and the melt pool behaves like a liquid, which can be modeled by the Navier–Stokes equations. The heat transport in the liquid is governed by diffusion and convection. Heat losses by convection and radiation away from the liquid's surface is negligible, particularly convection because the process takes place in vacuum conditions. Therefore, most of the heat will dissipate from the liquid melt pool into the solid below via conduction, ensuring that the melt pool solidifies. The conservation equations that model thermo-fluid incompressible transport consisting of melting and solidification are all similar to those equations used for LDED and explained in Chapter 4. However, those equations are listed again with the new notation to be customized for EB-PBF as expressed by [58]:

$$\nabla.\boldsymbol{u} = 0 \tag{5.62}$$

$$\frac{\partial \boldsymbol{u}}{\partial t} + (\boldsymbol{u}.\nabla)\boldsymbol{u} = -\frac{1}{\rho}\nabla p + v\nabla^2\boldsymbol{u} + g \tag{5.63}$$

$$\frac{\partial E}{\partial t} + \nabla(\boldsymbol{u}E) = \nabla.(k\nabla E) + Q \tag{5.64}$$

All parameters are defined in Chapter 4 or can be found in the nomenclature. The thermal energy density is (5.64) is expressed as [59]:

$$E = \int_0^T \rho C_p dT + \rho\Delta H \tag{5.65}$$

ΔH is usually a complex function of temperature for most metallic alloys but can be simplified to:

$$\Delta H(T) = \begin{cases} L & T \geq T_l \\ \dfrac{T-T_s}{T_l-T_s}L & T_s < T < T_l \\ 0 & T \leq T_s \end{cases} \tag{5.66}$$

The liquid fraction ξ in the cell can be expressed thus:

$$\xi(T) = \frac{\Delta H(T)}{L} \tag{5.67}$$

The thermal energy density in (5.65) can be written to have an effective specific heat capacity at constant pressure $\tilde{c}_p$:

$$E = \int_0^T \rho c_p dT + \rho \Delta H = \int_0^T \rho \tilde{c}_p dT \tag{5.68}$$

where

$$\tilde{c}_p = \begin{cases} c_p & T \geq T_l \\ c_p + \dfrac{L}{T_l - T_s} & T_s < T < T_l \\ c_p & T \leq T_s \end{cases} \tag{5.69}$$

Heat transfer in the solidified part and powder particles is by conduction and is modeled by the transient form of Fourier's equation:

$$\rho\left(\frac{\delta H}{\delta t}\right) = -\nabla.q + Q \tag{5.70}$$

when it can be written as:

$$\rho\left(c_p \frac{\partial T}{\partial t}\right) = \frac{\partial}{\partial x}\left(K_x \frac{\partial T}{\partial x}\right) + \frac{\partial}{\partial y}\left(K \frac{\partial T}{\partial y}\right) + \frac{\partial}{\partial z}\left(K \frac{\partial T}{\partial z}\right) + Q(x, y, z, t) \tag{5.71}$$

where the heat source Q is frequently modeled by the Gaussian distribution, which is sufficiently discussed in the physics and modeling of LPBF-AM. As was earlier stated, convection at the surfaces is negligible since the process takes place in vacuum conditions. Therefore, the prevailing thermal boundary condition is heat radiation away from free surfaces expressed by the radiation equation in Chapter 4.

5.3.4.3 Navier–Stokes Equation

The Navier–Stokes (NS) equations are fundamental fluid dynamics equations that model the motion of viscous fluids. Typically, the incompressible NS equation is used for modeling the EB-PBF process. The compressible NS equation is more complicated since compression of the fluid is included in the calculations. The materials of interest for EB-PBF are typically metals, and liquid metal is incompressible, which makes it appropriate to only consider the incompressible NS equation. While the equation is widely used in fluid dynamics, it is also used for modeling the EB-PBF process and other AM processes, as explained in Chapter 4. On a fundamental level, the NS equation is the mass and momentum conservation of fluids. The equation equates the force of the mass (due to density and acceleration) to the forces from pressure, gravity, and viscous forces. The NS equation is beneficial for AM because it is a universal representation of the fluid and can fit into a variety of applications, being essentially independent of the application. Though it has a wide range of applicability, the NS equation must be simplified (at least to some degree) to be useful [57].

The NS equation is very comprehensive, and unless simplified, is impossible (or nearly impossible) to solve explicitly. Typically, for the NS equations to be directly solved rather than modeled or numerically solved, assumptions are made to simplify the equation. The typical assumptions for a melt pool in EB-PBF, which is fundamentally the same as the LPBF process, are: the geometry (size/shape) of the melt pool is known, the velocity at the liquid–solid interface is zero, traction is known on the liquid arc surface, complex body forces (buoyancy, electromagnetic forces) are assumed to be negligible, and the viscosity is assumed to be known [60]. These simplifications allow for the NS equation to be solved directly, at least for low Reynolds numbers, as turbulent flows are also too complex [60]. While these simplifications ensure a solution is found, it does cause differences between numerical and practical results.

While the usefulness of the NS equation for fluid flow is obvious, the equation can be adapted to be used for the EB-PBF process. First, the equation can model phase transitions in the material, to a degree, with everything included in a single-phase continuum. The different phases are modeled by applying different material properties, typically viscosity and velocity to different areas. In addition to phases, the equations account for gravitational forces and advection–diffusion during melt pool motion. The advection–diffusion creates an energy density field due to the high energy and relatively low penetration (compared to the melt pool size) of the EB. The energy density field is what is used to determine the material temperature and phase. Typically during the EB process, the material gets to the point of melting rather than being processed in solid state, so different phases are required. Even after a portion of the powder melts, the entire area does not immediately act as a fluid. The amount of liquid present needs to exceed a certain amount before it behaves as a fluid. This amount of liquid depends on the material type and powder sizes present. The liquid motion itself is governed by the NS equation with heat transfer within the pool caused by both convection and conduction. The convection in the pool is modeled by the NS equations, which are incorporated into larger equations modeling the thermo-fluid process [61, 62].

While the NS equation is useful, it is difficult to use beyond simplified or small-scale models. The equation must be simplified to some degree to be solved analytically. Even with simplification, it is usually still easier to solve the equation numerically. While the basic equation can be used to solve the fluid flow of the EB-PBF process, it must be expanded to model heat flow and phase changes. The difficulty surrounding the NS equation is the main reason why it is not used in many EB-PBF models. Rather, most models use a numerical approximation which can be implemented more easily.

5.3.4.4 Numerical Methods for Solving Heat Transfer and Navier–Stokes Equation

There are few modeling methods for the NS equation coupled with heat transfer that has been customized for PBF processes, including EB-PBF. They are

1. Lattice Boltzmann method, and
2. Finite element methods.

The following sections mainly focus on the first method, as it is not much popular by the scientific community; however, it is an excellent method when it comes to the melt pool modeling of PBF processes.

Lattice Boltzmann method

The lattice Boltzmann (LB) method is a means of discretizing and approximating the more complex NS equation. Using the LB method, hydrodynamic processes can be modeled numerically and scaled up to a macro-size more easily than with NS. The LB method works, as shown in Figure 5.15, by creating a lattice where there are points connected by links to adjacent cells. The number of links for each point, or cell, depends on the shape of the lattice and whether the lattice is in two or three dimensions. On the lattice, some particles can either move or be at rest. If a particle is moving, it will be displaced along a link corresponding to its velocity [57].

Two particles cannot occupy the same link (except for some higher-order lattices) and two particles cannot occupy a cell. To replicate the NS equations on a macro-scale, the movement (or lack of movement) needs to be done in such a way that mass, momentum, and energy are conserved. To avoid spurious invariants (false equilibriums/constant states), three-particle collisions and rest particles had to be added to an earlier version of the method which only involved two particles, or binary, collisions (depicted in Figure 5.15). To better approximate fluid problems, a high-order lattice must be used. Typically, a face-centered hypercubic lattice (four-dimensional lattice, see Figure 5.16) is used, giving 24 links to adjacent cells. On the lattice, particles can move and collide, and boundary conditions can be applied [57].

LB models, as mentioned earlier, are an alternative numerical method for solving hydrodynamic problems, with the model itself a discretization of the Boltzmann equation. The LB method is used to solve the Boltzmann equation in the hydrodynamic limit for a particle distribution function in the physical momentum space. The Boltzmann equation is a way of representing, statistically, the behavior of a thermodynamic system, with the hydrodynamic limit meaning the equation has been scaled up from microscopic descriptions of matter to the macro-scale. The basic form of the Boltzmann equation is shown in (5.72). The terms on the right-hand side account for external forces, diffusion of particles, and collision of particles.

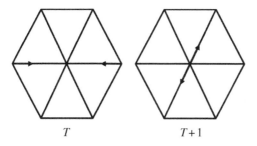

$$T \qquad\qquad T+1$$

Figure 5.15 Binary head-on collision of particles in a hexagonal grid model (known as Frisch, Hasslacher and Pomeau [FHP]) model (simplified LB method). Source: Redrawn and adapted from [57].

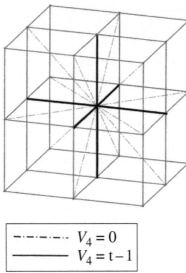

$$\dashrightarrow V_4 = 0$$
$$\text{———} V_4 = t - 1$$

Figure 5.16 Three-dimensional projection of a face-centered hypercubic lattice. Solid lines can carry two particles with speed v_4 along the fourth dimension. Source: Redrawn and adapted from [57].

Each of the terms changes based on the specific application of the equation, with differences between homogeneous materials and mixtures, liquids, and solids, etc.

$$\frac{\partial f}{\partial t} = \left(\frac{\partial f}{\partial t}\right)_{force} + \left(\frac{\partial f}{\partial t}\right)_{diffusion} + \left(\frac{\partial f}{\partial t}\right)_{collision} \tag{5.72}$$

Using the Boltzmann equation for a point distribution function (PDF) in the momentum space, one can determine the probability of finding a particle with a specific microscopic velocity at a particular point and time. The microscopic values are then integrated to get a macroscopic representation of the process (hydrodynamic limit). The LB method can be used for more than just solving for the velocity field, since the Boltzmann equation (used as the basis for the method) can be adapted for many different physical phenomena [58, 61].

Lattice Boltzmann Expansion for EB-PBF

The basic LB method is typically used for modeling fluid flow, since it is a numerical solution to the NS equation. However, the method can be expanded to model other aspects of the EB-PBF process. There are two major methodologies for using the LB method for EB-PBF.

1. The first is the multi-speed (MS) method. The MS-LB is an extension of the isothermal LB method, but it has extra discrete velocities and higher-order velocity terms. The thermal properties are regarded as an additional moment on the particles. The MS approach, though somewhat simple, can be unstable and typically requires the use of a single Prandtl number (specifically the ratio between kinematic viscosity and thermal diffusivity) to work [58, 62].

2. The next approach is the multi-distribution (MD) method. The MD-LB approach works by having the second particle distribution. Typically, one is used for density and momentum, while another distribution is used for thermal aspects (temperature, energy density, etc.). The benefit of the separate distributions is that the model can handle nearly arbitrary Prandtl numbers, which is beneficial for the high-energy EB process. Using the MD-LB method implies that a passive scalar approach is also being used. The passive scalar approach is based on the fact that temperature also follows the same definition and equation as a passive scalar, assuming viscous heat dissipation and compression work are negligible (typically excluded from LB methods) [58, 62].

Whichever method is used, the LB method is still divided into a stream and collide step. During the stream stage, the particles (making up the PDF) move along their respective links, which create the lattice structure. The model allows for particles to move and collide, though it also allows for particles to be at rest and not move. During the collide stage, the system is made to move toward a local equilibrium. For the surface of the material, there can be an open or closed thermal boundary condition. One of the conditions is needed to properly discretize the model. With open thermal conditions, a fixed temperature and no temperature gradient is set at the boundary. This allows for the initial unknown values for the rest of the lattice to be set by the local, constant equilibrium values from the boundary. With closed thermal conditions, there is no heat flux to the environment at all. This creates, essentially, an insulated surface [58].

Just as the NS equation, the LB method also accounts for different phases and interfaces. The different phases are assigned to different lattice cells to represent the process, see Figure 5.17.

While there are no specific interface cells between liquid and solid, there must be interface cells between liquid and gas and solid and gas. This is done because the gas phase is excluded (almost always) from the LB model. The interface cells are modeled using a free surface (free surface LB method) to simulate the moving interfaces between the different phases that can be presented during the EB process. The gas phase has significantly different properties when compared to the liquid or solid phases and this makes it difficult to integrate them into the model. Excluding the gas phase entirely from the calculations entirely is easier and makes

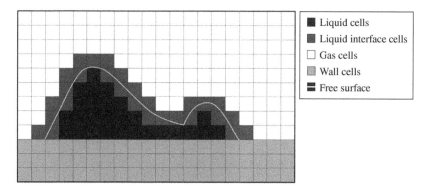

Figure 5.17 Phases assigned to different cells in the LB method. The specific example is of a deformed drop on a flat surface. Source: Redrawn and adapted from [58].

the process more reliable. This is accomplished by neglecting PDFs in the gas phase (i.e. no calculation is performed for that phase). While this does simplify things to some degree, it does mean that the interface cells must be reconstructed after the streaming step [62].

The interface cells, which make up the free surface (FS), are useful besides just isolating the gas phase. Since mass transfer has to occur across the interface cells, it is easy to have conservation of momentum and mass. The amount of material leaving the cell and its velocity is defined, which then defines mass and momentum transfer. A volume of fluid approach for each lattice cell (lattice cell has a fill level) is typically used to achieve this. The fill level determines the position of the FS but also affects EB absorption, with fuller cells absorbing more energy [63].

Since the gas phase is neglected, the only phases present in the PDF are liquid and solid. Whether a cell is liquid or solid depends on the energy density (related to temperature) of the cell. The energy density of a cell is changed by an applied heat source, which models the EB. The EB can be modeled only on the surface, but rather than a continuous spatial distribution on the surface, it is discretized to the lattice cells to fit with the LB method. The beam energy is included as a volumetric source in the top layers of the cells, since the absorption length of EB is larger than the thermal length (i.e. the energy is distributed within a large surface layer). There are two different ways to model the absorption, either exponential or constant absorption. Exponential and constant absorption profiles are shown in Figure 5.18.

The exponential absorption is typically used to lower the accelerating voltage (more common for EB melting), while the constant absorption is used for higher accelerating voltages. With the exponential function, the absorbance of the EB is high at the surface cells but decays exponentially, meaning most of the energy is deposited within the first few cells. With the constant function, the absorbance of the energy is the same for the entire depth until the EB energy is fully absorbed. This causes the penetration depth to be larger. With a larger cell size, the energy distribution shape does not change, but the depth of penetration increases [63].

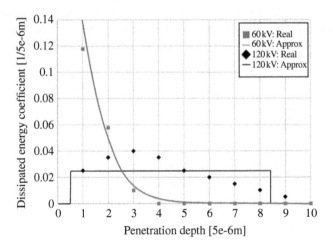

Figure 5.18 Exponential (60 kV) and constant (120 kV) absorption profiles. Source: Redrawn and adapted from [58].

Melting and Solidification of Powder by the LB Method

An important aspect that requires modeling using the LB method is the melting and solidification of the melt pool. The material can be in different phases depending on the processing
time and conditions. Whether a material is liquid or solid (or gas, though it is typically negated
due to complexity) is typically determined by the temperature. The evolution of a melt pool
during EB-PBF simulated using the LB method is shown in Figure 5.19.

By modeling the transition of the material, starting from powder, then melting and resolidifying, the effects of different process parameters on melt pool quality can be predicted. When modeling the melt pool, temperature-dependent properties such as viscosity should be known in
addition to the phases. However, since the temperature is not always continuously differentiable

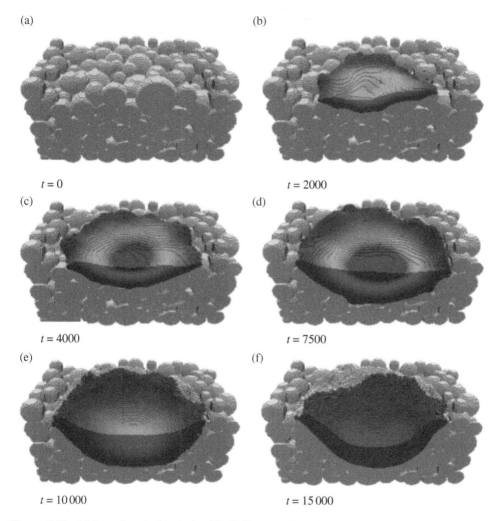

(a)

$t = 0$

(b)

$t = 2000$

(c)

$t = 4000$

(d)

$t = 7500$

(e)

$t = 10\,000$

(f)

$t = 15\,000$

Figure 5.19 Melt pool evolution during EB-PBF processing using the LB method at (a) 0 (b) 2 000
(c) 4 000 (d) 7 500 (e) 10 000 (f) 15 000 time steps, the beam pulse duration is 7500 time steps.
Source: Reprinted from [58] with permission from Elsevier.

with time, energy density is normally used instead. When the material is heated, the temperature will gradually increase but will reach a plateau as melting begins (in the case of pure metal; it is typically more complex for alloys). However, even with a freezing (or melting) range, the temperature will increase less quickly as energy will be used for melting (latent heat). The heating rate will change again once the material is fully melted (different specific heat). This process happens again in reverse as the material cools. The gradual and sudden change in the temperature value and heating rates make the use of temperature difficult. The energy density, on the other hand, is computed as a macroscopic value (rather than the temperature, which is microscopic) and it increases consistently. The energy density presented in Eq. (5.65) is related to temperature through an integral, and this process is easier and more stable than finding the temperature directly. The integral accounts not only for the temperature of the material as represented by the left term (based on specific heat) but also the latent enthalpy that occurs due to phase changes as indicated by the right term. While the model can be used for temperature and hydrodynamic effects, surface tension (temperature-dependent) is usually considered constant [58].

As mentioned earlier, the set of cell types is larger than just solid and liquid due to the FS approach used for the LB method. The different cell types are solid, liquid, gas, and solid and liquid interfaces. The different states and the transition between them are somewhat complicated due to the FS approach. Liquid and gas cells can change state but must do so through an intermediate liquid interface cell. This is to ensure that there is a closed interface layer, since gas evaporation is not modeled. The transition from a solid to a liquid cell depends on the energy density within that cell. As long as the energy density is higher than the value calculated for the specific material, the cell will change. A solid cell cannot convert to a gas cell and must first transition into a liquid cell (then liquid interface cell). Both solid and liquid cells can change into solid or liquid interface cells if they come into contact with gas cells. One issue with the transition from liquid to gas is that since the gas phase is not included in calculations, the PDFs for the LB method have to be constantly recomputed [58].

From [62], the amount of solid needed within a cell to designate it a solid is 55%. Solid cells are modeled by setting the velocity of that cell to zero. This effectively makes the cell act as a wall for hydrodynamic calculations of the remaining liquid. The surface tension of the liquid (typically not modeled) was approximated as a change in gas pressure acting on that cell. The more complex physics of wetting can also be included in the model, though they would be in a simplified form. In simple terms, the effects of wetting can be modeled by introducing an additional capillary force acting on the fluid [64]. In the model, the wetting angle between the liquid and powder can be varied between zero and pi. The wetting angle between the liquid and resolidified cell, which could be varied nonetheless, is held at zero (complete wetting).

The FS can be used to determine surface roughness from the simulation. To avoid more intensive measurements of the surface, [61] used a very quick approximation to determine if a surface was rough after processing.

In the simulation, evaporation is neglected, meaning no surface roughness is generated through that mechanism (a major mechanism during the actual process). Due to that, the melt pool surface is classified as uneven (rough) when the average maximum melt pool temperature is greater than 7500 K. This temperature is much higher than the actual process, but that is to compensate for the lack of heat loss through evaporation. The melt pool is considered porous if the calculated density is less than 99.5%. The process is deemed good if the process falls between those two bounds. While this approximation is simplistic, when correlated to empirical results, they match quite well [57].

In general, melting and solidification of the melt pool is a fairly straightforward process. When the energy density increases over a threshold, the material is liquid. Once liquid, fluid properties (velocity, viscosity, etc.) are assigned, and the material is effectively allowed to flow. The FS, which separates the PDF from the gas phase, acts as the melt pool surface and describes the volume change after solidification. The volume change occurs since the initial solid material existed in powder form, rather than being fully dense.

Simulation of Powder Bed Particle Distribution by the LB Method

The generation of the powder bed is done for the LB, so that phases can be assigned to the cells in the LB model initialization. Without this starting phase information, the process would either not work or not accurately model the process. All of the papers reviewed that used the LB method for simulation also had to use an additional process to generate a powder bed.

The actual generation of powder typically relies, to some degree, on empirical data. [58] analyzed actual powder size distributions to implement a realistic distribution in the simulation. The authors of [58] used an inverse Gaussian distribution, and three parameters govern the shape of the inverse Gaussian distribution, namely: the mean value, shape factor, and skewness factor. The mean value and shape and skewness factor were applied to the inverse Gaussian distribution to match the actual powder. This was done for initial tests, but the three factors were altered to examine the effects of different bulk powder properties on the EB process. The powder bed was generated using the *pe* framework from [65], representing the particles as rigid bodies and using that information for the generation of the cells for the LB method [58].

Another effort in [57] used a similar process as [58]. The particles are first generated above the bed (using the rain model) and allowed to free-fall onto the bed or previous layer. The particle size distribution is also approximated with an inverse Gaussian distribution. It is defined between a minimum and maximum particle size, with a skewness factor moving the peak of the distribution between them. The particle placement is imported into the LB method, defining the position and fill level of cells.

The rain model for powder bed generation is shown in Figure 5.20. First, the particle falls, and when it contacts a surface, it rotates (by gravitational forces) until it finds a lower energy position in contact with another particle. This allows particles to constantly adjust realistically to find the most favorable position on the bed. Once the bed is generated, particles can be removed to better match the actual density found in powder beds. Figure 5.20d shows a cross section of an actual powder bed and the adjusted rain model. It is observed that Figure 5.20c is a very close match.

The generation of the particles, or at least some approximation of it, is required for the LB method. Without it, the initial phases for the cells cannot be determined. Although modeling the particles gives a very good approximation of the actual powder bed, the process is intensive. More simplistic approximations could be made by modeling a single layer and applying the same results to each layer. However, the results may not be as accurate, but the simulation will be faster.

Additional Layers in the LB Method

The generation of particles is a known process, and doing it for another layer is somewhat trivial. No additional complexity is involved, and the simulation just takes more time. The only difference is that, rather than starting with a perfectly flat plate, the new powder layer is

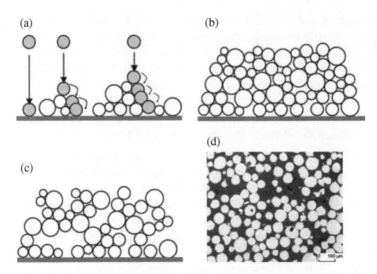

Figure 5.20 (a) Rain model schematic. (b) Generated powder bed. (c) Relative density is adjusted by removing particles. (d) Cross-section of actual powder bed. Source: Reprinted from [62] with permission from Elsevier.

generated on the solidified FS of the previous layer. While there is no fundamental change in the simulation process, it does allow for the effects of different layer thicknesses to be seen [58].

The efforts in [63, 66] used the same process as [58] but also used it for additional layers. These efforts were specifically interested in the thermal effects across multiple layers. [66] restricted the number of simulated layers to three, while [61, 67] only generated a single layer for the simulation. All three limited the number of layers due to the excessive computing time. Overall, the generation of additional layers is simple once a process has been established; however, the need for the extra computation time incurred is dependent on the specific application.

Lattice Boltzmann's Solution Methods

There are many different ways to solve the LB method; unfortunately, implementation is only possible by those highly experienced with coding. Consequently, most researchers use a commercially available LB solver. The most prevalent software used is waLBerla from Erlangen. It is an LB-based fluid flow solver that performs parallel simulations. The software can be used for fluid dynamic problems (LB method), but it can be expanded for other applications.

All of the efforts in [58, 61–[63] proceed with the simulation in a very similar fashion. A physics package (*pe*) is used to generate the particles. The powder bed is generated by the rain model, creating a powder bed with random packing. The model (not the software) is somewhat old, being published in 1987 [62]. Once generated, the particles are the boundaries (solid–gas interface) for the waLBerla software, which then models melting, fluid flow, heat transfer, and solidification during the EB process.

It is important to note that the LB method, for all the research efforts that utilized it, was used predominantly for simulating the melt pool flow during processing. The temperature distribution was found, but it was done so that more accurate material properties could be assigned.

Though the LB method is very good, it has its limitations. For modeling larger components and stress or strain distributions, FEMs are used.

Finite Element Methods

Though the LB method is popular, it is not the only method to model the EB-PBF process. Rather than basing the simulation on the NS equation, all fundamental equations governing the process can be included in a finite element (FE) software. As with the LB method, the process is simplified using FEM to reduce computation time. Approximations are made on the EB model and energy distribution, the material properties, the powder bed, and losses during the process. Unlike with the LB method, fluid flow is not simulated. The NS equation that would be needed to model the flow is too difficult and/or computationally intensive to solve. Additionally, the actual fluid flow is usually not required as many use the FE software for stress and temperature distributions rather than melt pool evolution. It should be noted that the process simulation of LDED, LPBF, and EB-PBF are largely the same using FEM except for the fact that convection is disregarded as a boundary condition in EB-PBF.

5.3.5   Case Studies

In the following, three case studies on the modeling of EB heat sources, temperature distribution, and residual stress distributions during EB-PBF are reviewed. All models in the following sections use FEM employed commercially available software. Abaqus [49, 68, 69] and Ansys [48] are used. In [66], an in-house code is developed using the FE C++ library deal.II. In [66], a code is developed since different solving methodologies were being tested, and the solving mechanisms in most commercial software are restricted.

Although the software packages differed, the solution methods are still the same. All of the fundamental equations that govern the process are included in the software (heat transfer, radiative losses, phase change, etc.), with only some specific material properties left as inputs. The boundary conditions are what the user applies to the body (both thermally and mechanically) to simulate the EB-PBF process. The boundary conditions are used to model the EB on the part, where the powder bed is fixed and the thermal conditions of the bed walls. The fundamental equations and boundary conditions are discretized and then solved using FEM in the software.

Overall, solving EB-PBF-AM process models using FEM has several benefits. First, the FEM is much simpler than the LB method. The LB method requires the generation of a powder bed to assign cell properties, since fluid flow is being modeled. Using FEM, the actual flow within the melt pool can be ignored completely. The powder bed can be assigned bulk properties modeling its behavior more simplistically. This way, the EB-PBF process is not accurately modeled; however, the melt pool evolution is not desired. The second benefit is that the residual stress and strain field of the part can be calculated. Part deformation during processing is a significant issue with EB-PBF and it cannot be effectively modeled using the LB method.

5.3.5.1   Micro- and Macro-Scale Modeling of the EB Heat Source

As with the LB method, the EB has to be included in the model as a heat source of some kind. Several efforts in literature use two major categories of models: micro-scale and macro-scale. The micro-scale model accurately models the EB on the powder bed and is found to be the most

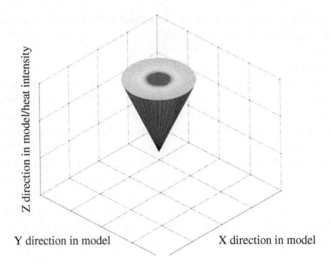

Figure 5.21 Gaussian EB model using the constant absorption profile. Source: Redrawn and adapted from [48].

common model used. For the micro-scale models, the EB is modeled as having a Gaussian in-plane distribution. This shape was used by papers [48, 63, 67, 68]. The Gaussian shape is shown in Figure 5.21.

The Gaussian distribution is fairly simple to model, and another benefit of the distribution is that it is also common for many lasers, which increases the usefulness of the model. The Gaussian distribution occurs in the $x-y$ plane, but as mentioned earlier, the EB has a significant penetration depth into the material. Due to this, the EB cannot be modeled simply as a surface heat source, and it must be modeled as a volumetric heat source. How the EB energy varies in z depends on the absorption profile used.

As discussed earlier, there are two specific absorption models: exponential and constant, and they are used to approximate two different types of penetration (see Figure 5.18). An absorption profile has to be developed, since accurately modeling a large number of reflections and absorption mechanisms with a complex, random powder bed is very difficult. The exponential absorption has a lower total penetration depth, which models more closely the absorption of 60-kV EB machines. With the exponential model, the energy at a specific depth is modeled by a decaying exponential function. This means more energy is absorbed in the top few layers. The constant absorption allows the EB energy to penetrate deeper into the material and more closely models higher accelerating voltages (120 kV). With constant absorption, the cell layers all absorb approximately the same amount of energy until there isn't any left [63, 68].

The exponential EB absorption is discretized in both x, y, and z, to account for the distribution along all axes. The basic equation for the discretization of the energy assumes that each cell absorbs all the energy allocated to it. However, some cells may be interfaces (making up the FS), partially filled, or even be gaseous (not included in the model or heat absorption) which needs to be accounted for. The energy distribution is calculated recursively not only on the initial energy attributed to each cell but on the quantity of energy that each cell can actually absorb. This allows for more energy to be distributed to other cells if a cell is only partially filled, which models actual absorption better. The fill level for the exponential function has a more tangible effect on the upper layers. The first layers, depending on the fill level, experience the greatest difference. However, since each subsequent cell absorbs a portion of the remaining, by around cells 3 or 4 (below the surface), the fill level effect (from the first cells) is effectively negligible. Only paper [63] used the exponential profile during simulations.

The function for the constant absorption is, as the name implies, constant, though it is still discretized in x, y, and z. As with the exponential, the actual absorption of each cell is recomputed to account for the cell type and fill level. The fill level has a more meaningful effect on the subsequent cells than with the exponential equation. Rather than subsequent cells absorbing more energy, all of the excess energy will be absorbed by more cells, increasing the penetration depth [63]. This profile is what creates a cone shape (pointing into the material) of heat generation. Beam intensity is larger closer to the center and closer to the top. Each x–y slice of the cone has a similar heat distribution, except the maximum temperatures reached will be lower. Paper [68] observed that the constant absorption profile fit best with the actual EB process. Both papers [48, 68] used the constant absorption profile exclusively during the simulation.

The macro-scale simulation was done by papers [49, 67]. Paper [49] modeled the production of an entire part, which will be otherwise difficult if done by the micro-scale model. The approximation used by the macro-scale approach is to assume that, since the EB process occurs very quickly, the surface temperature of a layer after scanning is approximately at a uniform temperature. Paper [67] also carried out macro-scale modeling, but only to compare it with the micro-scale model with actual simulations being limited in scale.

While this simplification does not allow one to see thermal buildups in areas based on scan pathing, it does simplify the process a great deal and makes simulating the full build process feasible. From the work in [49], it appears that for larger parts, the trade-off of speed versus accuracy can be useful.

5.3.5.2   Temperature Distribution During EB-PBF

The temperature distribution is important to model during the EB-PBF process for several reasons. First, the material must usually get up to a temperature where it will melt. If solid-state sintering is done, the material must be hot enough for long to ensure the necks between powders grow. In both cases, the temperature value is critical. Second, the temperature profile (heating and cooling) determines the resultant microstructure of the layer. Thirdly, temperature buildups and how the heat from a layer is distributed to previous layers will affect the initial microstructure that formed. All of this information needs to be known to get useful results from the simulation, but also to be able to apply the proper mechanical properties to specific areas of the part.

As mentioned earlier, the process can be modeled in two different ways. Paper [67] used both the micro- and macro-methods during simulation, see Figure 5.22a and b, respectively.

First, the process is modeled on a micro-scale level to accurately capture the EB absorption. This is done to ensure that the Gaussian approximation used works properly and the temperatures reached are reasonable. The second way is on a macro-scale to see the temperature evolution of the entire part. Most papers only perform the micro-scale analysis when only small sections of a few layers are being modeled, and higher accuracy is desired over computation time.

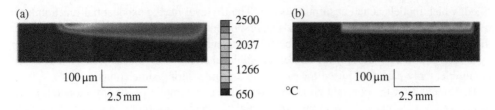

Figure 5.22 (a) Micro-scale simulation temperature distribution for a scan line. (b) Macro-scale temperature distribution for a 20-mm wide rectangle. Source: Reprinted from [67] with permission from Elsevier.

Other common aspects of all the models are the boundary conditions for the powder bed. For all models, the powder bed walls and bottom were assumed to be insulated and at a constant temperature (Dirichlet boundary). Most models accounted for radiative heat losses from the top surface, but not all. Papers [48, 67–69] all consider radiative heat losses. Paper [49] did not include them in the model to simplify it. Paper [49] compensated for the lack of losses by applying a lower uniform temperature to the surface. Paper [66] also did not include radiative heat losses. This paper was concerned with comparing different FEM solving strategies rather than getting actual results, which is why the simplification was deemed acceptable. Paper [68] incorporated convective heat losses into the model, since the goal was to model both EB and laser processing. During simulations for EB, the convective heat coefficient was simply set to zero to simulate a vacuum (no convective losses). In addition to the setup (boundary conditions), all of the papers accounted for changes in mechanical and thermal properties due to changes in temperature during processing. Mechanical property changes were especially important for simulations, including stress distributions. Finally, all of the models included, in some form or another, cooling between layers. This was done to accurately model the actual EB-PBF process, where it is typical to let the layer cool back down to its initial temperature before processing the next.

Heating and Cooling Rate

The heating and cooling rate are trivial to find once the temperature distribution has been found. Temperature is found not only over spatial variables but over time as well. The temperature change rate can be found using the temperature values for each time step. Heating and cooling rates are important primarily for determining the microstructure of parts.

5.3.5.3 Stress Distribution During EB-PBF

The stress distribution is used to find strain (based on part properties). The amount of deformation the part undergoes during processing is critical to the final dimensional tolerances. Most papers only found the residual stress within the part and did not proceed further and find the strain distribution. Nevertheless, the stress or strain field is derived from the temperature field and the mechanical properties [70]. The EB-PBF problem is a thermomechanical one, meaning the FEM solves both the thermal and mechanical problems. The stresses/strains are solved along with the temperature distribution using the same methodology, since mechanical properties are directly linked to the temperature. The entire methodology for solving the equations does not change, only the specific equations are different. Like with the thermal analysis, the stress field uses Dirichlet and Neumann boundary conditions on the body boundary.

As with the temperature field, the process of finding stress or strain is done in a layerwise approach. Displacement of elements can occur in x, y, and z directions. In all models, the walls of the powder bed are fixed, but the baseplate which is sitting in the powder is allowed to move. The baseplate has perfect bonding with the powder and the movement of one affects the other. Though still in powder form, the material is assumed to be sintered to some degree since it is at an elevated preheat temperature. The stress distribution and the material properties (based on the temperature distribution) are used to calculate strain [49, 66, 68, 69].

The strain field evolution for production simulated in [49] is shown in Figure 5.23. The temperature evolution of this build process is shown in the paper [49], and it is seen that the part does not have a uniform temperature during processing. In the first layer, the rectangular blocks cool quickly, with much of the heat being dispersed in the base plate. When the heat is dispersed, the base plate has a nonuniform temperature profile. As the process continues, the rectangular blocks stay at a relatively uniform temperature, while the base plate and the first few layers have the most variation. The amount of strain in those thermally nonuniform areas is apparent from Figure 5.23. The stress/strain in part is caused by the uneven expansion and contraction of the part. When the part expands or contracts nonuniformly, different areas will be under tension or compression. The most significant amount of stress/strain is expected in areas with temperatures or temperature profiles significantly different than other areas of the part; the results in Figure 5.23 confirm this. While [49] made simplifications to the model, with the uniform temperature layer being the most significant, the model was able to accurately predict results, as can be seen from the comparison to actual parts in Figure 5.24.

5.3.6   Summary

This chapter was devoted to the physics and modeling of LPBF and EB-PBF processes mainly using the concepts/equations discussed in Chapter 4. Fundamental equations governing both processes were discussed and customized for different modeling scenarios. Pros and cons of laser and electron beams were discussed.

Eight heat source models used for LPBF were introduced, and the heat transfer equation was introduced within the context of PBF processes. The dependencies of absorption factor and material properties to powder particle distribution were discussed.

The Lattice Boltzmann (LB) method was introduced to be very useful for modeling melt pool evolution during the EB-PBF process. It is a numerical approximation of the incompressible Navier–Stokes equation. The LB method accounts for phase changes in the material, to more accurately model the movement of material. A free surface is used in the model to isolate the gas phases and mimic the melt pool surface. The LB method requires the generation of a particle bed to assign phases to the cells during initialization. The LB method is limited in its use to modeling melt pool evolution, and it is typically not used for larger simulations or where the stress field is required. FEM is an alternative method for modeling the EB-PBF process. Fluid flow is not modeled during the FEM process, but residual stress and strain can be calculated.

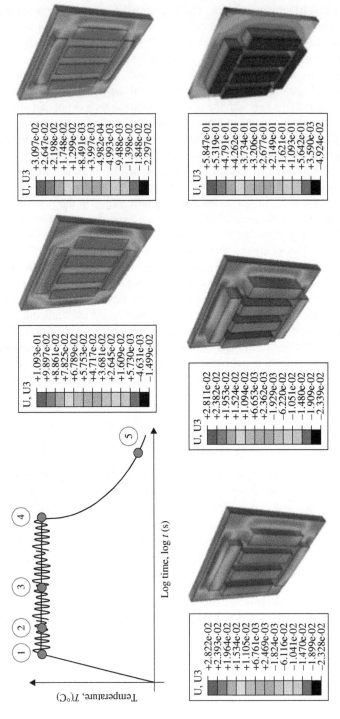

Figure 5.23 Strain evolution of a part during the build process. Source: Reprinted from [49] with permission from Elsevier.

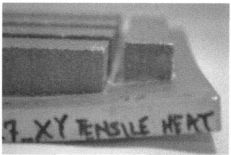

Figure 5.24   Comparison between the FEM model and produced parts. It uses the same scale as in Figure 5.23. Source: Reprinted from [49] with permission from Elsevier.

Both the LB and FEM require the user to assign boundary conditions to the problem. Mechanical boundary conditions need to be applied to define walls and other surfaces. Thermal boundary problems need to be assigned to simulate the EB and the insulating properties of the powder bed walls. Overall, there are a great many similarities between the two processes, though the specific applications of each differ.

References

1. W. E. King, A. T. Anderson, R. M. Ferencz, N. E. Hodge, C. Kamath, S. A. Khairallah, and A. M. Rubenchik, "Laser powder bed fusion additive manufacturing of metals; physics, computational, and materials challenges," *Appl. Phys. Rev.*, vol. 2, no. 4, p. 041304, 2015.
2. F. Klocke, C. Wagner, and C. Ader, "Development of an integrated model for selective metal laser sintering," *Proc. CIRP Int. Semin. Manuf. Syst.*, no. 1, pp. 387–392, 2003.
3. I. Yadroitsev, Selective laser melting: direct manufacturing of 3D-objects by selective laser melting of metal powders, LAP LAMBERT Academic Publishing, 2009.
4. Y. S. Lee and W. Zhang, "Mesoscopic simulation of heat transfer and fluid flow in laser powder bed additive manufacturing," Int. Solid Free Form Fabr. Symp. Austin, pp. 1154–1165, 2015.
5. T. Mukherjee, H. L. Wei, A. De, and T. DebRoy, "Heat and fluid flow in additive manufacturing—Part I: Modeling of powder bed fusion," *Comput. Mater. Sci.*, vol. 150, no. April, pp. 304–313, 2018.
6. I. Yadroitsev and I. Smurov, "Surface morphology in selective laser melting of metal powders," *Phys. Procedia*, vol. 12, pp. 264–270, 2011.
7. S. a. G. D. Correia, J. Lossen, M. Wald, K. Neckermann, and M. Bähr, "Selective laser ablation of dielectric layers," Proc. 22nd Eur. Photovolt. Sol. Energy Conf., pp. 1061–1067, 2007.
8. Y. L. Yao, H. Chen, and W. Zhang, "Time scale effects in laser material removal: A review," *Int. J. Adv. Manuf. Technol.*, vol. 26, no. 5–6, pp. 598–608, 2005.
9. E. Toyserkani, A. Khajepour, and S. Corbin, Laser cladding, CRC Press, 2004.
10. W. M. Steen and J. Mazumder, Laser material processing, Springer-Verlag London, 2010.
11. Z. Zhang et al., "3-Dimensional heat transfer modeling for laser powder bed fusion additive manufacturing with volumetric heat sources based on varied thermal conductivity and absorptivity," *Opt. Laser Technol., vol.* vol. 89, p. 109, 2019.
12. A. Foroozmehr, M. Badrossamay, E. Foroozmehr, and S. Golabi, "Finite element simulation of selective laser melting process considering optical penetration depth of laser in powder bed," *Mater. Des.*, vol. 89, pp. 255–263, 2016.

13. J. Goldak, A. Chakravarti, and M. Bibby, "A new finite element model for welding heat sources," *Metall. Trans. B*, vol. 15, no. 2, pp. 299–305, 1984.

14. A. V. Gusarov and J. P. Kruth, "Modelling of radiation transfer in metallic powders at laser treatment," *Int. J. Heat Mass Transf.*, vol. 48, no. 16, pp. 3423–3434, 2005.

15. H. C. Tran and Y. L. Lo, "Heat transfer simulations of selective laser melting process based on volumetric heat source with powder size consideration," *J. Mater. Process. Technol.*, vol. 255, no. May 2017, pp. 411–425, 2018.

16. L. Ladani, J. Romano, W. Brindley, and S. Burlatsky, "Effective liquid conductivity for improved simulation of thermal transport in laser beam melting powder bed technology," *Addit. Manuf*, vol. 14. pp. 13–23, 2017.

17. S. Liu, H. Zhu, G. Peng, J. Yin, and X. Zeng, "Microstructure prediction of selective laser melting AlSi10Mg using finite element analysis," *Mater. Des.*, vol. 142, pp. 319–328, 2018.

18. A. T. Sutton, C. S. Kriewall, M. C. Leu, and J. W. Newkirk, "Powder characterisation techniques and effects of powder characteristics on part properties in powder bed fusion processes," *Virtual Phys. Prototyp.*, vol. 12, no. 1, pp. 3–29, 2017.

19. Cox E., "A method of assigning numerical and percentage values to the degree of roundness of sand grains," *J. Paleontol.*, vol. 1, no. 3, pp. 179–183, 1927.

20. P. Schneiderhöhn, "Eine vergleichende Studie über Methoden zur quantitativen Bestimmung von Abrundung und Form an Sandkörnern (Im Hinblick auf die Verwendbarkeit an Dünnschliffen.)," *Heidelberger Beiträge zur Mineral. und Petrogr.*, vol. 4, no. 1–2, pp. 172–191, 1954.

21. M. Valdek, K. Helmo, K. Pritt, and M. Besterci, "Characterization of powder particle," *Methods*, vol. 7, pp. 22–34, 2001.

22. C. Wentworth, "Method of measuring and plotting the shapes of pebbles," *U.S. Geol. Surv. Bull.*, vol. 39, pp. 91–114, 1922.

23. V. C. Janoo, "Quantification of shape, angularity, and surface texture of base course materials," CRREL Spec. Rep. 98-1, no. January, p. 22, 1998.

24. E. Kundakc et al., "Thermal and molten pool model in selective laser melting process of Inconel 625," *Int. J. Adv. Manuf. Technol.*, vol. 95, pp. 3977–3984, 2018.

25. U. Ali et al., "On the measurement of relative powder bed compaction density in powder bed additive manufacturing processes," *J. Mater. Des.*, vol. 155, pp. 495–501, 2018.

26. S. S. Sih and J. W. Barlow, "The prediction of the emissivity and thermal conductivity of powder beds," *Part. Sci. Technol.*, vol. 22, no. 4, pp. 427–440, 2004.

27. S. I. Shahabad et al., "Heat source model calibration for thermal analysis of laser powder bed fusion," *Int. J. Adv. Manuf. Technol.*, vol. 106, pp. 3367–3379, 2020.

28. P. Promoppatum, S. Yao, P. C. Pistorius, and A. D. Rollett, "A comprehensive comparison of the analytical and numerical prediction of the thermal history and solidification microstructure of Inconel 718 Products made by laser powder bed fusion," *Engineering*, vol. 3, no. 5, pp. 685–694, 2017.

29. Y. Mahmoodkhani et al., "On the measurement of effective powder layer thickness in laser powder bed fusion additive manufacturing of metals," *Prog. Addit. Manuf.*, vol. 4, pp. 109–116, 2019.

30. T. Debroy and S. A. David, "Physical processes in fusion welding," *Rev. Mod. Phys.*, vol. 67, no. 1, pp. 85–112, 1995.

31. A. Block-Bolten and T. W. Eagar, "Metal vaporization from weld pools," *Metall. Trans. B*, vol. 15, pp. 461–469, 1984.

32. P. Mercelis and J. P. Kruth, "Residual stresses in selective laser sintering and selective laser melting," *Rapid Prototyp. J.*, vol. 12, no. 5, pp. 254–265, 2006.

33. C. Li, Z. Y. Liu, X. Y. Fang, and Y. B. Guo, "Residual stress in metal additive manufacturing," *Procedia CIRP*, vol. 71, pp. 348–353, 2018.

34. M. Shiomi, K. Osakada, K. Nakamura, T. Yamashita, and F. Abe, "Residual stress within metallic model made by selective laser melting process," *CIRP Ann. Manuf. Technol.*, vol. 53, no. 1, pp. 195–198, 2004.

35. D. Buchbinder, W. Meiners, N. Pirch, K. Wissenbach, and J. Schrage, "Investigation on reducing distortion by preheating during manufacture of aluminum components using selective laser melting," *J. Laser Appl.*, vol. 26, no. 1, p. 012004, 2014.

36. M. F. Zaeh and G. Branner, "Investigations on residual stresses and deformations in selective laser melting," *Prod. Eng.*, vol. 4, no. 1, pp. 35–45, 2010.

37. T. Vilaro, C. Colin, and J. D. Bartout, "As-fabricated and heat-treated microstructures of the Ti-6Al-4V alloy processed by selective laser melting," *Metall. Mater. Trans. A Phys. Metall. Mater. Sci.*, vol. 42, no. 10, pp. 3190–3199, 2011.

38. J.-P. Kruth, J. Deckers, E. Yasa, and R. Wauthlé, "Assessing and comparing influencing factors of residual stresses in selective laser melting using a novel analysis method," *Proc. Inst. Mech. Eng. Part B J. Eng. Manuf.*, vol. 226, no. 6, pp. 980–991, 2012.

39. X. Song et al., "Residual stresses and microstructure in Powder Bed Direct Laser Deposition (PB DLD) samples," *Int. J. Mater. Form.*, vol. 8, no. 2, pp. 245–254, 2015.

40. J. Zhang, Y. Zhang, W. Hoh, L. Wu, and H. Choi, "A multi-scale multi-physics modeling framework of laser powder bed fusion additive manufacturing process," *Met. Powder Rep.*, vol. 73, no. 3, pp. 151–157, 2018.

41. M. S. Daw, S. M. Foiles, M. I. Baskes, "The embedded-atom method: A review of theory and applications," *Mater. Sci. Rep.*, vol. 9, no. 7–8, 1993.

42. M. J. Matthews, G. Guss, S. A. Khairallah, A. M. Rubenchik, P. J. Depond, and W. E. King, "Denudation of metal powder layers in laser powder bed fusion processes," *Acta Mater.*, vol. 114, pp. 33–42, 2016.

43. S. A. Khairallah, A. T. Anderson, A. Rubenchik, and W. E. King, "Laser powder bed fusion additive manufacturing: Physics of complex melt flow and formation mechanisms of pores, spatter, and denudation zones," *Acta Mater.*, vol. 108, pp. 36–45, 2016.

44. H. K. Rafi, D. Pal, N. Patil, T. L. Starr, and B. E. Stucker, "Microstructure and mechanical behavior of 17-4 precipitation hardenable steel processed by selective laser melting," *J. Mater. Eng. Perform.*, vol. 23, no. 12, pp. 4421–4428, 2014.

45. A. Rubenchik, S. Wu, S. Mitchell, I. Golosker, M. LeBlanc, and N. Peterson, "Direct measurements of temperature-dependent laser absorptivity of metal powders," *Appl. Opt.*, vol. 54, no. 24, p. 7230, 2015.

46. X. Gong, T. Anderson, and K. Chou, "Review on powder-based electron beam additive manufacturing technology," *Manuf. Rev.*, vol. 1, pp. 1–12, 2014.

47. V. Adam, U. Clauß, D. Dobeneck, T. Krüssel, and T. Löwer, Electron beam welding – The fundamentals of a fascinating technology. pro-beam AG & Co. KGaA, pp. 1–98, 2011.

48. J. Romano, L. Ladani, J. Razmi, and M. Sadowski, "Temperature distribution and melt geometry in laser and electron-beam melting processes – A comparison among common materials," *Addit. Manuf.*, vol. 8, pp. 1–11, 2015.

49. P. Prabhakar, W. J. Sames, R. Dehoff, and S. S. Babu, "Computational modeling of residual stress formation during the electron beam melting process for Inconel 718," *Addit. Manuf.*, vol. 7, pp. 83–91, 2015.

50. H. Giedt and L. N. Tallerico, "Prediction of electron beam depth of penetration," *Weld. Res. Suppl.*, pp. 299–305, 1988.

51. S. I. Molokovsky and A. D. Sushkov, "Electron guns," in Intense electron and ion beams, Springer, Berlin, New York, pp. 125–155, 2005.

52. H. S. Fricker, "Why does charge concentrate on points?," *Phys. Educ.*, vol. 24, no. 3, pp. 157–161, 1989.

53. A. Khursheed, Scanning electron microscope optics and spectrometers, World Scientific, 2011.

54. P. Lulai, "Determination of filament work function in vacuum," 2001.

55. R. a. Serway and L. D. Kirkpatrick, Physics for scientists and engineers with modern physics, Cengage Learning, 1988.

56. Y. Tian, C. Wang, D. Zhu, and Y. Zhou, "Finite element modeling of electron beam welding of a large complex Al alloy structure by parallel computations," *J. Mater. Process. Technol.*, vol. 199, no. 1, pp. 41–48, 2008.

57. R. Benzi, S. Succi, and M. Vergassola, "The lattice Boltzmann equation: Theory and applications," *Phys. Rep.*, vol. 222, no. 3, pp. 145–197, 1992.

58. R. Ammer, M. Markl, U. Ljungblad, C. Körner, and U. Rüde, "Simulating fast electron beam melting with a parallel thermal free surface lattice Boltzmann method," *Comput. Math. Appl.*, vol. 67, no. 2, pp. 318–330, 2014.

59. E. D. T. Fakult and G. Doktor-ingenieur, "Simulation of selective electron beam melting processes Elham Attar," PhD thesis, 2011.

60. T. Zacharia, J. M. Vitek, J. a. Goldak, T. a. DebRoy, M. Rappaz, and H. K. D. H. Bhadeshia, "Modeling of fundamental phenomenon in welds," *Model. Simul. Mater. Sci. Eng.*, vol. 3, pp. 265–288, 1995.

61. M. Markl, R. Ammer, U. Rüde, and C. Körner, "Numerical investigations on hatching process strategies for powder bed-based additive manufacturing using an electron beam," *Int. J. Adv. Manuf. Technol.*, vol. 78, no. 1–4, pp. 239–247, 2015.

62. C. Körner, E. Attar, and P. Heinl, "Mesoscopic simulation of selective beam melting processes," *J. Mater. Process. Technol.*, vol. 211, no. 6, pp. 978–987, 2011.

63. M. Markl, R. Ammer, U. Ljungblad, U. Rüde, and C. Körner, "Electron beam absorption algorithms for electron beam melting processes simulated by a three-dimensional thermal free surface lattice boltzmann method in a distributed and parallel environment," *Procedia Comput. Sci.*, vol. 18, pp. 2127–2136, 2013.

64. E. Attar and C. Körner, "Lattice Boltzmann method for dynamic wetting problems," *J. Colloid Interface Sci.*, vol. 335, no. 1, pp. 84–93, 2009.

65. K. Iglberger and U. Rüde, "Large-scale rigid body simulations," *Multibody Syst. Dyn.*, vol. 25, no. 1, pp. 81–95, 2011.

66. D. Riedlbauer, P. Steinmann, and J. Mergheim, "Thermomechanical finite element simulations of selective electron beam melting processes: Performance considerations," *Comput. Mech.*, vol. 54, no. 1, pp. 109–122, 2014.

67. X. Tan et al., "An experimental and simulation study on build thickness dependent microstructure for electron beam melted Ti-6Al-4V," *J. Alloys Compd.*, vol. 646, pp. 303–309, 2015.

68. G. Vastola, G. Zhang, Q. X. Pei, and Y. W. Zhang, "Modeling and control of remelting in high-energy beam additive manufacturing," *Addit. Manuf.*, vol. 7, pp. 57–63, 2015.

69. B. Cheng and K. Chou, "Geometric consideration of support structures in part overhang fabrications by electron beam additive manufacturing," *CAD Comput. Aided Des.*, vol. 69, pp. 102–111, 2015.

70. H. J. Stone, S. M. Roberts, and R. C. Reed, "A process model for the distortion induced by the electron-beam welding of a nickel-based superalloy," Metall. Mater. Trans. A, vol. 31, no. September, pp. 2261–2273, 2000.

6

Binder Jetting and Material Jetting

Physics and Modeling

Learning Objectives

At the end of this chapter, you will be able to:

- Understand the basic governing physics of droplet formation in binder jetting and material jetting.
- Understand the basic governing physics of droplet–surface and droplet–powder bed interaction in material jetting and binder jetting processes, respectively.
- Learn about the numerical methods used to model material jetting and binder jetting.

6.1 Introduction

This chapter covers the governing physics of material jetting (MJ) and binder jetting (BJ) as both these processes start with the same step: droplet formation in the printhead. The impact of droplets on the substrate in MJ and the infiltration of the binder in the BJ powder bed process are described separately, along with their associated physics and available models.

Readers can refer to Chapter 2 for a detailed description of the working mechanism of BJ and MJ processes and the advantages and disadvantages associated with each process. As a summary, a printhead containing hundreds of small orifices ejects monodisperse droplets of a binding fluid with volumes in the order of picoliter onto a powder bed in the BJ process. The droplets are absorbed by the powder bed to fill the empty gaps between the powder particles.

Metal Additive Manufacturing, First Edition. Ehsan Toyserkani, Dyuti Sarker, Osezua Obehi Ibhadode, Farzad Liravi, Paola Russo, and Katayoon Taherkhani.

The binder acts as a glue and holds the cross section of each layer together while creating a bond between each layer and its immediate top and bottom layers. After the part is printed, it will be exposed to a relatively elevated temperature to cure the binder and create a green part. The green part is then sintered in order to remove the binder and fuse the particles together, so that a final part with improved mechanical properties is obtained.

The MJ process functions more or less similar to BJ in that the building blocks are made as a continuous body of liquid (jet) is broken into disconnected masses (droplet). The major difference between the two manufacturing systems is that these droplets are the main building material in MJ rather than just a binding agent. As droplets impact the substrate, they will normally spread to form semi-oval cross-sections under the effect of capillary, inertial, and gravitational forces until they are stable and then merge with other droplets to form a layer. The deposited material is solidified via evaporation, chemical reaction, or photo-polymerization to form the final product.

The manufacturing procedure in these two processes can be divided into four steps:

1. Droplet formation,
2. Droplet landing,
3. Droplet interaction with the substrate in MJ and with powder bed in BJ, and
4. Solidification.

With the diverse applications of these two manufacturing methods, it is important to understand the physical phenomena that control every step of the journey of droplets from formation to solidification. These physical phenomena, specifically the dynamics of droplet formation, have been extensively studied. Numerical models have also been developed to predict the jet rupture into droplets with high accuracy over the past few decades. The basics of these models are described in Sections 6.2 and 6.3. It is to be noted that the majority of these models are developed to describe the standard Newtonian fluids, which can only describe the behavior of some of the aqueous binders used in BJ. There have been efforts to model the effects of nonlinear rheological properties of fluids with long polymeric chains, widely used in both BJ and MJ [1]. However, the research on drop formation dynamics where the material of choice is liquid metal with high viscosity or highly viscous polymers containing metal particles is sparse and this topic needs to be further studied. These are the type of materials that need to be focused on to better understand the metal BJ-AM and metal MJ-AM and apply such knowledge to devise control algorithms to improve the quality of droplet generation, and ultimately the fabricated parts where these processes are used.

Different inkjet printing mechanisms are available, with continuous inkjet and drop-on-demand (DOD) inkjet being the most mainstream technologies. For a review of the history of inkjet technology, refer to [2, 3]. In the continuous inkjet printing process, a liquid column is ejected from an orifice under the influence of the backpressure. The liquid column is broken into a continuous stream of uniform droplets due to Rayleigh instabilities. Droplets are charged electrically, so that those droplets that are not intended to be jetted can be deflected and collected in a gutter and recycled if their properties are not changed due to interaction with the environment. The DOD inkjet printing is a more accurate process with less waste of materials and higher resolution in which a droplet is generated by applying pressure to the liquid chamber using either a thin heater (thermal DOD) or a piezoelectric (PZT) actuator (piezoelectric/mechanical DOD).

In thermal DOD, developed in the 1980s, the heater which is located near the orifice attached to the liquid chamber heats up when actuated by an electric voltage and evaporates the binder to create a bubble. The expansion of this vapor bubble generates a pressure large enough to eject the liquid out of the nozzle where a droplet is formed.

The piezoelectric DOD was developed in the late 1970s and has been the most widely used method of jetting for industrial applications, including AM. As such, PZT DOD systems are the focus of this chapter. The piezoelectric actuator attached to the liquid chamber expands or contracts based on the electric charge applied to it (inverse piezoelectric effect) and generates a pressure pulse. If this pressure is large enough to overcome the surface tension of the meniscus at the nozzle, a drop will be ejected. Different inkjet printing techniques are schematically demonstrated in Figure 6.1 [4].

6.2 Physics and Governing Equations

6.2.1 Droplet Formation

The formation of droplets due to fluid instabilities has long fascinated scientists as a classical problem of the fluid mechanics. This fascination has resulted in the generation of a great wealth of publications on the underlying physics of this phenomenon, starting with the work of scientists like Savart, Rayleigh, and Maxwell (see [3] for more details).

A column of the liquid ejecting from a nozzle starts to break into a droplet as the interfacial tensions try to minimize the surface area leading to capillary instabilities [5]. This process is highly affected by the properties of the liquid, such as viscosity and surface tension, as well as the printhead process parameters such as the shape and amplitude of pressure signals generated by the piezoelectric transducer. A typical droplet formation process in a PZT DOD printhead is described by Derby et al. in Figure 6.2. It is shown that the liquid column forms a droplet with a

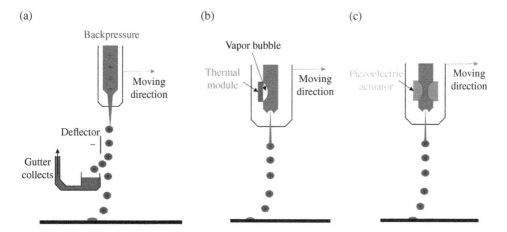

Figure 6.1 Schematic of (a) continuous inkjet printing showing the working mechanism of the deflector; (b) drop-on-demand thermal inkjet printing; (c) drop-on-demand piezoelectric inkjet printing. Source: Redrawn and adapted from [4].

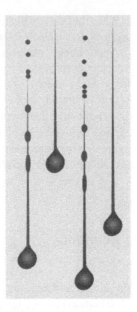

Figure 6.2 The droplet formation and flight in inkjet printing. The tail of the droplet might break into satellite droplets, which is not desirable. Source: Redrawn and adapted from [4].

leading tail, which is separated from the column very quickly at a region called the pinch-off region. At this point, the droplets have a semi-spherical shape and usually move with speed in the orders of m/s [3], and their diameter is generally the same as the nozzle diameter or slightly larger. Depending on the jetting and material properties, the tail might itself break into smaller droplets called satellites. These satellites might merge with the main droplet as it moves down its fly trajectory toward the substrate or reach the substrate after the main droplets. The latter has a detrimental effect on the resolution and quality of the print which is tried to be avoided through optimizing the process parameters such as nozzle stand-off distance, PZT voltage amplitude, and frequency among others [4].

Before describing the governing physics for the droplet formation, two important concepts should be explained. The droplet breakup, i.e., the pinching phenomenon, is considered a finite-time singularity and a self-similarity problem. These two concepts are very important and allow the droplet breakup to be explained by universal scaling laws and the Navier–Stokes equation. As described by Egger, the radius of curvature of droplet approaches zero near the singularity point at finite time while the strong forces acting upon the small volume of matter at the pinch-off lead to infinite velocity turning the droplet separation into a "singularity of the equations of motion" problem [3]. It is also mentioned that the main difference from the singularity at finite time that arises from surface tensions is the fact that there is an ability to explain breakage alone in terms of continuum mechanics without adopting the microscopic perception for explanation [3]. Regarding self-similarity, Egger elaborates on this concept that it occurs in a situation where the length scale is normally absent. If singularity happens, the solution length scale depends on small time steps [3].

6.2.1.1   Material Properties and Dimensionless Numbers

Viscosity, density, and surface tension of the jetting material directly affect the droplet formation process. As mentioned in Chapter 2, three important dimensionless numbers relate these
parameters together that are used to predict the dynamics of droplets:

1. Reynolds number (Re),
2. Weber number (We), and
3. Ohnesorge number (Oh).

These parameters are defined as:

$$Re = \frac{\rho v a}{\mu} \tag{6.1}$$

$$We = \frac{\rho v^2 a}{\gamma} \tag{6.2}$$

$$Oh = \frac{\sqrt{We}}{Re} = \frac{\mu}{\sqrt{\gamma \rho a}} \tag{6.3}$$

where ρ is the density (kg/m^3), $v = Q/\pi R^2$ is the velocity of droplet (m/s), in which Q is the
constant flow rate through a nozzle (m^3/s), a is the characteristic of length (m), μ is the viscosity
(N/(s.m^2)), and γ is the surface tension (N/m). Re is the ratio of the inertial and viscous forces,
and We is the ratio of inertia and surface tension. Oh number which is the ratio of We and Re
numbers, independent of the droplet speed, can be used to determine if a liquid is jettable based
on its viscosity, surface tension, density, and the geometry of the nozzle. Derby identifies
liquids with Oh in the range of 0.1–1 to be jettable. If $Oh \gg 1$, it means that the viscous forces
are dominant, and as such, the material is too viscous to break up into a droplet. On the other
hand, if $Oh < 0.1$, the generation of satellite droplets can be predicted. Refer to Figure 2.38 for
the printability plot based on the We and Re numbers. It is to be noted that the assumption here is
that the jetting material is a Newtonian fluid. The behavior of the droplet can be quite different
for non-Newtonian polymers and inks containing particle additives.

In some other references, other dimensionless numbers such as capillary number Ca and
gravitational Bond number G are used to describe the dynamics of droplet formation (in addition to Reynolds number); however, according to Basaran et al. [2], both sets of dimensionless
parameters can be used for this purpose as $We = Re \times Ca$ and $Oh = \frac{Ca}{Re^*}$ where Re^* is the prop-
erty-based Reynolds number. These numbers are expressed as:

$$Ca = \mu v / \gamma \tag{6.4}$$

$$G = \rho g a^2 / \gamma \tag{6.5}$$

$$Re^* = \frac{\sqrt{\rho a \gamma}}{\mu} \tag{6.6}$$

It was also mentioned that the energy exerted to the meniscus by the piezoelectric transducer (or any other form of backpressure generation) should overcome the surface tension of liquid at the nozzle to form a jet. According to Derby [4], a *We* number greater than 4 can be a suitable measure for the generation of sufficient energy to overcome the surface tension, and the minimum velocity of the droplet can be obtained from the following equation:

$$v_{min} = \left(\frac{2\gamma}{\rho R}\right)^{1/2}$$

(6.7)

6.2.1.2 Scaling Theories of Pinch-off

As mentioned previously, the droplet breakup is a finite-time singularity problem with identical conditions in the pinching zone no matter the global conditions of the experiment [2]. The pinching zone is shown in Figure 6.3a. The dynamics of droplet breakup at the singular point for Newtonian noncompressible fluids surrounded by a passive gas (ambient condition) can be described by three theories known as the scaling theories of pinch-off. The first scaling theory was proposed by Keller and Miksis [7] for inviscid (low viscous) liquid (inertial regime [I]). Later on, Papageorgiou [8] developed the second theory for high viscous liquids (viscous regime [V]). Finally, Eggers [9] introduced the last theory for inertial–viscous (I–V) regime under the assumption that all three influential forces, i.e., capillary, inertial, and gravitational, are in balance as the droplet breakup happens [2]. As the breakup point approaches, the initial I and V regimes are no longer able to describe the droplet formation behavior, and a transition of

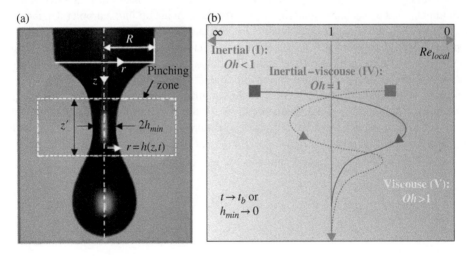

Figure 6.3 (a) The thinning of the jet on the onset of droplet formation in the pinching zone. (b) The trajectory of the path a low viscous ($Oh < 1$) and highly viscous ($Oh > 1$) liquid follow to reach the I–V regime. Source: Reproduced with permission from PNAS [6].

regime from I or V to I–V happens. Castrejón-pita et al., however, demonstrated that these transitions are not very direct and some temporary intermediate transitions take place, as shown in Figure 6.3b [6]. The description of the scaling theories described below is based on the work of Castrejón-pita et al. [6].

As shown in Figure 6.3a, a cylindrical coordinate system has been used to model the three above-mentioned regimes with r and z as radial and axial axes, respectively. If we consider z_0 as the location at which the column will break, z' will be the length scale of the pinching zone, $r = h$ (z, t) is the shape function describing the profile of the pinching zone, and h_{min} is the minimum radius of the liquid column.

Inertial regime (I): To make the time dimensionless, a capillary time is defined as:

$$\tau_c = (t_b - t)/t_I \tag{6.8}$$

where the inertial–capillary time is defined as

$$t_I = t_c = \sqrt{\rho R^3/\gamma} \tag{6.9}$$

and t_b is the moment when $h_{min} = 0$. Suppose the liquid has an ignorable viscosity ($h = 0$). In this case, pinching happens at an inertial regime where inertial and capillary forces balance, and inertial forces become dominant. In this situation, h_{min}, Re, and axial velocity (v) vary with τ as:

$$\frac{h_{min}}{R} \sim \frac{z'}{R} \sim \tau^{2/3} \tag{6.10}$$

$$Re\,(t) \sim \frac{1}{Oh}\tau^{1/3} \tag{6.11}$$

$$\frac{v}{v_c} \sim \tau^{-1/3} \tag{6.12}$$

where $z' = z - z_0$ and v_c is the characteristic velocity. In real situations, one cannot find a fluid with zero viscosity. As a result, for these situations ($Oh \ll 1$), the Reynolds number as a function of time can be approximated as:

$$Re\,(t) \sim \frac{\rho z' v}{\mu} \sim \frac{1}{Oh}\tau^{1/3} \tag{6.13}$$

Viscous regime (V): In the opposite case where the viscosity is the dominant force ($Oh = \infty$), viscous and capillary forces balance and h_{min} and $Re(t)$ vary with τ as:

$$\frac{h_{min}}{R} \sim \tau \tag{6.14}$$

$$\frac{z'}{R} \sim \tau^{\beta} \tag{6.15}$$

$$Re\left(t\right) \sim \frac{1}{Oh^2}\tau^{2\beta-1} \tag{6.16}$$

$$\frac{v}{v_c} \sim \tau^{\beta-1} \tag{6.17}$$

where $\tau_c = (t_b - t)/t_V$ while $t_V = t_c = \mu R/\gamma$ is the viscous–capillary time, and $\beta = 0.175$. Similarly, the viscosity value should be finite here; as such, $Re(t)$ for $(Oh \gg 1)$ can be approximated as:

$$Re\left(t\right) \sim \frac{\rho z'v}{\mu} \sim \frac{1}{Oh^2}\tau^{2\beta-1} \tag{6.18}$$

The difference between the necking and breakup behavior for a low viscous and high viscous liquid is shown in Figure 6.4.

Inertial–viscous regime (I–V): Equations (6.13) and (6.18) can only describe the pinching behavior at the early stages of thinning (as $\tau \to 0$, $Re(t)$ approaches 0 for I regime and ∞ for V regime). So near the breakup point, a third regime is required to explain the pinching dynamics wherein all three forces of viscous, inertial, and capillary balance. Here, it can be shown that:

$$\frac{h_{min}}{I_\mu} \sim \tau \tag{6.19}$$

$$\frac{z'}{I_\mu} \sim \tau^{1/2} \tag{6.20}$$

$$Re\left(t\right) \sim 1 \tag{6.21}$$

$$\frac{v}{v_c} \sim \tau^{-1/2} \tag{6.22}$$

where $v_c = \frac{I_\mu}{t_\mu} = \frac{\gamma}{\mu}$, $I_\mu \equiv \frac{\mu^2}{\rho\gamma}$ is the viscous length, and $t_c = t_\mu \equiv \frac{\mu^3}{\rho\gamma^2}$ is the viscous time.

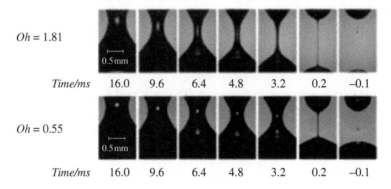

$Oh = 1.81$

0.5 mm

Time/ms   16.0   9.6   6.4   4.8   3.2   0.2   −0.1

$Oh = 0.55$

0.5 mm

Time/ms   16.0   9.6   6.4   4.8   3.2   0.2   −0.1

Figure 6.4 The pinch-off of liquid column as time passes for a liquid with relatively high viscosity ($Oh = 1.81$) and a liquid with relatively low viscosity ($Oh = 0.55$). Source: Reproduced with permission from PNAS [6].

6.2.2   Droplet–Substrate Interaction

When a single droplet impacts the substrate, its boundaries move until a steady state is reached. In the MJ process, reaching the stable condition might be more complex as the droplet will contact previously laid neighboring droplets. It should merge with them to form a layer followed by solidification to form a 3D structure.

Similar to droplet formation, the droplet–substrate interaction is controlled by capillary, inertial, and gravitational forces and can be explained using dimensionless numbers like Re, We, Oh, and G. In addition, the surface characteristics affect the final diameter of the droplet at equilibrium as well. For example, surface wettability highly influences the dynamics of droplet impact. Factors such as the chemical composition of the surface, temperature, surface texture among others affect the surface wettability. The contact angle (θ) is the parameter used to quantify the surface wettability. For example, when in contact with a water droplet, a surface with a value of $\theta < 90°$ is called *hydrophilic* while a surface with $\theta > 90°$ is called *hydrophobic* as shown in Figure 6.5. It can be observed that the lower wettability corresponds to lower spreading (smaller d_{con}).

The general physical phenomenon controlling the spreading and coalescence of droplets in MJ by Derby is an excellent source for understanding this process and is the source of information provided in this section. For further details, please refer to [4].

The presence of six regimes for droplet impact on dry surfaces is shown in Figure 6.6, as pointed out by Moghtadernejad et al. [10]. These six regimes are

1. Deposition,
2. Prompt splash,
3. Corona splash,
4. Receding breakup,
5. Partial rebound, and
6. Complete rebound.

Depending on material, surface, and ambient characteristics, some regimes might be present or not during a specific process.

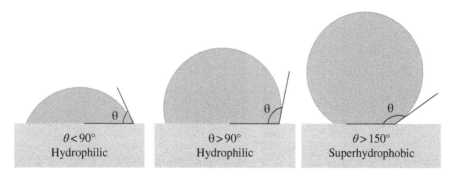

Figure 6.5   Schematic of surface wettability for a droplet of water as a function of contact angle. Source: republished under CC BY-SA 3.0 open access license [10].

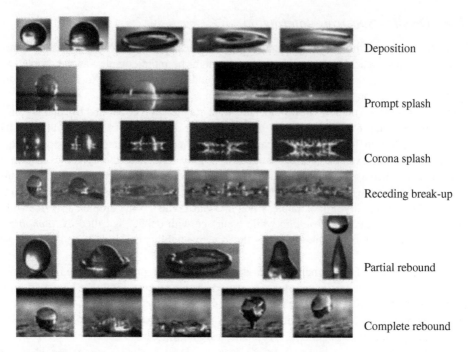

Deposition

Prompt splash

Corona splash

Receding break-up

Partial rebound

Complete rebound

Figure 6.6 Droplet impact regimes on dry surfaces. Source: Republished under CC BY-SA 3.0 open access license [10].

In Derby's work [4], the droplet behavior on the surface is divided into impact-driven and capillary-driven spreading as proposed by Schiaffino et al [30]. The dominant forces for impact-driven and capillary-driven scenarios are inertial forces and capillary forces, respectively. The gravitational forces can be neglected due to the typically small size of the droplets in PZT DOD where $G \to 0$.

The sequence of events in the deposition regime present in PZT DOD is illustrated in Figure 6.7 [4]. To describe the droplet–substrate interaction, a dimensionless time is defined as:

$$t^* = t\left(\frac{v}{d_0}\right) \tag{6.23}$$

| Droplet falling with momentum | Impact | Maximum spreading | Recoiling & oscillation | Capillary-driven spreading |

Figure 6.7 Droplet cross section changes as a function of time from impact to stability.

where d_0 is the droplet diameter (m) and v is the droplet velocity (m/s). As per this number, the following classification can be adopted:

1. $0 < t^* < 0.1$: the initial impact phase where the transfer of kinetic energy from vertical to radial direction happens [10].
2. $0.1 < t^* < 10$: this phase starts with impact-driven spreading. The conversion of kinetic energy to surface energy and heat is reported at the spreading phase [10]. The process continues by recoil due to surface tension, acting as a force to pull the droplet together, and maybe oscillation. This second part happens under the influence of viscous forces.
3. $10 < t^* < 100$: the capillary forces become dominant and control the spreading of the droplet.
4. $t^* > 1000$: at the final phase, equilibrium is reached.

Once the equilibrium is reached, the diameter of the semi-spherical droplet d_{con} can be calculated as [4]:

$$d_{con} = d_0 \sqrt[3]{\frac{8}{\tan \frac{\theta_{eqm}}{2} \left(3 + \tan^2 \frac{\theta_{eqm}}{2}\right)}} \tag{6.24}$$

where θ_{eqm} is the equilibrium contact angle (rad). The smaller the contact angle, the larger the d_{con}. To quantify the level of spreading, another important dimensionless number known as spreading ratio can be defined as:

$$S_{max} = d_{con}/d_0 \tag{6.25}$$

Various other equations have been proposed to calculate the value of S_{max} (see [10, 11]).

Derby also introduces an experimental dimensionless grouping to determine the onset of splashing as proposed by Stow and Hadfield:

$$We^{1/2} Re^{1/4} > f(R) \tag{6.26}$$

where $f(R)$ is a function of surface roughness (e.g., $f(R) \sim 50$ for smooth surfaces).

6.2.3 Binder Imbibition

Similar to the final dimensional footprint of a droplet spread onto a solid substrate in MJ, the final dimensions of the region inside a porous medium wetted by a binder droplet in the BJ system directly determine the resolution of the printed part as well as its quality. It could however be argued that the interaction of droplet and powder particles are more complicated than the interaction of droplet and solid substrate; as such, in the BJ process, not only the positioning of the droplets but also their imbibition should be carefully controlled [12].

The process starts with the droplets impacting the surface of the powder particulates, going through the steps not different from what described in Section 6.2.2 to form a semi-oval cross section while infiltrating into the porous medium known as drainage. Marston et al. describe two limiting cases characterizing the imbibition at this point: (1) constant drawing area (CDA) where the drainage happens while the contact line remains unchanged and the contact angle decreases (radius of curvature increases), and (2) decreasing drawing area (DDA) where the drainage happens while the contact line rescinds and the contact angle remains constant (radius of curvature decreases) [13]:

To calculate the corresponding penetration time for each case, Marston et al. introduce the following equations:

$$t_{CDA} = 1.35 \frac{V_0^{2/3}}{\varphi^2 R_{pore}} \frac{\mu}{\gamma \cos \theta_p} \tag{6.27}$$

$$t_{DDA} = 9 t_{CDA} \tag{6.28}$$

where V_0 is the initial drop volume (m^3), φ is the powder bed porosity (%), μ is the viscosity (N/(s.m^2)), γ is the surface tension (N/m), and θ_p is the apparent contact angle (rad). Also, R_{pore} is the effective pore diameter in the bed (m) that is defined by:

$$R_{pore} = 2\varphi / ((1 - \varphi) s_0 \rho_s) \tag{6.29}$$

where s_0 is the particle-specific surface area (m^2) and ρ_s is the packing density of the pores (kg/m^3). Another parameter used in various models as a measure of the effective pore diameter is

$$R_{eff} = \frac{\phi d_{3,2}}{3} \frac{\varphi_{eff}}{\left(1 - \varphi_{eff}\right)} \tag{6.30}$$

where φ is a shape factor, $d_{3,2}$ is the surface mean particle diameter (m), and

$$\varphi_{eff} = \varphi_{tap} \left(1 - \varphi + \varphi_{tap}\right) \tag{6.31}$$

where φ_{tap} is the tapped porosity that is one minus the ratio of the powder mass to the volume occupied by the powder after being tapped for a defined period.

As seen from Eqs. (6.27) and (6.28), the DDA is longer than CDA. This drainage step for both dry and pre-wetted powder is shown in Figure 6.8.

Various factors such as viscosity of the binder, printing parameters, droplet velocity, wetting properties of the powder, and powder bed packing density might affect the binder infiltration process [14]. At the microscopic level, the droplet fills the pore space under the influence of capillary and inertial forces, with the latter being the primary force. The capillary force expressed as capillary pressure (N/m^2) can be calculated as:

$$P_c = \frac{2\gamma \cos \theta}{R_{pore}} \tag{6.32}$$

The binder imbibition continues through the unsaturated/dry regions until the pressure across the liquid–gas interface balances the wetting forces [12]. It is worth noting that the saturation level, i.e., the level of empty space in the wetted region filled with the binder, is not constant

(a)   (b)   (c)

Figure 6.8 Infiltration of the droplet into (a) dry and (b) - (c) pre-wetted powder bed. Source: Republished with permission from Elsevier [13]

throughout this region as experimentally shown by Munuhe et al. [14] in Figure 6.9 where different local saturation levels are shown with different shades. The effective saturation can be calculated as:

$$S_{eff} = \frac{m_b \rho_{pb}}{m_b \rho_p \left(1 - P_f\right)}$$
(6.33)

where m_b is the mass of the deposited binder, ρ_p is the density of binder, m_b is the mass of the bound powder, ρ_{pb} is the powder bed density, and P_f is the packing fraction.

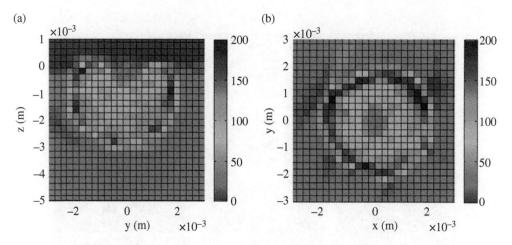

Figure 6.9 The wetted region imaged using micro-CT for one droplet dispensed on a bed of polymer powder: (a) side view and (b) top view at the distance 0.5 mm below the surface. Source: Republished with permission from Elsevier [14].

An important point, seen in Figure 6.9, is the fact that although the wetted region has a semi-spherical cross section as observed or hypothesized by most researchers, a crater might form near the surface. This is mainly attributed to the distribution of powder particles due to the high-velocity impact of binder onto powder bed (up to 10 m/s) under the effect of droplet inertia. Munuhe et al. [14] also attribute the creation of crater to capillary forces at the liquid–air interface during the drainage stage pulling the droplet inward and displacing the particles to form a more compact, less porous region near the surface, as shown in Figure 6.10b. This is amplified by the effect of drag and gravity forces (see Figure 6.10c). On the contrary, at the wetting front inside the powder bed, the inward capillary forces are normally subdued as the liquid further penetrates where the powder disturbance becomes minimal, as shown in Figure 6.10c.

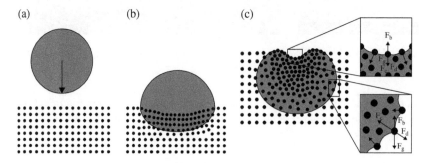

Figure 6.10 The creation of crater geometry: (a) droplet approaching the surface of powder bed, (b) impact and drainage step, and (c) equilibrium. F_c, F_b, F_g, and F_d are the capillary, buoyancy, gravitational, and drag forces respectively. Source: Republished with permission form Elsevier [14].

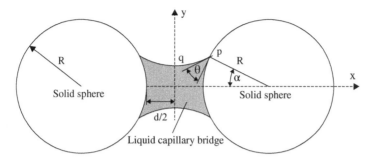

Figure 6.11 Schematic of liquid bridge between two identical spherical particles. Source: Republished with permission from Elsevier [15]

6.2.3.1 Geometrical Features of Liquid Bridge Among Particles

After binder injection in the powder bed, liquid bridges form between particles, resulting in particles agglomeration due to wetting and capillary forces [15]. To study the geometrical features of liquid bridges, a simple case is considered in which two identical spheres are adhered by liquid, as shown in Figure 6.11.

At the equilibrium, the shape of the liquid bridge state can be described by the Young–Laplace equation when the pressure difference ΔP (N/m^2) can be expressed as [15]:

$$\Delta P = \gamma \left(\frac{1}{r_1} + \frac{1}{r_2} \right) \tag{6.34}$$

where γ is the surface tension (N/m), and r_1 and r_2 are the principal curvature radii at the interface (m). This equation is derived in the absence of gravity. The following Young-Laplace equations can be established based on r_1 and r_2 [15]:

$$\frac{\Delta P}{\gamma} = \frac{1}{y \left(1 + \left(\frac{dy}{dx} \right)^2 \right)^{0.5}} - \frac{\frac{d^2 y}{dx^2}}{\left(1 + \left(\frac{dy}{dx} \right)^2 \right)^{1.5}} \tag{6.35}$$

where y is a function of x. An integration of the above equation yields the following equation [15]:

$$\frac{1}{\left(1 + \left(\frac{dy}{dx} \right)^2 \right)^{0.5}} = \frac{\Delta P}{2\gamma} y^2 + C \tag{6.36}$$

Consider points p and q in Figure 6.11, the first derivative of y with respect to x at point p is $\cot(\alpha + \theta)$ and at point q is zero [15]. Thus, C and ΔP can be calculated as [15]:

$$C = y_q y_p \frac{y_q - y_p \sin(\alpha + \theta)}{y_q^2 - y_p^2} \tag{6.37}$$

$$\frac{\Delta P}{2\gamma} = \frac{y_p \sin(\alpha + \theta) - y_q}{y_p{}^2 - y_q{}^2} \tag{6.38}$$

These equations can be solved numerically for a given boundary condition to identify the equilibrium shape of liquid bridges.

Figure 6.12 shows the penetration of a droplet with the impact velocity of 5 m/s into the powder bed from different views where the formation of liquid capillary bridges between the particles is visible.

6.3   Numerical Modeling

In addition to uncovering the governing physics of phenomena controlling the formation, falling, and spreading of droplets on the substrate (or droplet imbibition in powder bed), researchers have been trying to develop computational models to simulate such phenomena over the past few decades. These computational methods are important as they generate data that can be used in design and development of printing devices, as well as in providing feedback models in the control systems to improve the printing quality. Many different computational fluid dynamics (CFD) methods have been adopted for modeling the different steps of the inkjet process. However, a review of the literature shows that methods such as volume of fluid (VOF), level-set (LS), and lattice Boltzmann (LBM) have been used more frequently than others [5]. VOF and LS methods are both Eulerian numerical methods that function through locating and tracking free surfaces. These are suitable for modeling the changes in the interface of liquid–liquid systems (e.g., binder and air). In these computational models, the motion of the flow is described by Navier–Stokes equations (see Chapter 4), i.e., the conservation equations of mass, momentum, etc., are solved at a macroscopic scope. On the other hand, in the more recently developed LBM, which is a mesoscopic approach, the fluid is supposedly made of particle distributions for which a discretized Boltzmann equation is solved on a regular lattice [16]. To introduce the basics of these two different approaches in this section, the governing equations of both methods are described along with the boundary conditions based on the sample models provided as case studies. The application of LS method in modeling of droplet formation and LBM for modeling of droplet spread on a substrate is showcased. According to Hume et al., the LBM approach is less computationally intensive than the Eulerian methods with modeling of up to nine consecutive droplets reported using LBM in comparison with a maximum of four droplets modeled using other methods [17]. There is a rich literature on the numerical modeling of inkjet printing. Readers are encouraged to refer to many articles published in this area for more details and state-of-the-art approaches.

6.3.1   Level-Set Model[1]

Different numerical methods have been proposed to simulate the motion of the interface of two immiscible fluids. These methods can be classified into three groups of Lagrangian, Eulerian, and arbitrary Lagrangian–Eulerian (ALE) [19]. The level-set method is an Eulerian method

---

[1] This section is mainly adopted from the authors' previous work [18] with permission from Elsevier.

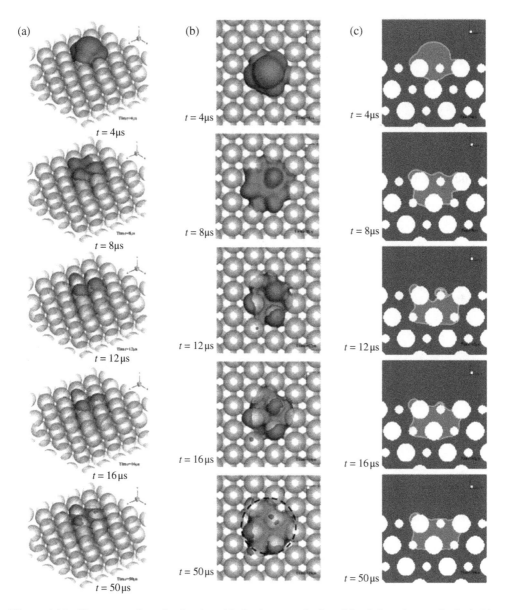

Figure 6.12 The penetration of a droplet with the impact velocity of 5 m/s into the powder bed. (a) isometric view, (b) top view, and (c) cross-section view. (c) The formation of liquid capillary bridges between the particles can be seen. Source: Adapted with permission from Elsevier [15].

using transient Navier–Stokes equations that have been used to predict the emerging and evolution of droplets of a Newtonian fluid by Wilkes et al. [20] for the first time.

The level-set method was introduced by Osher and Sethian [21] in 1980s to capture the interface of two fluids and the changes that the interface goes through due to motion. This Eulerian method is applicable to the analysis of the motion of interfaces, e.g., merging two droplets, in two or three dimensions.

In this method, the interface curve is defined by a level-set equation $\varphi(x, y)$ (in 2D space) dividing the computational region into two sections by assigning an arbitrary value to φ (e.g., 0) at the liquid–liquid boundary as described below:

$$\begin{cases} \phi < 0 & air \\ \phi = 0 & boundary \\ \phi > 0 & liquid \end{cases} \tag{6.39}$$

Density and viscosity of the two fluids change in the transition area according to the following equations using the level-set function:

$$\rho = \rho_{air} + \left(\rho_{liquid} - \rho_{air}\right)\phi \tag{6.40}$$

$$\mu = \mu_{air} + \left(\mu_{liquid} - \mu_{air}\right)\phi \tag{6.41}$$

where ρ denotes the density (kg/m^3) and μ denotes the viscosity (Ns/m^2). The movement of the fluids interface under the velocity field v (m/s) is described by:

$$\frac{\partial \phi}{\partial t} + v.\nabla\phi = 0 \tag{6.42}$$

In the LS method, the interface is considered to move normal to itself. Hence, only the normal component of the velocity field is important [22]. Replacing $v_n = v.\hat{n} = v.\frac{\nabla\phi}{|\nabla\phi|}$ with v in the above equation results in the following LS equation:

$$\frac{\partial \phi}{\partial t} + v_n|\nabla\phi| = 0 \tag{6.43}$$

The LS function should remain as a distance function ($|\nabla\varphi| = 1$) throughout the simulation process in order to depict the changes of interface accurately. In order to solve this problem, numerical reinitialization and stabilization terms are added to level-set equation so that it is set back to distance function from the boundary after each time step [20, 23]. The following equation presents the nonconservative reinitialized level-set function:

$$\rho\left(\frac{\partial \phi}{\partial t} + v.\nabla\phi\right) = \alpha\left[\varepsilon\nabla.\nabla\phi - \nabla.\left(\phi(1-\phi)\frac{\nabla\phi}{|\nabla\phi|}\right)\right] \tag{6.44}$$

where ε is the thickness of the transition layer (m), usually set as half of the typical mesh size, and α is the reinitialization factor and is approximately equal to the maximum value of the velocity field.

As mentioned above, the flow dynamics are based on the Navier–Stokes equation (see Chapter 4) and is represented as:

$$\rho\frac{\partial v}{\partial t} + \rho(v.\nabla)v = \nabla.\left[-P\bar{I} + \mu\left(\nabla v + (\nabla v)^T\right)\right] + \rho g + F_{st} \tag{6.45}$$

The continuity equation representing the mass conservation and incompressibility condition is shown as:

$$\nabla.v = 0 \tag{6.46}$$

In these equations, v is the velocity field (m/s), P is the pressure (N/m²), $\bar{I}$ is the identity matrix, μ is the viscosity (kg/(m·s)), g is the acceleration of gravity (m/s²), and F_{st} is the surface tension force which can be calculated as:

$$F_{st} = \nabla.\left[\gamma\left(I - nn^T\right)\delta\right] \tag{6.47}$$

where γ is the surface tension (N/m), and $n = \frac{\nabla\phi}{|\nabla\phi|}$ denotes the normal to the interface, and $\delta = 6|\varphi(1 - \varphi)\|\nabla\varphi|$ is the Dirac delta function.

Many researchers have adopted the LS method to model the droplet formation. The results of the work by Suh et al. for modeling a PZT DOD system based on the sharp-interface LS method are showcased here as a sample [24]. They have used their model to investigate the effect of dynamic contact angles (advancing or receding) on the flow behavior (Figure 6.13).

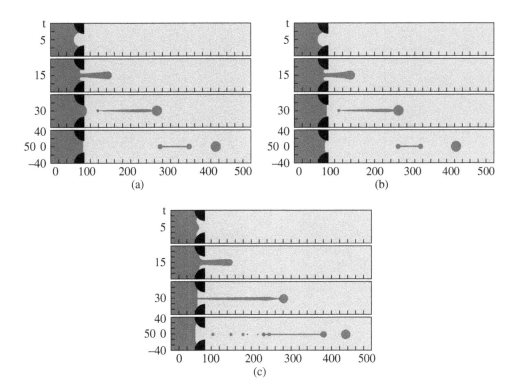

Figure 6.13 Droplet formation modeling using level-set method showing the effect of receding contact angle on droplet jetting. The value of receding contact angels is (a) 30°, (b) 90°, and (c) 150°. The value of the advancing contact angle is kept constant at 150°. Source: Reproduced with permission from Taylor & Francis [24].

6.3.2 Lattice Boltzmann Method

LBM method has been derived from lattice gas automata (LGA) method and has been developed into various approaches itself, as highlighted in Chapter 5. In this method, the LBM equation is based on "simplified, fictitious molecular dynamics in which space, time, and particle velocities are all discrete" [25]. For clarity, the descriptions of the model and case study provided below are mainly from the works of Raiskinmaki et al. [16], Tu et al. [25], and O'brien [26]. Readers are recommended to refer to these publications for background and detailed description of LMB. One of the improved LBM models known as a lattice Bhatnagar–Gross–Krook approximation (lattice BGK) introducing a single-relaxation-time collision has been identified as the most popular LBM method [26] and will be discussed in the following. LBM models work based on lattices made up of cells where they will be updated at each time step, and the flow dynamics are determined through the interaction of these cells. Various lattice structures exist in the LBM method denoted by the general expression DXQY, where X is the dimension and Y is the number of lattice vectors. A D2Q9 lattice sample is shown in Figure 6.14.

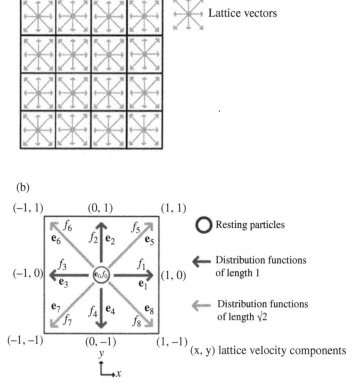

Figure 6.14 (a) Sample D2Q9 LBM. Source: Reproduced under CC BY-SA 3.0 open access license [26] (b) Lattice vectors in a D2Q9 cell. Source: Reproduced under CC BY-SA 3.0 open access license [27].

In LBM, the particle distribution functions $f_i(x, t)$ are used to model the arbitrary velocities where each $f_i(x, t)$ is the expected number of particles moving along a lattice vector e_i. The LBM simulation consists of two steps:

1. Streaming
2. Collision

During the streaming step, the particles move between the adjacent cells; for example, "cell$_{i,j}$'s distribution function for the lattice vector pointing downwards would be copied to cell$_{i+1,j}$'s distribution function for the lattice pointing downwards" [26] as shown in Figure 6.15 and described by Eq. (6.48). It is followed by the collision steps, where a particle collides with another one in an adjacent cell after moving to that cell, as shown in Figure 6.16. It should be noted that it only changes the distribution of cells without affecting their density or velocity. This step is described mathematically by Eq. (6.49):

$$f_i(x + e_i \Delta t, t + \Delta t) \tag{6.48}$$

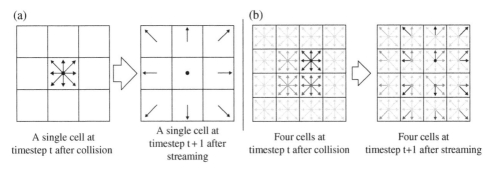

(a) A single cell at timestep t after collision | A single cell at timestep t+1 after streaming

(b) Four cells at timestep t after collision | Four cells at timestep t+1 after streaming

Figure 6.15 Streaming step in LBM. Source: Reproduced under CC BY-SA 3.0 open access license [27].

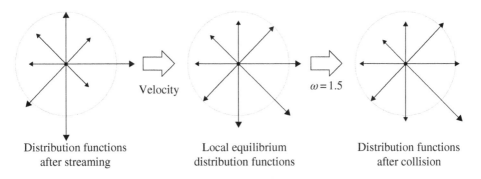

Distribution functions after streaming | Velocity | Local equilibrium distribution functions | $\omega = 1.5$ | Distribution functions after collision

Figure 6.16 Collision step in LBM. Source: Reproduced under CC BY-SA 3.0 open access license [27].

where e_i is a vector pointing to a neighboring lattice node, $f_i(x, t)$ is the density of particles moving in the e_i direction, τ is the BGK relaxation parameter, $f_i^{eq}(x, t)$ is the equilibrium distribution toward which the particle populations are relaxed, and g_i denotes the effect of external forces such as gravity or surface tension [28]. The following equation is the local equilibrium distribution:

$$f_i^{eq} = \rho w_i \left[1 + \frac{1}{c_s^2}(e_i.u) + \frac{1}{2c_s^4}(e_i.u)^2 - \frac{1}{2c_s^2}u^2\right] \tag{6.49}$$

where w_i is a weight factor depending on the length of the link vector e_i and c_s is the speed of sound in the fluid [28]. Combining Eqs. (6.48) and (6.49) results in LBM equation:

$$f_i(x + e_i\Delta t, t + \Delta t) = f_i(x, t) + \frac{1}{\tau}\left[f_i^{eq}(x, t) - f_i(x, t)\right] + g_i \tag{6.50}$$

ρ and u can be derived from the first and second velocity moments of the distribution function as:

$$\rho(x, t) = \sum_{i=0}^{Q-1} f_i(x, t) \tag{6.51}$$

$$\rho(x, t)u(x, t) = \sum_{i=0}^{Q-1} e_i f_i(x, t) \tag{6.52}$$

In the model developed by Raiskinmaki et al. [16] for droplet spread on surfaces with different roughness levels, a no-slip boundary condition (also known as the bounce-back boundary) has been applied to the liquid–solid interface. They have also selected a D3Q19 lattice that represents a 3D lattice with 19 velocities that are surrounding a center point; and the c_s as $1/\sqrt{3}$, the kinetic viscosity of the fluid as $v = (2\tau - 1)/6$, and the fluid pressure as $p = c_s^2\rho$. They have also considered the effect of momentum exchange between the neighboring particles through an attractive range force ($F_G(x)$) as well as adhesive forces between the liquid and solid ($F_W(x)$), thus the velocity moments of the distribution function can be rewritten as:

$$\rho(x)u'(x) = \rho(x)u(x) + F_G(x) + F_W(x) \tag{6.53}$$

where u' is the new fluid velocity when considering the effects of the short-range momentum exchange and adhesive forces. The simulation results for smooth and rough surfaces in [16] are shown in Figure 6.17 and are indicative of the usefulness of LBM in computational modeling of droplet dynamics. In Figure 6.17a, deposition on a smooth surface is simulated when the shape of droplet on the surface is mainly a symmetrical oval, whereas an irregular shape forms on a non-smooth surface as shown in Figure 6.17b.

(a)   (b)

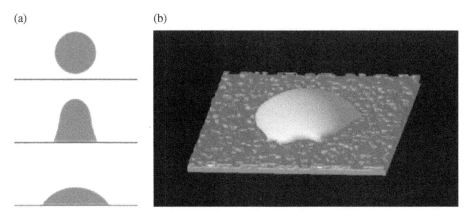

Figure 6.17 (a) 2D view of droplet spreading on a smooth surface from initial condition (topmost) to steady-state (lowermost); (b) 3D view of droplet spreading on a rough surface. Source: Republished with permission from Elsevier [16].

6.4 Summary

This chapter covered basic physics of binder jetting and material jetting AM processes within the context of droplet formation mechanism and the role of material properties and dimensionless numbers in such mechanism. In addition, introductory-level description of scaling theories of pinch-off, droplet–substrate interaction, and binder imbibition were provided. The chapter also shed some light on equations used for identifying geometrical features of liquid bridge among particles. Two methods of level-set and LBM for numerical modeling of BJ and MJ were also discussed.

References

1. E. Antonopoulou, O. G. Harlen, M. A. Walkley, and N. Kapur, "Jetting behavior in drop-on-demand printing: Laboratory experiments and numerical simulations," *Phys. Rev. Fluids*, vol. 5, no. 4, pp. 1–17, 2020.
2. O. A. Basaran, "Small-scale free surface flows with breakup: Drop formation and emerging applications," *Am. Inst. Chem. Eng. AIChE J.*, vol. 48, no. 9, p. 1842, 2002.
3. J. Eggers, "Nonlinear dynamics and breakup of free-surface flows," *Jens Eggers. Rev. Mod. Phys.*, vol. 69, no. 1833, pp. 865–929, 1997.
4. B. Derby, "Inkjet printing of functional and structural materials: Fluid property requirements, feature stability, and resolution," *Annu. Rev. Mater. Res.*, vol. 40, pp. 395–414, 2010.
5. D. Hernández-Cid, V. H. Pérez-González, R. C. Gallo-Villanueva, J. González-Valdez, and M. A. Mata-Gómez, "Modeling droplet formation in microfluidic flow-focusing devices using the two-phases level set method," Mater. Today Proc., 2020, in press.
6. J. R. Castrejón-pita, A. A. Castrejón-pita, S. Suresh, and K. Sambath, "Plethora of transitions during breakup of liquid filaments," *Proc. Natl. Acad. Sci.*, vol. 112, no. 15, pp. 4582–4587, 2015.
7. J. B. Keller and M. J. Miksis. "Surface tension driven flows." *SIAM J. Appl. Math.*, vol. 43, no. 2, pp. 268–277, 1983. https://doi.org/10.1137/0143018
8. D. T. Papageorgiou "On the breakup of viscous liquid threads." *Phys. Fluids* vol. 7, no. 7, pp. 1529–1544, 1995. https://doi.org/10.1063/1.868540
9. J. Eggers. "Universal pinching of 3D axisymmetric free-surface flow." *Phys. Rev. Lett.*, vol. 71, no. 21, 1993, p. 3458. https://doi.org/10.1103/PhysRevLett.71.3458

10. S. Moghtadernejad, C. Lee, and M. Jadidi, "An introduction of droplet impact dynamics to engineering students," *Fluids*, vol. 5, no. 3, p. 107, 2020.

11. C. Tang et al., "Dynamics of droplet impact on solid surface with different roughness," *Int. J. Multiph. Flow*, vol. 96, pp. 56–69, 2017.

12. T. Colton and N. B. Crane, "Influence of droplet velocity, spacing, and inter-arrival time on line formation and saturation in binder jet additive manufacturing," *Addit. Manuf.*, vol. 37, p. 101711, 2020.

13. J. O. Marston, J. E. Sprittles, Y. Zhu, E. Q. Li, I. U. Vakarelski, and S. T. Thoroddsen, "Drop spreading and penetration into pre-wetted powders," *Powder Technol.*, vol. 239, pp. 128–136, 2013.

14. T. Munuhe, A. Lebrun, L. Zhu, and R. Ma, "Using micro-ct to investigate nanofluid droplet sorption in dry powder beds," *Powder Technol.*, vol. 305, pp. 232–240, 2017.

15. H. Tan, "Three-dimensional simulation of micrometer-sized droplet impact and penetration into the powder bed," *Chem. Eng. Sci.*, vol. 153, pp. 93–107, 2016.

16. P. Raiskinmäki, A. Koponen, J. Merikoski, and J. Timonen, "Spreading dynamics of three-dimensional droplets by the lattice-Boltzmann method," *Comput. Mater. Sci.*, vol. 18, no. 1, pp. 7–12, 2000.

17. C. A. Hume and D. W. Rosen, "Low cost numerical modeling of material jetting-based additive manuracturing," Solid Free. Fabr. 2018 Proc. 29th Annu. Int. Solid Free. Fabr. Symp. – An Addit. Manuf. Conf. SFF 2018, pp. 1821–1831, 2020.

18. F. Liravi, R. Darleux, and E. Toyserkani, "Additive manufacturing of 3D structures with non-Newtonian highly viscous fluids: Finite element modeling and experimental validation," *Addit. Manuf.*, vol. 13, 2017.

19. V. Tirtaatmadja, G. H. McKinley, and J. J. Cooper-White, "Drop formation and breakup of low viscosity elastic fluids: effects of molecular weight and concentration," *Phys. Fluids*, vol. 18, no. 4, p. 43101, 2006.

20. E. D. Wilkes, S. D. Phillips, and O. A. Basaran, "Computational and experimental analysis of dynamics of drop formation," *Phys. Fluids*, vol. 11, no. 12, pp. 3577–3598, 1999.

21. S. Osher and Sethian, J. A. "Fronts propagating with curvature-dependent speed: Algorithms based on Hamilton-Jacobi formulations." *J. Comput. Phys.* vol. 79, no. 1, pp. 12–49, 1988.

22. K. Sarikhani, K. Jeddi, R. B. Thompson, C. B. Park, and P. Chen, "Effect of pressure and temperature on interfacial tension of poly lactic acid melt in supercritical carbon dioxide," *Thermochim. Acta*, vol. 609, pp. 1–6, 2015.

23. Y. Suh and G. Son, "A level-set method for simulation of a thermal inkjet process," *Num. Heat Transf. B Fundam.*, vol. 54, no. 2, pp. 138–156, 2008.

24. Y. Suh and G. Son, "A sharp-interface level-set method for simulation of a piezoelectric inkjet process," *Num. Heat Transf. B Fundam.*, vol. 55, no. 4, pp. 295–312, 2009.

25. J. Tu, G. H. Yeoh, and C. Liu, Computational fluid dynamics: A practical approach. Butterworth-Heinemann, 2018.

26. P. M. O'Brien, "A framework for digital watercolor." Texas A&M University, 2008.

27. N. Thürey, "A single-phase free-surface Lattice Boltzmann Method," 2003. Available: http://citeseerx.ist.psu.edu/viewdoc/summary?doi=10.1.1.396.7266

28. A. Valli, A. Koponen, T. Vesala, and J. Timonen, "Simulations of water flow through bordered pits of conifer xylem," *J. Stat. Phys.*, vol. 107, no. 1, pp. 121–142, 2002.

29. Ö. E. Yıldırım and O. A. Basaran, "Dynamics of formation and dripping of drops of deformation-rate-thinning and-thickening liquids from capillary tubes," *J. Nonnewton. Fluid Mech.*, vol. 136, no. 1, pp. 17–37, 2006.

29. Ö. E. Yıldırım and O. A. Basaran, "Dynamics of formation and dripping of drops of deformation-rate-thinning and-thickening liquids from capillary tubes," *J. Nonnewton. Fluid Mech.*, vol. 136, no. 1, pp. 17–37, 2006.

30. S. Schiaffino, A.A. Sonin AA, "Molten droplet deposition and solidification at low Weber numbers," Phys. Fluids, vol. 9, pp. 3172–87, 1997.

7

Material Extrusion

Physics and Modeling

Learning Objectives

At the end of this chapter, you will be able to:

- Understand the basic governing physics of material extrusion.
- Gain insight into analytical and numerical modeling of material extrusion.
- Learn about a few case studies on modeling and analysis of material extrusion.

7.1 Introduction

As described in Chapter 2, the filament-based material extrusion (ME), known by the commercial name of fused deposition modeling (FDM), is the dominant extrusion-based method for the fabrication of metal parts. In this method, filaments composed of polymer containing metal particles (known as highly filled [HF] filaments) are used. The filaments are directed toward the printhead using a pinch roller mechanism. Different components of the printhead are shown in Figure 7.1. The feedstock is melted in a heated chamber inside the printhead known as the liquefier. The melted feedstock is pushed out of the nozzle's capillary by the pressure exerted by the solid part of the filament. The printhead is attached to a gantry system and deposits the melted feedstock, also known as extrudate or bead, in a continuous fashion. The extrudate is solidified as it leaves the nozzle to form a line with a semi-oval cross section, also known as

---

Metal Additive Manufacturing, First Edition. Ehsan Toyserkani, Dyuti Sarker, Osezua Obehi Ibhadode,
Farzad Liravi, Paola Russo, and Katayoon Taherkhani.
© 2022 John Wiley & Sons Ltd. Published 2022 by John Wiley & Sons Ltd.

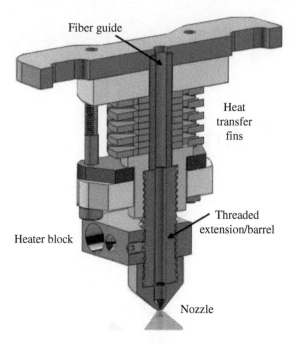

Figure 7.1 Components of a typical printhead in a ME system. Source: Reproduced with permission from Journal of Rheology [1].

"track." As the gantry system moves horizontally, a layer is formed from overlapped tracks encompassed by a frame outlining the contour of the layer. When the next layer is being deposited, the high temperature of extrudates heats the surface of the previous layer, which helps to form a bond between the molecules of different layers. As this process continuous, a 3D object is created. For metal ME (MME), there is a need to remove the polymer medium from metal–polymer composite (MPC) and sinter the green part, as explained in Chapter 2. The change in orientation and composition of HF composite filaments in extrusion, solidification, and sintering stages are shown in Figure 7.2.

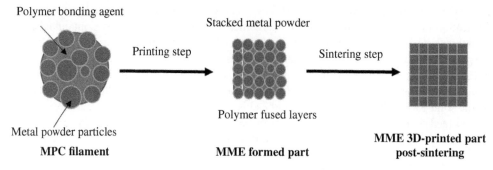

Figure 7.2 HF composite filament at different ME process stages: Metal–polymer composite (MPC) is heated to make metal ME (MME). Source: Redrawn and adapted with permission from Springer [2].

With the fused deposition modeling (FDM) being the most dominant class of material extrusion systems specifically for metals, it is important to understand the physics of this process at various stages of extrusion to be able to control the process, improve the parts' quality, establish the material–property relationships, and increase the process throughput. The ME physics is an under-researched area that has not been studied as thoroughly as other AM processes such as DED, PBF, and MJ that are explained in the previous chapters. Reviewing the literature shows that only a handful of analytical and numerical models describing this emerging process is available. Since FDM is a very complicated process featuring nonlinear thermal and rheological behavior, most available models have been developed based on many simplifying assumptions. One of these assumptions is that the feedstock is a filament made of a single-phase thermoplastic material. Obviously, that is not the case for metal ME in which HF composite filaments made of metal particles embedded in a polymer medium are used. In this chapter, available analytical and numerical models will be reviewed to provide a base for a better understanding of material extrusion with this fact in mind that none of these models have been customized for HF composite filaments. Extrusion of such feedstock, among other complicating factors, results in increased viscosity and consequently the need for increased pressure in the nozzle [3], which should be taken into account in the models. Modeling ME-AM is a recent research field, with the first models having been developed in the early 2000s [4]. With the increase in adoption of this technology for inexpensive metal processing in a non-laboratory environment, it is expected that more advanced models for HF composite filaments will be developed.

7.2   Analytical Modeling of ME

Some of the important physical processes that govern the ME process include heat transfer to the filament in the liquefier, fluid mechanics, and capillary forces inside the conical nozzle during the die swell, deposition, and solidification and binding of deposited beads [5]. Mackay describes these phenomena, as shown in Figure 7.3 [1]. The transfer of heat from the wall of the liquefier to the filament melts the material. The melted feedstock fills the nozzle's conical parts with the half-angle $\beta/2$, where shear and elongation flow caused by the backpressure from the filament acting as the piston push the melted material toward the capillary and eventually out of the orifice. The bead diameter increases as the stress applied to the material in the die is removed when the melted filament exits the nozzle. This phenomenon is known as the die swell effect. A track is deposited as the bead is extruded from the nozzle. Heat is transferred from the recently deposited bead to the previous layer and the adjacent track through conduction. The track eventually cools down and solidifies to form a 3D shape. These physical phenomena are reviewed in the following sections. For clarity, the majority of the work described in this section is based on articles by Bellini [4], Turner et al. [6], and Mackay [1].

7.2.1   Heat Transfer and Outlet Temperature

The liquefier is the part of the printhead in which the filament is melted. The liquefier is normally made up of a threaded barrel surrounded by a heater and a non-conductive protection layer. The nonlinear dependency of the rheological properties of the melted material on temperature and shear rate complicates the flow dynamics inside the liquefier [6]. On the other

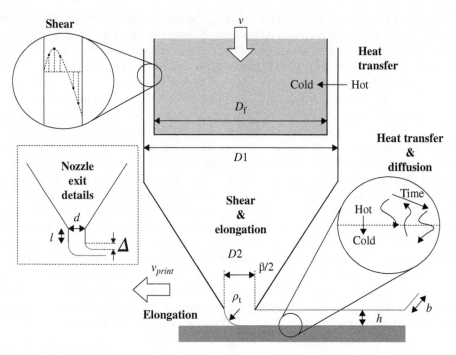

Figure 7.3 The schematic view of liquefier and nozzle in ME. Source: Adapted with permission from the Journal of Rheology [1].

hand, it is the most important component of the ME system, and as such, its full understanding is necessary.

Figure 7.4 shows a schematic view of the heat transfer region in ME. In this figure, $\dot{m}$ is the mass flow rate (kg/s), T_{in} is the filament temperature (K), T_{out} is the outlet temperature (K), and $T_{\dot{m}}$ is the temperature of the liquefier wall transferred to the filament through convection (K) [1].

The temperature of the melted filament T_{out} is lower than the temperature of the liquefier walls. This is shown by Phan et al. [7] by determining the average heat transfer coefficient and consequently calculating a Nusselt–Graetz ($Nu - G_z$) number by equating the first law

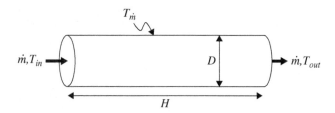

Figure 7.4 Schematic of heat transfer region in ME. Source: Adapted with permission from the Journal of Rheology [7].

of thermodynamics (Eq. 7.1) and Newton's law of heat transfer for convection heat loss (Eq. 7.2):

$$Q = \dot{m}C_p[T_{out} - T_{in}] \tag{7.1}$$

$$Q = h_a \Delta T_a \pi DH \tag{7.2}$$

where Q is the heat flow (W), C_p is the specific heat capacity (J/(kg.K)), and h_a is the average heat transfer coefficient of convection (J/(K.m^2)) determined using a given average temperature driving force ΔT_a, D is the barrel inside diameter (m), and H is the barrel length (m). In the above equation,

$$\Delta T_a = \{[T_w - T_{in}] + [T_w - T_{out}]\}/2 \tag{7.3}$$

By equating the Eqs. (7.1) and (7.2), h_a can be determined as:

$$h_a = \frac{2\dot{m}C_p}{\pi DH} \frac{T_{out} - T_{in}}{[T_{\dot{m}} - T_{in}] + [T_{\dot{m}} - T_{out}]} \tag{7.4}$$

Equation (7.4) can be rewritten in terms of Nu as a function of Graetz number G_z if we assume $Nu = \frac{h_a D}{K}$ is a measure of the rate of heat transfer where K is the thermal conductivity (W/(m.K)), $G_z = \frac{\dot{m}C_p}{HK} = Re \times Pr \times H/D$, where Re is the Reynolds number and Pr is the Prandtl number, and Θ is a dimensionless temperature ($\Theta = \frac{T_{out} - T_{in}}{T_{wall} - T_{in}}$).

$$Nu = \frac{2}{\pi}G_z\frac{\Theta}{2-\Theta} \tag{7.5}$$

Equation (7.5) can be used to estimate the outlet temperature and the maximum bead temperature that can be reached [8].

7.2.2 Flow Dynamics and Drop Pressure

As mentioned previously, it is assumed that the solid section of the filament in the liquefier acts as a piston exerting compressive force onto the melted material. This force should overcome the pressure drop in the nozzle in order to extrude the melt, which is under shear deformation, as it adheres to the internal walls of the liquefier and capillary system. This shear rate $\dot{\gamma}$ (1/s) can be calculated by:

$$\dot{\gamma} = -\frac{dv}{dr} \tag{7.6}$$

where v is the maximum velocity of filament at the center (m/s) and r is the radial dimension (m). The feedstock used in ME usually has shear-thinning behavior, i.e., the viscosity of the material is reduced as the shear stress increases. In order to model the shear-thinning behavior, the majority of researchers select the simple power law model (Eq. 7.7). It should be noted that this model has limitations, e.g., it can only predict the value of viscosity for a limited range of shear rates; nonetheless, it simplifies the modeling and calculations [6];

$$\tau = \left(\frac{\dot{\gamma}}{\phi}\right)^{1/m}$$

(7.7)

where ϕ and m are material constants (fluidity and flow exponent, respectively) and τ is the shear stress (N/m^2).

One of the first analytical models describing the dynamics of liquefier in ME relates the flow of melted material to the pressure drop in three different nozzle zones [4, 6], as shown in Figure 7.5. The assumptions of this model, as outlined in [9], include:

1. The filament is in a melt state when it leaves the liquefier and enters the nozzle;
2. The solid part of the filament acts as a piston and pushes the melted materials through the three nozzle zones;
3. The diameter of zone I is the same as the filament (no gap between the walls and filament);
4. The melted material in all three zones is under steady-state flow conditions;
5. The fluid is incompressible;
6. The no-slip boundary condition is applied to the walls;
7. Heat capacity (C_p) is considered to be constant.

In order to measure the pressure drop at each zone, Bellini et al. [4] have applied the power law to the momentum flux balance on a fluid element at each zone resulting in the following equations:

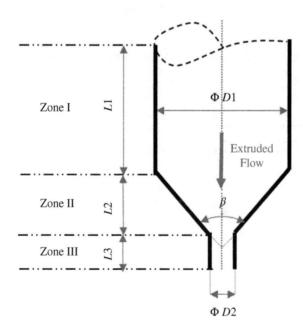

Figure 7.5 Three pressure drop zones in the nozzle. Source: Redrawn and adapted from [6].

$$\Delta P_{1_v} = 2L_1 \left(\frac{v}{\phi}\right)^{\frac{1}{m}} \left(\frac{m+3}{\left(\frac{D_1}{2}\right)^{m+1}}\right)^{1/m} \tag{7.8}$$

$$\Delta P_{2_v} = \left(\frac{2m}{3\tan(\beta/2)}\right)\left(\frac{1}{D_2^{\frac{3}{m}}} - \frac{1}{D_1^{\frac{3}{m}}}\right) \times \left(\left(\frac{D_1}{2}\right)^2 (m+3)2^{m+3}\right)^{1/m} \tag{7.9}$$

$$\Delta P_{3_v} = 2L_3 \left(\frac{v}{\phi}\right)^{\frac{1}{m}} \left(\frac{(m+3)(D_1/2)^{1/m}}{(D_2/2)^{m+3}}\right)^{1/m} \tag{7.10}$$

where β is the convergence angle of the liquefier (see Figure 7.5), v is the filament velocity at the liquefier inlet, and D_1, D_2, L_1, and L_2 are the diameter and length of zones shown in Figure 7.5.

If we assume that the melted filament has a uniform temperature throughout the nozzle equal to T_{out}, the following equation can calculate the heat flow rate from the liquefier walls to the filament:

$$Q = \dot{m}C_p[T_{out} - T_{in}] = (T - T_{in})\frac{\rho v A C_p}{2\pi \frac{D_1}{2} L_1} \tag{7.11}$$

where ρ is the density of filament (kg/m^3) and A is the nozzle cross-section area (m^2). However, the temperature of the flow is not constant and changes when a sudden change in flow conditions in the nozzle happens. As such, the non-isothermal properties of the flow should be taken into account. As mentioned previously, the viscosity is dependent on the temperature and shear rate; as such, the viscosity expression can be considered as:

$$\eta = H_T(T)\eta_0(\dot{\gamma}) \tag{7.12}$$

where $H_T(T)$ is a function expressing the temperature dependency and $\eta_0(\dot{\gamma})$ is the power law viscosity at a reference temperature T_α (at which the parameters ϕ and m should be determined). Bellini et al. [4] used the Arrhenius relation to model the materials' temperature-dependent dynamics:

$$H_T(T) = \exp\left[E_a\left(\frac{1}{T-T_0} - \frac{1}{T_\alpha-T_0}\right)\right] \tag{7.13}$$

where E_a is the energy of activation (J/mol), $H_T(T) = 1$ for T_α and $T_0 = 0$ for absolute temperatures T and T_α.

By substituting the temperature dependence and shear rate dependence terms in Eq. (7.12) with Eqs. (7.13) and (7.7), respectively, the pressure drop equations can be rewritten as:

$$\Delta P_1 = 2L_1 \left(\frac{v}{\phi}\right)^{\frac{1}{m}} \left(\frac{m+3}{\left(\frac{D_1}{2}\right)^{m+1}}\right)^{1/m} \exp\left[E_a\left(\frac{1}{T} - \frac{1}{T_\alpha}\right)\right] \tag{7.14}$$

$$\Delta P_2 = \left(\frac{2m}{3\tan(\beta/2)}\right)\left(\frac{1}{D_2^{\frac{3}{m}}} - \frac{1}{D_1^{\frac{3}{m}}}\right) \times \left(\left(\frac{D_1}{2}\right)^2 (m+3)2^{m+3}\right)^{1/m} \exp\left[E_a\left(\frac{1}{T} - \frac{1}{T_a}\right)\right]$$

$$(7.15)$$

$$\Delta P_3 = 2L_3\left(\frac{\nu}{\phi}\right)^{\frac{1}{m}}\left(\frac{(m+3)(D_1/2)^{1/m}}{(D_2/2)^{m+3}}\right)^{1/m} \exp\left[E_a\left(\frac{1}{T} - \frac{1}{T_a}\right)\right] \qquad (7.16)$$

$$\Delta P = \Delta P_1 + \Delta P_2 + \Delta P_3 \qquad (7.17)$$

With the total pressure drop (Eq. 7.17) known, the required compression force to extrude the melted material can be derived by:

$$F = \Delta P.A \qquad (7.18)$$

where the filament cross-section A (m^2) is assumed to be equal to the liquefier cross section [4]. The torque (Γ) and power (P_{el}) of each electrical motor that pushes the filament to the liquefier (assuming there are two motors) are then calculated as:

$$\Gamma = \frac{F}{2}R_r \qquad (7.19)$$

$$P_{el} = \omega_r\Gamma \qquad (7.20)$$

In above equations, R_r is the radius (m) and ω_r is the angular velocity of the actuating motor (1/rad).

Other researchers have tried to build on the work of Bellini et al. to develop models predicting the reality of flow dynamics in the liquefier and nozzle more accurately. Readers are encouraged to refer to these articles for the details of modeling. Mackay [1] has developed another model for predicting the volumetric flow rate in FDM by assuming that the diameter of the filament and liquefier in zone I of the nozzle are not equal, but a small gap of the size $B = \frac{D - D_f}{2}$ exists between them, as shown in Figure 7.6. The author argues that the polymer melt fills this gap and acts as a seal generating a large amount of pressure to push the bead outside the capillary. For a successful extrusion, the height of the melt polymer (H^*) should be less than the height of the liquefier (H). In another model developed by Osswald et al., it is assumed that the nozzle is not filled with melted material; rather the zone I is very small in size and is reduced to a thin film (as shown in Figure 7.7). The authors argue that this assumption that the nozzle is filled with melt under steady-state flow conditions requires a pressure field to extrude the bead that can only be achieved at relatively low filament velocities, which provides sufficient time for the filament to melt. Such an assumption is then not realistic, and the volume of the melt in the nozzle should be much less. In their model [9], it is assumed that the pressure within the melt film pushes the melt toward the capillary, known as the melting with pressure melt removal.

There are a few major assumptions to be able to use the above model [9]:

1. The material inside all zones is in the melt state.
2. The material is at the steady-state flow condition.
3. A pressure field to flow through the three zones should be given. Such a pressure field is required to compute the force, pushing the filament through the nozzle.

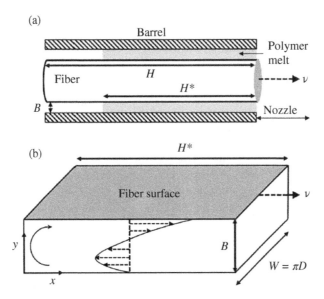

Figure 7.6 (a) The gap (B) between the filament and liquefier walls filled with melted polymer. (b) The pressure profile in the empty gap is shown. Source: Reproduced with permission from the Journal of Rheology [1].

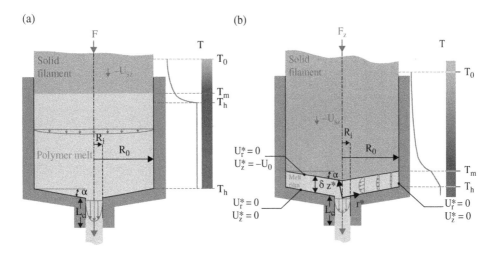

Figure 7.7 (a) Nozzle configuration in a conventional ME (FDM) model and (b) reduced sized zone I model. Source: Reproduced with permission from Elsevier [9].

This approach is only validated for more and less low filament velocities (~ <0.25 mm/s). At low velocities, the filament has adequate time to fully melt [9].

The model developers assume that the first zone tends to be smaller while being reduced within the conical transition zone to a small melt film. On the surface of the conical portion, a small melting film is formed. The film pressure pushes the melt through the capillary toward the middle and out. This type of melting process is so-called "melting with pressure melt removal." The higher the applied force, the higher the melting rate. By increasing the force, eventually, the filament motion would be jammed.

7.2.3 Die Swell

The diameter of the bead extruded from the nozzle is normally larger than that of the capillary zone. This increase in size after extrusion is even true for Newtonian materials; however, it is more obvious in non-Newtonian materials used in ME. For non-Newtonian materials, the deformation energy of polymer under stress is stored elastically and released as the bead exits the nozzle because the stresses applied to the melt are relaxed now that the flow has a free boundary. This phenomenon, known as "die swell," is shown in Figure 7.8.

As this phenomenon affects the resolution and quality of the print, it is important to be able to estimate it. The diameter of bead to the capillary ratio (χ) is a measure of the die swell, which can be calculated using the following equation for non-Newtonian fluids, known as Tanner's relation [10]:

$$\chi = \left\{ 1 + \frac{1}{2} \left[\frac{N_1}{2\sigma} \right]^2 \right\}^{1/6} + 0.13 \tag{7.21}$$

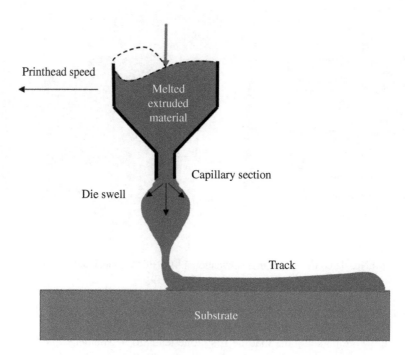

Figure 7.8 Die swell effect results in an increase in the diameter of the bead. Source: Redrawn and adapted from [6].

where $\frac{N_1}{2\sigma}$ is the recoverable shear strain, N_1 is the first normal stress difference, and σ is the shear strain.

According to Mackay [1], the requirements of Eq. (7.21) cannot be met in ME, since the ratio of the capillary length to its diameter should be ~20, which does not happen in ME. Furthermore, the addition of particles and fibers to the polymer to make a composite filament will affect the die swell level. Thus, there is a need for more research in this area to accurately estimate χ in MME [1, 6].

7.2.4 Deposition and Healing

As the bead is deposited on top of a previous track, its temperature is reduced through convection of heat to the surrounding area, which has a lower temperature, and through conduction to previous layers (see Figure 7.3). This transfer of heat to the previous layer is necessary as it increases the temperature of the material in that layer above its glass transition temperature (T_g) which facilitates the bonding of layers through reptation [1], which is the thermal motion of long and entangled macro-molecules present in polymer melts. The mechanical properties of the final product are dependent on the quality of this bonding [11]. The following equation describes the dependency of the properties of the printed part on time [1]:

$$D(t) = \frac{P(t)}{P(\infty)} = \int_{-\infty}^{t} R_h(t-\tau) \times \frac{d\phi(\tau)}{d\tau} d\tau \qquad (7.22)$$

where $P(\infty)$ is the value of the property after an infinite healing time, $R_h(t)$ is the intrinsic healing function, and $\phi(\tau)$ is the wetted areal fraction of a typical 2D system. The wetted areal function corrects the impingement of growing wetted areas with the assumption that it is circular in shape.

At the next step, the bead is spread as the nozzle moves to form a track. The adjacent tracks are also joined through their contact surface in a similar mechanism as the heat is transferred from the recently laid track to the adjacent track. Turner et al. [6] link the cross-sectional area A of the bead to the velocity of the nozzle using the following relationship:

$$A = \frac{Q_v}{v_{print}} \qquad (7.23)$$

where Q_v is the volumetric flow rate (m³/s) and v_{print} is the velocity of the printhead (m/s). The maximum velocity of the nozzle to obtain a stable bead and the minimum cross-sectional area are measured by the following equations, respectively:

$$v_{print} < \frac{Q_v \pi}{h_w^2} \qquad (7.24)$$

$$A_{min} = \frac{h_w^2}{\pi} \qquad (7.25)$$

where h_w is the distance between the nozzle and surface (m).

As discussed in Chapter 6, there are many models describing the spreading of droplets on different types of surfaces; however, there is limited research conducted on the spreading of a bead, specifically considering the beads should be deposited on the curved surface of previous tracks. Readers can refer to the article by Turner et al. [6] for a summary of some of these models developed considering many simplifying assumptions. This aspect of ME processes is not very well known and requires more research to be fully understood.

Finally, the tracks cool down, and bonding is formed between adjacent tracks through their contact areas. The final part properties are largely dependent on the thermal history of the tracks, which has been mainly investigated through experimental rather than analytical modeling. Turner et al. [6] have reviewed two of the most important models describing the thermal history of tracks. The first model has been developed by Thomas et al. [12], wherein the average temperature at each location in the width of the track is measured as:

$$T_{ave}(x, y, z) = T_{out}\left[1 + \sum_{m=1}^{\infty}\sum_{n=1}^{\infty}(a_{mn}\sin(\lambda_m y)\cos(\beta_n x)) \times \exp\left(-\alpha^2(\lambda_m^2 + \beta_m^2)t\right)\right] \quad (7.26)$$

where t is the time (s), T_{out} is the track temperature (K), α is thermal diffusivity of deposited material (m^2/s), and

$$a_{mn} = \frac{4T_L^*}{E_m^2 F_n^2 \lambda_m \beta_n}\sin\left(\frac{9\lambda_m H_t}{2}\right)\sin\left(\frac{\lambda_m H_t}{2}\right)\sin\left(\frac{\beta_n W_t}{2}\right) \quad (7.27)$$

$$E_m^2 = \frac{1}{2}\left(5H_t - \frac{\sin(10\lambda_m H_t)}{2\lambda_m}\right) \quad (7.28)$$

$$F_n^2 = \frac{1}{2}\left(\omega - \frac{\sin(\lambda_m \beta_n W_t)}{\beta_n}\right) \quad (7.29)$$

and the eigenvalues are the roots of the following equations:

$$\lambda_m \cot(5\lambda_m H_t) = -\frac{h}{K} \quad (7.30)$$

$$\beta_n \tan\left(\frac{\beta_n W_t}{2}\right) = \frac{h}{K} \quad (7.31)$$

In the above equations, H_t is the height (m) and W_t is the width of the deposited track (m), h is the coefficient of convective loss (J/(K.m^2)) that may be tuned to count for the effects of both heat convection with air and conduction with the substrate, and T_L^* is temperature (K). It should be noted that the conduction of heat to the substrate is omitted in this model.

Bellehumeur et al. [13] have proposed a simpler lumped capacity model with the following assumptions:

1. The heat transfer and convection coefficients are constant;
2. The length of the filament is semi-infinite;
3. The cross-sectional area of the filament is isothermal, and
4. The model is 1-D transient heat transfer.

In their model:

$$\rho C_p A \upsilon_{print} \frac{\partial T}{\partial x} = A \frac{\partial \left(K \frac{\partial T}{\partial x} \right)}{\partial x} - hP(T - T_\infty) \tag{7.32}$$

with the following analytical solution:

$$T = T_\infty + (T_0 + T_\infty) \exp \left(\frac{(1 - \sqrt{1 + 4\alpha\beta})\upsilon_{print} t}{2\alpha} \right) \tag{7.33}$$

where P is the filament cross-section perimeter, and

$$\beta = \frac{hP}{\rho C_p A \upsilon_{print}} \tag{7.34}$$

The experimental comparison of these two models shows that the performance of each model depends on the experimental condition and temperature.

7.3 Numerical Modeling of ME

In addition to the analytical models presented in Section 7.2, computational fluid dynamics (CFD) tools have also been used to model the extrusion of single-phase polymers. The available models are focused on different parts of the extrusion process, such as the liquefier dynamics, extrusion, deposition, and healing. Given the complex nature of the extrusion process involving multiple physics and nonlinear relationships, the numerical computations can quickly become expensive in terms of time. As such, simplifying assumptions not that different from those introduced for analytical modeling are made to make the models resolvable. Some of these models and their underlying assumptions are reviewed in this section.

One of the earliest numerical models of the liquefier has been developed by Yardimci et al. [5] using finite element analysis (FEA). Figure 7.9 shows the geometry of the liquefier, which is the computational domain in this work. The following axisymmetric conduction equation and boundary conditions have been presented as the modeling basis for heat transfer to study the convection at the liquefier in a polar coordinate [5]:

$$\frac{1}{r} \frac{\partial}{\partial r} \left(Kr \frac{\partial T}{\partial r} \right) + \frac{\partial}{\partial z} \left(K \frac{\partial T}{\partial z} \right) = 0 \tag{7.35}$$

$$-K \frac{\partial T}{\partial n} = h_1(T - T_1) \quad \vec{r} \in \Gamma_1 \tag{7.36}$$

$$-K \frac{\partial T}{\partial n} = h_2(T - T_2) \quad \vec{r} \in \Gamma_2 \tag{7.37}$$

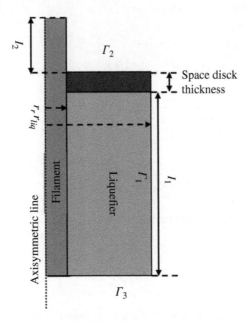

Figure 7.9   Schematics of the liquefier entrance. Source: Redrawn and adapted from [5].

$$T = T_0 \quad \vec{r} \in \Gamma_3 \tag{7.38}$$

$$\frac{\partial T}{\partial n} = 0 \quad \vec{r} \in \Gamma_4 \tag{7.39}$$

where r and z are the radial and axial coordinates, respectively, K is the spatially distributed thermal conductivity, T_0 is the temperature of liquefier, T_1 is the temperature of the liquefier side, T_2 is the entrance side temperature, h_i's are the convective heat loss (cooling) coefficients.

In order to calculate the location of the heat front, a 2D axisymmetric steady-state advection–conduction equation has been adopted, assuming that the temperature of the walls and the flow across the cross section are both constant. The non-dimensional solution to this problem results in:

$$\theta = 2 \sum_{n=1}^{\infty} \exp\left(-\lambda_n^2 z'\right) \frac{\mathcal{F}_0(\lambda_n r')}{\lambda_n \mathcal{F}_1(\lambda_n)}, \mathcal{F}_0(\lambda_n) = 0 \tag{7.40}$$

$$\theta = \frac{T_E - T_{\dot{m}}}{T_{in} - T_{\dot{m}}} \tag{7.41}$$

$$r' = \frac{r}{r_f}, \tag{7.42}$$

$$z' = \frac{\alpha z}{\nu r_f^2} \tag{7.43}$$

where T_E is the filament temperature at (r, z), $T_{\dot{m}}$ is the wall temperature, T_{in} is the temperature of filament at the liquefier entrance $(z = 0)$, r_f is the filament radius, α is the thermal diffusivity of the filament, ν is the filament feeding velocity, λ_n are "roots of zero-order Bessel function of its first kind $(\mathcal{F}_0)$, and $\mathcal{F}_1$ is the first-order Bessel function of the first kind. θ, r', and z' are the dimensionless temperature, radius, and length, respectively [5]."

The location of melt front, i.e., the location at which the melting temperature is reached in the middle of the filament, can be calculated using Eq. (7.41) as:

$$\theta_m = 2 \sum_{n=1}^{\infty} \frac{\exp\left(-\lambda_n^2 z'_m\right)}{\lambda_n \mathcal{F}_1(\lambda_n)} \tag{7.44}$$

Various other FEA models have been developed to simulate the flow of materials inside the nozzle while assuming the melt fills the entire liquefier, there is no gap between the walls and filament, and no-slip contact exists between the liquid phase and the walls. In a recent model by Serdeczny et al. [14], an FEA model considering the temperature and shear dependency of non-Newtonian semi-melted fluids is presented. In this model, instead of assuming the nozzle is filled with the melted feedstock, the free surface of the polymer is resolved. The geometry and boundary conditions of the model are shown in Figure 7.10.

As seen in the figure, the nozzle geometry is divided into five zones. D_f is the diameter of filament entering the nozzle at the feeding rate v and the inlet temperature T_{in} (equal to the room temperature). $T_{\dot{m}}$ is the wall temperature, which is constant for zones III–V. The assumptions made in developing this model include:

1. The density of the melt is constant with respect to pressure but is a function of temperature;
2. The viscosity is a function of temperature modeled using power law;
3. The heat exchange between the free surface of the polymer and air is negligible;
4. The filament remains at solid-state until reaching its glass transition temperature; and
5. The heat transfer from the free surface of melt to the filament due to radiation is included in the model.

In the following, the governing flow equations and boundary conditions for this model are presented:

$$\frac{\partial \rho}{\partial t} + \nabla \cdot \rho \boldsymbol{u} = 0 \tag{7.45}$$

$$\rho \frac{D\boldsymbol{u}}{Dt} = -\nabla p + \nabla \cdot \left(\eta \bar{\bar{S}}\right) + \rho \boldsymbol{g} \tag{7.46}$$

$$\rho C_p \frac{DT}{Dt} = \nabla \cdot (K \nabla T) + \frac{1}{2}\eta \left(\bar{\bar{S}} : \bar{\bar{S}}\right) \tag{7.47}$$

$$\bar{\bar{S}} = \nabla \boldsymbol{u} + (\nabla \boldsymbol{u})^T \tag{7.48}$$

where $\boldsymbol{u}$ is the velocity vector, η is the dynamic viscosity, $\boldsymbol{g}$ is the gravity acceleration vector, $\boldsymbol{g} = (0, 0, -g)^T$, and $\bar{\bar{S}}$ is the strain rate deformation tensor. Furthermore, the heat flux convective loss from the wall to the polymer is calculated using Eq. (7.2).

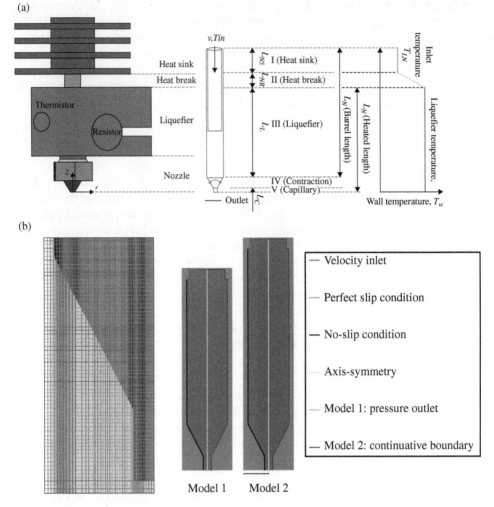

Figure 7.10 (a) Geometry and temperature zones of Serdeczny's model and (b) axisymmetric cylindrical mesh and boundary conditions for two scenarios: Model 1: pressure outlet, Model 2: continuative boundary. Source: Reproduced with permission from Elsevier [14].

This model is solved for various input conditions, a sample of which is shown in Figure 7.11. Interestingly, the model shows the upstream movement of the melted material to fill the gap between the filament and walls in zone II called the recirculation region by the authors. The model developers relate the stability of this region and the lack of excessive overflow leading to the system jamming to the relatively low temperature and high viscosity of the fluid in this region. The figure also shows the dynamic viscosity deviation across the nozzle for both models.

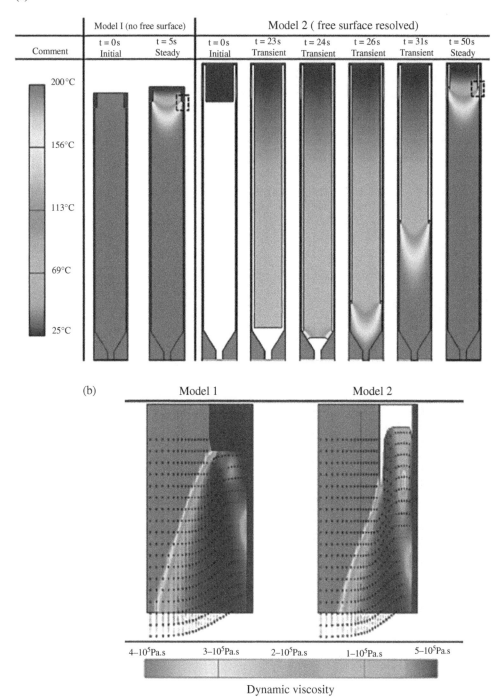

(a)

Comment	Model I (no free surface)		Model 2 (free surface resolved)						
	$t = 0\,$s	$t = 5\,$s	$t = 0\,$s	$t = 23\,$s	$t = 24\,$s	$t = 26\,$s	$t = 31\,$s	$t = 50\,$s	
	Initial	Steady	Initial	Transient	Transient	Transient	Transient	Steady	

200 °C

156 °C

113 °C

69 °C

25 °C

(b)   Model 1   Model 2

4–10⁵Pa.s   3–10⁵Pa.s   2–10⁵Pa.s   1–10⁵Pa.s   5–10⁵Pa.s

Dynamic viscosity

Figure 7.11 (a) Temperature distribution at the liquefier at different times for Model 1 with no free surface and Model 2 where the free surface is solved for the filament feeding rate of 40 mm/min and (b) dynamic viscosity in the recirculation region for Model 1 and Model 2. Source: Reproduced with permission from Elsevier [14].

7.4   Summary

This chapter briefly reviewed the physics and modeling of the material extrusion (ME) additive manufacturing process. It was mentioned that the number of models for metal ME, specifically in regard to bead positioning, spreading, and healing, is very limited. Several analytical and numerical models available in the literature were introduced and their formulations were presented.

Although most of the models are mainly customized for the conventional polymer-based ME process, with adequate consideration and experimentally identified metal filament properties, the polymer-based models would be good platforms for modeling of metal ME. It should also be noted that post-processing for green metal ME-made parts should follow the same steps (debinding and sintering) as for binder-jet AM processes.

References

1. M. E. Mackay, "The importance of rheological behavior in the additive manufacturing technique material extrusion," *J. Rheol. (N. Y. N. Y).*, vol. 62, no. 6, pp. 1549–1561, 2018.
2. S. Terry, I. Fidan, and K. Tantawi, "Preliminary investigation into metal-material extrusion," *Prog. Addit. Manuf.*, pp. 1–9, 2020.
3. J. J. Fallon, S. H. McKnight, and M. J. Bortner, "Highly loaded fiber filled polymers for material extrusion: A review of current understanding," *Addit. Manuf.*, vol. 30, p. 100810, 2019.
4. A. Bellini, Selçuk Güçeri, and M. Bertoldi, "Liquefier dynamics in fused deposition," *J. Manuf. Sci. Eng.*, vol. 126, no. 2, pp. 237–246, 2004.
5. M. AtifYardimci, T. Hattori, S. I. Guceri, and S. C. Danforth, "Thermal analysis of fused deposition," in 1997 International Solid Freeform Fabrication Symposium, University of Texas, Austin, 1997.
6. B. N. Turner, R. Strong, and S. A. Gold, "A review of melt extrusion additive manufacturing processes: I. Process design and modeling," *Rapid Prototyp. J.*, 2014.
7. D. D. Phan, Z. R. Swain, and M. E. Mackay, "Rheological and heat transfer effects in fused filament fabrication," *J. Rheol. (N. Y. N. Y).*, vol. 62, no. 5, pp. 1097–1107, 2018. doi: /10.1122/1.5022982.
8. U. Roy and P. K. Roy, "Advances in heat intensification techniques in shell and tube heat exchanger," in Advanced analytic and control techniques for thermal systems with heat exchangers, Academic Press, pp. 197–207, 2020.
9. T. A. Osswald, J. Puentes, and J. Kattinger, "Fused filament fabrication melting model," *Addit. Manuf.*, vol. 22, pp. 51–59, 2018.
10. S. Rafaï, D. Bonn, and A. Boudaoud, "Spreading of non-Newtonian fluids on hydrophilic surfaces," *J. Fluid Mech.*, vol. 513, pp. 77–85, 2004. doi: https://doi.org/10.1017/S0022112004000278.
11. Y. Yan, R. Zhang, G. Hong, and X. Yuan, "Research on the bonding of material paths in melted extrusion modeling," *Mater. Des.*, vol. 21, no. 2, pp. 93–99, 2000.
12. J. P. Thomas and J. F. Fodriguez, "Modeling the fracture strength between fused-deposition extruded roads," Solid Freeform Fabrication Proceedings, University of Texas, Austin, 2000.
13. C. Bellehumeur, L. Li, Q. Sun, and P. Gu, "Modeling of bond formation between polymer filaments in the fused deposition modeling process," *J. Manuf. Process*, vol. 6, no. 2, pp. 170–178, 2004. doi: https://doi.org/10.1016/S1526-6125(04)70071-7.
14. M. P. Serdeczny, R. Comminal, M. T. Mollah, D. B. Pedersen, and J. Spangenberg, "Numerical modeling of the polymer flow through the hot-end in filament-based material extrusion additive manufacturing," *Addit. Manuf.*, vol. 36, p. 101454, 2020.

8

Material Design and Considerations for Metal Additive Manufacturing

Learning Objectives

After reading this chapter, you should be accomplished with the followings:

- Relevant historical background on material science: its structure and properties relationship
- Manufacturing of metallic materials: conventional versus AM
- Fundamentals of solidification
- Factors affecting solidification in AM
- Phase transformation in solidification for different alloys used in AM
- Solidification defects in AM
- Environmental effect on the solidification process in AM

8.1 Historical Background on Materials

On earth, our life is surrounded and driven by materials. The way of our living, transportation, communication, accommodation, recreation, everywhere materials are dominated. Therefore, pre-historical human civilizations are classified according to the advancement of materials as Stone Age, Bronze Age, and Iron Age.

In the early stage of life on earth, people only use natural materials like stone, clay, skins, and wood, known as Stone Age. Bronze Age started at about 3000 BCE, when people got to know

Metal Additive Manufacturing, First Edition. Ehsan Toyserkani, Dyuti Sarker, Osezua Obehi Ibhadode, Farzad Liravi, Paola Russo, and Katayoon Taherkhani.
© 2022 John Wiley & Sons Ltd. Published 2022 by John Wiley & Sons Ltd.

about copper and the way to make it harder by alloying. In 1200 BCE, the application of steel and iron gave advantages in wars. The modern civilization, after 1850, discovers an economical steel manufacturing process that empowered the routes and constructs the current infrastructure of the industrial world.

Science and technology are enriching in all areas, and the progression of materials has a major role in this progress. The development of materials is quicker today than ever allowing enhancement on the function of existing products and technologies that have influenced all segments of our lives. Therefore, materials science and engineering is now an imperative subject of engineering that has a substantial influence on the world's economy. With new advancements in manufacturing, such as the birth of AM processes, material science has obtained another momentum to address challenges and opportunities associated with these advancements to transfer them from a fancy mode to mainstream manufacturing.

8.2   Materials Science: Structure–Property Relationship

Materials science is concerned with properties of solid materials, and the study of those properties and the way they are linked to material compositions and structures. It is based on the integration of solid-state physics, metallurgy, and chemistry because various material properties cannot be explained by the context of a single discipline. A fundamental concept of materials properties can be nominated for a massive diversity of consumptions, extending from structural steels to computer microchips. Materials science is then essential in the electronics, automotive, aerospace, consumables, telecommunications, data processing, nuclear, energy, and many more industries.

Generally, the structure of a material describes the array of core constituents. The structure could be a subatomic level, where electrons inside the discrete atoms interact with the nuclei. At the atomic level, structure involves groups of atoms or molecules in relation to one another. The greater structural level encompasses group of atoms together to make "microscopic element," where objects are detected under a microscope. The structure in the "macroscopic level" can be observed using the bare eye.

Properties can be defined as the way a material will respond to the environment. Understanding materials behavior and their difference in properties is possible with the knowledge of materials science. In materials science, besides the "structure" and "property" of materials, two other major components are "processing" and "performances." With respect to the relationship of these four constituents, the processing route will determine the final material structure,

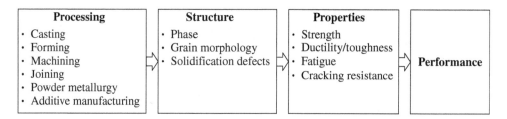

Figure 8.1   The relationship between four major components of materials science.

as schematically shown in Figure 8.1. The manufacturing processes for metallic materials involve casting, forming, machining, joining, powder metallurgy, and additive manufacturing, where products are given to a desired geometric shape either by liquid pouring in a mold, melting of feedstock materials layer by layer, or by solid-state forming.

Day by day, scientists strive to perceive the relationship among structural components of materials and their properties to develop new techniques for producing various superior materials compared to the natural materials like gold and copper. Properties of materials are correlated with the microstructure, which can be modified by changing the microconstituents' relative magnitude, known as phases. In the microstructure, phases are categorized according to their distinctive crystal structures, elemental composition, and properties. These properties impact the performance of materials in applications and alter the performance; the microstructure then needs to be modified. The modification of microstructure or alteration of crystal structures can be done by adding new elements or processing through mechanical and thermal treatments.

Microstructural modifications can be made in most of the manufacturing materials to change their mechanical properties. This is typically accomplished through thermal treatments, including heating and cooling under controlled environments. The most important factors of thermal treatments are (i) maximum temperature and heating rate, (ii) soaking time: the period during which the material would remain at the elevated temperature, (iii) cooling rate, and (iv) surrounding environment of the thermal treatment. Based on material types and compositions, the thermal treatment factors are considered to tweak the microstructure. Thermomechanical treatments, which are the combination of mechanical and thermal treatments, are used to yield properties that cannot be achieved using other techniques. Hot isostatic pressing (HIP), which exposes manufactured parts to simultaneously elevated temperature and isostatic gas pressure in a high-pressure containment vessel, may also be used to minimize porosity, while changing the phases.

8.3 Manufacturing of Metallic Materials

The manufacturing of materials by hand is as ancient as civilization. The Second Industrial Revolution in the eighteenth century started mechanization. Afterward, in England in the nineteenth century, the basic machinery for forming, shaping, and cutting was technologically advanced. From that time, materials manufacturing processes and machinery have expanded in varieties and numbers.

The sequence of conventional manufacturing methods to convert materials into products begins with extracting raw materials from minerals or fabrication using chemicals or natural substances. In this process, two stages normally exist: Stage 1: in this stage, industry generally makes metallic raw materials, where the crude ore is treated to enhance the material concentration, which is known as beneficiation. The typical beneficiation methods include crushing, roasting, magnetic separation, flotation, and leaching. Stage 2: Further processing steps such as melting and alloying are done to manufacture metal ingots, which will be machined into parts and finally assembled into a product. The detailed steps of the conventional manufacturing process, i.e., casting, are schematically shown in Figure 8.2.

In the casting process, the metal alloys are heated in a crucible until they melt. Once the heating temperature goes above the freezing temperature of a metal, it starts to become liquid. The metal alloy melts within a range of temperatures based on the composition of alloys. Heating above the melting point allows sufficient time for the alloy to cool throughout the pouring

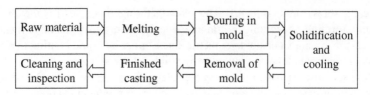

Figure 8.2 Conventional manufacturing processes: e.g., casting.

process. Once the liquid metal is poured into a mold, it quickly cools. When the particular metal or alloy temperature drops below its melting point, the solidification process starts. As the temperature drops gradually, the liquid metal loses energy and crystallization starts close to the mold walls, where they rapidly cool. These crystals together will form grains within the final structure. After becoming hard, the metal part is separated or broken from the mold at the end of the solidification process.

Distinct from traditional fabrication methods, AM is a technique of producing three-dimensional solid products of, in fact, any geometry through a digital model. AM follows an additive method, where materials are added in consecutive layers. AM is differentiated from conventional subtractive machining techniques based on the subtraction of materials through cutting or milling. AM possesses 30 years of history in the production of plastic components, and from 1995, the capability to manufacture metal components applicable to engineered products and modern tech manufacturing has been discovered. In this book, only metal AM technologies are discussed.

The manufacture of AM metal powder mostly comprises three stages, as illustrated in Figure 8.3 [1]. This figure highlights several further points related to these stages. These stages are:

I. Mining and extracting of ore to fabricate pure metal or alloy products (i.e., billet, ingot, and wire).
II. Powder production by either water atomization, plasma atomization, plasma rotate electrode process (PREP) atomization, rotating electrode process (REP) atomization, gas atomization, or electrode induction melting inert gas atomization (EIGA).
III. Powder sorting, classification, and validation.

The powder morphology has a substantial effect on bulk packing and flow behaviors. Spherical, regular, and equiaxed powders can organize and pack more competently than irregular powders; however, there are reports claiming that irregular powders behave well in terms of flowability in the AM powder bed and powder-fed processes for many applications. The powder morphology can significantly influence the density of final AM components. Very spherical powders are more advantageous to AM processes; on the other hand, this, in fact, reduces the use of possibly cheaper powder manufacture methods. Recent research outcomes demonstrate that the more irregular the powder shape, the inferior the product density [2].

Powder size distribution is also another vital parameter of metallic powders in AM. Generally, powder size is about 15–100 μm, depending on the various AM equipment, AM processes, and preferred geometrical resolution. The powder size distribution can impact the size of layer

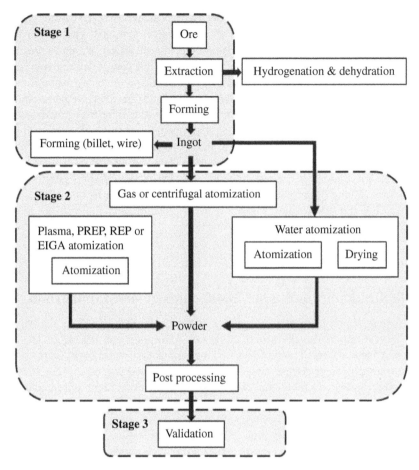

Figure 8.3 AM powder production steps [1]. Source: Reproduced with permission from Johnson Matthey Plc.

thickness and the finest aspect of the AM products. The detailed manufacturing way of using this metallic powder in different AM techniques to fabricate 3D products is described in Chapter 2.

8.4 Solidification of Metals: Equilibrium

In the conventional casting or laser/electron-beam AM techniques, after melting, the liquid metal converts to a solid form through the cooling process known as solidification. Thermodynamically, solid and liquid equally have the same energy at their melting stage; thus, both are steady at the melting point. During the cooling process, the liquid metal starts to lose energy, which is the latent heat. Consequently, the liquid temperature is dropped, which minimizes the average interatomic space between movable and disordered atoms. On further cooling, the interaction forces of atoms stop them from moving away from each other and finally an absolute transformation of liquid to solid occurs.

A controlled solidification protocol is employed in the extensively used casting process with a large ingot of metal alloy to a single crystal of solid. Transformation of a homogeneous liquid into a perfect homogeneous solid structure is enormously difficult to obtain in practice. This requires deep knowledge of the process at numerous levels that can be defined as the angstrom level, micron level, meso level, and macro level. At the angstrom level, no matter if the crystal is growing or not, the atomic movement and conformation of the interface between solid and liquid are essential to consider. Together with this, the early step of crystal growth and nucleus formation should be studied. It is important to study the redistribution of solute at the micron level, which will happen if the formed solid crystal structures have different chemical compositions compared to the liquid. This can result in undesirable segregation or extreme refinement of solid, which must consider the situation under which the solid/liquid interface becomes unstable to form cellular or dendritic structures or even crystal imperfections. The knowledge on solidification at the macro level is linked to the application, which covers the area of heat flow and fluid flow in a liquid metal during cooling.

8.5   Solidification in Additive Manufacturing: Non-Equilibrium

Since AM comprises rapid melting of metallic materials, knowledge of the theory of solidification plays an important role in predicting and monitoring the technique. The laser- or EB-based AM methods entail a localized moving heat source with a very short interaction time. For example, in laser powder bed fusion, the laser speed could be about 4 m/s, where the laser spot size is 100 μm. This implies that a process zone of 100 μm goes under a heating cycle when the maximum temperature in the process zone can be achieved in less than 25 μs. Given the material's thermal conductivity, the solidification stage followed by the cooling duration may take twice the heating time. Hence, the solidification and cooling rates are very high, which are generally ascribed by the surrounding heat conduction/convection regimes, mainly due to the substrate material. As the point heat source delivers extremely focused energy, it causes vastly localized heat flux in the melt pool zone, together with a massive temperature gradient in the deposited layers. The temperature gradient at the bottom of the substrate surface is higher compared to the top of the deposited surface. During the solidification process, the alloy partition coefficient drops, which results in the rejection of solute atoms at the solid–liquid boundary. The concentration of solute atoms rises until the solution reaches a steady condition. The solidification temperature at the solid–liquid boundary is influenced by the liquid composition, as well as the process speed and angle of this boundary with respect to the heat source centerline.

In AM, the metallurgy is controlled by the chemistry of raw materials and the thermal history exposed to the material during the manufacturing technique. For different AM techniques (i.e., LPBF and LDED), the heat transfer mechanism may differ; however, where a full melting occurs in laser or EB processed AM, they have the same metallurgical principles. Solidification followed by cooling to the ambient temperature controls the early phase distribution and grain morphology in the deposited layer. Moreover, the melt pool geometry is governed by the heat source speed, power, and size, which consequently controls the solidification kinetics. For the overall process, the thermal cycling and cooling path control different precipitation kinetics, phase growth, and grain growth [3].

In a few cases, the solidification characteristics in AM are apart from conventional techniques, e.g., casting. For AM, the melt pool size ranges from hundreds of micrometers to millimeters, irrespective of the final component size, and hence is relatively different from castings with bigger melt pool area [4]. In a small size melt pool, the macrosegregation is partial compared to the bigger size castings. However, in the LPBF process, the powder bed around the melt area exposes a relatively lower thermal conductivity (due to voids between particles) than the similar alloy material in a bulk form. For example, alumina shows a thermal conductivity between 25 and 30 W/m.K, whereas for powder bed at the ambient temperature, it is reported to be about 0.3 W/m.K [5]. For this reason, the conductive heat transfer frequently takes place through the earlier melted layers under the melt pools. In the LPBF process, the higher cooling rate is estimated to be between hundreds of Kelvin per second to 10^3 K/s [6] causes major deviations from near-equilibrium cooling conditions, generating much attention in the AM and rapid solidification communities. In the melt spinning process, the typical cooling rate is estimated to be between 10^5 and 10^6 K/s, which usually maintains the local interfacial equilibrium condition. Therefore, a similar hypothesis can be considered in the laser and EB-AM processes.

AM is comparatively a rapid solidification process [7]. During manufacturing, metallic AM products pass through a complex thermal cycle, including directional heat conduction, repeated heating and cooling, and rapid solidification. There are mainly two reasons for rapid solidification: (i) enormous undercooling of the melt and (ii) fast-moving temperature fluxes [8]. A common feature of the rapid solidification is accompanied by the robust liquid flow, like Marangoni convection, where flow velocities can be as high as 1–4 m/s. These factors will not support an assumption that the microstructural development and properties could be the same as conventional techniques. Moreover, in the rapid solidification mode, elemental partitioning is reduced, which extends the solid solubility that may cause the formation of metastable phases. Also, because of the directional heat conduction, a preferred directionality in grain growth may occur. The combined influence of rapid and directional solidification and phase transformation persuaded by continual thermal cycles has a substantial influence on the deposited parts' microstructure.

It has become evident that the rapid solidification in the melt pool offers the prospect of improving mechanical properties by allowing larger constitutional and structural latitude than is probable in the conventional manufacturing approach [9]. The development of these unique structures is associated with the liquid undercooling before nucleation and growth rate during solidification. Therefore, the cooling rate is another vital factor in achieving the desired undercooling that eventually minimizes the increased temperature due to recalescence [10]. Rapid solidification characteristics can appropriately conclude as the refinement of microstructure, an increase of solubility limits, lessening of microsegregation, and non-equilibrium of metastable phase formation [10]. The formation of distinctive microstructural features like grains, lamellae, and second-phase particles is generally reduced in the fast cooling compared to the standard casting methods [11]. Another possible outcome is the mitigation of the dendritic segregation at a point where compositional homogeneity can be attained [11].

Therefore, the solidification in AM stimulates the formation of grain size, shape, distribution, growth kinetics, elemental segregation and precipitation, phase transformation, and eventually the material properties. The vital factors in this process are associated with the multiscale multiphysics phenomena (i.e., fluid flow, heat and mass transfer, beam interaction with the material) that happen because of the localized heating and cooling in AM.

8.6   Equilibrium Solidification: Theory and Mechanism

Thermodynamics is a viable area to analyze the kinetics of solidification. It covers the evaluation of alloy phase composition, solidification way, fundamental alloy properties, i.e., partition coefficients, and solidus and liquidus line slopes. The free energy of a material is a function of pressure, temperature, and composition. Equilibrium condition is obtained if the Gibbs free energy is at a minimum level. Though the equilibrium does not exist in actual systems, under the hypothesis of thermodynamic equilibrium, the alloy composition in solid and liquid forms can be estimated from the equilibrium phase diagram. The structure of a material, which is a function of composition and temperature, can be identified from an equilibrium phase diagram using the assumption of a relatively slow transformation rate or a faster diffusion rate.

8.6.1   Cooling Curve and Phase Diagram

A cooling curve is a graphical presentation of the phase transition temperature with time for pure metals or alloys over a complete temperature range through which it cools. When a pure liquid metal cools, it first reaches the melting point (a unique temperature) and then solidifies, as schematically shown in Figure 8.4.

To understand the solidification of pure metals, a cooling curve with supercooling is adopted in Figure 8.4a, where the liquid metal cools as the specific heat drops from point A to B. This requires undercooling between points B and C. Since the nucleation starts at point C, the latent heat of fusion is evolved, resulting in the liquid's temperature rise. This process is recognized as recalescence (from point C to D). From this point, the solidification continues at a constant temperature (T_m) and is completed at point E. After that, the solid material starts to cool when this stage is recognized by the cooling rate.

Moreover, based on the atomic configuration, the metal in solid and liquid forms possesses different thermal conductivities. Therefore, a big difference between the slopes of the cooling curves is obvious in different regions. Before the solidification starts, the temperature falls relatively below the equilibrium cooling temperature, known as "supercooling", and is necessary to develop a thermodynamic driving force for the process. The term "supercooling" is also named "undercooling," where the liquid temperature is lowered below the melting point without transforming into a solid. For a pure metal, as shown in Figure 8.4b, the cooling curve shows a zero slope (constant temperature) throughout the solidification.

The cooling behavior of the alloy is quite different compared to the pure metal, as presented in Figure 8.4c. The alloy usually cools from a two-phase region, and the solid precipitates are postponed in liquid through a range of temperatures, which is called "mushy" zone. Unlike pure metals with consistent melting temperature, metal alloys show a mushy zone within solidus and liquidus temperatures, where solid and liquid phases concur throughout the solidification process [12], as shown in Figure 8.4d.

A group of cooling curves of an alloy at different alloy compositions are plotted in Figure 8.4e, where rather than showing a consistent temperature, the cooling curves progress with negative slopes. This region of the curve is non-linear because of the evolution of the latent heat, which arises due to the variation of the solid–liquid proportions during the solidification process. The temperature from which the slope first deviates corresponds to the formation of the first solid and is termed the liquidus temperature. The temperature below in which the alloy

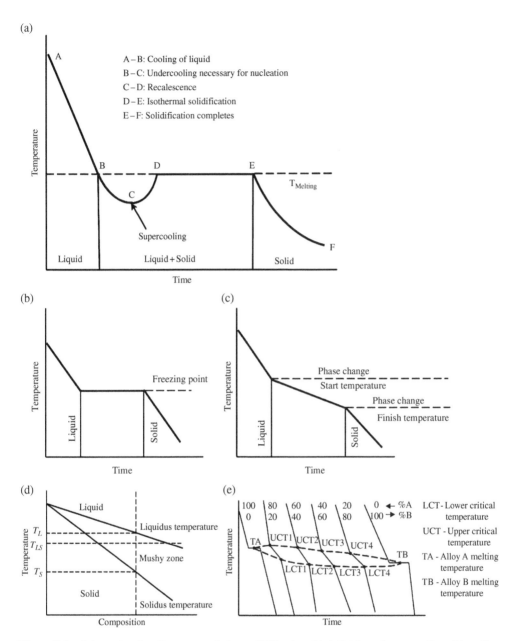

Figure 8.4 Schematic of cooling curves during solidification, (a) definition of supercooling, (b) cooling curve for pure metals, (c) cooling curve for alloys, (d) definition of mushy zone (Source: Reprinted by permission of the publisher Taylor & Francis Ltd. [12]), and (e) set of cooling curves to construct binary phase diagram.

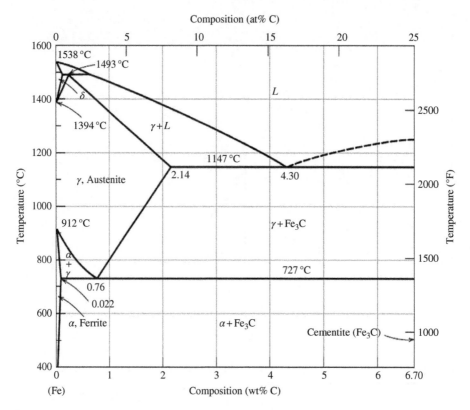

Figure 8.5 Fe–C phase diagram. Source: Openly accessible through creative commons license.

solidifies is known as the solidus temperature. Thus, a phase diagram can be generated by plotting the liquidus and solidus point temperatures with alloy compositions, as shown in Figure 8.4e.

A phase diagram is a useful tool to understand the formation of phases and their transformation throughout the heating and cooling practice with different alloy compositions. For better understanding, a typical phase diagram of Iron–Carbon (Fe–C) alloy is shown in Figure 8.5. At different C content, phase transformation temperatures of the alloy can be understood. This can be in the help of optimizing AM process parameters and modification of microstructure. For example, during the manufacturing of stainless steel using laser-based AM technology, it is crucial to optimize the laser input temperature to control the austenite and ferrite phase formation.

Another important tool is the continuous cooling transformation (CCT) diagram, which presents the knowledge on the type of phases that will form in alloys at different cooling rates. It is vital in AM techniques since each deposited layer goes through the repeated thermal cycles and eventually has different cooling rates. A CCT diagram for steel is shown in Figure 8.6, where it is evident that a complete martensitic structure forms at fast cooling conditions, whereas bainite, ferrite, and pearlite can occur at relatively slower cooling rates. In the AM process, when

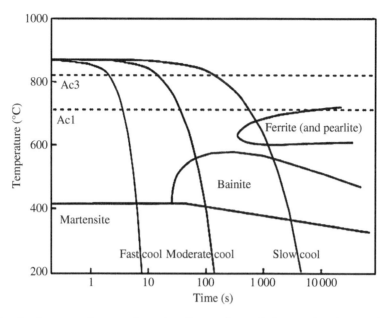

Figure 8.6 Continuous cooling transformation diagram for steel. Source: Openly accessible through creative commons license.

the thermal profile of each deposited layer is known, the CCT diagram can be used to estimate the formation of phase types.

8.7 Non-Equilibrium Solidification: Theory and Mechanism

The most significant part in the solidification of AM is the heating and cooling cycle, which may cause suppressed phase changes or supersaturated phases. In AM, when every layer passes through a repeated heating and cooling cycle, the temperature in the layer may reach above the phase transformation temperature. This will result in the multiple phase transformation or intricate microstructure, in addition to a residual stress formation. For example, a typical thermal profile for a single-layer Ti-6Al-4V alloy manufactured in AM is shown in Figure 8.7 [7]. The time–temperature profile shows fluctuations and reaches various phase-change temperatures, which eventually lead to the transformation of phases.

For better control of microstructure in AM parts, knowledge on the phase transformation during the fast solidification and repeated thermal cycle behaviors is essential. Some AM metal powders do not show good weldability from the beginning; therefore, the optimization of alloying elements and process parameters is required. The AM process optimization will require knowledge in the area of equilibrium and non-equilibrium phase diagrams, thermodynamics, thermo-physical quantities, laser/EB material interaction, diffusion kinetics, etc.

During the rapid cooling, some phases that generally form under equilibrium conditions may not arise; hence, there would be a chance for the occurrences of metastable phases. For example, the CCT diagram of Ti-6Al-4V alloy for both standard casting and AM techniques is

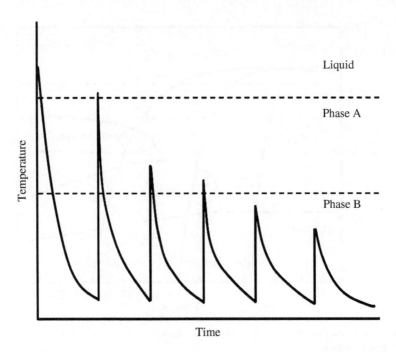

Figure 8.7 Time–temperature profile of a single-layer AM-manufactured Ti-6Al-4V alloy. Source: Redrawn and adapted from [7].

compared in Figure 8.8 [7]. The diagram shows a faster cooling for the alloy in AM, which forms martensitic structure in a metastable phase.

The characteristics of the metastable phases are based on the alloy chemical structure and thermal behavior of the cooling method. This aspect is graphically emphasized in Figure 8.9. From the time–temperature diagram, at cooling rate T_1, the primary phase is nucleated as phase I, whenever at faster cooling rate T_2, $(T_2 \gg T_1)$, another phase, phase II is nucleated by detouring phase I. To clarify the phase formation, a phase diagram is shown in Figure 8.10, where phase δ is evident at equilibrium condition. Due to the rapid cooling process, a metastable phase diagram may be created, which is highlighted by the dashed lines. In this cooling condition, if the delta phase cannot generate, a eutectic system may appear at lower temperature with different chemical compositions compared to the equilibrium phase diagram.

8.8   Solute Redistribution and Microsegregation

When a liquid solidifies from uniform composition, the resulting solid hardly shows uniformity in composition. This is because of the redistribution of the solute atoms in the liquid during solidification, which is a major problem in the process. Besides, the growth phenomenon, phase distribution, and segregations are related to the solute redistribution that is dependent on thermodynamics (i.e., phase diagram) and kinetics (i.e., diffusion, undercooling, fluid flow, etc.) [10].

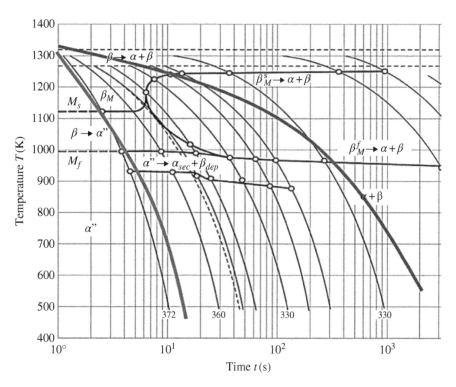

Figure 8.8 Critical continuous cooling transformation diagram for welded or cast and AM Ti-6Al-4V [7]. Source: Openly accessible through creative commons license.

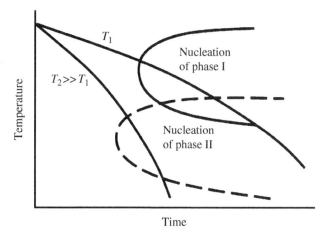

Figure 8.9 Time–temperature diagram presenting the nucleation onset of two dissimilar theoretical phases with unlike cooling behaviors.

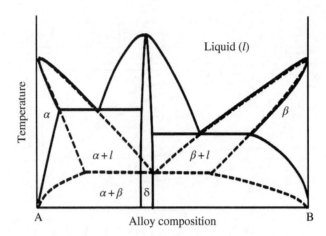

Figure 8.10   A comparative presentation of the theoretical equilibrium (solid lines) and metastable non-equilibrium phase diagram (dotted lines).

The microsegregation is defined as the ejection of solute from the freezing material, which afterward distributes heterogeneously and eventually affects the solidification mode. The alloy in liquid form contains a higher solute compared to the solidified condition. Therefore, during the solidification process, there is a chance of the high solute liquid to be trapped within the solidified structures. This causes microsegregation or banding of high and low solute alloys and substantial incoherence in material properties [10].

For example, the microsegregation is unavoidable in the solidification process of Nickel alloys, which may create unwanted phases, such as low melting point eutectic [8]. This may lead to strong performance deprivation. Nowadays, laser AM (e.g., LDED) possesses the potentiality of rapid forming and specific repair of single-crystal elements by allowing fast and precise addition of an exact quantity of materials for required positions with a minimum heat input at a maximum cooling rate. Therefore, it would be an effectual approach to lower and check microsegregation in a single-crystal material as a faster solidification technique.

The kinetics of solute redistribution and microsegregation is demonstrated in Figure 8.11. The phase diagram is partially shown in Figure 8.11a, to illustrate the compositional change during solidification. Whereas C_S and C_L indicate the solid and liquid compositions, respectively, at their interfaces, C_o is the nominal alloy composition, f_S and f_L are the fractions of solid and liquid, and k is the equilibrium distribution coefficient, which is defined by $k = C_S/C_L$. These terms adopt linear solidus and liquidus lines to continue with constant k through the process. The term k implies a significant value to describe the degree of specific elemental partitions within solid and liquid. When $k < 1$, the solute makes barriers to the liquid, and thus with a smaller value of k, the barriers to the liquid become more violent. For $k > 1$, the solute makes barriers to the solid in the solidification process.

The theory and mechanism of solute redistribution can be expressed using equilibrium and non-equilibrium models, considered in the vigorous conditions of solute redistribution. The factors that are assumed in equilibrium lever laws are (i) complete diffusion in both liquid and solid state, (ii) equilibrium at solid/liquid boundary, and (iii) no undercooling through the growth. Also, in the non-equilibrium lever law (known as Scheil equation), similar assumptions are considered, apart from considering negligible diffusion in the solid. In solidification,

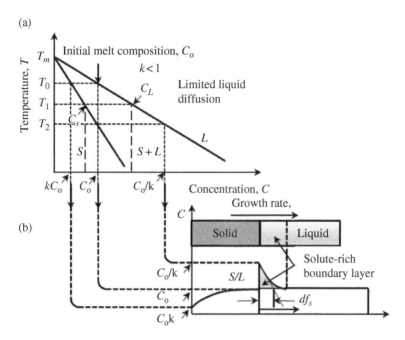

Figure 8.11 Solidification during inadequate diffusion in liquid and no diffusion in solid: (a) partial phase diagram and (b) solute-enriched boundary layer [10]. Source: reproduced with permission from John Wiley & Sons Inc.

these three factors signify the extreme situations of residual microsegregation. According to the equilibrium lever law, there are no concentration gradients in the liquid and solid during solidification and eventually no residual microsegregation in the solidified structure. On the other hand, the non-equilibrium situation possesses residual microsegregation in the solidified structure because of very minor solid diffusivity.

Generally, the solute diffusion coefficient of a solid phase is about four times lower than a liquid phase; therefore, a precise image of plane front solidification is expressed with a hypothesis that no diffusion happens in the solid phase and complete mixing of liquid occurs. According to Figure 8.11b, the early conformation is similar to the equilibrium condition, where the solid has an equal composition, i.e., kC_0. At the middle of the stage, the interface temperature has lowered to T_1; the solute concentration at the boundary continues along the solidus curve of the equilibrium phase diagram. However, as the solute is not endorsed to mix in the solid phase, an uneven concentration profile grows behind the progressing interface. The liquid concentration remains consistent with the hypothesis of thorough mixing. An equilibrium solute concentration at the developing solid–liquid interface may be of interest to develop an expression for the shape of the solute concentration in solid.

A small piece of material represented by the shaded area denotes the growth of a solid, with a volume fraction of df_S, which develops at any specified time. Considering the length df_S on the concentration vs. distance plot, the number of atoms ejected from the solid is equivalent to:

$$\#atoms\ ejected\ from\ solid = (N_LC_L - N_SC_S)Ldf_S \tag{8.1}$$

where N_S and N_L are the numbers of atoms per unit volume of solid and liquid phases, respectively. The number of solutes ejected from the solid must be considered in the liquid, demonstrating the growth of liquid concentration. The growth in solute performing the liquid is shown as:

$$Growth\ in\ solute\ in\ liquid = (1 - f_S)N_L L d C_L \tag{8.2}$$

Assuming the equal atomic volumes in solid and liquid, equating from (8.1) and (8.2):

$$(C_L - C_S)df_S = (1 - f_S)dC_L \tag{8.3}$$

The integration of the above-described equation develops a relation between C_S and the fraction conversion f_S,

The interfacial equilibrium condition $C_L = C_{S/k}$ is

$$C_S = kC_0(1 - f_S)^{k-1} \tag{8.4}$$

$$C_L = C_0 f_L^{k-1} \tag{8.5}$$

where it is known as the non-equilibrium lever rule or Scheil equation.

Three different categories of solute redistribution are presented in Figure 8.12, which can form with insignificant diffusion in the solid. For type 1, the liquid diffusion or convection-controlled mixing in the liquid is finished or the subsequent solute segregation is vigorous. On the contrary, for type 3, liquid diffusion is incomplete without convection-controlled mixing in the liquid and, eventually, the solute segregation is less vigorous. However, type 2 shows intermediate behavior of types 1 and 3, which ensure the solute segregation [10]. Therefore, based on the relations between the alloying elements, there will be remarkable importance on the diffusion rates, temperature, and various concentration profiles at the boundary during solidification.

8.9   Constitutional Supercooling

Usually, in the solidification process, the material in the liquid form first cools at the phase transformation temperature and then starts to solidify with the release of its latent heat [13]. The addition of the cooling practice, together with the latent heat release, maintains the consistent temperature of the system. On the other hand, the phenomenon "supercooling" or "subcooling" or "undercooling" of the liquid does not allow the solidification to occur at the minimal phase transformation temperature. As the cooling progresses, the temperature of the liquid continues to dropdown. However, the system becomes thermodynamically unsteady. Therefore, solidification can be commenced with a small amount of energy accomplished with the latent heat release, which eventually increases the temperature. This mechanism is familiar with the kinetics of solidification. Therefore, the real temperature (change in temperature between the solidification and the nucleation process) at which the solidification starts is termed "the degree of supercooling (or undercooling)." This occurs because of the alloying element segregation at the solid–liquid boundary [14]. The additional concentration of the elements

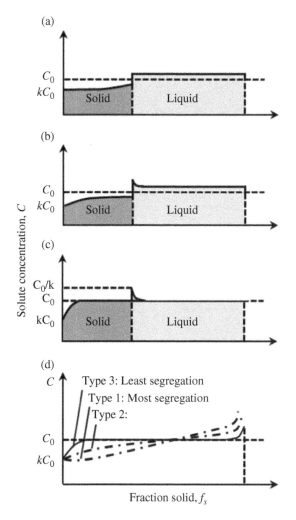

Figure 8.12 Solute distribution without diffusion in the solid and dissimilar diffusion in the liquid, (a) Type 1: complete liquid diffusion or mixing, (b) Type 2: limited liquid diffusion, some convection, (c) Type 3: limited liquid diffusion, no convection, (d) combination of Type 1, 2 and 3 [10]. Source: Reproduced with permission from John Wiley & Sons Inc.

lowers the melting temperature of the liquid. When this reduction is adequate to drop the melting temperature far below the actual temperature, then the liquid will be locally supercooled.

From the phase diagram shown in Figure 8.13a, it is obvious that the solid formation occurs through the rejection of solute in liquid. Therefore, the liquid enriches with the solute element through the solidification process, and their composition is followed along the liquidus lines. The rejected solute is conveyed far from the solid/liquid boundary via diffusion or convection in the liquid.

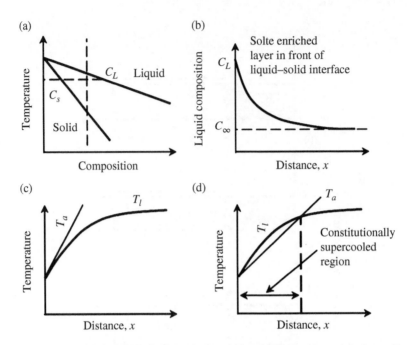

Figure 8.13 Schematic presentation of constitutional supercooling: (a) partial phase diagram, (b) composition profile, (c) temperature profile, and (d) shaded area shows the constitutional supercooled region.

When the solute rejection is too high to develop a higher growth rate at the solid/liquid boundary compared to the transportation rate of solute, then a boundary layer will be formed near the interface. Since solute enrichment lowers the liquidus temperature, the existence of solute boundary layer develops a gradient in the liquidus temperature, as shown in Figure 8.13b and c. The actual temperature gradually decreases with the increase of solute concentration in the liquid. For a comparatively lower thermal gradient, $\frac{dT_a}{dx} < \frac{dT_l}{dx}$, the tips of the projections meet liquid at lower liquidus temperature of the alloy, as shown in Figure 8.13d. This is the mechanism of constitutional supercooling where liquid starts to cool below the liquidus line due to the constitutional variation in the liquid.

Constitutional supercooling considers the interactions between the temperature gradient, interface speed, and the alloying element to come with a simple criterion to forecast the position of the melt, whether it is above or below the liquidus. In the former position, the melt remains steady to a lump on the interface that advances shortly into the melt. In the latter position, the melt is supercooled to solidify naturally on any interface lump [15].

8.10   Nucleation and Growth Kinetics

Nucleation is the early chapter of crystallization and is a key phenomenon in the theory of solidification. In the rapid solidification of supercooled alloy, the nucleation mechanism is far different than the conventional one; thus, it has stimulated plenty of research and fundamental knowledge investigations [16]. When the liquid is supercooled below the liquidus temperature of a metastable phase, solidification for that phase may occur thermodynamically; then, there would be competitive nucleation among the stable and metastable phases. However, for the highly supercooled pure metals and solid solution alloys, a single nucleation event may adequate to start and complete the solidification process having a faster crystal growth rate. But the growth of the supercooled material is always diffusion-controlled, which inherently entails a sluggish growth behavior. So, there is a doubt whether a single nucleation matter can lead the solidification system.

In laser- or EB-based AM techniques, heat sources are comparable to the traditional welding methods such as arcs, plasma, lasers, and electron beams, where very similar structures are formed in multipass welds [17]. The well-established structure–property relationship of the welded material can be of interest for metallic parts in AM. However, in some AM techniques, such as laser powder bed fusion, the processing speed is one to two orders of magnitude higher than the conventional welding. This may result in substantial changes in the microstructure formed. In a nutshell, the AM structures diverge from the traditional weld structures because of the circumstances, accuracy, and flexibility encountered in AM techniques with further layer-by-layer deposition and control of the structures.

Usually, solidification starts with the initiation of nucleation and afterward by growth mechanism [17]. Moreover, solidification can straightly begin through the growth mechanism, which is affected by the components of the additive powders and the substrate metal. Generally, in most AM techniques, the deposited material in every layer has similar chemical composition unless a powder-fed DED is utilized in which the powder stream may change from one composition to others on-demand. In these circumstances, due to a similar crystal structure, epitaxial growth can eliminate nucleation and let the natural growth happen without activation energy till below the liquidus temperature. On the other hand, when dissimilar materials are deposited during the fabrication of composites, or surface cladding of AM, then the nucleation phenomenon should be considered. In layer-by-layer AM processes, this may also happen when the first layer is deposited on the dissimilar substrate material. When the nucleation occurs at the solid/liquid boundary, the newly formed phase needs to conquer an energy obstacle that eventually controls the structure and property of the solidified part.

8.10.1   Nucleation

Nucleation is classified into two categories: (i) homogeneous nucleation and (ii) heterogeneous nucleation. Usually, these categories are proposed based on the position where the nucleation takes place. The homogeneous nucleation occurs with the formation of uniform nuclei all over the parent phase, while in the heterogeneous nucleation, nuclei may advance from the structural discontinuity, such as boundaries of the impurities, foreign particles, dislocations, and so on. The chance of nucleation happening at any position of the parent phase is similar through the phase. Unlike heterogeneous nucleation, the occurrence of homogenous nucleation inside of a uniform substance is challenging, which comprises the evolution of an interface boundary for the new phase. Another difficulty with inhomogeneous nucleation is the demand for high undercooling without forming foreign particles. A large proportion of driving force from the surface energy of the nucleus is not considered in homogeneous nucleation, which makes it uncommon in practice.

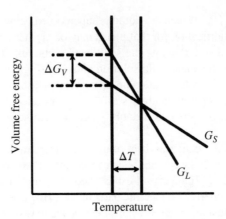

Figure 8.14 Schematic presentation on the relation between the Gibbs free energy and undercooling temperature.

Let us consider a relation between the Gibbs free energy and the undercooling temperature shown in Figure 8.14. When a liquid cools far below its melting point (T_m), the Gibbs free energy (G_S) for the solid becomes lower than the liquid (G_L), there would be an undercooling temperature ΔT and a driving force ΔG to result in an impulsive phase transformation [18]. The volume of free energy which is the change in the free energy in each volume can be expressed as:

$$\Delta G_V = G_L - G_S \tag{8.6}$$

In a simple way, the solidification of pure metal is considered in Figure 8.15, where nuclei of a solid phase are formed inside of a liquid-like packing arrangement of clustered atoms [19]. Moreover, the nucleus is presumed as spherical in shape with a radius r, which is shown in Figure 8.15a. Besides, Figure 8.15b presents two energy sources for the total free energy evolution that conduct a solidification transformation. The first source is the difference of free energy within the solid and liquid, or known as the volume free energy, Δ_V. It becomes negative when the temperature drops down the equilibrium, where it is quantified by the product of nucleus volume (i.e., $4/3\pi r^3$).

The second energy source comes by the development of a solid/liquid interface through the solidification process. This is a surface free energy γ with a positive value, which is quantified by the product of the nucleus surface area (i.e., $4\pi r^2$). Therefore, the complete energy change ΔG is the addition of those two energy sources,

$$\Delta G = \frac{4}{3}\pi r^3 \Delta G_v + 4\pi r^2 \gamma_{sl} \tag{8.7}$$

From a physics point of view, when solid particles in the liquid cluster together to form atoms, their free energy rises. After reaching the size of a critical radius $r*$, the growth of the clustered atoms begins with a decline of free energy. Contrarily, below the critical size, the cluster will shrink or dissolve. This critical size particle is known as an embryo, whereas

(a)   (b)

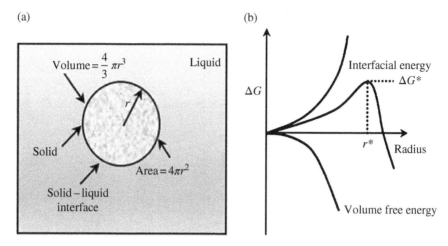

Figure 8.15 (a) The figure depicting the nucleation of a sphere-shaped particle in a liquid. (b) A plot of free energy against embryo/nucleus radius, where also presented the critical free energy change ($\Delta G*$) and the critical nucleus radius ($r*$) [19]. Source: Reproduced with permission from John Wiley and Sons.

with the greater size, it is called a nucleus. The free energy that arises at the critical radius is the critical free energy $\Delta G*$, which is the highest of the curve in Figure 8.15b. This is actually an activation free energy required to form a stable nucleus, or equally as an energy barrier in the nucleation process.

The criterion to begin a nucleation process can be theoretically derived as:

$$\frac{d(\Delta G)}{dr} = \frac{4}{3}\pi \Delta G_v \left(3r^2\right) + 4\pi \gamma_{sl}(2r) = 0 \tag{8.8}$$

And, the critical radius of nucleus:

$$r^* = \frac{2\gamma_{sl}}{\Delta G_v} \tag{8.9}$$

where the critical free energy for homogeneous nucleation is:

$$\Delta G^*_{hom} = \frac{16\pi \gamma^3_{sl}}{3\Delta G^2_v} \tag{8.10}$$

In heterogeneous nucleation or epitaxial growth, a nucleus in a liquid is formed in connection with a substrate. Therefore, the interfacial energies between the liquid, nucleus solid, and substrate metal control the geometry of the nucleus [20]. The total energy can be reduced by assuming the nucleus as the geometry of a spherical cap, as presented in Figure 8.16a. At equilibrium, the interfacial energy is equal to:

$$\gamma_{ML} = \gamma_{SM} + \gamma_{SL} \cos \varnothing \tag{8.11}$$

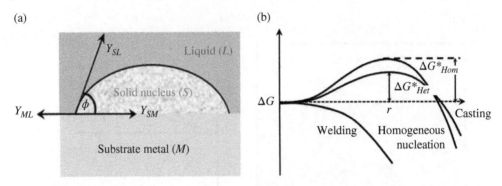

Figure 8.16 (a) Solid nucleus connected with substrate metal and liquid. (b) Graphic presentation for free energy change related to heterogeneous nucleation observed in casting and welding together with homogeneous nucleation [20]. Source: reproduced with permission from John Wiley and Sons.

where γ_{ML} represents the interfacial energy between the substrate and the liquid, γ_{SM} is the interfacial energy within the solid nucleus and substrate, and γ_{SL} is the interfacial energy between the solid nucleus and liquid. Also, φ is the contact angle of the nucleus. When the chemical composition of the substrate and the nucleus is identical, then the interfacial energy between them is negligible. Moreover, the interfacial energies among the substrate and liquid are similar to the interfacial energies between the nucleus and liquid. These assumptions give the following:

$$\gamma_{SM} \approx 0$$

$$\gamma_{ML} \approx \gamma_{SL}$$

$$\emptyset \rightarrow 0$$

The free energy change related to the creation of a solid spherical cap of radius r is shown by:

$$\Delta G_{het} = -V_S \Delta G_v + A_S \gamma_{SL} \tag{8.12}$$

$$\Delta G_{het} = S(\emptyset) \left[-\frac{4}{3} \pi r^3 \Delta G_v + 4\pi r^2 \gamma_{SL} \right] \tag{8.13}$$

where ΔG_V is the free energy change per unit volume, accompanying the nucleus development, V_s is the volume of the nucleus, As is the surface area of the interface between the nucleus and the liquid, and $S(\varphi)$ is the shape factor, which depends on the contact angle. By adopting differentiation of the above equation with respect to r, and considering the result as zero, the critical radius of the heterogeneous nucleation is presented as:

$$r^*_{het} = \frac{2\gamma_{SL}}{\Delta G_v} \tag{8.14}$$

The energy obstruction for the heterogeneous nucleation is then,

$$\Delta G_{het} = \frac{16\pi \gamma^3_{SL}}{3\Delta G^3_v} S(\emptyset) \tag{8.15}$$

In epitaxial solidification, φ is zero and so $S(\varphi)$ approaches zero, which makes ΔG_{het} zero as well. It emphasized that the energy barrier for the epitaxial solidification is negligible contrasted to the casting or other processes, as shown in Figure 8.16b. Because of this small energy barrier, initiation of nucleation is very fast in epitaxial solidification. Usually, in metal AM, the chemical composition of the solid and liquid is pretty similar, which makes γ_{SL} small, as well as the critical radius. On the other hand, this type of solidification demands incomplete or through melt-back of the substrate to expedite grain evolution from the existing ones. The melt-back of the earlier solidified layer is critical in the context of the continuity of the microstructure through the successively melted and solidified multiple layers.

8.10.2 Growth Behavior

After the initiation of the nucleation, the solid/liquid interface deeds as a growth front. The growth kinetics is dominated by the roughness of the solid/liquid interface in the atomic scale, which may be atomically rough (in metals) or atomically flat (in non-metals). The first one progresses with continuous growth, whereas the second one follows a lateral growth, including nucleation and dispersion of ledges [21].

Growth starts from the previously deposited layers through partial or complete melting, which eventually governs the crystallographic pattern [17, 22]. The intensified heat may penetrate further below the deposited layers, enabling the remelting process required to eliminate surface contaminants and the breakdown of oxides, thus offering a clean solid/liquid interface. The microstructure developed close to the melt boundary is controlled by the substrate material, whereas far away from the boundary, it is advanced through competitive growth [23].

A schematic presentation of the Walton and Chalmers model for competitive growth is shown in Figure 8.17 [24]. Usually, competitive growth takes place between dendrites with numerous crystallographic orientations and is commonly found in alloys of iron [25, 26], nickel

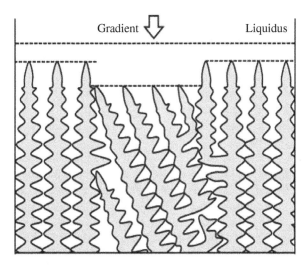

Figure 8.17 Schematic presentation of the Walton and Chalmers model showing competitive grain growth, adapted from [24].

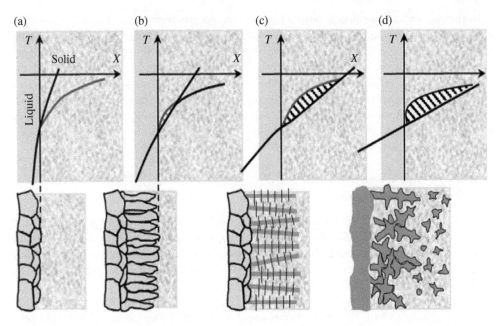

Figure 8.18 The graphics showing the growth characteristics and constitutional supercooling impact on solidified patterns. Source: adapted from https://www.tf.uni-kiel.de/matwis/amat/iss/kap_6/illustr/s6_2_1b.html.

[27], titanium [28], and tantalum [29]. Dendrites normally progress by the path of higher heat flow and lead to competitive growth in the structure.

The growth characteristics and constitutional supercooling impact on solidified patterns are depicted in Figure 8.18. When the solute-rich boundary layer creates a temperature gradient sharper than a critical gradient for constitutional supercooling, a stable planner interface growth would be introduced, as shown in Figure 8.18a [21]. If the constitutional supercooling is encountered, successive lumps at the solidification front may propagate with rapid growth to advance into long arms or cells, approaching parallel to the heat flow, developing a cellular microstructure, as presented in Figure 8.18b. However, with a smaller temperature gradient, a broader "mushy zone" is formed, which advances a dendritic growth with secondary or tertiary arms, as schematically presented in Figure 8.18c and Figure 8.18d, respectively.

In joining, the base metal at the boundary line plays as a nucleation site [10]. As the liquid metal within the melt pool is closely touching the thin layer of a substrate while splashing them totally, nucleation progresses without difficulties. During the autogenous joining, nucleation starts with the agglomeration of atoms in the liquid on the previously developed structures continuing with a similar crystallographic orientations. This type of growth phenomenon is shown in Figure 8.19 and is known as the epitaxial nucleation or epitaxial growth. The figure describes the grains' growth with $\langle 100 \rangle$ crystallographic orientations, which is common for the FCC or BCC crystal structures, where columnar grains advance in the $\langle 100 \rangle$ direction.

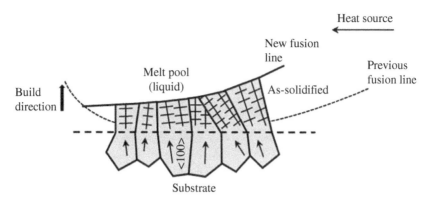

Figure 8.19 Epitaxial growth of the solidified metal adjacent the fusion line [10]. Source: Reproduced with permission from John Wiley & Sons Inc.

8.11 Solidification Microstructure in Pure Metals and Alloys

The solidification of a liquid implicates the chemistry of many physical aspects. The solid/liquid boundary layer acts as a front edge to generate latent heat, which is conveyed through thermal diffusion. Moreover, the solute ejection takes place in the liquid that contains impurity. The solidification process may be stable or unstable. The stable solidification, which is also told as Stefan problem [30], is defined as a smoothly growing interface, responsible for the thermal diffusion through the solid phase. In these circumstances, the surface tension and the dynamic movement at the interface are neglected, also with reduced perturbation. However, unstable solidification occurs when the metastable liquid cools far below the equilibrium temperature. In this situation, thermal diffusion from the interface occurs both by solid and liquid phases when the surface tension and dynamic movement are also considered. The local perturbation on the boundary layer will be enriched, and further protrusion in the liquid will progress with various structural patterns. The formation of grain pattern and the crystallographic texture is controlled by the melting process and the solidification of the liquid melt zone.

The melt area drives away heat through the substrate causing the melt pool to have a curve shape. Depending on the process parameters, i.e., heat source intensity, scanning speed, etc., the melt pool shape may vary from oval or tear dropped upon the substrate with semicircular or keyhole cross section. The geometric profile of the melt pool is significant as it affects the grain structure in the fusion zone. In the keyhole case, the beam goes down of the substrate with minimum heat input, which changes the conduction mode at high speeds, equally in electron and laser beam techniques. The conduction approach is desired for AM because of the unsteady keyhole that may cause undesired porosity in AM products. The significance of the melt pool geometry to make AM builds with desired structural features is a vital consideration.

There are four major solidification patterns: (i) planar, (ii) cellular, (iii) columnar, and (iv) equiaxed dendritic, which are shown in Figure 8.20. The various solidification structures grow depending on the driving force of constitutional supercooling, distribution of the solute at the boundary, as well as the characteristics of the elements containing in the solidified alloys. Generally, the devastating forces of constitutional supercooling do not exist in pure metal, which will then show a planer solidification approach [10].

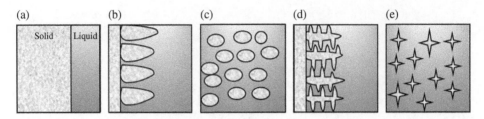

Figure 8.20 The schematic diagrams illustrate the modes of solidification pattern: (a) planar, (b) and (c) cellular, (d) columnar dendritic and (e) equiaxed dendritic.

The relation between the different solidification patterns with the degree of constitutional supercooling is shown in Figure 8.21. A planar solidification starts with a greater real temperature contrast to the equilibrium temperature of the liquid melt [31]. In the planer solidification, a stochastic protuberance may grow into an area of higher temperature and results in the breakdown of the protuberance (Figure 8.21a). Planar solidification is feasible, especially for single-crystal growth, where it demands either high purity metal or tremendously high degrees of temperature gradients or solidification rate.

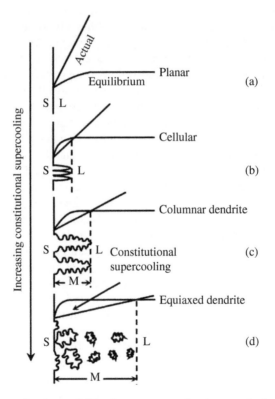

Figure 8.21 Occurrence of various solidification structures related to constitutional supercooling, [10]. Source: Reproduced with permission from John Wiley & Sons Inc.

In constitutional supercooling, the real temperature gradient is lower than the liquidus temperature gradient; a protuberance may mature in the undercooled melt and approaches as a cellular or dendritic pattern (Figure 8.21b). When the grain grows like a column without branching the arms, it will form a cellular structure. In contrast, the grains with secondary or tertiary arms will develop a dendritic structure (Figure 8.21c and Figure 8.21d). The cellular and the dendritic growth are estimated by the degree of constitutional supercooling and the complete stability, described by the critical solidification rate. However, planar solidification is independent of the temperature gradient.

The completion of the supercooling and starting of the solidification phenomenon are described in Figure 8.22. While the dendrite growth occurs throughout the melt, the melt temperature backs to the equilibrium by absorbing latent heat releasing from the solid/liquid boundary. Therefore, the knowledge on the mechanism of dendritic growth and their propagation to enhance solidification has become an intensely investigated subject.

Generally, the solidification starts homogeneously after adequate cooling or heterogeneously with the existence of a solid particle in the supercooled melt. Moreover, after nucleation, the consequent growth of the solid from the particle may be unstable, and based on the degree of supercooling, dendritic structures start to form. Dendrites are defined as prototypical structure growing from homogeneous initial states into compound spatio-temporal configuration distinct from equilibrium. The root of the word "dendrite" originates from the Greek term "dendron," meaning "tree". Similar to a tree, a dendrite shows an extremely branched, arborescent pattern, which is schematically presented in Figure 8.23.

Because of the unsteadiness of the interface, a dendrite structure comprises the primary stem, secondary, or tertiary branches, all advancing in particular crystallographic directions. The dendritic structures in metallurgy form the microstructure of metals or alloys, where it heavily

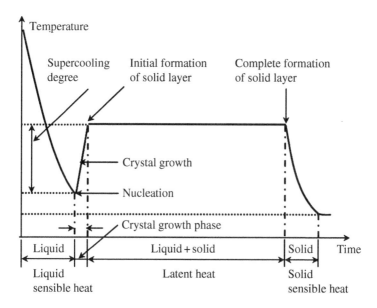

Figure 8.22 Change of the temperature in time, throughout the solidification progression of a supercooled liquid and the crystal growth. Source: Reprinted from [30] with permission from Elsevier.

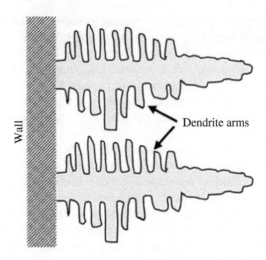

Figure 8.23 Schematic of the dendrite formation/growth.

controls the mechanical, chemical, and physical characteristics of the material. During the solidification, the growth of the perturbation at the solid/liquid boundary may be inhibited by the surface tension or kinetic phenomenon. The challenge between the steady influence of surface tension and the unsteady influence of supercooling is mainly the cause source for the formation of dendritic compound patterns.

The control of microstructure considering grains or phases has the potential in AM techniques by demonstrating the accurate speed of lasers or electron beams. The manipulation of the scan strategies can eventually manipulate the thermal gradient, solidification rate, and cooling rate to control the grain structure.

As the grain structure influences the mechanical properties, it is essential to create finer grains in the solidified material to achieve superior properties. However, in laser or EB metal AM, it is crucial to stop grain growth during remelting of the formerly deposited layers, which also brings heterogeneous nucleation and epitaxial grain development either in cellular or dendritic form. When the epitaxial grain growth of the columnar grains is prevented by the development of equiaxed grains close to the melt zone area, and the equiaxed grains are deeper than the remelted penetration depth, then the equiaxed grains govern the average grain size. It is demanded in laser- or EB-based AM that the occurrence of equiaxed grains is estimated by the density of heterogeneous nucleation sites, which may be feasible in the LPBF process with a proper powder feed rate.

8.12 Directional Solidification in AM

The most vital factor in controlling the mechanical properties in AM products is the solidification structure [12]. The connection with higher laser energy provides rapid heating and cooling, resulting in faster solidification during cooling. It is well known that the laser-based AM techniques generally deal with the higher heating/cooling rates (10^3–10^8 K/s) at the solid/liquid boundary even for a smaller sized melt pool (~1 mm). Moreover, the heat dissipation rate through the substrate is fast enough to develop a rapidly solidified structure. Therefore, the grain refinement is usually predicted in the distinctive AM structures because of their inadequate time for grain formation/progression.

However, the solute intensity or temperature gradients in the melt may produce surface tensions and subsequent Marangoni convection, resulting in an unstable solidification process. Moreover, because of the dynamic restraint of crystal growth, rapid solidification generally progresses along with the maximum heat flow. The mechanisms of non-equilibrium solidification and the propensity for directional propagation are instantaneous but competitive, which may generate diversity in crystal orientation with limited consistency. As a result, AM processed metallic products may possess intrinsic anisotropic characteristics.

For a known material composition, the solidification morphology is controlled by parameters, such as solidification velocity and temperature gradient. The developed structure may differ from planar to cellular and to dendrites, which normally turn into a finer structure until the regeneration of cellular structures with a growth rate near absolute stability. At velocities greater than this range, the banding acts and finally the planar interface is completely stabilized. The well-developed cellular structure normally grows without advancing sidearms, where their axes are aligned to the heat flux direction without considering any crystal orientation. However, dendrites are regarded as the growth of their arms along crystallographic orientations. Because of the anisotropy in solid/liquid interface energy and growth mechanism, cubic crystal dendrite will propagate along the $\langle 001 \rangle$ direction, indicating the heat flow direction.

Throughout the directional solidification and advances of columnar structures, the heat flux follows contrary to the growth direction. Which means that the growth rate of the fronts limits the solid/liquid boundary to propagate at this rate. During the alloy solidification, solute will stack on the boundary, while the distribution coefficient is normally lower than the unity and, eventually, this variation of concentration will impact the equilibrium solidification temperature. The supercooled zone, where equiaxed grains with various volume fractions may take place, depends on the thermal gradient and the solidification rate and finally directs to the columnar-to-equiaxed transition (CET).

8.13   Factors Affecting Solidification in AM

8.13.1   Cooling Rate

The structure of the AM products is controlled by the thermal behavior of the process, precisely by the temperature gradients, solidification, and cooling rates. The thermal stages of AM products are complex, having multiple repeated heating and cooling cycles. Moreover, cooling rates are higher due to the small localized heat input that normally moves at a high speed (e.g., 5 m/s), which creates a narrow heat-affected zone with finer grain structures. These phenomena are very uncommon in conventional processes [32, 33]. Generally, solidification/cooling rates are influenced by the heat input, which is manipulated by the laser or EB power, beam scan speed layer thickness, scanning strategies, and etc. For example, when the laser power is low and scanning speed is high, this combination normally delivers a smaller heat flux that results in a larger cooling rate. On the contrary, with a higher laser power and a lower scan speed, heat input would be intensified to melt a larger substrate area and eventually results in a slower cooling rate.

8.13.2   Temperature Gradient and Solidification Rate

Through the solidification, columnar grains advance along the path of a higher temperature gradient in the melt pool [22]. Various melt pool geometries impact the temperature gradient and eventually influence the grain structure in terms of types, size, texture, and shape. For example, the spherical melt pool generates curved and tapered columnar structures because of the deviation in the thermal gradient path from the pool border. However, the comet-featured melt pool creates conventional and wide columnar structures, where the path of the maximum thermal gradient does not shift notably through the process.

Let us consider the angle between the direction of grain growth and beam scan speed (SS) is θ, then the constant nominal growth rate R_N would be $R_N = SS. \cos \theta$. In a cubic crystal structure, $\langle 100 \rangle$ direction designates the main dendrite growth direction. This favored growth direction makes the local growth rate R_L greater than the nominal rate R_N. Another angle φ is considered between the normal direction of the melt pool border and the $\langle 100 \rangle$ direction to make a link between R_L and R_N, which is $R_L = R_N/\cos \varphi$. The relationship states that the local growth rate becomes larger with misaligned crystals with respect to the direction of a higher temperature gradient.

Several factors like moving heating source power (P), beam scan speed, substrate material temperature (T), and beam spot diameter (d) control the G/R_L ratio [17], where G is the local temperature gradient. The value of G is lower with higher values of T. The lower value of G must be recompensed by a lower P value. Therefore, P must be decreased when T increased to have columnar growth through the solidification. With an increased scan speed, R_L increases without a significant impact on G, resulting in a lower value of G/R_L ratio. However, as the higher scan speed causes a smaller melt area with greater temperature gradients, the increased scan speed may increase the G/R_L ratio. A larger energy intensity or a wider beam diameter is inconvenient to work with because of the prompt decreased beam intensity in the area far from the centerline in a single mode laser with a Gaussian intensity distribution that restricts uniformity in the microstructure. Therefore, when a larger d is unavoidable, P needs to be intensified to confirm enough heat flux for the substrate remelting to continue with epitaxial grain growth. To arise the substrate material temperature by a pre-heating procedure somewhat enable the melting with a larger volume, but at the same time, lowers the processing window.

Moreover, the stability of the solid/liquid boundary is dominated by the thermal and supercooling behavior. Considering the constitutional supercooling, the instability at the interface is expressed as: $\frac{G}{R} > \frac{\delta T}{D}$, G represents thermal gradient, R as the growth rate, δT is the temperature range for solidification, and D is the diffusion coefficient of solute material in the melt.

To understand the solidification morphology and their structure, a solidification map is designed with G and R, in the form of their product as $G.R$ and ratio as G/R. Figure 8.24 describes the influence of G/R and $G.R$ on the solidification structure, where G/R governs the type of solidification pattern and $G.R$ controls the size of the structure [10]. The solidification structure, which may be planar, cellular, columnar dendritic, or equiaxed dendritic, normally occurs at high and low G/R values when normally a higher G/R results in a planer structure and a lower G/R results in an equiaxed structure. Moreover, the size of these four structures decreases with the larger value of $G.R$ (cooling rate).

The size of the solidification structure can be estimated using the product $G.R$. Therefore, using the cooling rate or solidification time, the arm spacing of columnar or equiaxed grains can be measured, which can be stated as [10]:

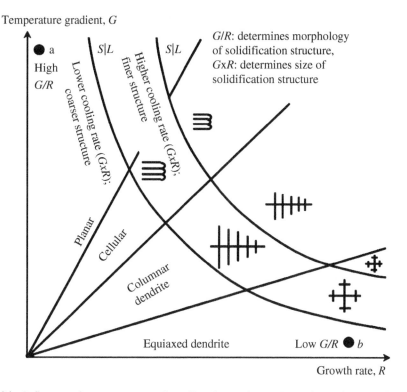

Temperature gradient, G

G/R: determines morphology of solidification structure,

$G{\times}R$: determines size of solidification structure

Figure 8.24 Influence of temperature gradient G and growth rate R on size and morphology of the solidified structure [10]. Source: Reproduced with permission from John Wiley & Sons Inc.

$$y = at_f^n = b(\varepsilon_c)^{-n} \tag{8.16}$$

where t_f is the solidification time, ε_c is the cooling rate, and a, b and n represent material constants that should be identified based on the experimental data.

The mathematical relation depicts that with the slower cooling rate and the extended growth time, coarser dendritic structures are formed. The surface energy of the solidified material may be lowered with the formation of coarse dendrite arms, as the coarse arms have less surface area per unit volume. This is because the slower cooling allows a longer time for growth and hence forming coarser dendritic arms. On the other hand, a faster cooling rate does not allow a longer time for growth and hence producing a finer structure.

The driving force, which is required for dendrite to grow properly, comes from the undercooling. There is a difference in temperature among the liquidus and the dendrite slant that makes the undercooling temperature, as stated by [10]:

$$\Delta T_{tot} = \Delta T_c + \Delta T_T + \Delta T_K + \Delta T_R \tag{8.17}$$

where ΔT_C, ΔT_T, ΔT_K, and ΔT_R are the undercooling temperatures accompanied with solute diffusion, thermal diffusion, attachment kinetics, and solid/liquid boundary curvature, respectively.

Through the solidification of most metallic materials, ΔT_T, ΔT_K, and ΔT_R are negligible; hence, the solute diffusion undercooling ΔT_C leads the process. Therefore, different solidification structures from planar to cellular, columnar dendrite, or equiaxed dendrite are basically formed because of the supercooling at the solid/liquid interface.

However, the morphology of the microstructure developed in the direct laser energy deposition process for different geometries is analyzed by Liu et al. [34], as shown in Figure 8.25. In the first case, the geometry of a single track is considered in Figure 8.25a, where a higher thermal gradient and faster cooling rate by heat dissipation through the bottom of the substrate are the potential reason for the formation of columnar structure. In the first case, the geometry of a single track is considered in Figure 8.25a, where columnar structures formed at the lower end of the melt as an outcome of higher thermal gradient and faster cooling by heat dissipation through the bottom. With expanding the depth of the melt, the temperature gradient and the cooling rate are lowered, thus forming the equiaxed grains at the upper part of the melt. In the second case, multi-tracks are considered in Figure 8.25b. Here, the successive scans create overlapped areas and remelt the previously formed equiaxed grains, which then solidify as columnar structures. In the overlapped areas, because of the lower melt depth, the greater temperature gradient and faster cooling rate stimulate the epitaxial growth. In the third case, a geometry of multi-layer deposition is shown in Figure 8.25c. It is already stated that during the remelting of the overlapped areas, the previously formed equiaxed grains remelt partially and act as nuclei to

(a)

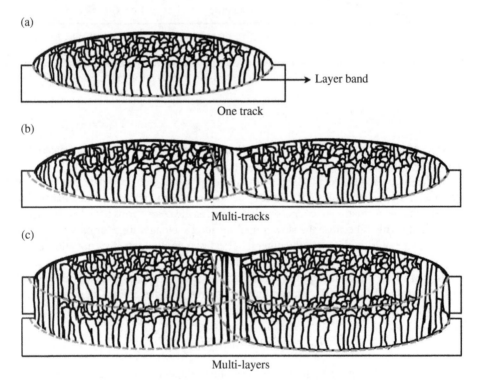

One track

(b)

Multi-tracks

(c)

Multi-layers

Figure 8.25 Formation mechanism of grains in AM: (a) single track, (b) multiple tracks, (c) multiple layers during laser DED. Source: reprinted from [34] with permission from Elsevier.

continue epitaxial grain growth and subsequent columnar structures. Thus, the grain morphology in newly formed layer grows into similar to the last one and repeats in each layer till the last deposited one.

Therefore, regardless of the complications of AM, the solidification microstructure is mostly resulted by temperature gradient, solidification rate, and undercooling temperature. The correlation between these basic solidification parameters can significantly endorse the knowledge within process and structure. Besides the dendritic structure reported in AM, there may be a considerable amount of precipitates in some precipitation strengthened alloys such as Ni-based superalloys. This happens because of the very fast cooling in the localized melt area, which creates a non-equilibrium condition where diffusion is limited. Another reason is the inadequate time for the alloying constituents to diffuse back into the solidified structure. Therefore, the concentration of the residual melt increases with alloying elements and promotes the eutectic solidification to take place at the end of solidification, resulting in precipitates.

8.13.3 Process Parameters

The solidified microstructure and the phase formations are considerably controlled by the input process parameters. Together with the higher specific energy and the faster deposition rate, the liquid melt will be at a higher temperature for an extended time to lower the temperature gradients. Therefore, the grains are allowed to grow coarsened and mainly equiaxed. On the other hand, the minor specific energy is understood by applying a faster scan speed, and hence no adequate time for the grain growth. Moreover, the geometry of the melt pool becomes narrow at a faster scan speed; therefore, the temperature gradients are higher, resulting in the formation of columnar grains. Layer thickness is also another influencing factor, which depends on additional parameters, i.e., power, speed, specific energy, and powder flow rate, to dominate the microstructure in AM products. When the specific energy is brought down, the energy required per unit area to melt down the powder is lowered. This calls for a need to lower the layer thickness. However, the thicker layer causes slower cooling and results in a coarser microstructure. Therefore, the structural development is complex to layer thickness because of the dominance of heat conduction through the substrate, which controls the cooling rate and consequent microstructures.

8.13.4 Solidification Temperature Span

Usually, the broader solidification temperature span creates a greater solid/liquid or mushy zone, which is mostly responsible for the solidification cracking, as the liquid cannot allow load. This temperature range is altered by several factors, such as the existence of impurities like sulfur and phosphorous in ferrous or nickel alloys and some specific alloying elements to enhance mechanical properties. In ferrous alloys, there is a propensity for sulfur and phosphorus to segregate at the grain boundaries and alloying with the element to form low melting point compounds; such as FeS, which also prolongs the mushy zone. Simultaneously, the nearby grain material will be solidified, which then creates a gap with the grain boundary liquid due to thermal stresses. The eutectic temperature range also enhances the extent of the mushy zone.

8.13.5   Gas Interactions

In the area of rapid prototyping and additive manufacturing of metallic components, it is very important to ensure an appropriate gas environment to fabricate quality products. In AM techniques, argon and nitrogen are generally used to offer an inert atmosphere and satisfy the high-tolerance criteria. In the case of electron beam powder bed fusion (EB-PBF), a true vacuum in the chamber is required. Helium is also used for shielding during laser DED to enhance temperature temporal behavior.

The impact of the shielding gas on the quality of metal powders during laser DED, combined with the AM input parameters, plays a major role in the growth of the final structure and properties of the manufactured products. Although nitrogen is a reactive gas, it can be a suitable shielding gas for those materials, which never react with nitrogen. Generally, the metallic materials such as Fe, Cr, and Ti react with nitrogen to form nitrides. Improper shielding or vacuum can cause air contamination, from where nitrogen may be introduced in the liquid melt [10]. For example, in ferrous alloys, nitrogen stabilizes the austenite phases and hereby raises the retained carbon and solute content, which afterward initiates the solidification cracking. Simultaneously, the structure is changed with the existence of iron nitrides, which has a needle-like structure with brittle interfaces susceptible to cracking.

8.14   Solidification Defects

Generally, AM-made products include defects that influence the mechanical properties of the components. The reasons for the defects are recognized as: (i) un-melted or partially melted powders, (ii) deficiency of fusion, (iii) delamination between subsequent layers due to residual stresses, and (iv) interactions of gases during manufacturing. Therefore, the possible defects, their formation and growth mechanisms, and potential ways of exclusions are reviewed below:

8.14.1   Porosity

Usually, in AM, the pore formations are connected with the process input parameters such as laser power or beam scan speed (see Chapters 2 to 7). Also, the porosities are categorized as (i) powder induced and (ii) process induced.

Three major mechanisms cause porosity in AM products. First, in some AM techniques such as LPBF, at very high operating powers, melting may be accomplished throughout a keyhole mode [17]. When the keyholes are not developed and stabled properly, they can turn into an unstable mode, which then frequently forms and collapses, making voids inside the melt. The shape of the keyhole porosity can differ based on the keyhole's geometry, as evident in Figure 8.26a.

Second, during the atomization of metallic powder, gas can be trapped inside powder particles, creating microscopic gas pores during the process. Consequently, the powder caring pores can be transferred to the final fabricated parts, as presented in Figure 8.26b. Moreover, gas pores can be generated because of the potential of gas attraction/solution by the alloy melt. In laser-based AM, usually, an inert gas such as argon is used to avoid impurity from the liquid melt or solidified parts, while electron beam AM is carried out in vacuum or underneath helium. The inert gases remain insoluble in the liquid melt; therefore, the pores will be kept in the

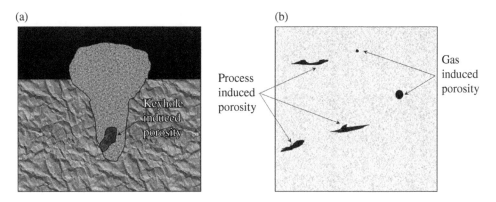

Figure 8.26 Schematic of (a) Keyhole porosity. Source: Adapted from [35]. (b) Process and gas-induced porosity. Source: Redrawn and adapted from [36].

solidified parts except they can outflow the liquid melt. However, the non-inert gas like nitrogen has the characteristics to lower or remove porosity from stainless steel melt by mixing with liquid melt before solidification. Therefore, removing the shielding gas and carrying out the laser AM under vacuum might be a way to lower the porosity formation in the solidified products. Third, the deficiency in fusion imperfections can result from insufficient infiltration of the upper layer melt into the substrate or the formerly solidified layers. These lack of fusion areas are detectable by un-melted or partially melted powders nearby the pores.

To understand the mechanisms of defect formations during LPBF, the process window for a known and fixed layer thickness and hatch spacing can be classified into four different zones, as shown in Figure 8.27 [37]. The zone I, termed "fully dense zone," which is free from porosity. Zones II and III, which are termed "over melting zone" and "incomplete melting zone," respectively, have measurable porosities. The porosities introduced by process input parameters in zone II are resulted from the exceeded energy, whereas those of zone III are caused by inadequate energy flux. Last, the "overheating zone" is generated by very slow scan speed and higher laser power. With extra energy, it is challenging to intricate any samples under this process environment. The geometry of the porosity is also a sign of understanding the formation mechanism. Actually, the spherical defects are typically accomplished with gas bubbles during higher energy laser application to a destabilized liquid interface.

Porosities generated by inadequate melting are typically observed near the boundaries, which are geometrically irregular in shape. Their amounts are greatly persuaded by operating parameters such as laser power, layer thickness, scanning speed, and hatch spacing. In the deficient fusion porosity, the top surface of the earlier layer may not be melted to develop a coherent bond due to the unsatisfactory dissipation of laser energy through the powder layers. Another potential reason for the poor fusion porosity is the entrapment of gas bubbles between the layers during processing. These gas bubbles then result in an unsteady scan track with non-uniform evaporation. The distinctive uncertainty of the scan tracks therefore causes intermittent failure of the liquid melt adjacent to the vapor cavity and forms periodic voids. There is a hypothesis that metallic vapor together with the shielding gas probably enters into the melt tracks, where, during cooling, the metallic vapor becomes solidified, leaving the shielding gas as pores.

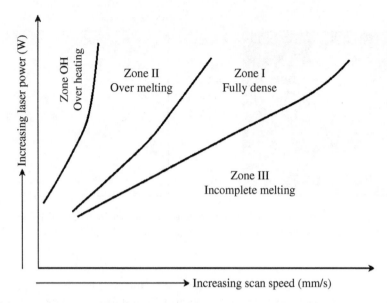

Figure 8.27 Process window for LPBF manufactured Ti-6Al-4V alloy. Source: Redrawn and adapted from [37].

Similar to the process input parameters, the unsteadiness of the scan track and pore formation can be minimized by the proper adjustment of the beam [38]. There is a chance to release more gas during the slow solidification process, usually at higher energy density or lower scan depth with lower energy densities. Using a pulsed or modulated laser beam, it is possible to regulate the energy indulgence into the powder layer or substrate material, thus manipulating solidified structures, porosity, and other defects. The steadiness of the scan track is significantly impacted by the pulse because the pulse needs to revive the scan track once it fails. The adequate over-lapping ratio of the two scan tracks can provide effective removal of porosity formed in early pulse, which eventually minimizes the process-induced cavities. Hence, it is suggested to reduce the lay-off time of the pulse, lower than the solidification time required for the melt. In that situation, it is necessary to use a high-duty cycle and a greater extent of overlap area in the melt zone through the pulse transition. During pulsed-wave methods, the consideration of short pulse periods and lower energies is necessary to maintain a small melted area.

8.14.2 Balling

Usually, the structural characteristics of AM-fabricated parts comprise the outer surface structure, together with the inner grain structure. The balling phenomenon is considered as the unusual melt pool segregation/breakout that may take place on the surfaces of the laser additive manufactured parts, especially laser powder bed fusion. Through the processing, the laser beam scans the surface linearly and the melting occurs along a row of powders, which then creates a constant liquid track like a tubular shape. The breakdown of the tube into the spherical-shaped (balling effect) metallic agglomerates drops the surface energy of the melt track till the ultimate equilibrium condition is aroused. The balling effect can result in an intermittent scan track with poor bonding and can be an obstruction to a constant deposition of powder on early deposited layers. These phenomena can result in porosity or even delamination because of weak interlayer joining together with induced thermal stresses.

There are three types of balling mechanisms: (i) laser scanning on cold powder offers "first line scan balling" because of the greater temperature gradients enforced on the melt, (ii) applying larger scan speed promotes "shrinkage induced balling" because of a substantial capillary unsteadiness, (iii) the "splash induced balling," which forms a greater extent of micrometer-sized balls, predominates at higher laser power with slow scan speed, and extensive period of lower viscosity liquid.

The mechanism of balling phenomenon due to inadequate liquid formation is described in Figure 8.28. In powder-based AM techniques, powder particles absorb energy by the mechanisms of bulk coupling and powder coupling. Initially, a thin layer of distinct powder absorbs energy, influenced by the powder properties. This elevates the temperature on the particle surface, forming a liquid phase through the surface melting of powders. Consequently, the heat flows mostly in the direction of the center of the persisted powders till the steady-state melting temperature is achieved. The volume of liquid formation is influenced by the melting temperature, which is regulated by the laser power and scan speed through line scanning. For example, with a known scan speed and a lower laser power, the solidus temperature drops, forming a smaller volume of liquid melt. This makes a higher viscosity in the liquid–solid mix that, in turn, impends the liquid flow and particle reordering. This eventually drops the general rheological performance of liquid and solid particles that are adjacent and in touch together. Subsequently, the liquid melt in each exposing spot area apts to combine into a discrete coarsened sphere nearly with the same size as the laser beam's diameter. In these circumstances, no effective bonding is occurred between the adjacent balls since inadequate liquid volume stops the satisfactory growth of continuous connecting metal agglomerates. In fact, the result would be multiple irregular shape discrete solidified zones when one looks at the build plate from the top. In addition, lower laser power is responsible for the inadequate undercooling temperature of the liquid melt. Therefore, the formation of coarsened and irregular dendrite structure in solidified balls generates characteristics that are mechanically weak, thus undermining the part quality.

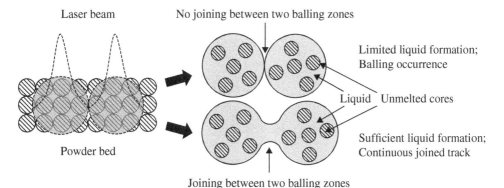

Figure 8.28   Schematic of balling incident appeared by coarsened sphere-shaped sintered agglomerates initiated by inadequate liquid formation. Source: Reprinted from [39] with permission from Elsevier.

With adequate liquid content, the balling mechanism is depicted schematically in Figure 8.29. When a satisfactory volume of the liquid phase is grown with both greater laser power and scan speed, the melt converts an incessant tubular melted track because of the short exposure time of laser input on every area underneath the moving beam. But the melt pool track would be in an unsteady state; hence, the surface energy will keep dropping to reach an ultimate

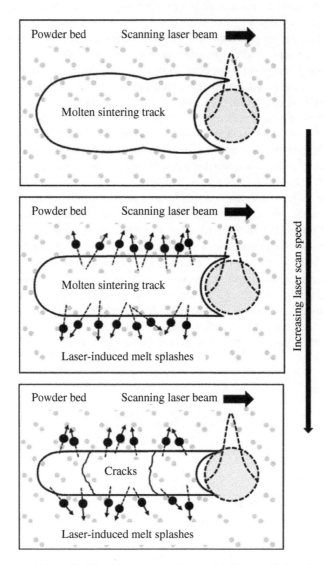

Figure 8.29 Representation of balling phenomenon characterized by small shaped balls on the sintered surface produced by melt splashes through laser scanning. Source: Reprinted from [39] with permission from Elsevier.

equilibrium state. When the scan speed increases, the energy intensity by the laser input drops, which then lowers the surrounding temperature and subsequently the diameter of the tubular melt pool track. Therefore, the melt zone unsteadiness rises considerably. Under these circumstances, the dropped surface energy promotes the spattering of liquid droplets from the melt pool track surface. As a result, many micrometer-sized spherical spatters are solidified near the sintered surface, ensuing the balling phenomenon. These irregular shape-solidified zones may cause a manufacturing issue by manifesting the potential of recoater jamming during LPBF.

However, in a multi-layer deposited sample, the balling effect may be excluded by reducing the powder layer thickness. By applying a deoxidizing agent, oxide films can be satisfactorily eliminated from the melt surface, thereby cleaning the balling zone joining system. A proper dispersion and leveling of the melt through oxide removal ensures proper wetting characteristics, and therefore effective layer-by-layer joining of melted powder particles without any obvious balling effect.

8.14.3 Cracking

Usually, three categories of cracks are detected in metal AM processed products. The solidification cracking in AM products occurs along the grain boundaries, similar to what would normally happen in the conventional welding. The solidifying melt passes through a contraction from solidification shrinkage and thermal contraction. The substrate or the earlier solidified layers show a lower temperature than the solidifying layers. As a result, a higher shrinkage develops in the solidifying layers, impeded by the earlier deposited layers, forming tensile residual stresses. When the extent of residual stresses surpasses the strength of the solidifying layer, cracks may be generated throughout the boundaries.

8.14.3.1 Solidification Cracking

Solidification cracking takes place at the last step of solidification, at the location where dendrites have become complete grains and are detached by a small volume of liquid known as a grain boundary. There are three basic theories of solidification cracking: (i) the shrinkage–brittleness theory, (ii) the strain theory, and (iii) the comprehensive theory that comprises the related philosophies from theories (i) and (ii). The comprehensive theory emphasizes the formation of cracks in a material, where the liquid melt isolates the grains or solid–solid bonds occur between grains. Therefore, the solidification cracking occurs with the surpassing of tensile stresses/strains above the critical range.

The possible reasons for solidification cracking in melted metals are depicted as: (i) temperature range of solidification, (ii) volume and dispersion of liquid melt at the end of solidification, (iii) the early solidification phase, (iv) the surface tension at the grain boundary melt, (v) the grain morphology, (vi) the ductility of the solidifying metal, and (vii) the propensity of the weld metal to contract and the amount of restriction. All these aspects are ultimately connected with the metal composition. Here, the first two aspects are influenced by microsegregation, which is controlled by the cooling rate through solidification like the primary phase formation.

8.14.3.2 Intergranular Cracking

Intergranular cracking arises at the grain boundaries during the last step of the solidification, where solidifying and cooling material possesses higher tensile stresses compared to the strength of the solidified metal. Moreover, the substrate material, which also passes through a thermal cycle developing expansion and contraction on a small scale. Intergranular cracking is worsened by the intensification of thermal power or thickness of the substrate.

8.14.3.3 Reheat Cracking

Reheat cracking is a common phenomenon in low-grade ferrite steels containing Cr, Mo, V and W to enrich corrosion and high-temperature properties. After manufacturing through welding or AM techniques, heat treatment is usually carried out to release stress and minimize the propensity to hydrogen cracking or stress corrosion cracking. However, the major issue is that the cracking may take place at the heat-affected region through reheating.

The propensity to crack formation (crack susceptibility (CS)) is also related to the alloy content. For example, in the case of steel, when the CS value is equivalent to or larger than zero, there may be a possibility for reheat cracking [10].

$$CS = \%Cr + 3.3 \times (\%Mo) + 8.1 \times (\%V) - 2 \tag{8.18}$$

During the AM processes, the temperature in the heat-affected zone close to the fusion line is raised to the austenitizing temperature, where *Cr*, *Mo*, and *V* carbides dissolve into austenite regions. Faster cooling of the heat-affected zone may allow inadequate time to transform carbides, resulting in martensite with supersaturated alloy content. When the reheating is done to release the stress, the alloy content starts to reprecipitate along the high-energy austenite grain boundaries. This stimulates the crack formation because of residual thermal stresses.

8.14.3.4 Liquation Cracking

Liquation cracking typically occurs in the mushy zone (MZ) or partially melted zone (PMZ) of the solidified build product, schematically shown in Figure 8.30. The alloy containing low melting temperature carbides results in melting in PMZ, during fast heating even under the liquidus temperature. When cooling starts to take place, the PMZ undergoes the tensile stress because of solidification shrinkage together with the thermal contraction from the solidified layers. In these circumstances, the liquid melt pool around the grain boundaries or carbides may perform as the crack initiation locations.

Therefore, the possible reasons for liquation cracking are summarized as: (i) wider mushy zone, formed because of greater difference between liquidus and solidus temperatures, as observed in nickel-based superalloys, (ii) greater solidification shrinkage due to a larger size melt pool, such as Ti-6Al-4V alloy, (iii) greater thermal contraction because of a large coefficient of thermal expansion, as observed in aluminum alloys.

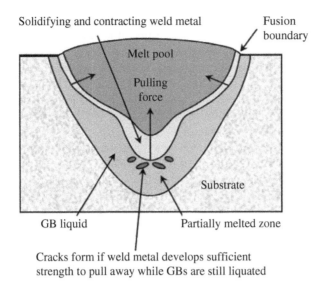

Solidifying and contracting weld metal

Fusion boundary

Melt pool

Pulling force

Substrate

GB liquid

Partially melted zone

Cracks form if weld metal develops sufficient
strength to pull away while GBs are still liquated

Figure 8.30 The mechanism of liquation cracking in the melt pool area. Source: Adapted from [10] (reproduced with permission from John Wiley & Sons Inc.).

8.14.4 Lamellar Tearing

Lamellar tearing is caused because of the combined effect of localized internal stresses and the substrate material with lower ductility. The substrate material normally reveals with nonmetallic inclusions. The tearing is activated by de-bonding of nonmetallic inclusions such as silicates or sulfides in the substrate metal close to the heat-affected zone, where there is no retrieval of grains or reabsorption of precipitates for the homogenization of microstructure. This region of the substrate also receives greater thermal stresses because of the higher heat input during the AM processes.

8.15 Post Solidification Phase Transformation

The earlier sections have focused on the influence of process input parameters on the occurrence of AM structures that grow through the solidification, which is a crucial stage in the fabrication of AM products [17]. After completing the solidification and cooling far down the solidus temperature, the metal will continue to cool down to the room temperature and then endures the solid-state transformation. Moreover, reheating of previously deposited layers can again possess phase transformation. Here, the phase transformation that takes place through AM processing is presented for both cases of heat treatable and non-heat-treatable alloys.

8.15.1 Ferrous Alloys/Steels

The phase formation and the microstructural growth in AM (laser powder bed fusion [LPBF] and laser powder-fed [LPF], a class of laser directed energy deposition [LDED]) fabricated steels are totally dissimilar compared to the conventionally manufactured cast steels. The basic difference is the cooling behavior and the application of thermo-mechanical approaches after casting to achieve uniform material properties.

A wide variety of steels have been manufactured through AM, i.e., austenitic stainless steel, precipitation hardened steel, maraging steel, and soft magnetic high silicon steel. The very usual microstructure of AM processed austenitic stainless steels is cellular and columnar dendrites. Equiaxed structure is hardly observed because of the higher temperature gradient during the metal AM. The solidified microstructure is mainly cellular, and their size increases with the increase of deposit depth by accumulating heat through the AM techniques. Austenitic steels regularly have tiny amount of ferrite distant from austenite.

During solidification, solute rejects at the boundary and enhances the intercellular areas with chromium and molybdenum, resulting in the ferrite formation. However, the ferrite content drops at a faster cooling rate due to the inadequate time for solute restructuring at higher rates. The phase formation of different steels before and after the heat treatment is shown in Table 8.1. Austenitic steels show comparatively higher thermal expansion coefficients, thus being susceptible to solidification cracking. For austenitic steels, the propensity to solidification cracking is reduced with primary δ-ferrite phase, compared to austenite phase.

8.15.2 Al Alloys

The production of various aluminum alloys using AM techniques is still scarce. This is due to the fact that many aluminum alloys are recognized to be barely weldable due to solidification cracking. Another reason might be that unlike other alloys, aluminum is easy to machine and cast with lower product costs. Therefore, the production of aluminum parts in AM may not have much commercial advantage yet, while the AM adoption for Al alloys is mainly dependent on applications and geometrical features of parts. Usually, higher functioning alloys show their strength through precipitation strengthening. For example, several hardenable alloys, i.e., EN AW-7075 of the Al-Zn 7xxx series, have a strongly reactive element like zinc, resulting in violent melt pools, spatter, and porosity, which make it difficult to use in AM. Under vacuum conditions, if the alloying elements have vapor pressure very dissimilar to Al, e.g., Mg and Li, they prone to evaporate favorably. Moreover, Al shows greater reflectivity with the wavelengths typically applied in LPBF and LPF, which is then another barrier for Al products to manufacture through AM. The low viscous Al melt limits the AM process with smaller melt pool sizes.

However, the most promising side to fabricate Al products using AM technique is their higher thermal conductivity, which minimizes thermally produced stresses, as well as the necessity of support structures. Moreover, higher thermal conductivity permits greater processing speeds. The very familiar Al alloys obtainable in AM are the hardenable AlSi10Mg (EN AC-43000) and the eutectic $AlSi_{12}$ (EN AC-44200). As a high-strength alloy, a hardenable Al–Mg–Sc alloy is recommended by Schmidtke et al. ($AlMg_{4.5}Sc_{0.66}$) [61], where Sc allows precipitation strengthening by Al_3Sc precipitates, which also refines the grain. Al–Sc alloys have received significant attention from the aerospace industry recently due to their high mechanical strength and attractive elongation rate. Airbus has filed multiple patents as of 2021 on Al–Sc powders deployed to LPBF. Different Al alloys in AM and their phase formation during solidification are summarized in Table 8.2.

Table 8.1 Phase formation during solidification of different types of AM steels.

Alloy	Manufacturing techniques	Heat treatment	Microstructure	References
316L	LPBF		Primary austenite (fcc γ) with small amount of retained ferrite (bcc δ)	[40]
316L	LPF (LDED)		Austenite phases	[41]
316L	LPF (LDED)		Austenite and ferrite phases	[26]
304	LPBF		Austenite phase	[42, 43]
17-4 steel	LPBF		Austenite, martensite, and metal carbides	[44]
17-4 steel	LPBF		Austenite and martensite	[45–47]
17-4 steel	LPF (LDED)	SP, aging 755K and 866K	Retained austenite → martensite (after SP) retained austenite (866K), martensite (755K)	[48, 49]
18Ni (300)	LPF(LDED)		Martensite phases and a transformation of austenite to Fe–Ni martensite.	[50]
420 SS	LPBF		Austenite reversion	[51]
SS SC420	LPF (LDED)		Tempered martensite with fine carbides and ferrite through the grain boundary	[52]
H13 tool steel	LPBF	Preheating at 200 and 400 °C	Martensitic phase and retained austenite (before and after preheating at 200 °C), preheating 400 °C→bainite	[53]
H13 tool steel	LPF (LDED)		Martensite	[54, 55]
HY100	LPBF		Martensite	[56–58]
4140	LPBF		Martensite and bainite	[59]
4340	LPBF		Martensite	[60]

SP → Shot Peening.

Usually, pure aluminum does not show low melting point solute enrich liquid at grain boundaries to result in solidification cracking. However, the highly alloyed aluminum shows a low melting point solute-rich liquid to fill or heal any initial crack that might have been formed. However, within the above-mentioned two circumstances, there may occur a composition with the solute enriched liquid, which then form thin incessant grain boundary layers resulting in greater susceptibility to cracking.

In the weldable Al alloys, the finer equiaxed grains are known to be less prone to solidification cracking. The equiaxed grains may have distortion by accommodating contraction strains, which make them ductile. Liquid supplying and the remedial of initial cracks can also be an additional efficient approach for fine-grained materials. Moreover, finer grain materials with bigger grain boundary may have less rigorous segregation of low melting solutes. Therefore, the propensity of the weld metal to contract and the level of restraint are the reasons to influence solidification cracking.

Table 8.2 Phase formation during solidification of different types of AM Al alloys.

Alloy	Manufacturing techniques	Heat treatment/ condition	Microstructure/phases	References
Al-5Si-1Cu-Mg	LPF (LDED)		Vermicular Si, Fishbone-shaped θ-Al_2Cu, Blocky π-$Al_8Mg_3FeSi_6$, and Irregular Q-$Al_5Mg_8Cu_2Si_6$.	[62]
Al 2139	EBF	Solution annealing: 800 K $\pm$ 3 K/2 h Aging: 433 K $\pm$ 0.5 K/12, 18, 24, 36, 48, 72 h	Ω-Al_2Cu precipitates	[63]
AlSi12	LPBF		Si particles (400 nm) in Al matrix	[64]
AlSi10Mg	LPBF		Eutectic silicon network in α-Al matrix	[65]
A357	LPBF	Stress relieve: 300 $\pm$ 1 °C$\rightarrow$air cooling, Solution heat treatment$\rightarrow$ 535 $\pm$ 3°C$\rightarrow$salt bath for 0.25, 1, 4, 24 and 150 h$\rightarrow$water quench	Generally, Si is observed in the microstructure as like continuous segregation across the cellular Al in the as-built part after etching.	[66]

The possibility of AM processed Al alloy parts to meet cracking is very robust. This could have been attributed from: (i) greater solidification temperature span, (ii) higher coefficient of thermal expansion, and (iii) larger solidification shrinkage [67]. The laser processed AM Al alloy also meets liquation and solidification cracks, such as laser-welded Al components. The addition of higher alloying elements in heat-treatable alloys may precipitate lower melting point eutectics, which then create liquation cracks. However, the liquation cracking can be lowered in LPBF manufactured Al alloy by reducing scattered energy concentration from the substrate.

Solidification cracking is barely found in Al alloy manufactured using continuous-wave Nd: YAG laser; however, the opposite is correct for pulsed-wave Nd:YAG systems. Also, the heat treatable 2000 and 6000 Al alloy series are more prone to solidification cracking than the work hardening 5000 alloy series through laser processing [38]. In Al alloys, the solidification cracking is metallurgically guided by the temperature limit of dendrite consistency and the existing liquid level during cooling. The tendency of solidification cracking enhances with a wider solidification temperature range, which is directly correlated with solidification strains.

Moreover, the faster cooling accompanying with less energy density confirms that the mushy zone persists broadly in the laser manufactured components. The faster cooling rate speeds up the formation of large thermal shrinkage strains and expands the stress gradient that stimulates the crack growth rate. The remaining liquid at the grain boundaries of the mushy zone may act as a thin film, resulting in strain concentration and solidification cracks. In addition, the time needed for the remaining liquid to seal the initiated cracks is reduced in the faster cooling rates.

In the LPBF process, the non-equilibrium and rapid solidification may result in insufficient diffusion that eventually lower the liquidus and solidus temperatures. Subsequently, a broader temperature range and larger solidification cracking propensity are usually through the LPBF process of Al alloys, i.e., 0.8% Si in Al–Si; 1–3% Cu in Al–Cu; 1–1.5% Mg in Al–Mg; and 1% Mg_2Si in Al–Mg–Si alloys [67]. However, the addition of some alloying constituents with a focus on narrowing down the solidification temperature limit may change the melt pool composition to minimize cracking. The solidification cracking tendency in LPBF processed Al is considered to be a process responsive.

In laser-processed AM, there exists an ideal energy density to develop crack-free, entirely dense products. Therefore, the solidification cracking starts at energy densities larger than the ideal value because of the following: (i) lower liquid viscosity, (ii) lengthy liquid period, and (iii) subsequent higher thermal stresses. On the other hand, with lesser energy densities, a disorganized solidification front and a major balling phenomenon may result in the crack formation because of the higher unsteadiness of the liquid due to Marangoni convection, nonlinear capillary forces and inconsistent wetting angles.

The alloying constituents and impurities are segregated along the grain boundaries during the solidification through the microsegregation process, resulting in the liquation layers that also cause the temperature to further cool down. To reduce the origination of stresses and cracks, the adequate liquid is needed to seal the cracks and remove the strain generated through the solidification. Thus, crack admittance by strain development competes with crack remedial through refilling by remaining liquid. Although the crack growth rate rises with strains, the replenishing and remedial approach of the remaining liquid is regulated by its fluidity. The eutectic required to avoid cracking differs with composition and cooling rate. For example, the 5000 series alloys do not show a tendency to solidification cracking because of the higher Mg content. However, the heat-treatable alloys show a greater propensity to solidification cracking for the larger amount of alloying elements, which develop low melting components and wider temperature ranges.

8.15.3 Nickel Alloys/Superalloys

The superalloy refers to an alloy with several key properties: tremendous mechanical strength, excellent thermal creep property, better surface steadiness, corrosion, and oxidation resistance at elevated temperature, etc. In current years, Ni-based superalloy components have been manufactured using various AM techniques, for example, LPBF, EB-PBF, and BJ. The chemical composition of different Ni alloys, i.e., IN625, IN718, and Stellite, is presented in Table 8.3.

8.15.3.1 Inconel 625

Inconel 625 (IN625) is a Ni-based solid solution strengthening superalloy and is greatly strengthened by Mo and Nb contents [68, 69]. The reaction between Nb and Mo makes the alloy matrix harden while offering a greater strength without passing through the heat treatment. Inconel 625 has application in aerospace, marine, chemical, and petrochemical industries, possessing superior properties comprising strength at elevated temperatures, better

Table 8.3 Chemical composition of different superalloys.

	Ni	Cr	Fe	Mo	Nb	C	Mn	Si	P	S	Al	Ti	Co	W
IN625	58	20–23	5	8–10	3.15–4.15	0.1	0.5	0.5	0.015	0.015	0.4	0.4	1	1
IN718	55–55	17–21	balance	2.8–3.3	4.75–5.5	0.08	0.35	0.35	0.015	0.015	0.2–0.8	0.65–1.15	1	11–13
Stellite 1	1	28–32	1			2–3		1.2					57	11–13
Stellite 6		27–32				0.9–1.4							Base	4–6
Stellite 21		26–29		4.5–6.0		0.2–3.0							Base	

creep resistance, excellent fatigue property, resistance to oxidation and corrosion, and accessible processability.

However, the microstructure of AM-processed IN625 has austenitic phases, where no carbides and any other phases are recognized. In the laser-processed AM technique, the beam travels very fast (>1000 mm/s) and makes the solidification time short (<1 ms). The atomic restoration speed ahead of the short period liquid/solid solidification is greater than the diffusion speed. Therefore, the faster solidification makes the solute atoms to be trapped and creates the well-known "solute trapping." The majority of the solute atoms like Cr, Mo, and Nb are trapped in the Ni matrix and phase transformation is difficult to take place. Similarly, carbides are also hard to accumulate and precipitate. IN625 alloy in AM and its phase formation during solidification are summarized in Table 8.4.

8.15.3.2 Inconel 718

Inconel 718 (IN718) is known as a Ni-based superalloy with higher quantities of Nb, Cr, and Fe [77]. As a major structural component, Ni has applications in the aerospace and energy/resources industries because of their good oxidation resistance, creep, and mechanical properties at elevated temperatures. IN718 is reinforced by the precipitation of consistent secondary phases, i.e., γ' and γ'', in a face-centered cubic (FCC) γ matrix, which is on the (100) plane variants. Here, γ' shows a FCC, L12, crystal structure, and composition of $Ni_3(Al,Ti)$, whereas γ'' shows a body-centered tetragonal (BCT), DO_{22}, crystal structure, and a composition of Ni_3Nb. Other significant phases are topological close-packed (TCP) Laves $(Ni,Cr,Fe)_3(Nb, Ti,Mo)$, MC carbides (Nb,Ti)C, and δ (Ni_3Nb), which possess an orthorhombic crystal structure and precipitates inconsistently on (111) planes. In IN718, the precipitated δ phase does not impact the strength of the products but influences the grain size.

Knowledge of the solidification phenomenon of IN718 is crucial to be familiar with the phase progression in the AM-processed alloys. The detected microstructural development with various phases in IN718 is schematically shown in Figure 8.31. The solidification structure of IN718 begins with $L \rightarrow \gamma$ reaction (1359 °C) resulted in the $L \rightarrow \gamma + MC$ kind of eutectic reaction at a temperature of 1289 °C. The liquid phase ends at 1160 °C, from where Laves phase starts. The δ-phase precipitates are formed through a solid-state reaction, which occurs at 1145 ± 5 °C, and other phases, γ'' and γ' precipitate at 1000 ± 20 °C.

The existence of both precipitates and intermetallic components in Ni alloys is crucial to their microstructures [17]. To better understand various phase formations depending on the cooling behaviors, the transformation–time–temperature (TTT) diagram for IN718 alloy is presented in Figure 8.32.

The summarized list of microstructure types for various Ni-based alloys processed by different AM techniques is provided in Table 8.5. The chemical constituents as well as the thermal behaviors influence the microstructure. IN718 does not instantly form the γ' phase through the LPBF process; however, the phases γ' and γ'' are formed via heat treatment post-processes [80, 85]. On the other hand, René 142 forms γ' precipitate through EBM technique without having post-process heat treatment. The relationships between γ, γ', and γ'' phases comprising precipitate shape, volume fraction, and their arrangements govern the build qualities. Therefore, the microstructure of Ni-based alloys is essential to improve through compositional variants,

Table 8.4 Post solidification phases of In625.

Alloy	Process	Microstructure/phases	Heat treatment (HT)/post-processing	Microstructure/phases after post-processing	References
IN625	LPBF	Dendritic/cellular microstructure of austenite phases	—	—	[70]
IN625	LPBF	Columnar dendrite, austenite phases	Annealing 700, 1000, 1150 °C/1 h→air cooling	Microstructure is same after HT, at 1000 and 1150 °C, NbC and MoC carbides particulates distribute around the grain boundary	[71]
IN625	Laser cladding-wire	Cellular grains, columnar dendrites, and equiaxed grains			[72]
IN625	BJ-sintered at 1280 °C/4 h	In γ-Ni matrix, precipitation of δ phase (Ni$_3$Nb), and NbC, M$_6$C carbides	Solution treatment: 1150 °C/2 h Aging:745 °C/20h, 60h	Solution treatment-dissolve δ phases and carbides. Aging precipitation of Ni$_3$Nb phases and carbides of M$_{23}$C$_6$ and M$_6$C	[73]
IN625	Laser cladding-on Gr22 and 316L steel substrate	Dendritic microstructure, γ-Ni matrix, segregated particles γ''-Ni$_3$Nb, M$_6$C (M = Mo, Cr, Ni, and Nb) and M$_2$C (M = Mo and Nb) carbides	Heat treatment: (i) 520 °C/48, 336 h (ii) 800 °C/24, 336 h	520°C/48 h-increased content of MC (M = Mo, Nb, and Si) and M$_6$C (similar to as-deposited clad). 520°C/336h-laves phases ((Fe, Cr, Ni, Si)$_2$(Nb, Mo)), 800°C/24 h, 336 h-phases of laves and needle like δ-Ni$_3$(Nb, Mo).	[74]
IN625	LPF (LDED)	Columnar dendritic microstructure, γ-Ni matrix	Annealing:700,800, 900,1000,1100 °C and 1200 °C/1 h→air cooling	From 700 to 1000 °C, no change in microstructure, At 1100 °C-mixture of equiaxed and dendritic microstructure, At 1200 °C-fully equiaxed structure. Precipitated phases, γ' [Ni$_3$Al], γ'' [Ni$_3$Nb] and δ [Ni$_3$Nb] in γ-Ni matrix.	[33]

Table 8.4 (continued)

Alloy	Process	Microstructure/phases	Heat treatment (HT)/post-processing	Microstructure/phases after post-processing	References
IN625	BJ	Sintering at 1200 and 1240 °C: NbC in γ-Ni matrix. Sintering at 1280 °C: In γ-Ni matrix, precipitation of δ phase (Ni$_3$Nb), and NbC, M$_6$C carbides. Sintering at 1290 and 1300 °C: NbC and phases enriched in Nb, Cr and Mo in γ-Ni matrix.			[75]
In718	EBM	Columnar grain, γ-Ni matrix	In-situ heat treatment: Solution treatment: 980 °C/1 h; Aging (precipitation): 720 °C/8 h, Aging (coarsening): 620 °C/10 h	Precipitation of both γ' and γ'' phases, Some needle like δ phases and carbides of Nb and Ti.	[76]

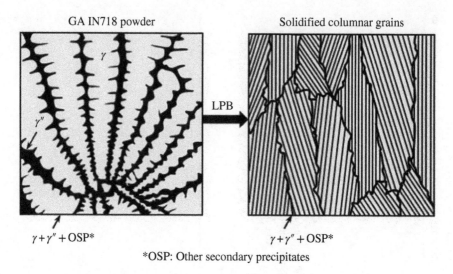

Figure 8.31 Schematic presentation of the microstructural development and phase formation in LPBF processed IN718 alloy [78]. Source: Open access, reproduced with permission from authors.

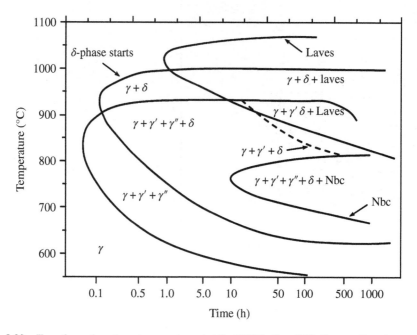

Figure 8.32 Transformation–time–temperature plot for IN718 alloy [79]. Source: Openly accessible.

Table 8.5 Post solidification phases of IN718.

Alloy	Process	Microstructure/phases	Heat treatment/post-processing	Microstructure/phases after post-processing	References
In718	EBM	Columnar grain, γ-Ni matrix	In-situ heat treatment: Solution treatment: 980 °C/1 h; Aging (precipitation): 720 °C/8 h, Aging (coarsening): 620 °C/10 h	Precipitation of both γ' and γ" phases, Some needle like δ phases and carbides of Nb and Ti.	[76]
In718	LPBF	γ austenite matrix with cellular microstructure	Solution heat treatment: 1065 °C/1 h → air cooling, Aging 760 °C/8 h → furnace cooling to 650 °C/2 h → holding at 650 °C/8 h.	Precipitation of γ', γ" and δ phases in γ matrix	[80]
In718	LPF (LDED)	γ-Ni matrix	Aging: 720 °C/8 h → furnace cooling to 620 °C/8 h → air cooling	Three kinds of precipitates, Laves phases, γ" and γ' phases	[81]
In718	LPF (i) CW (ii) QCW	γ-Ni matrix for both conditions, (i) Columnar dendrite (ii) Equiaxed dendrite	Aging: 720 °C/8 h → furnace cooling to 620 °C/8 h →air cooling	Precipitates of NbC and laves phases	[82]
In718	LPBF	γ-Ni matrix with cellular solidification structure, also presence of γ' and γ" phases.	(i) Solution and double aging (HT): ST: 980 °C/4 h → air cooling to RT Aging: 720 °C/8 h → furnace cooling to 620 °C/8 h → air cooling to RT (ii) Hot Isostatic Pressing (HIP) + HT: 1180 °C at 100 MPa for 4 h (iii) HT + Shot Peening (SP): For SP, 0.006–0.010 Almen intensity (iv) HIP + HT + SP	(i) Formation of δ and laves phases (ii) recrystallized grains and annealing twins are evident. (iii)	[77]
IN718	LPBF	γ-Ni matrix	Homogenization: 954–982 °C/1 h→air cooling, Solution: 1093 °C ± 14/1–2 h→air cooling, Aging: 718 °C ± 8/8 h → air cooling	Due to heat treatment, grains become bigger precipitates of δ phases, laves phases and carbides are occurred.	[83]
IN718	LPF (LDED)	γ-Ni matrix with columnar microstructure, laves phases.	Homogenization: 1100 ± 14 °C/1.5 h → air cool, Solution: 980 °C/1.5 h → air cool, Aging: 720±8 °C/8 h → furnace cool → 620 ± 8 °C/8 h → air cool	After homogenization and solution, Laves → δ, Aging: γ' and γ" phases precipitate in γ matrix.	[84]

process mechanism, and post-processing techniques to reach similar or superior properties of their equivalents manufactured by conventional methods.

Unlike conventional techniques such as casting and welding, AM is categorized by faster cooling behavior and repeated thermal cycles, which consequently influence the development of microstructure. For example, the γ' phase in Ni-based alloy precipitates at a temperature higher than 1223 K. The distinct layers are exposed to distinct thermal cycles through processing; therefore, inside of a sample, the formation of γ' shape is position-oriented. The bottom of the sample is subjected to greater thermal cycles and extended periods for growth compared to the top of the sample, which results in greater precipitates at the bottom, unlike the top surface. In laser-processed René 41 alloy, the morphology and dispersion of γ' phases have an effect on the dendritic structures. The γ' precipitates are greater in shape within dendrite cores compared to the inter-dendritic areas because of supersaturations and refined segregations of the solutes. Therefore, the eutectic Laves phases and NbC have a larger amount of Nb in the matrix compared to the inter-dendritic zones. Also, higher precipitation takes place in γ regions near the eutectic phases through repeated thermal processes. This microstructural heterogeneity of the build samples inevitably results in mechanical heterogeneity and therefore demands post-processing heat treatment to obtain homogeneous microstructure.

Intermetallic compounds like Laves Ni_3Nb-δ, Nb-rich MC are usually found in inter-dendritic areas or grain boundaries of Ni-based alloys, which undesirably affect the mechanical properties. Laves phase, which occurs in Nb-rich melt with long-chain structure controlled by Nb segregation and liquid melt distribution, is often found to produce hot cracking. In addition to the cooling rate, Laves phase formation is also influenced by the solidification structure reliant on the proportion of temperature gradient and the growth rate. Smaller dendrite arm spacing with higher cooling rate and lower G/R ratio is helpful for resulting in distinct Laves phase particles. On the contrary, the larger dendrite arm spacing with a lower cooling rate and higher G/R ratio have tendency to develop incessantly dispersed larger particles of Laves phase.

8.15.3.3   Stellite

Stellite alloys show better corrosion and wear resistance at a wide variety of interactions and environments in industries like aerospace, oil and gas, forging, and power production. Usually, Stellite products are manufactured through conventional casting processes as well as powder metallurgy techniques adopting compaction stages of isostatic pressing, die pressing, or automatic pressing. Because of the intricacy of these techniques, the production of Stellite components is inadequate by the geometry of the mold, which also makes them limited in the number of applications.

Moreover, the post-processing (i.e., surface grinding, cutting, shaping) of these materials is challenging because of their higher hardness and wear-resistant properties. Compared to the conventional techniques, AM reduces the time and manufacturing costs of molds needed to produce Stellite components through traditional methods. Moreover, AM causes effective usage of the powder material required for a particular component to build and recycling the remains. Therefore, using AM technique, the production of Stellite components could be advantageous to wear-resistant materials by; (i) allowing them in applications by reducing component weight, improving consistency with surface alterations and (ii) minimizing the step of post-machining.

Table 8.6 Different types of Stellite.

Co-based super alloy-Stellite			
	Stellite 1	Stellite 6K	Stellite 694
	Stellite 3	Stellite 12	Stellite 706
	Stellite 4	Stellite 21	Stellite 712
	Stellite 5	Stellite 31	Stellite F
	Stellite 6	Stellite 190	Stellite Star J
	Stellite 6B		

The various types of Stellites that are commercially obtainable are shown in Table 8.6. The characteristic microstructure of Stellite contains hard carbides distributed through a cobalt-rich solid solution matrix.

The typical microstructure of Stellite 12 with cobalt matrix dendrites together with interdendritic eutectic carbides is shown in Figure 8.33 [86]. There are two types of inter-dendritic structures: one with lamellar eutectic carbides (Figure 8.33a), and the other showing blocky eutectic carbides (Figure 8.33b).

Stellite 12, which is a hypo-eutectic alloy, forms a solid solution cobalt matrix through solidification.

When the temperature drops, the amount of Co in liquid is also lowered and then the eutectic state is attained. The residual liquid reacts with the eutectic structure comprising carbides and Co-based matrix. Moreover, the alloy possesses blocky eutectic carbides. Laser-processed AM follows faster melting and solidification practice, and the overlapping trends of multiple tracks and layers will result in remelting of the earlier solidified layers, which may cause divergence in microstructure development.

For Co-based superalloys, the major strengthening mechanisms include solid solution strengthening through the dispersion of Cr, W, and other elements in Co matrix, whereas M_7C_3, $M_{23}C_6$, and other carbides act as a major function in precipitation strengthening between

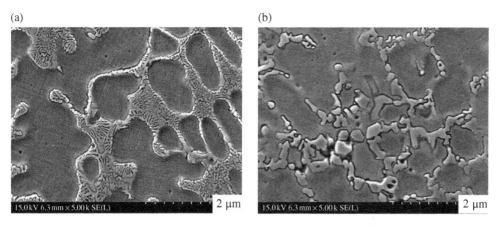

Figure 8.33 Microstructure of Stellite 12 manufactured through laser-based AM techniques, (a) lamellar eutectic carbides and (b) blocky eutectic carbides. Source: Reprinted by permission from Springer Nature [86].

the dendrite region. Therefore, the higher hardness of the carbides may enrich the hardness value and wear property of the alloys. For AM-processed Stellite 12, M_7C_3 is the primary carbide, which is in a metastable condition and decomposes to discharge Cr, C, and W elements at higher temperatures, and facilitates to form $M_{23}C_6$ and M_6C carbides. For different Stellite alloys, the phase formation during the manufacturing techniques and their transformation due to heat treatment are summarized in Table 8.7.

8.15.4 Titanium Alloys

Titanium alloys have vast applications in aerospace, chemistry, ship manufacturing, and other industrial sectors due to their superior properties, i.e., greater strength-to-weight ratio, high corrosion resistance, and compatibility with composite structure in the application of structural parts [88]. The advancement of titanium is because of their implementation, multi-purpose applications, and low price. Ti-6Al-4V (Ti64) as a first pragmatic alloy was effectively designed and manufactured in the United States in 1954. Nowadays, Ti64 is extensively being used in the aerospace and medical industries with profound research programs. A massive investigation is currently ongoing for the use of AM titanium products for different applications, such as Ti64 for lightweight components used in airplanes, Ti64 for personalized bone implants, Ti-5Al-5V-5Mo-3Cr (Ti5553) for landing gear parts, and many more.

The metallurgy of titanium alloys is directed by the phase transformation that occurs in pure metal at 882 °C. Pure titanium shows alpha (α) phase (hexagonal close packed-hcp structure) below the temperature, whereas, above the temperature, the phase is beta (β) (body-centered cubic-bcc structure).

The primary outcome of alloying additions to titanium is the modification of the conversion temperature and formation of two-phase structures, having both α and β phases. Commercially obtainable titanium alloys are categorized based on the impact of α and β phases, which consists of, (i) α alloys, (ii) near- α, (iii) α-β, (iv) near β, and (v) β alloys. In summary:

1. **α Alloys**: Generally, these alloys are not heat treatable and typically weldable. These alloys show low to medium strength, better notch toughness, fairly good ductility, and superior properties at cryogenic atmosphere. Moreover, the α or near-α alloys show higher temperature creep property and oxidation resistance.
2. **α-β Alloys:** These are the heat treatable and weldable alloys, with the possibility of losing ductility near the weld area. They have medium to higher strength levels. Also have better hot forming properties, with inferior cold-forming properties. Likewise, creep strength is not good enough to compare with α-alloys.
3. **β Alloys**: β or near-β alloys easily heat treatable, mostly weldable, show higher to medium temperature levels. In a solution-treated state, the alloy possesses superior cold formability.

Various types of stabilizing constituents on Ti alloy and their impact on phase transformation are schematically presented in Figure 8.34.

α Stabilizers: These types of components have large solubility in α-phase which usually increases the transformation temperature and are known as α stabilizers. The effect of α-stabilizing elements on the titanium phase diagram is presented in Figure 8.34a. Very common α stabilizers to Ti alloy are Al, O_2, N_2, or C. The addition of O_2 to pure Ti has the potential

Table 8.7 Post solidification phases of Stellite.

Alloy	Process	Microstructure/phases	Heat treatment/post-processing	Microstructure/phases after post-processing	References
Stellite1	Laser cladding (LPF, LDED)	Eutectic lamellas (carbides + γ-Co), mixed with dendrites of chromium/tungsten enrich hard carbides			[87]
Stellite12	LPF (LDED)	(a) Cobalt matrix dendrites and interdendritic eutectic carbides are the primary microstructure, blocky and lamellar eutectic carbides and M_7C_3 and $M_{23}C_6$ are the primary carbides.	(i) Solution treatment: 1250 °C/4 h-air cooling, (ii) Aging:950 °C/12 h-furnace cooling (iii) Solution + aging: heating with 5 °C/min to holding temperature of 1250 °C/4 h → air cooling, and then heating to 950 °C/12 h → furnace cooling.	i. Solution treatment caused dissolution of some carbides. ii. Aging: • carbides are precipitated throughout the cobalt matrix. • lamellar eutectic carbides transformed to blocky carbides. iii. A greater amount of blocky carbides and globular and rod-like carbides are formed in the sample after solution plus aging treatment.	[86]

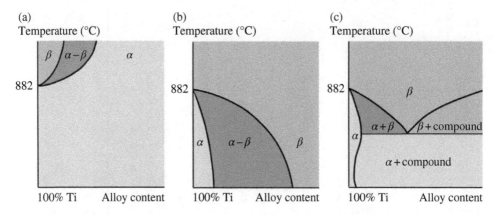

Figure 8.34 Effect of (a) α-stabilizing, (b) β-isomorphous, and (c) on β-eutectoid elements on titanium phase diagram. Source: openly accessible.

to make a variety of grades with higher strength. Al is a commercially used stabilizer, which also acts as the main constituent of many commercial alloys. It can effectively strengthen the α-phase at a higher temperature about 550 °C. Despite the lower density, Al has further suitable characteristics, its quantity is limited to 8% by weight due to the development of a fragile Ti–Al compound. Besides this, the α-phase can be reinforced with tin or zirconium, which have substantial solubility equally in α and β phases. They do not prominently affect the transformation temperature and therefore known as a neutral stabilizer.

β **Stabilizers:** Components that reduce the transformation temperature, easily disperse in and reinforce the β phase, and show lower α phase solubility are termed β stabilizers. They can be classified into two classes based on their constitutional activities with Ti:

- **β-isomorphous elements:** It is evident that the hardening influence of four isomorphous β stable components drops thoroughly in the sequence of Mo, V, Nb, and Ta, where Mo is more influential than V, to boost the strength of the alloy. A higher quantity of Al (>7 wt. %) is detrimental because of the existence of the intermetallic compound Ti_3Al. Zr as a neutral element can increase the strength of Ti extensively without raising the density of the alloy [89]. Also, Si can be present as a solid solute in Ti alloys or can be present as silicide, where a very small quantity of Si can improve the room-temperature properties. Moreover, Si is recently added to enhance the elevated temperature behaviors of Ti alloys. Effect of β-isomorphous elements on the phase diagram is shown in Figure 8.34b.
- **β-eutectoid elements:** β-eutectoid components have limited solubility in β Ti, and therefore intermetallic compounds formed through eutectoid decomposition of the β phase. A characteristic phase diagram is presented in Figure 8.34c. Common β-eutectoid elements are iron, chromium, and manganese, where eutectoids are decomposed as titanium–iron, titanium–chromium, and titanium–manganese. These steps are slow enough to form intermetallic compounds through conventional manufacturing and heat treatment; therefore, for practical reasons, the performance of iron, chromium, and manganese can be compared to that of beta-isomorphous components.

Variation in the phase composition is a significant basis of microstructural heterogeneity in AM-produced metallic products. For example, in $\alpha + \beta$ titanium alloys (Ti-6Al-4V), usually, three phases are stated, i.e., α phase, β phase, and α' martensitic phase. Because of the intricate phase transformation methods, it is challenging to calculate the phase composition in AM Ti alloys precisely. The cooling rate and the manufacturing temperature are the two key process parameters to affect the ultimate phase structures in laser or EB-AM made parts. Therefore, the microstructural differences are resulted from repeated thermal inputs from sequential buildups. The solidification phases in Ti-6Al-4V alloy, at different AM techniques, are summarized in Table 8.8.

A schematic quasi-vertical portion of a ternary titanium structure, including both α and β stabilizing components, is presented in Figure 8.35. The composition variations consisting of five kinds of alloy are also shown in the upper portion of the figure. The dashed line indicated M_s/M_f denotes the start and finish of martensitic formation and generally is very close to one another. The Ti-6Al-4V, which is a α–β alloy, contains 6 wt.% aluminum as an α stabilizer and 4 wt.% vanadium as a β stabilizer. The β-transus (T_β) temperature for this alloy is at about 980 °C in equilibrium conditions, and higher than this temperature, the alloy contains 100% β phase. This temperature is important as the heat treatments are usually taken at below or above the T_β.

During cooling from the melting temperature, 1670 °C, the liquid melt solidifies into β phase and then β grains begin to grow. The β grains of AM-fabricated Ti-6Al-4V alloy are usually columnar in shape and advance epitaxially, which is vertical to the build layers, because of thermal gradients generated from heat source. If the temperature drops far below the β transus temperature (>995 °C), for Ti-6Al-4V alloy, the β phase begins to transform into α phase (hcp structure) considering the Burgers link $\{0001\}\alpha\|\{110\}\beta$, $\langle1120\rangle\alpha\|\langle111\rangle\beta$ [97]. The cooling rates are generally observed in AM techniques, the α phase nucleates at the previous β grain boundaries. In AM-processed alloys, α phase can be formed both at grain boundaries and within the β grains as laths. Upon progressing in a parallel way, these laths create well-known "α colonies."

When the cooling rate is rapid enough to finish the occurrence of the entire α colony, the nucleation of α phase instantaneously happens inside the β grains, as well as along the β grain boundaries. However, if the α laths advancing from these places meet each other, a "basket wave" shaped structure will form. Also, at extremely higher cooling rates, there may occur diffusion less martensitic transformation in AM products. Therefore, the cooling rate is significant as it decides the diffusion retrained α phase growth, either as α grain boundary or α laths. In laser- and EB-based AM processes, the α may convert to α laths organized as Widmanstätten or basket weave pattern (one type of martensite).

Through the AM processes, the layers experience repetitive heating and cooling cycles and hence the ultimate temperature of individual layers drops constantly from T_β, as depicted in Figure 8.36.

Based on the temperature variations, the cooling system for every solidified layer can be split into three zones. In zone A, when the ultimate temperature is more than T_β, the earliest fine α phase lamella will convert into β phase entirely and again convert into α through cooling. In zone B, for an ultimate temperature below T_β, the fine α phase lamella will inadequately convert into β phase, the remaining α phase then coarsens, and some newly developed α phase will precipitate from β, throughout the cooling process. Therefore, the volume of residual α phase lamella raises as the ultimate temperature drops progressively. In zone C, when the ultimate temperature drops below T_α (the temperature wherein there is no observable phase conversion), the absolute structure grows as presented in the figure.

Table 8.8 Post solidification phases of Ti alloys.

Alloy	Process	Microstructure/phases	Heat treatment/condition	Microstructure/phases	References
(1) Ti-6Al-2V-1.5Mo-0.5Zr-0.3Si (2) Ti-6Al-4V	LPF (LDED)	For both alloys, $\alpha + \beta$ micro-structure with fine basket weave structures of lamellar α phase	Stress relief annealing: 600 °C/2 h $\rightarrow$ air cooling	Fine α phases	[88]
Ti-6Al-4V	DMD/LPF (LDED)	Martensitic structure	Annealing: 550 °C HIP:100 MPa/899–954 °C/2 h $\rightarrow$ slow cool to 427 °C	Grain boundary alpha and inter-granular coarse alpha plates	[90]
Ti-6.5Al-3.5Mo-0.3Si-xZr (x = 1.5, 2.5, 3.5, 4.5 wt%)	LAM (LDED)	The alloys in this study have equiaxed grain morphology with ($\alpha + \beta$) phase.	Effect of Zr content: 1.5, 2.5, 3.5, 4.5 wt.%	With higher Zr contents, the grains are refined and the microstructures have no prominent change, although the crystal lattice para-meters of α phase and β phase are to some extent intensified.	[91]
Ti-6Al-4V	LPBF	α' martensite microstructure	Heat treatment: 600, 730, 800, 850, 900, 950, 1000 °C $\rightarrow$ 2 h $\rightarrow$ Furnace cooling, 800, 850, 900, 950 °C $\rightarrow$ 2 h $\rightarrow$ Air cooling	The α/α' remain unchanged at 650 °C/2 h and 750 °C/2 h At 730 °C, initial martens-ite decomposed to $\alpha + \beta$ lamellar structure	[92]
Ti-6Al-4V	1. LPBF 2. LPF (LDED) 3. EBM	1. acicular α' 2. acicular α' in columnar prior-β with grain boundary α 3. lath-shaped α(decomposed α')			[93–95]

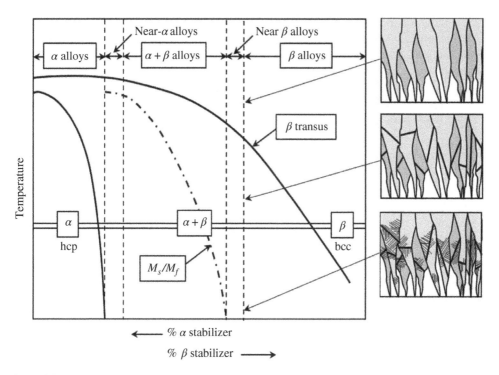

Figure 8.35 The graphical presentation of ternary titanium alloys having both α and β phases. Source: redrawn and adapted from [96].

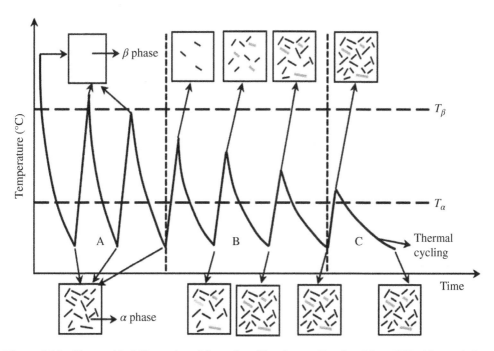

Figure 8.36 The graphical illustration of thermal profiles. Source: Reprinted from [88] with permission from Elsevier.

Multiple studies have been conducted to shed some light on the effect of faster cooling on the microstructure development in Ti64 alloy by Ahmed and Rack, where improved Jominy and quench test techniques were adopted [98]. The subsequent morphologies at various cooling rates are presented in Figure 8.37, and the consequent values in Table 8.9. The

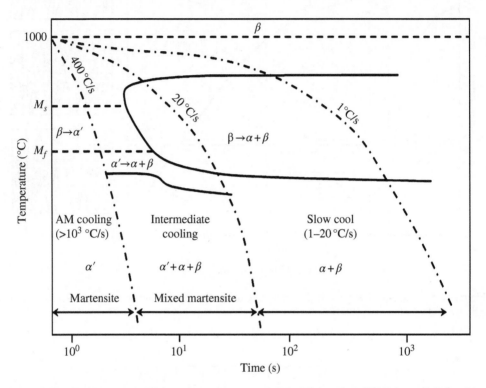

Figure 8.37 Continuous cooling transformation curve for Ti-6Al-4v alloy [99]. Source: Reprinted by permission of the Taylor & Francis Ltd.

Table 8.9 The estimated cooling rates necessary to obtain various morphologies in Ti-6Al-4V alloy through heat treatment [98].

Cooling rate (k/s)	Start temperature	Transformation	Remarks
CR > 410	Ms = 848 K	Martensitic (α') Pocket like or acicular morphology	Diffusion-less process ($\beta \rightarrow \alpha'$)
20 < CR < 410	1223–1273 K	Massive (α_m)	Competitive diffusion-less and diffusion control process ($\beta \rightarrow \alpha_m$)
CR < 20	~1173 and ~1223 K [29, 30]	Widmanstäten (α)	Diffusion control process
CR < 20	1243–1273 K	Allotriomorphic (grain boundary α~)	($\beta \rightarrow \alpha$ with Widmanstäten morphology or at grain boundary)

development of a complete martensitic structure is noticed at cooling rates 410 and 20 °C/s. This conversion is gradually switched by diffusion influenced Widmanstätten α phase at slower cooling rates.

8.16   Phases after Post-Process Heat Treatment

8.16.1   Ferrous Alloys

In laser manufactured 17-4 PH steel, austenite reversion occurs during the aging treatment. The austenite phases in laser-processed AM steel transform to martensite during heat treatment. This is believed to happen due to the stress release, which permits the austenite to transform martensite through post-treatment cooling. Austenite reversion is also common in laser-fabricated maraging steel, where during aging, Ni-rich returned austenite shell forms around the retained austenite areas [100].

8.16.2   Al Alloys

The precipitation-hardened AlSi10Mg alloy in the AM process does not show any precipitates because of the rapid solidification, except for some Si segregation near the grain boundaries. After solution treatment, Si starts to form as Si particles in α-Al matrix. The application of water quenching and peak-hardening followed by solution treatment results in globular Si particles with the needle-like Mg_2Si precipitates. Therefore, the microstructural anisotropy is diminished through disappearing dendrites, melt pool edges, and heat-affected zones. Moreover, for the 7xxx aerospace alloys, Sc is the element to enhance the strength of Al alloy in solid solution. From the heat treatment results, it is evident that the formation of completely coherent nano-sized Al_3Sc precipitate can increase the strength for about 50 MPa, per 0.1% Sc addition [101].

8.16.3   Ni Alloys

Faster solidification of laser-deposited thin layers that proceed with higher cooling rates results in directional grain growth, microsegregation of highly intense refractory components like Nb and Mo, and generation of non-equilibrium phases comprising carbides and Laves Phases. However, the precipitation-reinforced γ' and γ'' phases are obstructed. For this reason, post-process heat treatments are essential in developing the microstructure of AM-manufactured products to satisfy the functional requirements.

For IN718, the most commonly applied heat treatments are homogenization or solution-aging treatment, which can stimulate the diffusion process through segregation of some components and dissolve some phases of carbides and Laves in the austenite matrix. Moreover, aging can promote precipitation of γ' and γ'' elements. After the heat treatment process, the plate-like δ-Ni_3Nb precipitates form along the grain boundaries, as well as within the grains. However, γ' and γ'' precipitates are very fine to determine evidently.

For cobalt-based superalloys, solid solution and the precipitation strengthening are the major reinforcing mechanisms. Cr, W, and other components, which disperse on the cobalt matrix, can act as solid solution strengtheners. M_7C_3 is the major carbides of Stellite 12, which is a metastable phase, fabricated through Laser AM techniques. This carbide phase will decompose to Cr, C, and W components during heat treatment and promote the formation of $M_{23}C_6$ and M_6C carbides.

8.16.4   Ti Alloys

Ti-6Al-4V alloy possesses excellent stability in strength, ductility, fatigue, and fracture properties, except the creep property at temperatures above 300 °C. The structural development of this alloy is greatly influenced by heat treatment. For example, at a temperature of about 800 °C, the alloy contains 15% β phase. However, at room temperature, the alloy is dominated by the α phase, but at a temperature about 995 °C, which is above the β transus, endures as a single β phase.

In laser-processed AM techniques, the finer acicular α' phase in the as-manufactured Ti-6Al-4V alloy results in an inadequate outcome through a conventional heat-treatment process. The metastable α' phase is very fine and contains greater densities of dislocations and twins, which impede the grain growth during heat treatment, directing to a finer $\alpha + \beta$ lamellar structure. During the heat treatment (~400 °C), in the first step, α phase starts to nucleate along the boundary of the acicular α' phase, which pushes the vanadium in the boundary of the newly formed α grain. After that, the β phase forms in the area of higher vanadium content among the α phase laths. Therefore, the heat treatment temperature is important to retain the refined microstructure. Moreover, the appropriate temperature and the holding time may relieve the residual stress. Consequently, the stress-relief heat treatment may contribute to two major structural modifications through decreasing the dislocation density and the breakdown of α' phases. The lamellar $\alpha + \beta$ structure, which exists as a colony shape, starts to become coarsen α lamellae with increasing the heat treatment temperature. This way, some α grains become globular and reduce morphological anisotropy.

8.17   Mechanical Properties

The progressions in AM over the past few years have made the mechanical properties of AM-made parts comparable with the conventionally made counterparts for some materials and process parameter combinations. The defects like porosity influence crack generation and decline mechanical properties. Therefore, a higher density above 99.5% is usually the first priority for AM technique optimization, particularly if the AM-made parts are used in critical missions' applications such as jet engine components, etc. In addition to other effects, the product density is very much reliant on the given energy volume to the material being made by AM.

An inadequate energy flux may give rise to un-melted material, which minimizes the density by the occurrence of irregular voids. On the other hand, an excessive energy input may cause instability in the melt pool dynamics that in turn may enter the process to a keyhole mode that normally decreases the density due to the formation of pores through the trapped gas. These types of pores are usually spherical in shape as they are likely resulted from the trapped gasses (inter gas and evaporation of materials) due to instabilities in the melt pool and keyhole phenomenon. Therefore, the greater density which can be achieved with AM for different materials suggests a discrete optimization guideline for AM parameters to achieve the optimum volume energy. The rationale for optimizing parameters to maximize the density by reducing porosities is the fact that the size and distribution of pores have a major impact on mechanical properties.

8.17.1   Hardness

Hardness is important to illustrate the contained mechanical strength of AM products. It is known that the micro-hardness is associated with the microstructure of AM-processed alloys and follows the Hall–Petch relationship. It is believed that the hardness value depends on the cross-sectional area, where with a larger cross-sectional area, the hardness value drops due to microstructure coarsening. The basis of this microstructure coarsening is the greater thermal input in a bigger area, which makes a slower cooling process. Therefore, the heterogeneity in hardness depends on the thermal history of the individual layer. Prospective improvement using different process parameters with respect to the cross-sectional area may improve the heterogeneity in hardness.

8.17.1.1   Hardness of AM-Processed Ferrous Alloys

The hardness development in AM-processed ferrous alloys depends on the solidification behavior which is influenced by process input parameters. For multilayer deposited steels, the micro-hardness value drops from the very initial deposited layers, which afterward enhances along the upper layers. This phenomenon is because of the repeated heating of the former layers and letting them time to be annealed to some extent. Additionally, this inconsistency is ascribed to the time dependency of the cooling rate in the liquid melt and comparatively the slow solidification rate in the middle area. Therefore, greater hardness values are typically obtained both at the top and bottom of the AM parts in contrast with the central area.

In low alloy steels, i.e., 41XX series steels, the content of alloying components as well as the amount of carbon controls the phase formation, which eventually affects the hardness values. In high carbon-containing steel, through the rapid cooling in the AM process, hard martensite phases are developed, which contribute to hardness development.

Because of the higher manufacturing cost of AM at the moment, compared to the traditional one, the application of plain carbon and low alloy steels are not very common in AM applications. Most of the steels found in AM are tool steels, considering specific repair purposes, where higher strength and wear resistance are vital. M2, P20, and H13 tool steels attain higher strength because of their tendency to develop martensite even at comparatively lower cooling rates. Moreover, these alloys have a higher quantity of carbon, which accelerates the development of carbides and enhances strength and hardness. Generally, the structure of AM-processed tool steels show martensitic matrix with a few amount of carbide precipitates and retain austenite. After the AM process of tool steels, post-process heat treatment is necessary, which improves ductility and toughness through the tempered martensitic structure with carbide precipitates.

On the other hand, austenitic stainless steels do not show secondary phases or even undergo any solid-state phase transformation. Therefore, the strength and hardness are governed by the finer solidification structure, as well as the chemical composition of the alloy.

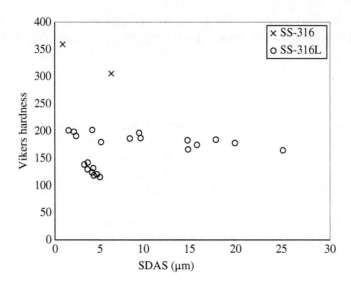

Figure 8.38 Micro-hardness values with respect to the secondary arms spacing (SDAS) for AM-manufactured stainless steels [41, 102–104].

The hardness of LDED-processed austenitic stainless steels, such as 316 and 316L, as a function of Secondary Dendritic Arm Spacing (SDAS) is shown in Figure 8.38. Both alloys have a similar chemical composition with an identical dendritic structure. However, a slight change in carbon content (316L $\leq$ 0.03 wt.% C, and 316 $\leq$ 0.08 wt.% C) between 316L and 316 shows a significant change in hardness.

8.17.1.2 Hardness of AM-Manufactured Al Alloys

The most familiar Al alloys fabricated through the AM technique contain greater volumes of Si, which stimulate eutectic solidification. Compared to other Al alloys, the eutectic Al-Si alloys, such as Al-12Si, and AlSi10Mg have lower absorptivity of Al for a broad range of wavelengths, which facilitate to manufacture these alloys using AM techniques.

The precipitation-hardened Al alloys depend on the post-process heat treatments to attain improved properties, such as higher strength and hardness through finer solidification structure. The hardness value of AM-fabricated alloys with respect to the SDAS is shown in Figure 8.39. The combined effect of higher cooling rates and alloying element like Si promote the formation of finer SDAS, which result in higher hardness in AM-manufactured alloys.

Some post-processing technique may influence the hardness property in AM-processed Al alloys. From various research works, it is noted that solutionizing and aging treatment may drop the hardness value, compared to the as-deposited condition. This is believed to happen due to the influence of microstructural change caused by solutionizing and aging treatment, which coarsen the Si particles.

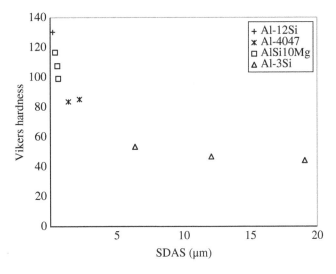

Figure 8.39 The micro-hardness values with respect to the secondary arms spacing (SDAS) for various aluminum alloys manufactured through AM technique [44, 105–107].

On the other hand, HIP as a post-process technique also has an impact on the hardness of AM-manufactured alloys. HIP usually results in the microstructural coarsening as well as releasing residual stresses, which drop the hardness of products.

8.17.1.3 Hardness of AM-Manufactured Nickel Alloys

The faster cooling rate in AM-processed Ni alloys leads the strengthening elements such as Mo and Nb to be persisted in Ni alloy matrix. Therefore, a greater lattice distortion occurs by these point defects, which also enhanced the hardness of the AM-fabricated samples compared to the conventional cast alloys. As IN625 is a solid solution-strengthened superalloy, phase transformation occurs through heat treatment. At low-temperature annealing, i.e., 700 °C, the release of residual stress lowers the hardness. However, annealing between 800 and 900 °C forms δ (Ni3Nb) precipitate, which improves the hardness. δ is an orthorhombic phase having a greater mismatch with the Ni matrix and thereby develops hardness. However, annealing above 1000 ° C dissolves δ phases in the Ni matrix and reduces the lattice distortion together with hardness.

8.17.1.4 Hardness of AM-Manufactured Ti Alloys

Generally, the AM-processed Ti-6Al-4V alloy possesses different structural features after various post-processes. For example, LPBF-processed Ti-6Al-4V alloy shows finer martensitic structure with acicular laths after stress relieving. In contrast, post-process Hot Isostatic Pressing (HIP) results in a coarser structure of lamellar α and β, as shown in Figure 8.40.

Ti-6Al-4V alloy endures a phase change from body-centered cubic β phase to a structure comprising hexagonal α-phase and few amounts of β-phase, when the temperature is about 1000 °C. The solid-state transformation may result in measureable structures within grains,

(a)   (b)

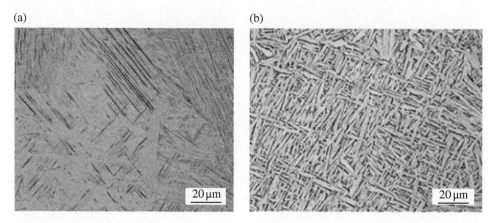

Figure 8.40 Microstructure of Ti-6Al-4V manufactured by LPBF, (a) after stress releasing and (b) after post-process Hot Isostatic Pressing (HIP). Source: Reprinted from [108] with permission from Elsevier.

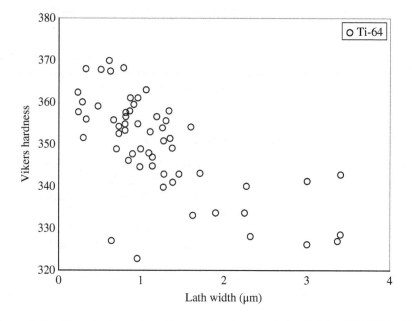

Figure 8.41 Micro-hardness plot with respect to the alpha lath width for Ti64 [109–115].

and based on the cooling behavior during the conversion of temperature, the α-phase may result in diverse morphology. There is a relationship between the α-lath width and the hardness, as obvious in Figure 8.41.

Some post-processing treatments, such as heat treatment and annealing, may release residual stress with some coarsening effect of α-phases, which eventually raise the ductility and toughness, compromising strength and hardness of AM-manufactured Ti-6Al-4V alloy. The precipitation hardening and solid solution strengthening both cause the mechanical property to increase.

8.17.2   Tensile Strength and Static Strength

Generally, the static strength of the products is governed by the density as well as the structure obtained through the AM process. In comparison with the conventional manufacturing technique (i.e., casting), the microstructure in AM products is finer, which yields greater static strength. However, the microstructure of AM-processed products is anisotropic regarding the build direction and frequently shows more or less distinct texture. As a result, the tensile behavior is also anisotropic, which depends on the orientation. Therefore, the tensile and the fracture strength in the as-built samples are larger in the building direction than perpendicular to that.

8.17.2.1   Tensile Behavior of AM-Fabricated Ferrous Alloys

Tensile behavior of the AM-manufactured steel often satisfies the required specifications for technical usages. Formation of finer grains precedes with a substantial rise in the yield strength and ultimate tensile strength. As for the ductility of the AM samples, a minor amount of porosity results in ductile fracture with elongation values similar to wrought alloys. On the other hand, a greater amount of porosity influences brittle failure, which considerably lowers the elongation. The tensile property of the AM-manufactured steels compared with the conventional steels is shown in Figure 8.42.

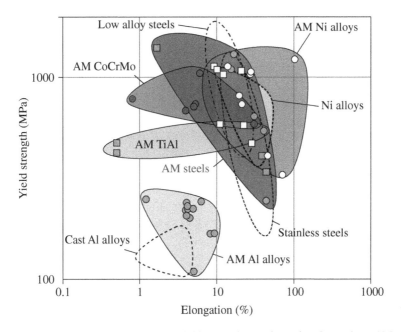

Figure 8.42   The property window presents yield strength vs. elongation for various AM-processed alloys and conventionally fabricated ferrous alloys [116, 117]. Source: Open access, reproduced with permission from authors.

It is obvious that the tensile properties vary with the manufacturing techniques controlling the obtained microstructure, also influenced by process input parameters. The AM-manufactured precipitation strengthening steels are relatively soft, as there is no precipitation formation due to inadequate time of faster solidification. Comparable to conventional manufacturing, these AM-made steels show better tensile properties through post-processing heat treatments. In martensitic grade steels, the amount of retained austenite and austenite reversion phenomenon impacts the tensile properties.

8.17.2.2   Tensile Behavior of AM-Fabricated Al Alloys

In AM aluminum alloys, the tensile properties follow the same phenomenon as obvious in AM ferrous alloys regarding the structure and yield strength relationship. The finer grains developed in AM-processed Al alloys promote increased strength in the as-built condition. However, the precipitation strengthens AM-fabricated AlSi10Mg alloy which shows similar tensile properties like solution strengthen $AlSi_{12}$ alloy. During heat treatment of AM-fabricated AlSi10Mg alloy, the earlier fine-grain becomes coarser and also the precipitation formation takes place. The coarse grains deteriorate the tensile strength, whereas precipitates strengthen the alloy. The AM-manufactured scandium-containing alloy aids to retain fine grain structure and also completely coherent precipitates after aging heat treatment. However, in AA 2139 (Al–Cu,Mg) alloy, loss of Mg occurs during AM, which afterward lowers the precipitate volume as well as drops the tensile strength.

8.17.2.3   Tensile Properties of AM-Manufactured Ni Alloys

The tensile behavior of AM-fabricated Ni alloys is comparable to the wrought one, while the ductility is low because of the precipitation of γ' and γ'' phases in austenite matrix, and δ phases around the grain boundaries.

8.17.2.4   Tensile Properties of AM-Manufactured Ti Alloys

In AM, Ti-6Al-4V is the most comprehensively studied group of alloys. As Ti is an appropriate material for different AM techniques, the process input parameters influencing microstructures and tensile behaviors have been considered systematically, specifically for Ti-6Al-4V alloys. The yield strength of Ti-6Al-4V alloy, manufactured at different AM techniques, is plotted in Figure 8.43. The better tensile strength is correlated with the finer martensitic structure resulted from rapid cooling. Finer grains always enhance the yield strength and the ductility. It is already established that the deformed hexagonal lattice of α' marensite is stronger compared to lamellar α, which is due to the finer lath width, without reducing the ductility.

Typically, AM-fabricated Ti-6Al-4V alloy has better tensile strength than cast or wrought alloys, i.e., $\alpha + \beta$ alloys, but lower ductility than pure Ti, because of the impeding of the twinning deformation phenomenon.

The tensile property of AM-processed Ti alloys depends on the microstructure and hence influenced by thermal cycling through manufacturing and post-process heat treatment. The wrought alloy, after annealing at 705 °C (1300 °F), according to Metallic Materials Properties

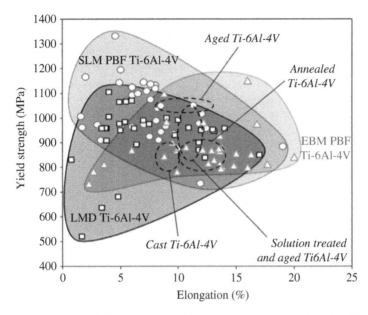

Figure 8.43 The property window presents yield strength vs. elongation for Ti-6Al-4V alloy manufactured using both AM and conventional technique [116, 117]. Source: Open access, reproduced with permission from authors.

Development and Standardization (MMPDS)[118], develops equiaxed microstructure. Because of the faster cooling of AM, the deposited Ti-6Al-4V alloy possesses a martensitic phase with columnar structure and compared to generally anneal wrought alloy shows higher tensile strengths with limited ductility and fracture strengths.

From many research studies, it is concluded that AM-manufactured Ti-6Al-4V alloy is equivalent to cast or wrought products. It is important to note that the tendency of anisotropy is divergent for tensile strength and ductility using different techniques. The lower ductility of Ti-6Al-4V alloy manufactured using LPBF and LDED techniques is ascribed to the development of brittle martensite phases. Ductility is the anisotropic property, which is different on the horizontal and vertical direction of the built and can be improved through post-process heat treatment.

8.17.3 Fatigue Behavior of AM-Manufactured Alloys

Fatigue is the most crucial failure factor for the operative design of numerous engineering materials. In this process, cyclic loading generates cracks by accumulating the plastic deformation directing to the initiation and propagation of cracks, which finally results in fatigue failure well below the material's ultimate strength. Therefore, the complete process can be subdivided into three stages: (i) nucleation of cracks, (ii) growth of cracks, and (iii) rapid fracture. The area of the surfaces where bending occurs frequently, as well as the geometric variations due to machining marks, surface flaws, notches, etc., create stress concentration, which initiates cracking. In the early stage, the crack grows on the particular crystallographic planes with the greater resolved shear stress, and its kinetics is influenced by surface topography, structural morphology, and crystallographic texture. Through the propagation of cracks, the stress intensity factor, K, becomes greater and leads to growth on different planes. Generally, a microcrack advances along the maximum shear stress plane resulting in macroscopic growth on the maximum tensile plane. When the maximum stress intensity, K_{max} reaches the plane strain fracture toughness, KIC, the steady crack growth changes to the unsteady approach. Therefore, when the tensile stresses become higher, the crack grows large enough to cause fatigue failure. Fatigue life is expressed by the load cycle number (N_f) until the failure occurs. Generally, fatigue lives are classified into three categories: (i) low cycle fatigue where fatigue life is below 10^3 cycles, (ii) medium cycle fatigue where fatigue life is in the range of $10^3 < N_f < 10^5$, (iii) high cycle fatigue with $N_f > 10^5$.

The fatigue property also depends on the isotropic behavior of the AM-fabricated samples. Usually, fatigue strengths are greater in the horizontal built direction than the vertical direction in the LPBF process. The LPBF-processed samples show a better Paris slope, unlike the equivalents manufactured by EBM, resulting in a greater fatigue crack progression rate. The fatigue life of the as-fabricated AM samples is considerably lower than the wrought ones, which is due to the surface roughness and internal defects influencing the cracks to initiate. The fatigue property of AM products can be enhanced by post-processing heat treatment, hot isostatics pressing, and surface quality improvement. Besides surface feature and internal defects, fatigue crack growth also depends on the crystallographic direction of grains that hold the crack tip, the quantity of grain boundaries around it, and the type of residual core stress. However, some AM-fabricated products may show less anisotropy because of the variation in the process parameters, which may result in a more homogeneous microstructure and proper distribution of residual stresses. Moreover, post-processing like surface grinding and heat treatment may significantly enhance the mechanical properties of AM-fabricated parts to be similar or even superior to conventional counterparts.

The appearance of the fatigue fracture surface is usually defined as brittle because of the plastic deformation. Fatigue fractures are easy to differentiate from other brittle fractures as they are gradual and create characteristic marks. The fracture surface macroscopically looks like "beach," "clam-shell," or "conchoidal," indicating intervals between the fatigue loading cycle.

Fatigue results are presented using $S–N$ curve, where the employed stress (S) is plotted against the number of cycles. There are several factors that influence the fatigue properties, such as material characteristics, AM techniques, process parameters, surface roughness, defects, residual stresses, post-processing techniques, and loading conditions.

8.17.3.1   Factors Influencing Fatigue Behavior in AM

Like other mechanical parts, most of the parts fabricated by AM pass through the cyclic loading during applications, making the fatigue property a key concern. Although AM possesses numerous benefits, predicting the fatigue property of the components manufactured using this process is quite challenging. This is because several factors influence fatigue performance, including microstructural features like grain morphology, defect type and shape, build direction, surface roughness, residual stress, and post-processing heat treatment. Several factors are briefly discussed below.

In AM, the characteristic features and defects are reliant on the process input parameters (laser power, beam scan speed, hatch spacing, etc.) as well as the thermal history (heating and cooling cycle, thermal gradient) through the manufacturing process. Besides the impact on microstructural and crystallographic texture, changing these parameters influences the quantity, category, as well as the distribution of the interior defects. The well-known defects to introduce fatigue cracks in AM components are the pores and voids generated from entrapped gas and/or lack of fusion, as well as inadequate fusion.

There is a controversy about the build orientation anisotropy on the mechanical properties of AM products. The difference in thermal profile experienced in different build directions control the morphology of the microstructure, defects, and eventually the ductility.

The manufacturing way of the AM techniques, which pass through the extreme thermal gradients, higher energy density, and most significantly rapid solidification, develops residual stresses in the final products. As known, the tensile residual stresses are disadvantageous to fatigue property; their impact may be reduced or prevented through suitable process parameters, corrected build direction, or releasing them by applying post-processing heat treatment (PPHT). Annealing is a possible PPHT to enhance fatigue life by releasing tensile residual stresses. HIP is another post-processing treatment that usually applies to shrink or close pores and voids, thereby dropping residual stresses and enhancing the component's densification and fatigue property.

From the view of mechanical and the structural performance of AM-processed materials, most of the experiments are performed using monotonic loading, which shows different performance during cyclic loading, and hence fatigue design or evolution. Therefore, for the better performance of AM products in cyclic loading, knowledge on the AM process parameters, the related thermal behavior, as well as the post-process treatment is considerably important.

An advanced sample design can be applied for fatigue testing to minimize the limitations often observed in the conventional fatigue test specimens. For example, the typical plate-type samples during the completely reversed fatigue test show buckling along the compression portion of the loading cycle. Therefore, the fatigue test outcomes using this type of sample are in the tension–tension condition, executing mean stress. The modification of the sample geometry may increase the buckling resistance. The conventional subtractive fabrication technique may restrict the sample geometry, which would be facilitated in the AM process. Because of the advantage of geometric flexibility, samples can be designed for homogeneous distribution of stress or strain through the gauge section to avoid stress or strain gradients during the shear loading.

Moreover, the other crucial factors for fatigue design are the loading conditions similar to the applications and atmospheres comparable to service.

8.17.3.2 Fatigue Performance of AM-Manufactured Ferrous Alloys

The fatigue property of 316L alloy is influenced by its monotonic strength. The fatigue behavior of LPBF-processed 17-4 PH steel at various directions is shown in Figure 8.44. The building orientation shows a significant impact on the fatigue life of 17-4 PH steel.

Usually, the horizontally built samples show higher fatigue property because of the structural configuration along the loading direction. However, in the vertical samples, defects are detrimental as they create stress accumulation during loading.

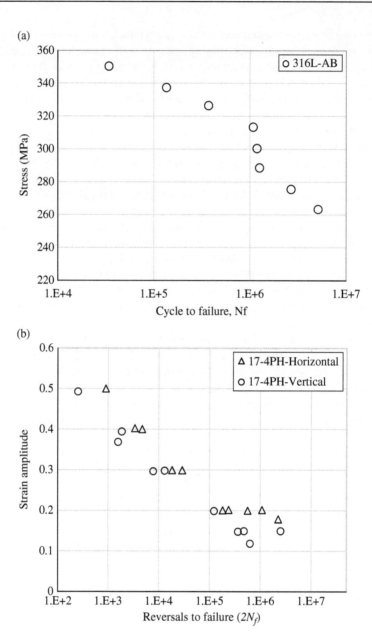

Figure 8.44 The fatigue behavior of (a) 316L and (b) LPBF-manufactured 17-4 PH steel at horizontal and vertical build direction [119, 120].

Fatigue life of the ferrous alloys is influenced by the alloying element and the post-processing technique. This is due to the following reasons: (i) lower carbon-containing ferrous alloys develop low hardness and therefore possess better fatigue life compared to the hardened steel, (ii) higher nickel content from Fe–Ni lath martensite, which accelerates ductility, and eventually fatigue life. Moreover, post-processing heat treatment can expand the fatigue life through the recovery of ductility and toughness of the alloy.

8.17.3.3 Fatigue Behavior of AM-Fabricated Al Alloy

Similar to the fatigue property of conventional metallic parts, the occurrence of structural flaws such as porosity or shrinkage cavities controls the fatigue life of AM-manufactured components. In laser-fabricated parts, the formation of lower melting porosity adjacent to the layers drops the actual load-bearing zone vertical to the layers and develops stress concentration, following a decrease of static and dynamic strength in the vertical direction. Therefore, porosity has the greatest worsening impact on the fatigue property of components when the shape and the number of pores surpass the threshold values.

Although Al alloys have different physical properties than ferrous and Ti alloys, in the area of thermal conductivity, surface reflectivity, and melt viscosity, the AM-fabricated properties are comparable. For example, in eutectic AlSi12 and AlSi10Mg alloy, the faster cooling rate develops a finer lamellar dendritic network of eutectic phases. The formation of strengthening phases of Mg_2Si as well as the distribution of Si particles in the Al matrix contributes to the tensile strength of the AM alloy compared to the conventional sand-cast or die-cast alloys. Throughout the plastic deformation, the dendrite structure acts as load-bearing elements by breaking through the dendritic arms, followed by the ultimate delamination of the Al matrix. The rapid cooling rate results in the growth of residual pores and eventually forms initial cracks at vital pores. Therefore, the joining of several cracks causes a rapid fracture while dropping the ductility.

The surface feature of the AM-manufactured Al alloys is very dominant in fatigue behavior. Therefore, the stress-released AlSi10Mg alloy shows better fatigue strength because of the enhanced ductility of the deformed microstructure. For example, the combined effect of pre- and post-heating of laser-processed AlSi10Mg alloy at 300 °C directs the homogenization of microstructure through the development of spheroidal eutectic of Si-particles, as well as reduction of crack origination and or growth, which ultimately improve the fatigue strength and the ductility. On the other hand, the build orientation has a minor impact on the laser-processed AlSi10Mg alloy's fatigue property. Moreover, due to the post-heating, the metallurgical defects like cracks and porosity would be reduced because the laser beam imposes on the previously heated powder bed resulting in lower cooling rates, and hence less stress development causes minor distortion.

The application of heat treatment usually reduces residual porosity, and thereby fatigue inconsistency. In the high cycle fatigue (HCF), the fatigue property is influenced by the resistance to the crack origination rather than growth. Materials performance is connected with each other regarding the fatigue scheme under consideration. It is well known that materials that have an improved tensile strength are stronger in the low cycle fatigue (LCF) scheme, which is different in HCF.

8.17.3.4 Fatigue Property of AM-Manufactured Nickel Alloys

Wrought IN718 contains δ phases, whereas the AM-fabricated IN718 comprises Laves phases. It is known that the wrought IN718 shows a better fatigue crack resistance than the AM-processed IN718 because of the absence of detrimental Laves phase.

There exists some controversial myth regarding the δ phases, such as: (i) δ phases are enriched with niobium, which makes the nearby matrix depletion of niobium content that after heat treatment results in lack of γ'' phase. Therefore, that region becomes weak. (ii) δ phase obstructs the crack growth during creep and fatigue failure showing an intergranular fracture.

The effect of Laves phases on the fatigue crack origination and transmission depends on their morphologies, which are different at various stress levels. To illustrate the phenomenon, a schematic is shown in Figure 8.45, which depicts the breaking up of Laves phase and their separation from the surrounding austenitic matrix. Through the fatigue progression, a plastic zone forms nearby the crack tip. The size of the plastic zone is associated with the stress concentration of the crack tip. At the initial stage of the fatigue crack generation, the lower stress concentration results in a smaller plastic zone, which makes it difficult to break Laves phases. Therefore, the fatigue cracks are stopped by Laves phases and make them easier to propagate through the austenite matrix. Thus, the Laves phases act as a vital role in impeding the crack growth, resulting in high-cycle fatigue property in AM-manufactured IN718 compared to wrought IN718.

On the other hand, at the higher stress amplitude, the Laves phases become weak and most of them break near the crack propagation region, and others detach from the austenite matrix, creating microscopic holes in the boundary between them. Therefore, the fatigue cracks grow along the broken interface resulting in the deterioration of fatigue property of AM-manufactured Ni alloy.

8.17.3.5   Fatigue Behavior of Additive-Manufactured Ti Alloy

The fatigue property of LPBF-manufactured Ti-6Al-4V alloy at different conditions is shown in Figure 8.46, comparing with the wrought alloy. From the plot, it is evident that heat treatment may considerably progress the fatigue life. This is due to the release of the residual stresses which develop through the manufacturing process. Stress accumulation on the printed rough surface, as well as the un-melted particles on the surface, may originate crack initiation zones, which afterward promote premature failure.

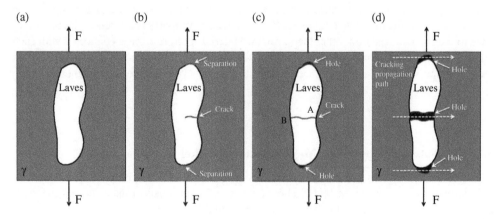

Figure 8.45   The schematic shows the breaking up of Laves phase and split-up from the surrounding austenitic matrix. Source: Reprinted from [81] with permission from Elsevier.

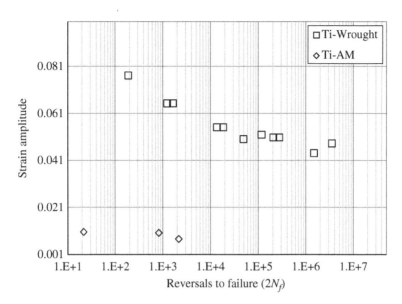

Figure 8.46 The fatigue plot shows strain amplitude vs. reversals to failure for both wrought and AM-manufactured Ti-6Al-4V alloy. Source: Redrawn and adapted from [119].

A comparison of the fatigue property with the wrought alloy specifies that AM-manufactured alloys have a shorter fatigue life. It is already identified that the fatigue property of AM alloy is greatly influenced by internal defects, which perform as micro-notches and result in stress accumulation. The application of post-processing treatment is important to remove or shrink pores to a far smaller size, which makes them unable to influence fatigue behavior.

The AM-manufactured alloys show a very fewer cyclic softening than the wrought alloy under loading. This behavior is because of poor ductility of the AM alloy and so as for shorter fatigue life.

Fracture toughness demonstrating the property of a material to repel fracture also depends on the build orientation of the sample. In the horizontally oriented samples, cracks grow through the columnar grains, whereas, in the vertically made samples, cracks propagate along the grain boundary. The poor toughness value in the LPBF-processed Ti-6Al-4V alloy is because of the acicular morphology of martensite, which is brittle compared to α/β duplex structure formed in EB-Melted Ti-6Al-4V alloy. Moreover, the residual stress developed in LPBF-processed alloy can result in anisotropy in the fracture toughness, which can be improved through post-processing such as heat treatment, Hot Isostatic Pressing, and stress relieving treatment.

8.18 Summary

At the end of this chapter, the following paragraphs recap the learning objectives:

Materials Science: Structure–Property Relationship
The area of material science can be categorized into four parts, as (i) structure, (ii) property, (iii) processing, and (iv) performance. The properties of materials used in AM are linked with

the composition and structure of materials controlled by the processing techniques and, all together, influence the performance of the material in the application.

Manufacturing of Metallic Materials

The manufacturing of metallic raw materials starts from the ore extraction, through melting and alloying to manufacture metal ingot. In conventional manufacturing techniques such as casting, raw materials are passed through melting, pouring in mold, and afterward solidification and cooling, whereas, in AM, 3D objects generated by computer-aided design (CAD) are fabricated by adding layer-by-layer of materials, followed by local melting and solidifying as near-net-shape.

Solidification

Both conventional and AM techniques follow the solidification scheme, which is the transformation of the liquid metal to solid form through the cooling process. In AM, because of the utilization of moving point power sources with focused energy in the localized area, the short interaction time results in a faster solidification rate compared to the conventional technique.

Equilibrium Solidification: Theory and Mechanism

Considering the hypothesis of thermodynamic equilibrium, the alloy composition in solid and liquid states can be assessed from the equilibrium phase diagram in conventional methods. The structure of a material influenced by composition and temperature can be identified from an equilibrium phase diagram using the assumption of a relatively slow transformation rate or a faster diffusion rate.

Non-Equilibrium Solidification: Theory and Mechanism

The AM process elevation will require knowledge on equilibrium and non-equilibrium phase diagrams, thermodynamics, thermo-physical quantities, diffusion kinetics, etc. During rapid cooling, some phases that generally form under equilibrium conditions may not arise; hence, there would be a chance for the occurrences of metastable phases.

Solute Redistribution and Microsegregation

When a liquid solidifies from uniform composition, the resulting solid hardly shows uniformity in the composition. In addition, the growth phenomena, phase distribution, and segregations are related to the solute redistribution and dependent on thermodynamics (i.e., phase diagram) and kinetics (i.e., diffusion, undercooling, fluid flow, and so on).

The microsegregation is defined as the ejection of solute from the freezing material, which afterward distributes heterogeneously and eventually affects the solidification mode. The alloy in liquid form contains higher solute compared to the solidified condition. Therefore, during the solidification process, there is a chance of the high solute liquid to be trapped within the solidified structures. This causes microsegregation or banding of high and low solute alloys and substantial incoherence in material properties.

Constitutional Supercooling

The phenomena "supercooling" or "subcooling" or "undercooling" of the liquid does not allow the solidification to occur at the minimal phase transformation temperature. As the cooling progresses, the temperature of the liquid continues to drop down. However, the system becomes thermodynamically unsteady, and therefore solidification can be commenced with a small amount of energy accomplished with latent heat release, which eventually increases the temperature. This mechanism is known as the kinetics of solidification.

The actual temperature (change in temperature between the solidification and the nucleation process) at which the solidification starts is termed "the degree of supercooling (or undercooling)."

Nucleation and Growth Kinetics

There are two kinds of nucleation, (i) homogeneous, and (ii) heterogeneous nucleation, which are categorized based on the position where the nucleation takes place. The homogeneous nucleation occurs with the formation of uniform nuclei all over the parent phase, while in heterogeneous nucleation, nuclei may advance from the structural discontinuity, like boundaries of the impurities, foreign particles, dislocations and so on.

After the initiation of the nucleation, the solid/liquid interface deeds like a growth front. Growth starts from the previously deposited layers through partial or complete melting, which eventually governs the crystallographic pattern.

Solidification Microstructures in Pure Metals and Alloys

There are four major solidification patterns, (i) planar, (ii) cellular, (iii) columnar, and (iv) equiaxed dendritic. The various solidification structures grow based on the driving force of constitutional supercooling, distribution of the solute at the boundary, as well as the characteristics of the elements containing in the solidified alloys. Generally, the devastating forces of constitutional supercooling do not exist in pure metal, showing a planer solidification approach.

A planer solidification starts when the actual temperature of the melt is larger compared to the equilibrium liquidus. In constitutional supercooling, the actual temperature gradient is lower than the liquidus temperature gradient; a protuberance may mature in the undercooled melt and approach as a cellular or dendritic pattern. When the grain grows like a column without branching the arms, it will form a cellular structure. In contrast, the grains with secondary or tertiary arms will develop a dendritic structure. The cellular and the dendritic growth are estimated by the degree of constitutional supercooling and the complete stability described by the critical solidification rate.

Directional Solidification in AM

The solute intensity or temperature gradients in the melt may produce surface tension and subsequent Marangoni convection, resulting in an unstable solidification process. Because of the dynamic restraint of crystal growth, the rapid solidification generally progresses along the maximum heat flow. The non-equilibrium solidification behavior and the propensity for the directional propagation, the mechanisms which are instantaneous but competitive in nature, may generate diversity in crystal orientation with a localized consistency.

Factors Affecting Solidification in AM

Cooling rate

In AM, cooling rates are higher due to the localized heat input and high process speed being used and also influenced by the heat input, manipulated by laser power, beam scan speed, and powder flow rate in the case of LDED.

Temperature Gradient and Solidification Rate

Solidification morphology is understandable using a solidification map, designed with a temperature gradient (G), and growth rate (R) in the form of their product as $G.R$ and ratio as G/R. G/R governs the type of solidification pattern, and $G.R$ controls the size of the structure. The solidification structure may be planar, cellular, columnar dendritic, or equiaxed dendritic when

G/R value decreases. Moreover, the size of these four structures decreases with the larger value of *G.R* (cooling rate).

Process Parameters
The solidified microstructure and the phase formations are considerably controlled by the input process parameters, i.e., power, speed, and specific energy that dominate the microstructure in AM products.

Solidification Temperature Span
The broader solidification temperature span creates a greater solid/liquid or mushy zone, which is the most responsible for the solidification cracking.

Gas Interactions
In 3D printing techniques, argon and nitrogen are generally used to offer an inert atmosphere and satisfy the high-tolerance criteria. As a reactive gas, nitrogen can be a suitable shielding gas for those materials that never react with nitrogen.

Solidification Defects
Porosity
In laser- and EB-based AM techniques, the melting may be accomplished throughout a keyhole mode at very high operating energies. When the keyholes are not melted properly, they can turn into an unstable mode, which then frequently forms and collapses, causing voids to be formed inside the melt. Also, through the atomization of metallic powder, gas can be captured by the powder particles, which then creates microscopic gas pores.

Balling
The balling phenomenon is considered as the unusual broken melt pools that take place on the surfaces of laser-manufactured parts. Through the processing, the laser is scanned linearly and the melting occurs along a row of powders in LPBF, which then creates continuous liquid tracks similar to a tubular shape. The breakdown of the tube into the spherical-shaped (balling effect) metallic agglomerates drops the surface energy of the melt track till the ultimate equilibrium condition is aroused.

Cracking
The solidification cracking in AM product occurs along the grain boundaries like welding. The solidifying melt passes through a contraction from solidification shrinkage and thermal contraction. The substrate or the earlier solidified layers show low temperature compared to the solidifying layers. Hence, a higher contraction develops in the solidifying layer, which is also impeded by earlier deposited layers, forming induced residual stresses. When the extent of tensile stress surpasses the strength of the solidifying layer, cracks may be generated along the boundaries.

Lamellar Tearing
Lamellar tearing causes because of the combined effect of localized internal stresses and the lower ductility substrate material.

Post Solidification Phase Transformation
Ferrous Alloys/Steels
The very typical microstructures of AM-processed austenitic stainless steels are cellular and columnar dendrites. Austenitic steels regularly have a tiny amount of ferrite distant from austenite.

Al Alloys

The most familiar Al alloys obtainable in AM are the hardenable AlSi10Mg, which in the LPBF process shows a eutectic silicon network in the α-Al matrix.

Nickel Alloys/Superalloys

LPBF and LDED manufactured IN625 alloy shows columnar dendritic microstructure in γ-Ni matrix. From the LPBF and BJ fabrication processes of IN625, there observed the precipitates of δ phase (Ni_3Nb) and NbC, M_6C carbides.

IN718 also shows γ-Ni matrix with cellular solidification structure, and presence of both γ' and γ'' phases.

In LPF/LDED-processed Stellite 12, cobalt matrix dendrites, and inter-dendritic eutectic carbides are the core microstructure, while blocky and lamellar eutectic carbides and M_7C_3 and $M_{23}C_6$ are the main carbides.

Titanium Alloys

LPBF-manufactured Ti-6Al-4V alloy shows α' martensitic structure, which after heat treatment decomposes to $\alpha + \beta$ lamellar structure.

References

1. J. Dawes, R. Bowerman, and R. Trepleton, "Introduction to the additive manufacturing powder metallurgy supply chain," *Johnson Matthey Technol. Rev.*, vol. 59, no. 3, pp. 243–256, 2015, doi: https://doi.org/10.1595/205651315X688686.
2. G. Egger, P. E. Gygax, R. Glardon, and N. P. Karapatis, "Optimization of powder layer density in selective laser sintering," 10th Solid Free. Fabr. Symp., pp. 255–263, 1999.
3. W. J. Sames, F. A. List, S. Pannala, R. R. Dehoff, and S. S. Babu, "The metallurgy and processing science of metal additive manufacturing," *Int. Mater. Rev.*, vol. 61, no. 5, pp. 315–360, Jul. 2016, doi: https://doi.org/10.1080/09506608.2015.1116649.
4. R. J. Hebert, "Viewpoint: metallurgical aspects of powder bed metal additive manufacturing," *J. Mater. Sci.*, vol. 51, no. 3, pp. 1165–1175, 2016, doi: https://doi.org/10.1007/s10853-015-9479-x.
5. S. S. Sih and J. W. Barlow, "The prediction of the emissivity and thermal conductivity of powder beds," *Part. Sci. Technol.*, vol. 22, no. 4, pp. 427–440, 2004, doi: https://doi.org/10.1080/02726350490501682.
6. C. A. Brice and N. Dennis, "Cooling rate determination in additively manufactured Aluminum Alloy 2219," *Metall. Mater. Trans. A Phys. Metall. Mater. Sci.*, vol. 46, no. 5, pp. 2304–2308, 2015, doi: https://doi.org/10.1007/s11661-015-2775-x.
7. W. E. Frazier, "Metal additive manufacturing: A review," *J. Mater. Eng. Perform.* 2014, doi: https://doi.org/10.1007/s11665-014-0958-z.
8. Y. J. Liang, X. Cheng, and H. M. Wang, "A new microsegregation model for rapid solidification multi-component alloys and its application to single-crystal nickel-base superalloys of laser rapid directional solidification," *Acta Mater.*, vol. 118, pp. 17–27, 2016, doi: https://doi.org/10.1016/j.actamat.2016.07.008.
9. F. H. Froes, Y. Kim, and F. Hehmann, "Rapid solidification of Al, Mg and Ti," *JOM*, vol. 39, no. August, pp. 14–21, 1987.
10. S. Kou, Metallurgy, Welding Metallurgy, 2nd edition, vol. 822, no. 1–3, John Wiley & Sons, 2003.
11. B. Cantor, W. T. Kim, B. P. Bewlay, and A. G. Gillen, "Microstructure – cooling rate correlations in melt-spun alloys," *J. Mater. Sci.*, vol. 26, no. 5, pp. 1266–1276, 1991, doi: https://doi.org/10.1007/BF00544465.
12. D. D. Gu, W. Meiners, K. Wissenbach, and R. Poprawe, "Laser additive manufacturing of metallic components: Materials, processes and mechanisms," *Int. Mater. Rev.*, vol. 57, no. 3, pp. 133–164, 2012, doi: https://doi.org/10.1179/1743280411Y.0000000014.

13. A. Y. Uzan, Y. Kozak, Y. Korin, I. Harary, H. Mehling, and G. Ziskind, "A novel multi-dimensional model for solidification process with supercooling," *Int. J. Heat Mass Transf.*, vol. 106, pp. 91–102, 2017, doi: https://doi.org/10.1016/j.ijheatmasstransfer.2016.10.046.

14. W. Kurz and D. Fisher, Fundamentals of solidification, Trans Tech Publications, Switzerland, Germany, UK, USA, 1986.

15. M. E. Glicksman, "Principles of solidification: An introduction to modern casting and crystal growth concepts," no. 2, Springer, New York, pp. 1–520, 2011, doi: https://doi.org/10.1007/978-1-4419-7344-3.

16. Y. Z. Chen, F. Liu, G. C. Yang, and Y. H. Zhou, "Nucleation mechanisms involving in rapid solidification of undercooled Ni80.3B19.7melts," *Intermetallics*, vol. 19, no. 2, pp. 221–224, 2011, doi: https://doi.org/10.1016/j.intermet.2010.08.010.

17. T. DebRoy et al., "Additive manufacturing of metallic components – Process, structure and properties," *Prog. Mater. Sci.*, vol. 92. pp. 112–224, 2018, doi: https://doi.org/10.1016/j.pmatsci.2017.10.001.

18. H. Zhao, "Microstructure Heterogeneity in Additive ManufacturedTi-6Al-4V," Thesis, 2017.

19. W. Callister and D. Rethwisch, Materials science and engineering: an introduction, 8th edition, vol. 94, John Wily & Sons, Inc, 2007.

20. S. Das, "Physical aspects of process control in selective laser sintering of metals," *Adv. Eng. Mater.*, vol. 5, no. 10, pp. 701–711, 2003, doi: https://doi.org/10.1002/adem.200310099.

21. D. A. Porter and K. E. Easterling, "Phase transformations in metals and alloys, 3rd edition," CRC Press, p. 138, 2014, doi: https://doi.org/10.1146/annurev.ms.03.080173.001551.

22. A. Basak and S. Das, "Epitaxy and microstructure evolution in metal additive manufacturing," *Annu. Rev. Mater. Res.*, vol. 46, no. 1, pp. 125–149, 2016, doi: https://doi.org/10.1146/annurev-matsci-070115-031728.

23. S. A. David and J. M. Vitek, "Correlation between solidification parameters and weld microstructures," *Int. Mater. Rev.*, vol. 34, no. 1, pp. 213–245, 1989, doi: https://doi.org/10.1179/imr.1989.34.1.213.

24. Y. Z. Zhou, A. Volek, and N. R. Green, "Mechanism of competitive grain growth in directional solidification of a nickel-base superalloy," *Acta Mater.*, vol. 56, no. 11, pp. 2631–2637, 2008, doi: https://doi.org/10.1016/j.actamat.2008.02.022.

25. M. Garibaldi, I. Ashcroft, M. Simonelli, and R. Hague, "Metallurgy of high-silicon steel parts produced using selective laser melting," *Acta Mater.*, vol. 110, pp. 207–216, 2016, doi: https://doi.org/10.1016/j.actamat.2016.03.037.

26. A. Yadollahi, N. Shamsaei, S. M. Thompson, and D. W. Seely, "Effects of process time interval and heat treatment on the mechanical and microstructural properties of direct laser deposited 316L stainless steel," *Mater. Sci. Eng. A*, vol. 644, pp. 171–183, 2015, doi: https://doi.org/10.1016/j.msea.2015.07.056.

27. X. Zhao, L. Liu, and J. Zhang, "Investigation of grain competitive growth during directional solidification of single-crystal nickel-based superalloys," *Appl. Phys. A Mater. Sci. Process.*, vol. 120, no. 2, pp. 793–800, 2015, doi: https://doi.org/10.1007/s00339-015-9290-1.

28. T. Wang, Y. Y. Zhu, S. Q. Zhang, H. B. Tang, and H. M. Wang, "Grain morphology evolution behavior of titanium alloy components during laser melting deposition additive manufacturing," *J. Alloys Compd.*, vol. 632, pp. 505–513, 2015, doi: https://doi.org/10.1016/j.jallcom.2015.01.256.

29. L. Thijs, M. L. Montero Sistiaga, R. Wauthle, Q. Xie, J. P. Kruth, and J. Van Humbeeck, "Strong morphological and crystallographic texture and resulting yield strength anisotropy in selective laser melted tantalum," *Acta Mater.*, vol. 61, no. 12, pp. 4657–4668, 2013, doi: https://doi.org/10.1016/j.actamat.2013.04.036.

30. M. A. Jaafar, D. R. Rousse, S. Gibout, and J.-P. Bédécarrats, "A review of dendritic growth during solidification: Mathematical modeling and numerical simulations," *Renew. Sustain. Energy Rev.*, vol. 74, pp. 1064–1079, 2017, doi: https://doi.org/10.1016/j.rser.2017.02.050.

31. A. Weisheit, A. Gasser, G. Backes, T. Jambor, N. Pirch, K. Wissenbach, Direct laser cladding, current status and future scope of application. In: Majumdar J., Manna I. (eds) Laser-assisted fabrication of materials. Springer series in materials science, vol 161, Springer, Berlin, Heidelberg, 2013. https://doi.org/10.1007/978-3-642-28359-8_5. pp. 221–240.

32. M. Gäumann, S. Henry, F. Cléton, J.-D. Wagnière, and W. Kurz, "Epitaxial laser metal forming: Analysis of microstructure formation," *Mater. Sci. Eng. A*, vol. 271, no. 1–2, pp. 232–241, 1999, doi: https://doi.org/10.1016/S0921-5093(99)00202-6.

33. G. P. Dinda, A. K. Dasgupta, and J. Mazumder, "Laser aided direct metal deposition of Inconel 625 superalloy: Microstructural evolution and thermal stability," *Mater. Sci. Eng. A*, vol. 509, no. 1–2, pp. 98–104, 2009, doi: https://doi.org/10.1016/j.msea.2009.01.009.

34. C. M. Liu, X. J. Tian, H. B. Tang, and H. M. Wang, "Microstructural characterization of laser melting deposited Ti-5Al-5Mo-5V-1Cr-1Fe near β titanium alloy," *J. Alloys Compd.*, vol. 572, pp. 17–24, 2013, doi: https://doi.org/10.1016/j.jallcom.2013.03.243.

35. W. E. King et al., "Observation of keyhole-mode laser melting in laser powder bed fusion additive manufacturing," *J. Mater. Process. Technol.*, vol. 214, no. 12, pp. 2915–2925, 2014, doi: https://doi.org/10.1016/j.jmatprotec.2014.06.005.

36. W. J. Sames, F. Medina, W. H. Peter, S. S. Babu, and R. R. Dehoff, "Effect of Process Control and Powder Quality on Inconel 718 Produced Using Electron Beam Melting," 8th Int. Symp. Superalloy 718 Deriv., pp. 409–423, 2014, doi: 10.1002/9781119016854.ch32.

37. H. Gong, K. Rafi, H. Gu, T. Starr, and B. Stucker, "Analysis of defect generation in Ti-6Al-4V parts made using powder bed fusion additive manufacturing processes," *Addit. Manuf.*, vol. 1, pp. 87–98, 2014, doi: https://doi.org/10.1016/j.addma.2014.08.002.

38. M. Pastor, H. Zhao, and T. Debroy, "Pore formation during continuous wave Nd:YAG laser welding of aluminium for automotive applications," *Weld. Int.*, vol. 15, no. 4, pp. 275–281, 2001, doi: https://doi.org/10.1080/09507110109549355.

39. D. Gu and Y. Shen, "Balling phenomena in direct laser sintering of stainless steel powder: Metallurgical mechanisms and control methods," *Mater. Des.*, vol. 30, pp. 2903–2910, 2009, doi: https://doi.org/10.1016/j.matdes.2009.01.013.

40. Z. Sun, X. Tan, S. B. Tor, and W. Y. Yeong, "Selective laser melting of stainless steel 316L with low porosity and high build rates," *Mater. Des.*, vol. 104, pp. 197–204, 2016, doi: https://doi.org/10.1016/j.matdes.2016.05.035.

41. K. Zhang, S. Wang, W. Liu, and X. Shang, "Characterization of stainless steel parts by laser metal deposition shaping," *Mater. Des.*, vol. 55, pp. 104–119, 2014, doi: https://doi.org/10.1016/j.matdes.2013.09.006.

42. K. Abd-Elghany and D. L. Bourell, "Property evaluation of 304L stainless steel fabricated by selective laser melting," *Rapid Prototyp. J.*, vol. 18, no. 5, pp. 420–428, 2012, doi: https://doi.org/10.1108/13552541211250418.

43. H. Yu, J. Yang, J. Yin, Z. Wang, and X. Zeng, "Comparison on mechanical anisotropies of selective laser melted Ti-6Al-4V alloy and 304 stainless steel," *Mater. Sci. Eng. A*, vol. 695, pp. 92–100, 2017, doi: https://doi.org/10.1016/j.msea.2017.04.031.

44. S. Cheruvathur, E. A. Lass, and C. E. Campbell, "Additive manufacturing of 17-4 PH stainless steel: postprocessing heat treatment to achieve uniform reproducible microstructure," *JOM*, vol. 68, no. 3, pp. 930–942, 2016, doi: https://doi.org/10.1007/s11837-015-1754-4.

45. L. Facchini, N. Vicente, I. Lonardelli, E. Magalini, P. Robotti, and M. Alberto, "Metastable austenite in 17-4 precipitation-hardening stainless steel produced by selective laser melting," *Adv. Eng. Mater.*, vol. 12, no. 3, pp. 184–188, 2010, doi: https://doi.org/10.1002/adem.200900259.

46. A. Kudzal et al., "Effect of scan pattern on the microstructure and mechanical properties of Powder Bed Fusion additive manufactured 17-4 stainless steel," *Mater. Des.*, vol. 133, pp. 205–215, 2017, doi: https://doi.org/10.1016/j.matdes.2017.07.047.

47. T. LeBrun, T. Nakamoto, K. Horikawa, and H. Kobayashi, "Effect of retained austenite on subsequent thermal processing and resultant mechanical properties of selective laser melted 17-4 PH stainless steel," *Mater. Des.*, vol. 81, pp. 44–53, 2015, doi: https://doi.org/10.1016/j.matdes.2015.05.026.

48. B. AlMangour and J. M. Yang, "Understanding the deformation behavior of 17-4 precipitate hardenable stainless steel produced by direct metal laser sintering using micropillar compression and TEM," *Int. J. Adv. Manuf. Technol.*, vol. 90, no. 1–4, pp. 119–126, 2017, doi: https://doi.org/10.1007/s00170-016-9367-9.

49. B. AlMangour and J. M. Yang, "Improving the surface quality and mechanical properties by shot-peening of 17-4 stainless steel fabricated by additive manufacturing," *Mater. Des.*, vol. 110, pp. 914–924, 2016, doi: https://doi.org/10.1016/j.matdes.2016.08.037.

50. S. L. Campanelli, A. Angelastro, C. G. Signorile, and G. Casalino, "Investigation on direct laser powder deposition of 18 Ni (300) marage steel using mathematical model and experimental characterisation," *Int. J. Adv. Manuf. Technol.*, vol. 89, no. 1–4, pp. 885–895, 2017, doi: https://doi.org/10.1007/s00170-016-9135-x.

51. P. Krakhmalev, I. Yadroitsava, G. Fredriksson, and I. Yadroitsev, "In situ heat treatment in selective laser melted martensitic AISI 420 stainless steels," *Mater. Des.*, vol. 87, pp. 380–385, 2015, doi: https://doi.org/10.1016/j.matdes.2015.08.045.

52. G. A. Ravi, X. J. Hao, N. Wain, X. Wu, and M. M. Attallah, "Direct laser fabrication of three dimensional components using SC420 stainless steel," *Mater. Des.*, vol. 47, pp. 731–736, 2013, doi: https://doi.org/10.1016/j.matdes.2012.12.062.

53. R. Mertens, B. Vrancken, N. Holmstock, Y. Kinds, J. P. Kruth, and J. Van Humbeeck, "Influence of powder bed preheating on microstructure and mechanical properties of H13 tool steel SLM parts," *Phys. Procedia*, 2016, vol. 83, pp. 882–890, doi: https://doi.org/10.1016/j.phpro.2016.08.092.

54. J. Mazumder, A. Schifferer, and J. Choi, "Direct materials deposition: Designed macro and microstructure," *Mater. Res. Innov.*, vol. 3, no. 3, pp. 118–131, 1999, doi: https://doi.org/10.1007/s100190050137.

55. J. Choi and Y. Chang, "Characteristics of laser aided direct metal/material deposition process for tool steel," *Int. J. Mach. Tools Manuf.*, vol. 45, no. 4–5, pp. 597–607, 2005, doi: https://doi.org/10.1016/j.ijmachtools.2004.08.014.

56. J. J. S. Dilip, G. D. J. Ram, T. L. Starr, and B. Stucker, "Selective laser melting of HY100 steel: Process parameters, microstructure and mechanical properties," *Addit. Manuf.*, vol. 13, pp. 49–60, 2016, doi: https://doi.org/10.1016/j.addma.2016.11.003.

57. X. Yue, J. C. Lippold, B. T. Alexandrov, and S. S. Babu, "Continuous cooling transformation behavior in the CGHAZ of naval steels," *Weld. J.*, vol. March, pp. 67s–75s, 2012.

58. X. Yue, "Evaluation of heat-affected zone hydrogen-induced cracking in high-strength steels," *J. Chem. Inf. Model.*, vol. 53, no. January, pp. 1689–1699, 2013, doi: https://doi.org/10.1017/CBO9781107415324.004.

59. W. Wang and S. Kelly, "A metallurgical evaluation of the powder bed laser additive manufactured 4140 steel material," *JOM*, vol. 68, no. 3, pp. 869–875, 2016, doi: https://doi.org/10.1007/s11837-015-1804-y.

60. E. Jelis, M. Clemente, S. Kerwien, N. M. Ravindra, and M. R. Hespos, "Metallurgical and mechanical evaluation of 4340 steel produced by direct metal laser sintering," *JOM*, vol. 67, no. 3, pp. 582–589, 2015, doi: https://doi.org/10.1007/s11837-014-1273-8.

61. K. Schmidtke, F. Palm, A. Hawkins, and C. Emmelmann, "Process and mechanical properties: Applicability of a scandium modified Al-alloy for laser additive manufacturing," *Phys. Procedia*, 2011, vol. 12, no. PART 1, pp. 369–374, doi: https://doi.org/10.1016/j.phpro.2011.03.047.

62. J. Li et al., "Phase evolution of a heat-treatable aluminum alloy during laser additive manufacturing," *Mater. Lett.*, vol. 214, pp. 56–59, 2018, doi: https://doi.org/10.1016/j.matlet.2017.11.111.

63. C. Brice, R. Shenoy, M. Kral, and K. Buchannan, "Precipitation behavior of aluminum alloy 2139 fabricated using additive manufacturing," *Mater. Sci. Eng. A*, vol. 648, pp. 9–14, 2015, doi: https://doi.org/10.1016/j.msea.2015.08.088.

64. D. Dai et al., "Influence of scan strategy and molten pool configuration on microstructures and tensile properties of selective laser melting additive manufactured aluminum based parts," *Opt. Laser Technol.*, vol. 99, pp. 91–100, 2018, doi: https://doi.org/10.1016/j.optlastec.2017.08.015.

65. H. Asgari, C. Baxter, K. Hosseinkhani, and M. Mohammadi, "On microstructure and mechanical properties of additively manufactured AlSi10Mg_200C using recycled powder," *Mater. Sci. Eng. A*, vol. 707, pp. 148–158, 2017, doi: https://doi.org/10.1016/j.msea.2017.09.041.

66. J. H. Rao, Y. Zhang, X. Fang, Y. Chen, X. Wu, and C. H. J. Davies, "The origins for tensile properties of selective laser melted aluminium alloy A357," *Addit. Manuf.*, vol. 17, pp. 113–122, 2017, doi: https://doi.org/10.1016/j.addma.2017.08.007.

67. X. Cao, W. Wallace, J. P. Immarigeon, and C. Poon, "Research and progress in laser welding of wrought aluminum alloys. II. Metallurgical microstructures, defects, and mechanical properties," *Mater. Manuf. Process.*, vol. 18, no. 1, pp. 23–49, 2003, doi: https://doi.org/10.1081/AMP-120017587.

68. C. P. Paul, P. Ganesh, S. K. Mishra, P. Bhargava, J. Negi, and A. K. Nath, "Investigating laser rapid manufacturing for Inconel-625 components," *Opt. Laser Technol.*, vol. 39, no. 4, pp. 800–805, 2007, doi: https://doi.org/10.1016/j.optlastec.2006.01.008.

69. V. Shankar, K. Bhanu Sankara Rao, and S. Mannan, "Microstructure and mechanical properties of Inconel 625 superalloy," *J. Nucl. Mater.*, vol. 288, no. 2, pp. 222–232, 2001, doi: https://doi.org/10.1016/S0022-3115(00)00723-6.

70. L. E. Criales, Y. M. Arısoy, B. Lane, S. Moylan, A. Donmez, and T. Özel, "Laser powder bed fusion of nickel alloy 625: Experimental investigations of effects of process parameters on melt pool size and shape with spatter analysis," *Int. J. Mach. Tools Manuf.*, vol. 121, 2017, doi: https://doi.org/10.1016/j.ijmachtools.2017.03.004.

71. S. Li, Q. Wei, Y. Shi, Z. Zhu, and D. Zhang, "Microstructure characteristics of Inconel 625 superalloy manufactured by selective laser melting," *J. Mater. Sci. Technol.*, vol. 31, no. 9, pp. 946–952, 2015, doi: https://doi.org/10.1016/j.jmst.2014.09.020.

72. X. Xu et al., "Research on microstructures and properties of Inconel 625 coatings obtained by laser cladding with wire," *J. Alloys Compd.*, vol. 715, pp. 362–373, 2017, doi: https://doi.org/10.1016/j.jallcom.2017.04.252.

73. A. Mostafaei, Y. Behnamian, Y. L. Krimer, E. L. Stevens, J. L. Luo, and M. Chmielus, "Effect of solutionizing and aging on the microstructure and mechanical properties of powder bed binder jet printed nickel-based superalloy 625," *Mater. Des.*, vol. 111, pp. 482–491, 2016, doi: https://doi.org/10.1016/j.matdes.2016.08.083.

74. D. Verdi, M. A. Garrido, C. J. Múnez, and P. Poza, "Microscale evaluation of laser cladded Inconel 625 exposed at high temperature in air," *Mater. Des.*, vol. 114, pp. 326–338, 2017, doi: https://doi.org/10.1016/j.matdes.2016.11.014.

75. A. Mostafaei, E. L. Stevens, E. T. Hughes, S. D. Biery, C. Hilla, and M. Chmielus, "Powder bed binder jet printed alloy 625: Densification, microstructure and mechanical properties," *Mater. Des.*, vol. 108, pp. 126–135, 2016, doi: https://doi.org/10.1016/j.matdes.2016.06.067.

76. W. J. Sames et al., "Feasibility of in situ controlled heat treatment (ISHT) of Inconel 718 during electron beam melting additive manufacturing," *Addit. Manuf.*, vol. 13, pp. 156–165, 2017, doi: https://doi.org/10.1016/j.addma.2016.09.001.

77. B. Farber et al., "Correlation of mechanical properties to microstructure in Metal Laser Sintering Inconel 718," *Mater. Sci. Eng. A*, vol. 712, pp. 539–547, 2018, doi: https://doi.org/10.1016/j.msea.2017.11.125.

78. A. Mostafa, M. Medraj, M. Jahazi, V. Brailovski, and I. Picazo Rubio, "Structure, texture and phases in 3D printed IN718 alloy subjected to homogenization and HIP treatments," *Metals (Basel).*, vol. 7, pp. 196–217, 2017, doi: https://doi.org/10.3390/met7060196.

79. R. G. Thompson, J. R. Dobbs, and D. E. Mayo, "The effect of heat treatment on microfissuring in alloy 718," *Weld. J.*, vol. 11, pp. 299–304, 1986.

80. G. H. Cao et al., "Investigations of γ′ γ″ and δ precipitates in heat-treated Inconel 718 alloy fabricated by selective laser melting," *Mater. Charact.*, vol. 136, pp. 398–406, 2018, doi: https://doi.org/10.1016/j.matchar.2018.01.006.

81. S. Sui, J. Chen, E. Fan, H. Yang, X. Lin, and W. Huang, "The influence of Laves phases on the high-cycle fatigue behavior of laser additive manufactured Inconel 718," *Mater. Sci. Eng. A*, vol. 695, pp. 6–13, 2017, doi: https://doi.org/10.1016/j.msea.2017.03.098.

82. H. Xiao, S. Li, X. Han, J. Mazumder, and L. Song, "Laves phase control of Inconel 718 alloy using quasi-continuous-wave laser additive manufacturing," *Mater. Des.*, vol. 122, pp. 330–339, 2017, doi: https://doi.org/10.1016/j.matdes.2017.03.004.

83. Z. Xu, C. J. Hyde, C. Tuck, and A. T. Clare, "Creep behaviour of inconel 718 processed by laser powder bed fusion," *J. Mater. Process. Technol.*, vol. 256, pp. 13–24, 2018, doi: https://doi.org/10.1016/j.jmatprotec.2018.01.040.

84. K. Yuan et al., "Influence of process parameters and heat treatments on the microstructures and dynamic mechanical behaviors of Inconel 718 superalloy manufactured by laser metal deposition," *Mater. Sci. Eng. A*, vol. 721, pp. 215–225, 2018, doi: https://doi.org/10.1016/j.msea.2018.02.014.

85. J. Strößner, M. Terock, and U. Glatzel, "Mechanical and microstructural investigation of nickel-based superalloy IN718 manufactured by selective laser melting (SLM)," *Adv. Eng. Mater.*, vol. 17, no. 8, pp. 1099–1105, 2015, doi: https://doi.org/10.1002/adem.201500158.

86. B. Ren, M. Zhang, C. Chen, X. Wang, T. Zou, and Z. Hu, "Effect of heat treatment on microstructure and mechanical properties of Stellite 12 fabricated by laser additive manufacturing," *J. Mater. Eng. Perform.*, vol. 26, no. 11, pp. 5404–5413, 2017, doi: https://doi.org/10.1007/s11665-017-2984-0.

87. M. Alimardani, V. Fallah, A. Khajepour, and E. Toyserkani, "The effect of localized dynamic surface pre-heating in laser cladding of Stellite 1," *Surf. Coatings Technol.*, vol. 204, no. 23, pp. 3911–3919, 2010, doi: https://doi.org/10.1016/j.surfcoat.2010.05.009.

88. G. C. Li, J. Li, X. J. Tian, X. Cheng, B. He, and H. M. Wang, "Microstructure and properties of a novel titanium alloy Ti-6Al-2V-1.5Mo-0.5Zr-0.3Si manufactured by laser additive manufacturing," *Mater. Sci. Eng. A*, vol. 684, pp. 233–238, 2017, doi: https://doi.org/10.1016/j.msea.2016.11.084.

89. Y. Zheng, R. E. A. Williams, S. Nag, R. Banerjee, H. L. Fraser, and D. Banerjee, "The effect of alloy composition on instabilities in the β phase of titanium alloys," *Scr. Mater.*, vol. 116, pp. 49–52, 2016, doi: https://doi.org/10.1016/j.scriptamat.2016.01.024.

90. B. Dutta and F. H. (Sam) Froes, "The Additive Manufacturing (AM) of titanium alloys," *Met. Powder Rep.*, vol. 72, no. 2, pp. 96–106, 2017, doi: https://doi.org/10.1016/j.mprp.2016.12.062.

91. Z. Jiao et al., "The effects of Zr contents on microstructure and properties of laser additive manufactured Ti-6.5Al-3.5Mo-0.3Si-xZr alloys," *J. Alloys Compd.*, vol. 745, pp. 592–598, 2018, doi: https://doi.org/10.1016/j.jallcom.2018.02.079.

92. X. Y. Zhang, G. Fang, S. Leeflang, A. J. Böttger, A. A. Zadpoor, and J. Zhou, "Effect of subtransus heat treatment on the microstructure and mechanical properties of additively manufactured Ti-6Al-4V alloy," *J. Alloys Compd.*, vol. 735, pp. 1562–1575, 2018, doi: https://doi.org/10.1016/j.jallcom.2017.11.263.

93. D. Herzog, V. Seyda, E. Wycisk, and C. Emmelmann, "Additive manufacturing of metals," *Acta Mater.*, vol. 117, pp. 371–392, 2016, doi: https://doi.org/10.1016/j.actamat.2016.07.019.

94. K. Yamanaka, W. Saito, M. Mori, H. Matsumoto, and A. Chiba, "Preparation of weak-textured commercially pure titanium by electron beam melting," *Addit. Manuf.*, 2015, doi: https://doi.org/10.1016/j.addma.2015.09.007.

95. B. E. Carroll, T. A. Palmer, and A. M. Beese, "Anisotropic tensile behavior of Ti-6Al-4V components fabricated with directed energy deposition additive manufacturing," *Acta Mater.*, 2015, doi: https://doi.org/10.1016/j.actamat.2014.12.054.

96. H. M. Flower, "Microstructural development in relation to hot working of titanium alloys," *Mater. Sci. Technol.*, vol. 6, no. 11, pp. 1082–1092, 1990, doi: https://doi.org/10.1179/mst.1990.6.11.1082.

97. G. Lütjering and J. C. Williams, "Special properties and applications of Titanium," Titanium, 2007, pp. 383–415.

98. T. Ahmed and H. J. Rack, "Phase transformations during cooling in α+β titanium alloys," *Mater. Sci. Eng. A*, vol. 243, no. 1–2, pp. 206–211, 1998, doi: https://doi.org/10.1016/S0921-5093(97)00802-2.

99. J. Tong, C. R. Bowen, J. Persson, and A. Plummer, "Mechanical properties of titanium-based Ti–6Al–4V alloys manufactured by powder bed additive manufacture," *Mater. Sci. Technol. (United Kingdom)*, vol. 33, no. 2. pp. 138–148, 2017, doi: https://doi.org/10.1080/02670836.2016.1172787.

100. E. A. Jägle, P. P. Choi, J. Van Humbeeck, and D. Raabe, "Precipitation and austenite reversion behavior of a maraging steel produced by selective laser melting," *J. Mater. Res.*, vol. 29, pp. 2072–2079, 2014, doi: https://doi.org/10.1557/jmr.2014.204.

101. F. Palm and K. Schmidtke, "Exceptional grain refinement in directly built up Sc-modified AlMg-alloys is promising a quantum leap in ultimate light weight design," In ASM Proceedings of the International Conference: Trends in Welding Research, 2013.

102. G. L. Knapp et al., "Building blocks for a digital twin of additive manufacturing," *Acta Mater.*, vol. 135, pp. 390–399, 2017, doi: https://doi.org/10.1016/j.actamat.2017.06.039.

103. T. Amine, J. W. Newkirk, and F. Liou, "Investigation of effect of process parameters on multilayer builds by direct metal deposition," *Appl. Therm. Eng.*, vol. 73, pp. 500–511. 2014, doi: https://doi.org/10.1016/j.applthermaleng.2014.08.005.

104. K. Shah, I. ul Haq, A. Khan, S. A. Shah, M. Khan, and A. J. Pinkerton, "Parametric study of development of Inconel-steel functionally graded materials by laser direct metal deposition," *Mater. Des.*, vol. 54, pp. 531–538, 2014, doi: https://doi.org/10.1016/j.matdes.2013.08.079.

105. C. P. Paul, A. Jain, P. Ganesh, J. Negi, and A. K. Nath, "Laser rapid manufacturing of Colmonoy-6 components," *Opt. Lasers Eng.*, vol. 44, pp. 1096–1109, 2006, doi: https://doi.org/10.1016/j.optlaseng.2005.08.005.

106. M. Javidani, J. Arreguin-Zavala, J. Danovitch, Y. Tian, and M. Brochu, "Additive manufacturing of AlSi10Mg alloy using direct energy deposition: Microstructure and hardness characterization," *J. Therm. Spray Technol.*, vol. 26, pp. 587–597, 2017, doi: https://doi.org/10.1007/s11666-016-0495-4.

107. C. Yan, L. Hao, A. Hussein, P. Young, J. Huang, and W. Zhu, "Microstructure and mechanical properties of aluminium alloy cellular lattice structures manufactured by direct metal laser sintering," *Mater. Sci. Eng. A*, vol. 628, pp. 238–246, 2015, doi: https://doi.org/10.1016/j.msea.2015.01.063.

108. J. S. Keist and T. A. Palmer, "Development of strength-hardness relationships in additively manufactured titanium alloys," *Mater. Sci. Eng. A*, vol. 693, pp. 214–224, 2017, doi: https://doi.org/10.1016/j.msea.2017.03.102.

109. B. K. Foster, A. M. Beese, J. S. Keist, E. T. McHale, and T. A. Palmer, "Impact of interlayer dwell time on microstructure and mechanical properties of Nickel and Titanium alloys," *Metall. Mater. Trans. A Phys. Metall. Mater. Sci.*, vol. 48, pp. 4411–4422, 2017, doi: https://doi.org/10.1007/s11661-017-4164-0.

110. S. M. Kelly and S. L. Kamper, "Microstructural evolution in laser-deposited multilayer Ti-6Al-4V builds: Part 1. Microstructural characterization," *Metall. Mater. Trans. A Phys. Metall. Mater. Sci*, vol. 35, pp. 1861–1867, 2004, doi: https://doi.org/10.1007/s11661-004-0094-8.

111. N. Hrabe and T. Quinn, "Effects of processing on microstructure and mechanical properties of a titanium alloy (Ti-6Al-4V) fabricated using electron beam melting (EBM), part 1: Distance from build plate and part size," *Mater. Sci. Eng. A*, vol. 573, pp. 264–270, 2013, doi: https://doi.org/10.1016/j.msea.2013.02.064.

112. H. Galarraga, R. J. Warren, D. A. Lados, R. R. Dehoff, M. M. Kirka, and P. Nandwana, "Effects of heat treatments on microstructure and properties of Ti-6Al-4V ELI alloy fabricated by electron beam melting (EBM)," *Mater. Sci. Eng. A*, vol. 685, pp. 417–428, 2017, doi: https://doi.org/10.1016/j.msea.2017.01.019.

113. A. R. Nassar, J. S. Keist, E. W. Reutzel, and T. J. Spurgeon, "Intra-layer closed-loop control of build plan during directed energy additive manufacturing of Ti-6Al-4V," *Addit. Manuf.*, vol. 6, pp. 39–52, 2015, doi: https://doi.org/10.1016/j.addma.2015.03.005.

114. P. Wang, X. Tan, M. L. S. Nai, S. B. Tor, and J. Wei, "Spatial and geometrical-based characterization of microstructure and microhardness for an electron beam melted Ti-6Al-4V component," *Mater. Des.*, vol. 95, pp. 287–295, 2016, doi: https://doi.org/10.1016/j.matdes.2016.01.093.

115. M. Jamshidinia, M. M. Atabaki, M. Zahiri, S. Kelly, A. Sadek, and R. Kovacevic, "Microstructural modification of Ti-6Al-4V by using an in-situ printed heat sink in Electron Beam Melting® (EBM)," *J. Mater. Process. Technol.*, vol. 226, pp. 264–271, 2015, doi: https://doi.org/10.1016/j.jmatprotec.2015.07.006.

116. S. Gorsse, C. Hutchinson, M. Gouné, and R. Banerjee, "Additive manufacturing of metals: a brief review of the characteristic microstructures and properties of steels, Ti-6Al-4V and high-entropy alloys," *Sci. Technol. Adv. Mater.*, vol. 18, pp. 584–610, 2017, doi: https://doi.org/10.1080/14686996.2017.1361305.

117. J. J. Lewandowski and M. Seifi, "Metal Additive Manufacturing: A Review of Mechanical Properties," *Annu. Rev. Mater. Res.*, 2016, vol.46, pp. 151–186, doi: https://doi.org/10.1146/annurev-matsci-070115-032024.

118. "MMPDS-02: Metallic Materials Properties Development and Standardization," Anti-Corrosion Methods Mater., 2007, doi: 10.1108/acmm.2007.12854eac.002.

119. R. Molaei and A. Fatemi, "Fatigue Design with Additive Manufactured Metals: Issues to Consider and Perspective for Future Research," in *Procedia Engineering*, vol. 213, pp. 5–16, 2018, doi: https://doi.org/10.1016/j.proeng.2018.02.002.

120. S. Leuders, T. Lieneke, S. Lammers, T. Tröster, and T. Niendorf, "On the fatigue properties of metals manufactured by selective laser melting - The role of ductility," *J. Mater. Res.*, vol. 29, pp. 1911–1919, 2014, doi: https://doi.org/10.1557/jmr.2014.157.

9

Additive Manufacturing of Metal Matrix Composites

Learning Objectives

At the end of this chapter, it is expected that you would:

- Understand the concept of Metal Matrix Composites (MMCs).
- Learn about applications of AM technology in the fabrication of MMCs and possible challenges.
- Differentiate categories of MMCs in AM, e.g., ferrous matrix composites, titanium matrix composites, aluminum matrix composites, nickel matrix composites.
- Understand the major factors affecting the properties of AM-made MMCs.

9.1 Introduction

In recent years, the application of lightweight materials in modern technology expands because of their excellent mechanical properties. Among them, the lightweight aluminum (Al) and titanium (Ti) alloys with relatively low wear and hardness values have limited applications in their extensive use in the automotive and aerospace sectors. To expand their applications, these alloys are reinforced with strengthening materials, making them capable of meeting industrial requirements. The particulate-reinforced alloys are known as composites, defined as combining two or more basic materials with considerably different physical and chemical properties. Combining the two or more constituents produces a new material with very different characteristics

Metal Additive Manufacturing, First Edition. Ehsan Toyserkani, Dyuti Sarker, Osezua Obehi Ibhadode,
Farzad Liravi, Paola Russo, and Katayoon Taherkhani.
© 2022 John Wiley & Sons Ltd. Published 2022 by John Wiley & Sons Ltd.

from the individual components. As the individual component remains distinct and separates in the final structure, composites can be distinguished from the mixture of materials and solid solutions. Generally, composites are comprised of three parts; (i) matrix as an incessant phase, (ii) reinforcing elements as inhomogeneous or distributed phases, and (iii) an interfacial boundary or binder. Through the appropriate selection of component materials and manufacturing techniques, the mechanical performance of composites as well as their material and structural characteristics can be selectively designed. The new material may be selected for desirable properties such as lower thermal expansion coefficient, higher wear resistance, greater compressive strength, tensile strength, flexibility, and hardness.

9.2   Conventional Manufacturing Techniques for Metal Matrix Composites (MMCs)

There is a wide variety of manufacturing techniques available for MMCs. Based on the processing temperature of the matrix material, the conventional techniques of MMCs are generalized into four as follows:

  i. **Liquid-phase methods**: mixing liquid metal matrix and ceramic reinforcements; melt infiltration; squeeze casting or pressure infiltration; reaction infiltration; and melt oxidation process.
 ii. **Solid-phase methods**: powder metallurgy techniques such as pressing followed by sintering, forging, and extrusion; higher energy and higher rate methods; diffusion bonding; and plastic deformation processes such as friction stir welding and superplastic forming.
iii. **Solid/liquid dual-phase methods**: rheo-casting/compo-casting; and variable co-deposition of multiphase materials.
 iv. **Deposition methods**: spray deposition; chemical vapor deposition (CVD); physical vapor deposition (PVD); and spray forming processes.

Among the several accessible methods to combine metals/alloys with ceramics in the production of ceramic–metal composites, the melt infiltration process is upheld as extremely easy and cost-effective. Melt infiltration techniques are fast, enabler of key features of lower-cost materials, enabler of the production of intricate shaped parts, and suitable for upscaling. This is also an appropriate technique for producing interpenetrating phase composites. This is because forcing the outer liquid material into porous ceramics is a complicated method, based on numerous physical and chemical facts. In the process of infiltration of ceramic preforms, liquid metals are flowing through the pores due to backpressure or propelled by self-supporting capillary force. The pressure applied does not help metal to wet ceramics. The pressure can be directed using advanced techniques of squeeze, gaseous, and die casting or forced castings such as electromagnetic, centrifugal, or ultrasonic. However, the threshold pressure required to force the liquid material through the pores depends on their surface tension, wetting angle, and pore size. For the reactive liquid melt infiltration, the reactant oxide transforms to product oxide, which influences the volume fraction of metallic phases in the resulted composites, estimating from the available space. Therefore, liquid melt techniques tend to accomplish composites with considerably higher ceramic fractions. Moreover, preform porosities are adapted carefully to develop composites with preferred phase composition. As the infiltration process requires time and reaction based on the preform size and porosity, short-length infiltration contributes to better shaping microstructure development and composite properties.

These conventional methods are employed to fabricate simple geometrical MMC parts, which are not suitable for intricate design. Moreover, composites are fabricated using conventional methods (i.e., casting and powder metallurgy) where the key concerns are related as follows: (i) in conventional castings, the requirements of molds, as well as post-process techniques (heat treatment/machining), make the process costly, (ii) the lower operating temperature creates poor interfacial bonds within reinforced elements and the matrix that may weaken the properties of the composites, (iii) microstructural inconsistency is another limitation because of the inhomogeneous distribution of reinforcing materials, which is the result of Van der Waals forces among the nearby particles. Therefore, it is necessary to adopt advanced processing techniques to obtain a uniform distribution of ceramic and metallic phases, improved grain morphology and interface structure, thereby creating superior physical and mechanical properties compared with the traditional composites.

9.3   Additive Manufacturing of Metal Matrix Composites (MMCs)

Generally, MMCs are fabricated using metals or alloys strengthened with elements of different materials targeting higher specific strength and toughness, greater thermal steadiness, superior wear resistance, enhanced fatigue property, and the potential to hold those properties for specific applications. Since AM is accomplished with multidirectional and free form production, it could be a promising manufacturing technique for multifaceted performs in the composite area. The possibility of fabricating composites in AM techniques creates the opportunity for their use in regions where lighter and lower cost materials both have the preference, such as in the automotive, aerospace, and biomedical sectors.

Figure 9.1 reviews the general AM methods of MMCs. Based on the nature of the reaction within phases throughout composites' fabrication, the techniques can be generalized in two significant classes: 1) ex-situ, and, 2) in-situ methods. In some applications, deposition processes are employed to coat the reinforcing elements, which are made by e.g. the powder metallurgy process. In the ex-situ of nonreactive process, the reinforcements are either premixed with the matrix powder or fed distinctly to the melt during the AM process. The fabrication method contains reactive processes of in-situ MMCs between the reinforcing elements and the preliminary materials. These reactions give the probability of forming dense products with desired microstructures and properties for higher-performing applications. Compared with the ex-situ composites, filler materials of in-situ composites are smaller, thermally steady, and homogeneously dispersed throughout the matrix. Therefore, in-situ composites are better consistent as they are not absorbing gas or having oxidation, which may improve interfacial strength and enhance mechanical properties of the composite products.

The design freedom accompanying the AM techniques exhibits an excellent capability to manufacture novel structures with unique geometry. Moreover, this technique allows manufacturers to integrate reinforcing materials into the structure to fabricate parts with the required properties. Compared with the conventional castings, the laser-processed composite melt possesses nonequilibrium solidification due to the faster heating and cooling on a localized area, which eventually causes the finer structure and consistent distribution of strengthening materials. In addition, AM can decrease operating times and expense through the manufacturing of

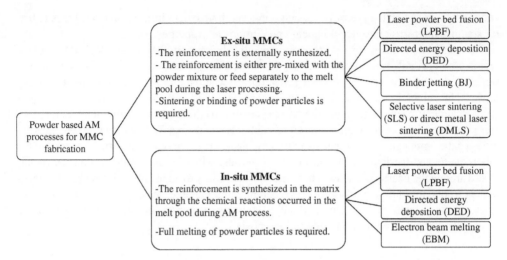

Figure 9.1 The production route of metal matrix composites in AM. Source: Reprinted by permission from [1].

near-net shapes by least machining and permitting the fabrication of composites with intricate geometry. Therefore, AM-fabricated MMC parts are excellent candidates to meet the requirements in the automotive, aerospace, and other sectors. Most of the AM techniques allow the production of MMCs to strengthen with particles or fibers. On a different note, the AM sheet lamination method is not suitable for producing complex-shaped MMCs. This chapter emphasizes the fabrication of MMCs through powder-based AM techniques.

9.4 AM Challenges and Opportunities

In heat-based AM processes where the heat source could be a laser, EBM, plasma, etc., the robust Marangoni convection generated by higher thermal capillary forces results in instability in the liquid metal. The melt flow is usually possessed with gravity force, buoyancy force, and surface tension of the melted region. Subsequently, the redistribution of strengthening materials and their consequent dispersions are drastically affected by the liquid flow.

The filler/reinforcing materials, present all over the matrix, are usually concentrated near the solidification cell boundaries, which develop higher filler content areas such as honeycomb morphology. The well-distributed higher-melting particulate fillers can act as nucleation sites causing grain refinement. According to the composite hypothesis, well-distributed nano-fillers have a significant effect by increasing nucleation sites and creating equiaxed grains, which show weak crystallographic texture in contrast with the unreinforced materials. Therefore, through the increased nucleation, the columnar structures in composites do not prolong over the layer.

The volume of the reinforcing elements has a substantial impact on the composites' size and shape. Moreover, different morphology of reinforcing elements such as globular, rod, flower-shaped, cubic, and dendritic structures can occur based on the heat source parameters. For example, for a laser-based moving heat source, the various range of laser power, intensity distribution, spot size, powder feed rate, scanning strategies, and scanning speed will influence the reinforcing elements that could be dissolved partly, entirely, or even remain undissolved throughout the matrix. Therefore, the process input parameters demonstrate an important factor in the characteristics of composites and the interfacial strength within the reinforcing materials and matrix.

MMCs reinforced with ceramic particulates are not always in the preferred form. There are still challenges on several factors such as porosity, instability in the melt pool, post-processing expenses, which need to be addressed. Moreover, the variation in physical and mechanical properties between particulates and matrix may generate micro-cracks at the particulate/matrix interface. These cracks deteriorate the functionality of the manufactured parts and, in turn, initiate failure.

Usually, the reinforcing elements are directly mixed with matrix materials to develop ex-situ composites. However, in those cases, the boundary within the reinforcements and the composite matrix creates a weaker zone due to the thermal coefficient mismatch. Moreover, when ceramic filler materials are added during ex-situ techniques, thin oxide films are produced on the reinforcements' exteriors, which deteriorate the bonding strength with the matrix and eventually initiate cracking.

9.5   Preparation of Composite Materials: Mechanical Mixing

The method of composite manufacturing using AM can be accomplished through either the material deposition or a hybrid process, where the mixture of various materials can be done before AM. A uniform spreading of the reinforcing materials in a mixture of matrix-reinforcing elements is essential for its consistent spreading throughout the bulk composite.

Mechanical mixing/alloying is a process of developing metal/alloy-reinforced powders, which is afterward used in various manufacturing techniques of compression molding, die casting, extrusion, remelting, and solidification. The dispersion of reinforcing elements depends on the processing techniques and the size of reinforcing elements. The mechanical alloying results in finer grains in a solid state by incorporating particles that enhance the solid solubility range and accomplish a distinct nonequilibrium condition.

The alloy powder properties and chemical compositions can be improved through the appropriate grinding process. Several factors affect the resulting particle size and properties, such as type of milling machine, particle size, the ratio of balls and powders, milling speed, and milling duration. The particle size in mechanical alloying is important as it governs the homogeneity of alloying within two materials. However, the particle refinement and uniformity of the composite structure depend on time, mechanical energy, and the material's strain hardening. Generally, the reinforcing materials are smaller to enable agglomeration with the matrix material. During agglomeration, the mechanical bonding between the particles may help to enter its lattice and eventually limit the dislocation motion, thereby intensifying the strength of the material.

In mechanical alloying, the rotational speed of the ball mill influences the milling process through the transmission of kinetic energy to powder particles from grinding balls during the collision of balls with powders. The kinetic energy Ekin (J) is calculated by Abdellaoui and Gaffet [2], on the basis of dynamic fundamentals as follows:

$$E_{kin} = \frac{1}{2}m_b v_c^2 \tag{9.1}$$

where m_b is the mass of the ball (kg) and v_c the collision velocity (m/s).

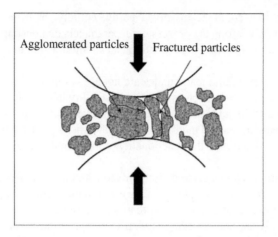

Figure 9.2 Schematic illustration of the collision between grinding ball and powder particles during mechanical alloying. Source: Reprinted from [3] with permission from Elsevier.

This technique uses a high-erosion mill to enable the process, where higher rotational speed through milling and extended impact during alloying results in hardening of the powder particles, thus distorting grain structure. A schematic presentation of mechanical alloying of powder mixtures is shown in Figure 9.2, where the rotational motion of powder particles and grinding balls results in agglomeration and fracture of particles. The rigorous plastic deformation from mechanical alloying develops different types of equilibrium, nonequilibrium, and the mixture of phases containing properties of distinct materials [4].

During the mechanical alloying of ductile powders, cold welding is a phenomenon that may happen due to severe plastic deformation. Therefore, lubricants, which act as surface-active agents, are introduced to the powder mixture at the time of the milling process to reduce powder agglomeration. Moreover, the lubricants can reduce the surface tension of the particles. In the milling process, the energy E required for particle size reduction is presented as:

$$E = \gamma_E \cdot \Delta S \tag{9.2}$$

where γ_E is the specific surface energy (J/m^2) and ΔS is the expansion of the surface area (m^2). The requirement of lower surface energy emphasizes a shorter milling period and results in finer particles.

The movement of the powder materials in the mills depends on the rotational speed of the ball mill. In the milling process, the grinding balls are located at the mill liner because of inertial and centrifugal forces. During the rotation of the mills, grinding balls are moved to some height and fall under gravity. The dropping of the grinding balls results in the crushing of the particles. To understand the impact of the dropping phenomena of powder particles, the kinematic trajectory motion is sketched in Figure 9.3.

In this sketch, some assumptions are as follows: r_b is the ball mill radius (m), A is a particle point, α is the separating angle (angle within OA and vertical direction), ω is the rotating speed (rad/s), and t_0–t_5 is the various times in the trajectory (s). For a particle at point A, the centrifugal force F is equal to the gravitational force in the opposite direction. Therefore, considering the gravity and ignoring the particle friction, F can be presented as:

$$F = m_p g \cos \alpha \tag{9.3}$$

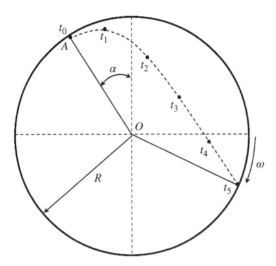

Figure 9.3 Geometrical presentation of the particle motion trajectory in an outer layer of the ball mill. Source: Reprinted from [5] with permission from Elsevier.

where m_p is the mass of the particles (kg), and g is the gravitational acceleration (m/s^2). In addition, considering the centrifugal force concept, F can be presented as (N):

$$F = m_p v_p^2 / r_b = m_p r_b \omega^2 \tag{9.4}$$

where v_p is the velocity of the particle (m/s). From Eq. (9.3) and (9.4), the rotational speed is as follows:

$$\omega = \sqrt{\frac{g}{r_b}} \cos \alpha \tag{9.5}$$

Therefore, the separating angle α, can be calculated as:

$$\alpha = \arccos \frac{r_b \omega^2}{g} \tag{9.6}$$

The aforementioned equations illustrate that the separating angle depends on the rotational speed and radius of the mill and the gravitational acceleration. The higher separating angle because of greater rotational speed may cause slipping of the particles from the ball mill interior affecting the mixing/alloying process.

9.6 Different Categories of MMCs

The physical and mechanical performance of MMCs is influenced by properties of matrix materials, distribution of reinforcing elements, the bond strength between matrix materials and reinforcing elements, and obviously the manufacturing methods. Several AM techniques exist to manufacture MMCs, which meet extensive application containing different matrix materials and reinforcements. AM processing of MMCs creates new prospects for the manufacture of multifunctional composites in contrast with conventional techniques. The following section will present various particulate reinforced ferrous, titanium, aluminum, and nickel matrix composites manufactured through AM techniques.

9.7 Additive Manufacturing of Ferrous Matrix Composites

Generally, MMCs fabricated by AM techniques are narrowed down to ferrous alloys as 316 stainless steel (SS) and H13 tool steel [6]. For composites of ferrous alloys, the most used strengthening elements are the phases of carbides (SiC, TiC, WC), nitrides (TaN, TiN), borides (TiB, TiB_2, WB), metal oxides (Al_2O_3), and carbon fibers. Among these reinforcing elements, TiC and TiB_2 do not consider the relative density of the printed composites such as Hydroxyapatite (HA) as adverse. The presence of second-phase carbides within the ferrous alloy matrix can slow down the grain growth and ensure thermal stability, which enhance mechanical properties. The manufacturing techniques and properties of several ferrous matrix composites are described further:

9.7.1 316 SS-TiC Composite

316L is a promising austenitic SS to manufacture composites because of the ductile matrix property. Overall, it is an attractive material to be reinforced with strengthening particles. Among the different types of reinforcing elements, TiC is most widely applied with ferrous alloys due to their higher melting temperature, lower density, better corrosion property, higher hardness, thermodynamic characteristics, and thermal steadiness. The hardness and the wear properties of MMCs vary on different structural constraints such as volume fraction and magnitude of the filler materials, dispersion of the inserted elements, and interfacial joining within the matrix and the filler material. If there are no defects during manufacturing, the higher volume of strengthening elements results in enhanced/increased density, hardness, and the composites' elastic modulus.

9.7.1.1 Solidification Phases of 316 SS-TiC Composite: Theory of Fabrication

The reaction within Ti and C to form TiC is considered by the higher activation energy. This reaction takes place vigorously because of their exothermic nature. The possible reaction of TiC formation and the associated variations in the enthalpy and Gibbs free energy are thought to communicate with two temperature extents as illustrated as follows [7]:

$$Ti + C \rightarrow TiC \tag{9.7}$$

When $T < 1939$ K,

$$\Delta H = -184\,571.8 + 5.042T - \left(2.425 \times 10^{-3}T^2\right) - \left(1.958 \times \frac{10^6}{T}\right) \tag{9.8}$$

$$\Delta G^{\circ} = -184\,571.8 + 41.382T - 5.042T \ln T + \left(2.425 \times 10^{-3}T^2\right) - \left(9.79 \times \frac{10^5}{T}\right) \quad (9.9)$$

and when $T \geq 1939$,

$$\Delta H = -160\,311.5 + 24.79T - \left(2.732 \times 10^{-3}T^2\right) - \left(1.862 \times \frac{10^6}{T}\right) \quad (9.10)$$

$$\Delta G^{\circ} = -160\,311.5 - 186.9T - 24.79T \ln T + \left(2.732 \times 10^{-3}T^2\right) - \left(9.31 \times \frac{10^5}{T}\right) \quad (9.11)$$

The negative value of ΔG° in the aforementioned equation implies a higher chance of TiC formation.

The strengthening element TiC can be produced in situ through direct reaction within elemental material Ti and graphite. Because of the reaction of Ti with other materials at higher temperatures, in-situ formed strengthening elements comprising Ti can be developed instantly.

The formation mechanism of the in-situ 316L-TiC composite is schematically depicted in Figure 9.4. Throughout the layer-by-layer laser scanning process, the feedstock materials are heated vigorously due to higher energy immersed. Subsequently, at higher operating temperature, the phases in 316 steel and Ti become melted. In the liquid melt, when the amount of molten Ti is significantly greater, Ti and C react exothermically because of the materials' exothermic character and the higher temperature made by greater energy density that stimulates the reaction process.

Throughout the solid–liquid reaction, TiC nucleates because of Ti diffusion into C, which then precipitates as TiC above temperature 1403 K. When the laser beam takes off from the scanning, the liquid melt will start to solidify quickly. Moreover, the solubility drops with the temperature fall, and many small embryos develop from the melt by reciprocal influence within Ti and C particles. These embryos play as the heterogeneous nucleation sites to speed up TiC nucleation through rapid solidification. TiC, which has a higher melting temperature of

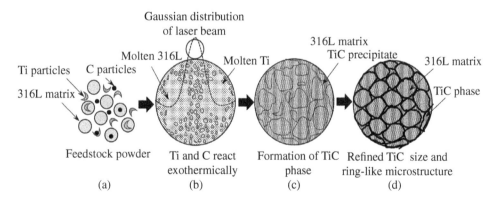

Figure 9.4 Formation mechanism of TiC reinforced 316L matrix composite, (a) feedstock powder, (b) Ti and C react exothermically, (c) formation of TiC phase, (d) refined TiC size and ring-like microstructure. Source: Reprinted from [7] with permission from Elsevier.

3480 K, forms by combining C with Ti in the liquid melt. By that time, precipitation or nucleation of TiC advances continuously through grain-boundary diffusion and develops in-situ TiC strengthened 316L SS composite.

In laser-based AM processes, TiC reinforcing elements are observed at the grain boundaries as well as inside the columnar dendrites. This is because of the segregation of TiC elements, which happens due to a higher solidification rate. The generation of strong chaotic flows within the liquid melt results in a comparative intensification of heavy particles among the nearby turbulences. Moreover, chemical inhomogeneity can influence the concentration contour, whereas the constitutional supercooling of the grown structures may result in variation in the solidification approach and therefore discrepancy in the volume of TiC precipitates. In addition, small interaction time and rapid heat transfer rates can generate less amount of TiC precipitates. Through the multiple scanning of the laser over the previously melted layers, the higher energy accumulated due to prolong heating can increase the volume of TiC.

9.7.2   316 SS-TiB₂ Composite

In 316–TiB$_2$ composites, the reinforcing materials are distributed homogeneously throughout the matrix, which also join each other and form an uninterrupted ring-like feature. The faster cooling accompanying the LPBF technique impacts the ultimate composite structure. The higher cooling rate as 10^6 K/s constrains the TiB$_2$ grain growth because of the insufficient time for grain advancing. This results in the most demanded finer TiB$_2$ particles dispersion throughout the matrix. On the other hand, the strong temperature gradient perceived through the laser AM process develops a surface tension gradient among the middle and the periphery of the melt. This gradient introduces Marangoni convection to expedite the relocation of TiB$_2$ particles by inhibiting their accumulation and regulating their distribution across the solidified matrix. As known, TiB$_2$ melts incompletely, whereas the 316 matrix melts completely. The forces applied to TiB$_2$ elements will move them in the direction of Marangoni flow. With the adequate melting of matrix materials, the repulsive force generates within the TiB$_2$ particles. The combination of repulsive force and the Marangoni convection leads to developing an uninterrupted ring-like feature, where TiB$_2$ elements are homogeneously dispersed throughout the matrix.

9.7.3   H13–TiB₂ Composite

The application of laser AM in the production of tools facilitates the accomplishment of intricate shaped parts using the digital method sequence. Moreover, the implementation of AM minimizes the cost of manufacturing tools, cuts the production period, and lowers manpower because of robotic application.

The laser-processed H13–TiB$_2$ composite is distinguished with α-Fe and TiB$_2$ phases, except austenite. The development of these structures is believed to influence the Gaussian distribution of heat during laser melting. The faster heating and solidifying cycle encourages fine equiaxed grains with consistent distribution of TiB2 reinforcing elements along the grain boundaries of the H13 matrix. During the laser melting, where a complete liquid formation occurs, the strengthening phases are generated by the dissolution mechanism, defined as the heterogeneous nucleation of TiB$_2$ and consequent grain development.

9.7.4   H13–TiC Composite

The faster heating and cooling cycle of the laser helps TiC structure occurrence by reducing the time required for TiC grains to grow. At higher temperatures, the intensified Marangoni flow brings capillary forces causing the liquid to flow. Therefore, the shear and rotational forces that induce around the TiC particles play a possible driving force for homogeneous dispersion of particles, which inhibits agglomeration. In addition, at lower volumetric energy density, the torque is decreased, which increases clustering in particles.

9.7.5   Ferrous–WC Composite

MMCs, strengthening with ceramic particles, have been fabricated using different AM processes that have gained interest in AM ceramic arenas. Similarly, ceramic particles-reinforced ferrous composites show better performance. WC is reflected as a high-quality ceramic reinforcing element having a higher melting temperature and excellent wettability with many ferrous alloys. Additionally, WC can keep the room temperature hardness as high as 1400 C. Obviously, the selection of reinforcing elements and the matrix is vital for increasing wear resistance; thus, the interface can play a significant role. As structural materials, the Fe-based alloys are reinforced with ceramic elements, which is a promising approach to improve the wear property.

The cross section of a WC-reinforced ferrous matrix composite is shown in Figure 9.5, where eutectic phases are noticed between grains. In Fe–WC composite, a significant amount of WC reinforcing elements are dissolved in the Fe-based matrix and release W and C atoms in the liquid melt. Some released W and C atoms react with ferrous alloy and form carbides, which distribute near the grain boundaries. In addition, the gradient interface is known as MC_3 (M = W, Fe, Cr, Ni), which develops between WC reinforcing element and the matrix because of the in-situ reaction during laser AM techniques. It is acceptable to believe that the development of gradient interface helps advance stronger bond strength between WC and the Fe-matrix. Moreover, the interface gradient's size and shape change considerably with the laser input parameters, such as laser power, intensity, and spot size.

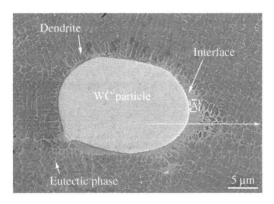

Figure 9.5   Microstructure showing the morphology of matrix, interface, and WC reinforcement in MMC. Source: Reprinted from [8] with permission from Elsevier.

Fe–WC composite's performance mechanism has a comprehensive link concerning densification, gradient interface, microstructural morphology, and hardness development in several ways. Firstly, when the densification is low at a higher scanning speed, there is a possibility for the material to be removed from the matrix due to the abundance of pores. However, the lower scan speed can result in higher densification, hence resulting in superior wear property. Secondly, the developed gradient interface from the in-situ reaction plays a significant role in enhancing wear property. It is well understood that the degradation of wear performance in the composite is the reason for the weak interfacial bond between reinforcing elements and the matrix. Therefore, the interfacial layers without pores and cracks guarantee coherent bonding between composite materials. In addition, the thicker interfacial layers promote strong bonding strength, making it difficult to wear away WC elements during the application, and thereby improve wear property. Moreover, the composite's higher hardness can impede the adhesive process such as scuffling and removing material, giving rise to a lesser wear rate and better wear property.

9.7.6   Ferrous–VC Composites

The formation mechanism of ferrous matrix composites strengthens with vanadium carbides (VCs) through the LPBF process, as schematically shown in Figure 9.6. In this technique, the greater energy from the laser power source along with the pressure/flow induced in the laser-stimulated liquid melt pool is adequate to decompose micron-sized VC through a complete melting-solidifying mechanism. As described in the schematic, the laser energy quickly heats up the mixture of 316L and VC, creating a molten area of fully dissolved V_8C_7/316L liquids. The V_8C_7 component can easily decompose and release V and C, because of their finer particle size, greater surface tension, and large heat input from the laser source. In addition, the microscopic pressure working on the ultrafine VC in the liquid melt pool would accelerate the activity throughout the rapid melt process.

Successively, the melt pool goes through a faster solidification process after the laser source moves away from the melt. Throughout the rapid solidification process, the higher melting phase VC and VC_x will tend to form through heterogeneous nucleation of VC nuclei and their successive grain growth. Due to subsequent nucleation and progression of austenite grains, ultimately, the matured VC_x strengthening elements will be dispersed at grain boundaries and in austenite grains. When the VC_x phases become bulk, they will be driven toward the grain boundaries with subsequent grain growth. However, the very small VC_x phases are kept within the austenite grains. As laser AM possesses inherently a fast solidification process, few amounts of V and C will get adequate time to be diffused in austenite solid solution.

Moreover, the nucleant and the grown VC_x reinforcing elements homogeneously integrating into the 316L alloy structure restrict the growth of austenite grains. Therefore, the nucleant ceramic elements supplied resistance to grain progression due to the pinning phenomenon acting through austenite solidification. A bulk of bigger-sized VC_x elements are progressed and gathered near the austenite grain boundaries, which also block the further movement of grain boundaries.

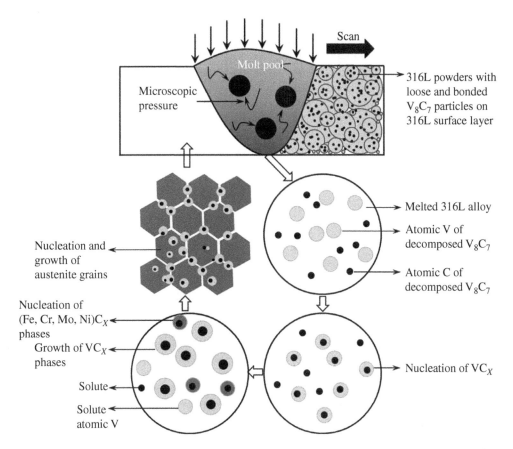

Figure 9.6 The schematic diagram shows that the formation mechanism of ferrous matrix composites strengthens with VC in LPBF technique. Source: Reprinted from [9] with permission from Elsevier.

9.8 Additive Manufacturing of Titanium-Matrix Composites (TMCs)

Extensive research programs on AM of Ti and Ti alloys are being conducted worldwide due to their outstanding characteristics. Different AM techniques are adopted to fabricate Ti-matrix composites (TMCs) for advancing the wear resistance property and high-temperature sustainability of commercially used pure Ti. Moreover, Ti intermetallic alloys do not have considerable room temperature ductility, which requires a next step for the fabrication of Ti intermetallic composites to possess improved microstructure, resulting in enhanced mechanical and thermal properties. Therefore, various reinforcing elements can be added to advance the mechanical properties by forming Ti intermetallic matrix composites.

Generally, compared with the pure Ti and Ti alloys, there exists less attention on AM of TMCs. Several processing complications make the manufacture of TMCs challenging. In AM, spherical-shaped powders are generally required for optimum processing; hence gas or plasma-atomized powders are frequently used in AM. On the other hand, very few composite powders in spherical shape are available. Moreover, the Ti-matrix and the reinforcing elements may show considerably different melting temperatures, making the laser techniques more complicated because of their dissimilar viscosities and melting/solidification characteristics. This demands the proper optimization of the melting and solidification process of Ti-matrix with reinforcing elements to fabricate composite parts with higher relative density.

The frequently used elements that offer efficient reinforcement for the production of TMCs include Cr_3C_2, TiC, TiN, TiO_2, $Si_3 N_4$, SiC, TiB_2, TiB, Al_2O_3, and Ti_5Si_3 [10]. In addition, boron particles and carbon as nanoparticles, nanotubes, and fibers are taken as effective additives to fabricate reinforcing compounds. Among the various reinforcing elements, TiC and TiB are regarded as in-situ types, while TiB_2 and SiC are considered ex-situ compounds. In-situ compounds react with the matrix to form reinforcing phases throughout the fabrication process of TMCs. The formation mechanism and properties of some TMCs with different reinforcing elements are discussed further.

9.8.1   Ti–TiC Composite

In laser-based AM of Ti–TiC composites, different contents of Ti and TiC powders are formulated by mixing and ball milling to ensure superior densification. Through the laser AM, TiC elements are formed by dissolution or precipitation mechanism, including heterogeneous nucleation of TiC and subsequent grain development. The ex-situ construction of Ti-TiC and the dispersion of TiC reinforcement depend on processing conditions, such as the functional energy density. At higher energy densities, the TiC reinforcing element disperses as a network array with dendritic morphology, whereas at lower energy densities, TiC becomes finer and distributes consistently. The lower energy density can also modify TiC shape from coarser dendritic to accumulated whisker shape and lastly as lamellar structure. These microstructural developments and refinements are connected with the higher scanning speeds, which create faster cooling rates and eventually inadequate time for the growth of TiC elements in the composites.

9.8.2   Ti–TiB Composites

Even though numerous particles are used for the reinforcements of TMCs, ceramic particles such as TiB and TiC have gained much attention because of their good chemical affinity with the matrix, which stimulates superior matrix/particle wetting and complete interfaces. TiB is a steady element, even with a very small solubility of boron in titanium (<0.001 at.%), which is far low compared with carbon (22 at.%) and nitrogen (1.8 at.%) [11]. Moreover, the quite similar thermal expansion coefficients of Ti and TiB lower the possibility of residual stresses in AM composite parts. Therefore, TiB can play a role to enhance elastic modulus and strength of composites, even with smaller additions, contrasted with other elements such as TiC, TiN, and TiB_2.

9.8.2.1   Solidification Phases of Ti–TiB Composite: Theory of Fabrication

A schematic presentation of the in-situ TiB formation through the reaction within Ti and TiB_2 during the synthesis of TMC is shown in Figure 9.7. Powder mixture shows well dispersion of

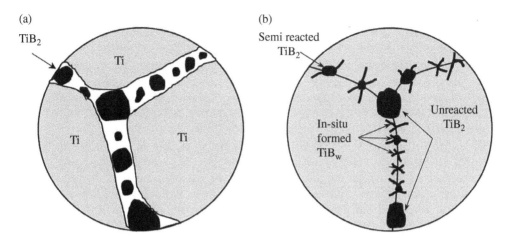

Figure 9.7 Schematic presentation for the formation of TiB phase from in-situ reaction within Ti and TiB$_2$ particles, (a) powder mixture with well dispersion of TiB$_2$, surrounded by bigger Ti powders prior sintering, (b) in-situ reaction develops needle-like TiB, semi-reacted TiB$_2$, and unreacted TiB$_2$. Source: Reprinted by permission of the publisher Taylor & Francis Ltd, [11].

TiB$_2$ particles adjoining bigger Ti powders prior to the sintering. The in-situ reaction develops TiB, together with semi-reacted TiB$_2$ and unreacted TiB$_2$.

In-situ Ti-TiB composites are manufactured from Ti-TiB$_2$ powders through the optimization of process parameters. TiB$_2$ has a greater melting temperature compared with CP-Ti, which necessitates a higher energy density. The processed Ti–TiB$_2$ composites show two phases of α-Ti and TiB$_2$. The in-situ reaction occurs between TiB$_2$ and Ti through the diffusional removal of boron (B) from TiB$_2$ or by melting and successive solidification of TiB.

$$Ti + TiB_2 \rightarrow 2TiB \tag{9.12}$$

Based on Eq. (9.12), the reaction between Ti and TiB$_2$ to form TiB is influenced when the amount of B is below the limit to form TiB$_2$ corresponding to the Ti-B phase diagram. The reinforcing element TiB with a needle-shaped morphology can distribute consistently throughout the Ti-matrix (Figure 9.7b). Due to the faster cooling rate and the presence of B content, Ti will have finer lamellar α phase morphology. Moreover, the presence of B reduces the martensitic transformation temperature; therefore, α-Ti matrix is formed without forming α' martensitic structure. A comparative study between AM-made Ti–TiB composite and conventional cast one reveals that TiB reinforcing elements in AM are finer than the cast equivalent, because of the higher cooling rate, and therefore, restrict further grain growth.

9.8.2.2 Formation Mechanism of Three-Dimensional Quasi Continuous Network of TiB

The structural morphology of TiB reinforcing element depends on the laser input parameters [12]. The three-dimensional quasi-continuous network (3DQCN) such as microstructure of TiB depends on the laser power. In addition, the lower laser power provides inadequate energy to activate robust Marangoni convection, liquid capillary force, and flowability to rearrange TiB reinforcing elements in Ti matrix. Therefore, the liquid melt pool of Ti-TiB solidifies as follows: liquid (liquid melt of Ti-TiB) → liquid + primary TiB → primary TiB + eutectic TiB + βTi→ primary TiB + eutectic TiB + αTi [13]. However, the higher laser power results in robust Marangoni convection, greater liquid capillary force, and elevated flowability to rearrange TiB elements to form QCN structure. Moreover, the increased laser power precedes with better material interaction in the melt and gives rise to eutectic solidification as follows: liquid (liquid melt of Ti-TiB) → liquid + eutectic TiB + βTi → eutectic TiB + αTi.

The formation mechanism of 3DQCN such as microstructure of TiB is schematically described in Figure 9.8, which can be classified into three steps. Step 1: laser power on the substrate material will create the melt pool, which will capture powders delivered through the powder stream. Because of the higher input energy, Ti will react with B to form TiB. The recently developed TiB, together with the other delivered powders, will be melted into the liquefied material. Step 2: When the input energy is moved to the next location, the liquid melt starts to solidify. Initially, TiB nucleates and progresses as long whiskers offering heterogeneous nucleation positions for liquid Ti. Afterward, Ti melt will nucleate heterogeneously at walls of TiB whiskers and homogeneously from liquid to develop βTi. As a result, βTi nuclei grow, pushing TiB into the liquid, which will not be submerged by βTi nuclei, due to the poor solubility of TiB in βTi. Step3: TiB whiskers accumulate at Ti grain boundaries, developing QCN-like structure. Meanwhile, βTi converts to αTi.

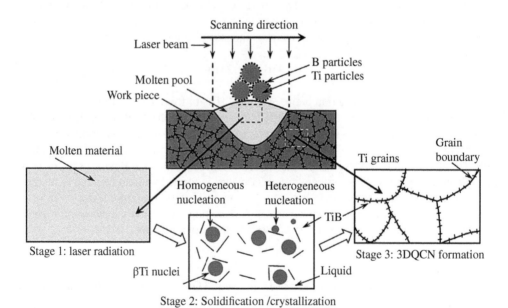

Figure 9.8 Schematic depiction of the formation mechanism of quasi-continuous network-like microstructure. Source: Reprinted from [12] with permission from Elsevier.

9.8.3   Ti–Hydroxyapatite (Ti–HA) Composites

Although Ti alloys have good properties for biomedical applications, their osseointegration still has space for potential enhancement using bioactive ceramic reinforcements such as hydroxyapatite (HA).

HA is a calcium phosphate ($Ca_{10}(PO_4)_6(OH)_2$) compound having similar composition and morphology of human bone tissue. Nano-hydroxyapatite (nHA) has gained much attention as biomaterials in prosthetic applications because of the similarity in size, crystallography, and chemical constituents such as human bone tissue.

9.8.3.1   Solidification Phases of Ti–HA Composite: Theory of Fabrication

The decomposition of HA takes place at higher temperatures and has a substantial impact on the final structure of the Ti–HA composites. The possible reactions that take place through the LPBF process are presented by the following equations [14, 15]:

$$Ca_{10}(PO_4)_6(OH)_2 \rightarrow Ca_{10}(PO_4)_6(OH)_{2-x} + xH_2O(gas) \tag{9.13}$$

$$Ti + 2H_2O(gas) \rightarrow TiO_2 + 2H_2 \tag{9.14}$$

$$Ca_{10}(PO_4)_6(OH)_2 + TiO_2 \rightarrow 3Ca_3(PO_4)_2 + CaTiO_3 + H_2 + \frac{1}{2}O_2 \tag{9.15}$$

$$Ca_3(PO_4)_2 + \frac{1}{2}O_2 + 3Ti \rightarrow 3CaTiO_3 + 2P \tag{9.16}$$

$$5Ti + 3P \rightarrow Ti_5P_3 \tag{9.17}$$

The phase formation mechanism of Ti–nHA composite in the LPBF process can be summarized as follows: After decomposition of HA, Ca/P elements react with the α phase resulting in the formation of Ti_5P_3, $Ca_3(PO_4)_2$, $CaTiO_3$, and Ti_xO phases. Moreover, the remaining phases of Ti_5P_3 and $Ca_3(PO_4)_2$ disperse randomly throughout the α matrix. The formation of $CaTiO_3$ phase yields a small amount of $Ca_3(PO_4)_2$ as residue. The low absorption rate of oxygen during the manufacturing process causes the occurrence of Ti_xO phase, rather than TiO_2. However, the oxygen from HA acts as an interstitial atom, which diffuses into Ti matrix and forms titanium oxides. A saturation then starts by faster solidification of the LPBF process and hence forms Ti_xO phase.

In CP-Ti, α structure is identified as the long lath-shaped grains. Relatively, the small acicular-shaped grains are formed with the presence of nHA into the Ti matrix. The in-situ reaction within Ti and HA can form composites with lath shapes of Ti grains to acicular and quasi-continuous spherical grains of Ti-HA, as shown in Figure 9.9. The microstructural development using HA is connected with the melt pool kinetics and the heterogeneous nucleation aspects. The lower content of HA can shorten the acicular-shaped grains without much densification. Whereas the higher amount of HA constrains the crystal growth and allows the formation of quasi-continuous spherical grains. Therefore, the microstructure of LPBF manufactured composites becomes refined progressively with the incorporation of HA. In summary, the grain morphology undergoes a unique approach as follows: long lath-shaped grains $\rightarrow$ short acicular-shaped grains $\rightarrow$ quasi-continuous circle-shaped grains.

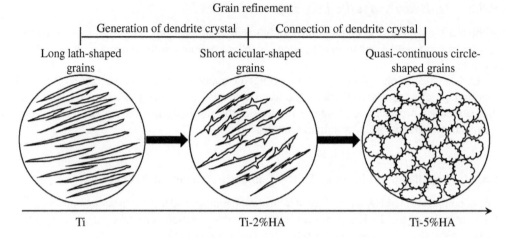

Figure 9.9 Schematic presentation of the microstructural development in pure Ti, and Ti-2%HA and Ti-5%HA composites manufactured through LPBF techniques. Source: Reprinted from [16] with permission from Elsevier.

9.8.4 Ti-6Al-4V-Metallic Glass (MG) Composites

The incorporation of Zr-based metallic glass (MG) reinforcement can create a new direction to modify the microstructure of TMCs, through stabilizing the β phase and refining both β and α′ phases; thereby can impart higher strength and better ductility in the TMC parts.

Generally, the solidification of Ti-6Al-4V liquid melt follows a planner mode, as there exists a minor temperature difference between the liquidus and the solidus points. Conversely, the addition of MG to Ti-6Al-4V alloy may interrupt the equilibrium processes, prompting a conversion from planner to dendritic solidification form. This type of conversion takes place when the liquid phase obtains a thermal gradient below the critical value near the solidification front. The critical thermal gradient $\frac{\delta T}{\delta X_{crit}}$, is presented in Eq. (9.18) [17].

$$\frac{\delta T}{\delta X_{crit}} = - \frac{C_0}{D_L/R} \frac{1-k}{k} \frac{\delta T_L}{\delta C} \tag{9.18}$$

where C_0 is the solute concentration, D_L is the solute diffusion coefficient, R is the solidification rate, and k is the solute partition coefficient, associated with the solidification limit among the liquidus and solidus temperatures. The slow diffusion coefficient of Cu and Ni in Ti melt compared with Al and V could affect the first part of Eq. (9.18). In addition, the extended liquidus/ solidus temperature difference owing to the addition of Zr/Al can impact the second part of Eq. (9.18). However, both effects can raise the critical thermal gradient, which leads to dendritic solidification of the MG and Ti-6Al-4V composite mixture. It is eminent that the dissimilarities between the density and the viscosity of Ti-6Al-4V and MG have significance effect on the subsequent solidified microstructure. Because of the different density of Ti-6Al-4V and MG melts, a gravity force will act on MG melt to settle down the melt. The dynamic viscosity, μ, for a liquid (kg/(m.s)) is presented by Eq. (9.19) [18].

$$\mu = \frac{16}{15}\sqrt{\frac{m}{\sigma_B T}}\gamma \qquad (9.19)$$

where m is the atomic mass (amu), γ is the surface tension (N/m), T is the temperature (K), and σ_B is the Boltzmann constant (J/K). Although Ti-6Al-4V and MG liquids have similar surface tension as ~1.5 N/m, the greater m value of MG results in a higher μ, with regard to Ti-6Al-4V liquid corresponding to Eq. (9.19). Theoretically, owing to a higher density and viscosity, the MG melt leans to separate from the Ti-6Al-4V melt. In addition, the β phase becomes stable by Cu and Ni and then passes through solidification, having the higher melting temperature. The schematic presentation of the Ti-6Al-4V/MG composite formation is shown in Figure 9.10.

The dendrite formation discards solute elements (Cu, Ni, and Al) through the solid/liquid interface to supercooled melt, which enriches the melt pool with MG contents leading to the initiation of amorphous phases. In addition, the faster cooling process restricts the growth of the amorphous phases, resulting in the MG as nanobands. However, the MG nanobands can pass a partial crystallization during continual heating of layer-to-layer deposition. Thereby, a hard/soft nanostructured Ti-6Al-4V/MG composite is formed with partially crystallized MG reinforcing elements (hard phase), embedded in β grains (soft phase).

9.8.5 Ti-6Al-4V + B₄C Pre-alloyed Composites

TMCs manufactured from the mixture of Ti-6Al-4V + B$_4$C powders with different volume contents of B$_4$C follow the in-situ chemical reaction of

$$5Ti + B_4C \rightarrow 4TiB + TiC \qquad (9.20)$$

The composite consists of a dual structure of α/β Widmanstätten Ti matrix, enclosing with nonuniformly distributed TiB whiskers, as obvious in Figure 9.11. On the other hand, TiC is

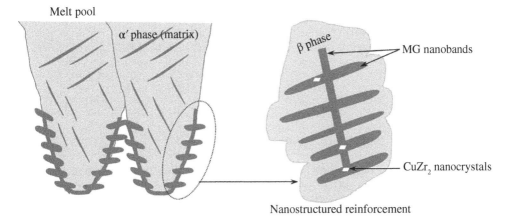

Figure 9.10 Schematic microstructure of the as-printed Ti-6Al-4V/MG composite. Source: Reprinted from [19] with permission from Elsevier.

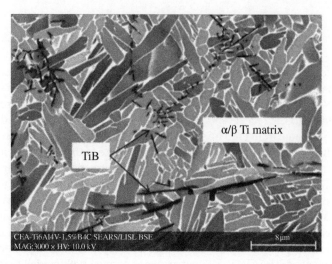

Figure 9.11 SEM micrograph showing Ti-6Al-4V- 3% B$_4$C composite. Source: Reprinted from [20] with permission from Elsevier.

difficult to detect in the composite structure because of its finer size or greater carbon solubility in α-Ti solid solution [21].

The addition of Ti$_4$B can influence the solidification process in many ways. Based on the Ti-B phase diagram, with a lower mass content of B, β-Ti can form as the first solid phase during solidification. Through the growth of β-Ti phase, boron gradually rejects to liquids and causes a micro-segregation of B near β grain boundaries, which ends with the completion of eutectic solidification. When eutectic temperature rises, fine TiB whiskers precipitate on some preferred position of β grain and stop grain growth.

Moreover, the variation of β-Ti grains may induce macro-segregation of boron at some stage of solidification. This particular refinement occurs just under the top of every additive layer, which is supposed to belong to non-melted zones. The residual segregation of boron occurs near the melt pool, resulting in a partially remelted area through the subsequent layer, which can improve the grain refinement acts.

The hardness development in Ti-6Al-4V+B$_4$C composite is influenced by the formation of TiB whiskers. Another key contributor in the hardening of this composite is carbon in α-Ti solid solution. Moreover, the network-like TiB structure around the β-Ti grains enhances the stiffness of the composite.

9.8.6   Ti-6Al-4V + Mo Composite

To manufacture TMCs, when Ti-6Al-4V powder is mixed with different percentages of Mo powder, a novel microstructure develops in the composite structure. In contrary to a complete α′ martensitic structure of Ti-6Al-4V alloy, Ti-6Al-4V-Mo composite contains β-Ti matrix with nonuniformly distributed Mo particles.

The alloying additions in Ti may stabilize α or β phase or even do some invisible control on phase equilibrium. The Mo equivalent, as presented in Eq. (9.21), can illustrate the overall effect of the β stabilizers. During the formation of Ti-6Al-4V alloy, the β phase converts to α′ martensite phase, because of faster cooling. It is already established that Mo addition can fully conceal the transformation, thereby completely retains the β phase. Mo can control the martensite transformation in two ways: primarily, Mo can drop the critical cooling rate to keep the β phase; secondarily, Mo as a β stabilizer significantly lowers the martensite start temperature.

$$Mo_{eq} = 1.0Mo + 0.67V + 0.44W + 0.28Nb + 0.22Ta + \ldots - 1.0Al \, (\text{wt.\%}) \qquad (9.21)$$

Thereby, Mo addition distracts the equilibrium of Ti-6Al-4V alloy, by expanding the solidification range. In addition, Mo has a lower diffusion coefficient in β-Ti, compared with Al and V, which results in a higher critical gradient. All these effects together create unsteadiness/ nonequilibrium in planner solidification and generate cellular solidification.

9.8.7 Structure and Properties of Different TMCs

The summary of microstructure developed in different types of AM-made TMCs and their mechanical and wear properties are listed in Tables 9.1 and 9.2, accordingly.

9.9 Additive Manufacturing of Aluminum Matrix Composites

The lightweight aluminum matrix composites (AMCs) are reinforced with hard ceramic particles showing better performance in aerospace, aircraft, and automotive industries because of higher stiffness, greater specific strength, and better wear properties. Generally, the superior

Table 9.1 Summary of microstructures formed in TMCs.

Composite matrix-vol%	AM techniques	Microstructural features
Ti-TiC	LPBF	Homogeneous distribution of TiC in Ti matrix, TiC shape is modified from coarsened dendritic to nanoscale lamellar at faster cooling rate.
Ti-TiB$_2$	LPBF	Consistent distribution of needle-like TiB in Ti matrix.
Ti-B$_4$C	LPBF	In-situ reaction takes place within Ti and B$_4$C and forms Ti-(TiB-TiC) with residual B$_4$C.
Ti-Si$_3$N$_4$	LPBF	In-situ reaction occurs within Ti and Si$_3$N$_4$, forms Ti$_5$-Si$_3$-TiN with uniform distribution of TiN. The size and shape of TiN changed with energy density from the irregular polyangular shape to spherical and finally coarsened dendrite.
Ti-SiC	LPBF	In-situ reaction between Ti and SiC and forms Ti$_5$Si$_3$-TiC. With increased energy density, TiC changed from slender dendrite to network structure and finally coarsened dendrite.
Ti-HA	LPBF	Various microstructures are observed: • long lath-type grains of Ti, • short acicular grains of Ti-2%HA, • quasi-continuous spherical grains of Ti-5%HA.

Source: Reprinted from [10] with permission from Elsevier.

Table 9.2 Mechanical and wear property of CP-Ti and different kinds of TMCs manufactured by LPBF.

Composite matrix-wt.% reinforcements	E (J/mm³)	H (GPa/Hv)	Wear rate (mm³/Nm)	Tensile strength (MPa)	Elongation (%)	References
CP-Ti	120	4.0 (GPa)/261 (Hv)	2.8×10^{-6}	757	19.5	[22] [18]
Ti-5TiC	142	—	—	914	18.3	[23]
Ti-7.5TiC	120	553 Hv	2.6×10^{-7}	—	—	[24]
Ti-12.5TiC	120	577	2.3×10^{-7}	—	—	[24]
Ti-15TiC	90–360	44.2–90.9 (GPa)	1.8×10^{-7}–7.0×10^{-7}	—	—	[25]
Ti-17.5TiC	120	487 Hv	3.75×10^{-7}	—	—	[24]
Ti-22.5TiC	120	412 Hv	6.5×10^{-7}	—	—	[24]
Ti-2CrB₂	—	320–380 Hv	7.5×10^{-4}–13.3×10^{-7}	—	—	[26]
Ti45Al2Cr5Nb-(1–3 wt. %) TiB₂	125	9.9–10.5 GPa	—	—	—	[27]
Ti45Al2Cr5Nb-1TiB₂	51–89	8.6–9.4 GPa	—	—	—	[28]
Ti-5B₄C	108–208	418–577 Hv	—	—	—	[29]
Ti-24.55Si₃N₄(Ti₅Si₃-TiN)	167–666	1083–1358 Hv	6.84×10^{-5}–9.73×10^{-5}	—	—	[30]
Ti-22.8SiC(Ti₅Si₃-TiC)	80–320	906–980 Hv	1.42×10^{-4}–2.65×10^{-4}	—	—	[31]
Ti-2HA	100	424 Hv	—	289	1	[16]
Ti-5HA	100	601 Hv	—	—	—	[16]
Ti-6Al7Nb-2vol%HA	—	501–518 Hv	—	657	—	[32]
Ti-6Al7Nb-5vol%HA	—	571–625 Hv	—	—	—	[32]

Source: Reprinted from [10] with permission from Elsevier.

mechanical properties of AMCs are affected by the size and the dispersion style of the reinforcements. It is well known that reinforcing with nanoscale particles, called nanocomposites, can result in considerably higher strength and ductility. The most common manufacturing methods of ceramic particle-strengthened AMCs are mechanical alloying, powder metallurgy, and stir casting.

For AMC, frequently used ceramic reinforcements are TiB_2, TiC, SiC, Al_2O_3, Si_3N_4, and B_4C, which follow the conventional manufacturing methods of stir casting, melt infiltration, powder metallurgy, and spray deposition. Although these methods may possess few advantages, their major limitations include poor wettability, nonuniform particle distribution, and brittle interface. These actually degrade the fatigue strength and longevity of AMC parts significantly. Moreover, fabrication of nanoscale composite with proper dispersion of reinforcements is challenging because of the Van der Waals attraction forces, which act to agglomerate particles, resulting in inhomogeneous microstructure.

However, the application of nano- or micro-particle-reinforced AMCs can diminish crack formation and offer better mechanical properties by converting the columnar grains to equiaxed grains. The homogeneous dispersion of reinforcements characterizes a proper strengthening phase, which also refines the grain. Reinforcing elements, which show good wettability, chemical stability, higher hardness and stiffness, excellent wear property, can be compatible with the aluminum matrix to manufacture composites with improved properties. In contrast with the whisker and fiber-strengthened MMCs, particle-fortified MMCs have benefits over processing routes, manufacturing expenses, microstructure shapes and distributions, which are alternatives of conventional processing techniques. However, careful care must be given to safety issues associated with the use of nanoscale particles.

9.9.1 Al–Fe₂O₃ Composite

Fe_2O_3 is another low-cost reinforcement, which minimizes the energy required to fabricate AMC in laser-based AM techniques. The in-situ reaction between the mixture of Al and Fe_2O_3 can produce Fe_3Al and Al_2O_3 in Al matrix, as highlighted in the following equation:

$$8Al + 3Fe_2O_3 \rightarrow 2Fe_3Al + 3Al_2O_3 + heat \tag{9.22}$$

The stoichiometric reaction shows that the laser beam ignites the additional heat and stimulates thermite interactions through the AM process. This exothermal excess heat of in-situ reaction lowers the laser energy required to melt powder mixtures, followed by solidifying the composite melt. The amount of Fe_2O_3 has a substantial impact on the hardness, density, and surface roughness. It can control the energy discharge in the reaction and, thereby, form the surface profile. The higher content of Fe_2O_3 does not affect the composites' densification property but can improve the hardness through excellent interfacial bonding and uniform distribution of reinforcements within the matrix.

9.9.2 AlSi₁₀Mg–SiC Composite

The selection of proper reinforcing elements for Al matrix gives the scope to design AMCs with superior strength and stiffness. For example, the lower hardness and poor wear property of Al alloys can be improved through the incorporation of hard particles such as SiC, Al_2O_3, and TiB_2 into the matrix. In the fabrication of $Al_4SiC_4 + SiC$ composite, in-situ reaction occurs between $AlSi_{10}Mg$ and SiC powder mixtures, as presented in the following equation:

$$AlSi_{10}Mg + SiC \rightarrow Al_4SiC_4 + SiC \tag{9.23}$$

The formation mechanism of Al_4SiC_4 phase through the in-situ reaction within Al and SiC includes development of irregular nuclei on the interface of Al and SiC. As the reaction continues, the nuclei start to grow and become plate-like structures. The uniform distribution of these plate-like Al_4SiC_4 structures has a key role in improving the mechanical property of AM-made composites.

In laser AM, one of the limitations that Al possesses is their lower absorption capability. The addition of SiC as a higher energy absorptive can enhance the energy absorption rate and thereby form adequate liquid with steady melt pools. Therefore, the intensified energy absorption rate, which raises the melt pool temperature, also improves the solid–liquid wettability. Moreover, SiC particle size impacts the densification of the composites, as the specific surface area between Al melt and SiC varies with the size of SiC reinforcement. With coarser SiC, there is a chance to melt fewer particles through the AM process. Therefore, the inadequate interface involving SiC and Al matrix impedes the in-situ reaction. In addition, the coarser SiC limits the specific surface area between the matrix and the reinforcement, which eventually reduces the wettability and, afterward, lowers the density of the composite. Moreover, the irregular SiC can affect the microstructural homogeneity and density of the composite. On the other hand, finer SiC can result in a higher density of the composite by accelerating the in-situ reactions between them. Therefore, the finer SiC can improve the wear performance and hardness property of the composite through the higher density and uniform microstructure.

9.9.3 $AlSi_{10}Mg$–TiC Composite

During the laser processing of AMC reinforced with TiC particles, the temperature gradient between the center and the border of the melt pool creates surface tension and subsequent Marangoni flow. All these effects make capillary forces, which act on TiC particles generating torque around the particles. The operating torque then rotates the TiC particles in the melt and thereby provides its redistribution. Furthermore, the concentration of the capillary forces is influenced by the amount of temperature gradient. The inadequate energy input may lower the temperature gradient and the associated capillary force concentration, which results in the inhomogeneous distribution of TiC particles in the melt. In addition, TiC particles are always propelled by torque and accumulate concurrently near the center of the Marangoni flow, thereby creating a ring pattern of TiC. However, when a satisfactory amount of Al melt occurs within the molten pool, the repulsive force occurs between TiC reinforcements. Therefore, the mutual effect of Marangoni flow and repulsive forces contributes to the development of the TiC ring structure in the composite matrix.

9.9.4 $2024Al$–TiB_2 Composite

In laser AM, the most focused aluminum alloys are $AlSi_{10}Mg$ and Al-12Si, which possess higher mechanical properties compared with the conventionally manufactured ones. The other

Al alloys become difficult to manufacture using AM processes because of their lower flowability and excessive reflectivity of alloy powders during laser processing. Moreover, they are susceptible to form large columnar grains through the solidification process, which deteriorates their properties by generating periodic cracks.

The particle-reinforced Al matrix composites using LPBF is an effective method to enhance the mechanical property of the Al alloys. Among various reinforcing elements, TiB_2 shows good compatibility with Al alloys having excellent wettability, hardness, stiffness, wear performance, and substantial chemical stability.

2024Al is one of the demandable alloys in aerospace applications, which has higher strength, lower density, and heat treatable property. Usually, the microstructure of 2024Al alloy consists of columnar grains aligned parallel to the build direction and continuing through the multiple layers. The addition of TiB_2 in 2024Al alloy causes columnar grains to transform to equiaxed. This is because TiB_2 acts as a grain growth inhibitor in Al matrix. In addition, TiB_2 creates heterogeneous nucleation sites in front of the solidification face to stimulate fine equiaxed structure.

9.9.5 $AlSi_{10}Mg–TiB_2$ Composite

The fabrication of in-situ nano-TiB_2-enhanced $AlSi_{10}Mg$ composite exhibits a very coherent interface with Al matrix, indicating good interfacial bonding. The solidified microstructure includes both nonequilibrium and equilibrium (eutectic) phases of Al-Si through faster cooling of AM process.

In this alloy, Al is supersaturated with Si, which is greater compared with the equilibrium content of Si in Al (~1.5%). The faster cooling rate of AM techniques can freeze Si in Al. Therefore, Al will solidify first in a supersaturated nonequilibrium condition containing a higher concentration of Si. When solidification continues, the increase of Si content in Al-Si melt moves the system in eutectic condition. Thus, Al-Si eutectic occurs near the boundaries of the previously formed Al cells. In addition, TiB_2 forms at Al grain boundaries because of the following reasons: (i) the higher melting temperature of TiB_2 (3230 °C) does not allow them to become melt, (ii) TiB_2 does not react with Al, Si, and Mg, which indicates their good chemical stability. Therefore, during solidification, TiB_2 is driven to the boundaries, where eutectic Al-Si forms too.

9.9.6 $AA7075–TiB_2$ Composite

TiB_2-reinforced AA7075 alloy has an irregular dispersion of reinforcing elements in the matrix. The faster solidification results in nonequilibrium phase transformation and inter-dendritic segregation of solutes. The laser-processed $AA7075$-TiB_2 composite forms $Mg(Zn, Cu, Al)_2$ and $Al_6(Cu, Fe)$ intermetallic phases along the grain boundaries and within the grains as well.

9.10 Additive Manufacturing of Nickel Matrix Composites

Nickel (Ni) and Ni superalloys have extensive applications in the aerospace industry (airplane jet engines), energy (power plant turbines), petrochemical, and nuclear energy sectors because of their outstanding properties such as higher corrosion resistance, fatigue property, and lower thermal growth. Some reinforcing elements such as titanium carbide, tungsten carbide, chromium carbide, and carbon nanotubes are added to strengthen nickel superalloys for further enhancement of mechanical performances.

9.10.1 Inconel 625–TiC Composites

Titanium carbide (TiC) has excellent properties, such as greater hardness, higher melting temperature, lower density, and improved mechanical strength. However, the brittle structure of TiC limits their applications as monolithic ceramic. For this reason, TiC-strengthened nickel matrix composites reflect good performance in high-temperature refractory, wear, and various surface engineering applications.

In contrast with many other metals, Ni shows a lower wetting angle with TiC, which results in major enhancements in interfacial bonding between TiC reinforcing elements and Ni matrix [33]. Moreover, TiC-strengthened Ni matrix composites are taken as a better replacement of WC-Co-based wear-resistant alloy. This is because the graphite phase in Ni-TiC-C composite acts as an in-situ lubricant during the friction process and thereby improves wear performance.

Ni alloy's typical microstructure, i.e., Inconel 625, contains columnar dendrites along the build direction, which relates to both primary and secondary dendrites along the vertical direction. The microstructure of Inconel 625-TiC composites has similar dendrite morphology like Inconel 625 alloy except with shorter dendrite arm spacing. Moreover, TiC reinforcing elements are distributed randomly throughout the composites.

Because of the higher melting temperature of TiC (3200 °C), some unmelted particles are possible to get within the dendrite and inter-dendrite regions. Therefore, TiC can act as a heterogeneous nucleation site for the γ phase of the alloy. Moreover, the faster cooling rate, which may be enhanced by unmelted particles, will lower dendrite spacing in the Inconel 625–TiC composite. The associated interstitial and substantial components of nickel alloy possessing self-diffusion can generate robust mixing of liquid melt in the melt pool during laser AM techniques, which will greatly influence the diffusion of components about some degrees compared with the normal diffusion. Therefore, when TiC is dissolved through the laser processing, Ti and C are homogeneously dispersed in the liquid melt for the period of building. In addition, TiC can be considered in tremendous heterogeneous nucleation positions for metal carbide (MC) phases.

Moreover, (Nb, Ti)C is the extensively prevailing phase in Nb-Ti-C structure. Therefore, instead of forming TiC_x nearby TiC elements, MC phases containing higher amounts of Nb develop at the transformation zone in TiC/matrix boundary. When there is no morphological dissimilarities in MC at transition areas near the unmelted TiC elements in dendrite and inter-dendrite regions, the period of occurrence of this type of MC is generally shorter compared with γ and rod-like MC phases. Based on the literature, the cooling behavior of Inconel 625 through the solidification process is as follows: $L \rightarrow L + \gamma \rightarrow L + \gamma + NbC \rightarrow \gamma + NbC$.

Because of the partial melting of TiC elements, additional Ti and C particles would be in the liquid melt. In place of developing NbC at inter-dendritic region, MC phases (typically contain Nb, Ti with higher content of Mo) as rod-like morphology precipitate out, then the cooling behavior of the alloy modifies as: $L \rightarrow L + \gamma \rightarrow L + \gamma + MC \rightarrow \gamma + MC$ (M = Nb, Ti and Mo).

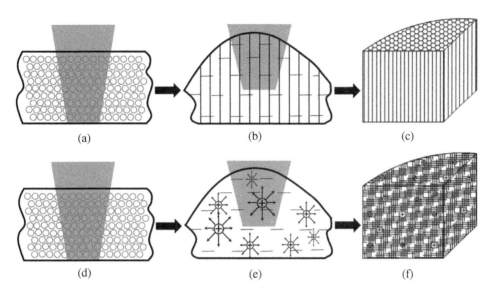

Figure 9.12 The schematic diagram illustrates the mechanism of Inconel–TiB$_2$ composite formation. Source: Reprinted from [34] with permission from Elsevier.

9.10.2   Inconel 625–TiB$_2$ Composite

The solidification mechanism of Inconel 625–TiB$_2$ composite in the LPBF process is schematically shown in Figure 9.12. During LPBF, a TEM$_{00}$ laser beam provides Gaussian energy supply on the powder bed making the melt pool within a short time period. Throughout the process, as the beam moves away from the liquid melt, a rapid cooling starts from the direction of the bottom to top of the melt with a solidification angle with respect to the processing speed (see Chapters 2–5 for more info), which originates long dendrite grains along that direction (Figure 9.12c).

However, TiB$_2$ particle remains stable because of the variation of melting temperature between Inconel 625 and TiB$_2$ (1350 °C for Inconel 625 and 3230 °C for TiB$_2$). On the other hand, Marangoni convention relocates TiB$_2$ particle in the melt leading a reaction with Mo from Inconel 625 at higher energy density. Consequently, a driving force radiates from particles to the matrix in the melt pool. During Inconel 625–TiB$_2$ composite processing, the complete thermal gradient from liquid melt bottom to top moves to an isolated area near TiB$_2$ particles. As a result, randomly oriented dendritic grains are formed in Ni matrix after solidification (Figure 9.12f).

9.11   Factors Affecting Composite Property

9.11.1   Mixing of Matrix and Reinforcing Elements

At the starting of the composite manufacturing steps, both the matrix powder and the reinforcing elements are prepared by either mixing or ball milling with different volume percentages of reinforcing materials. During mixing, powder particles are blended together, unlike ball milling, where stainless steel balls are introduced through the milling. Moreover, the ratio between ball weight and powder weight is necessary to maintain. To keep the mixing and milling environment protective, an inert atmosphere is commenced together with temperature control.

The reduction of crystallite size by milling may generate severe plastic deformation, such as crystal defects, which enhance both the strain and the lattice's core energy. The lattice then discharges energy through dislocation relocation with minimum energy level, preceding the formation of low-angle grain boundaries. When the milling time is intensified, further concentrated deformation takes place. Therefore the dislocation density is amplified with the number of sub-grain boundary misorientations. These misorientations are afterward converted to high-angle grain boundaries and ultimately into finer grains. Therefore, the existence of hard, small-scale reinforcing elements in a milling method considerably constrains dislocation activity via Orowan bowing mechanism [3], which intensifies dislocation density and hereby speeds up the grain refinement scheme. At some stages of milling, due to the saturation in the dislocation density, the crystal lattice achieves the smallest level to stay unchanged. However, at a longer milling time, the mixing materials break into tiny pieces. The plastic deformation, as well as the stress concentration near the reinforcing particles, increases the dislocation density. The accumulated strain may initiate and extend cracks, which afterward fracture particles and lead to a homogeneous distribution of reinforcing materials.

9.11.2   Size of Reinforcing Elements

The size and volume of the reinforcing elements have a significant impact on the microstructure, densification, hardness, toughness, and wear property of the composites. Composites reinforced with coarser elements possess poor wear property, compared with the finer elements-reinforced composites. During laser processing, the redisposition rate of finer TiC reinforcing elements is higher, leading to a better dispersion of TiC elements within the matrix and the greater densification of the solidified parts. In addition, the reduced particle size expands the surface contact within particles and the matrix and eventually results in proper melting and solidification.

Thereby, the finer TiC-reinforced composites have strong interfacial bonds, better densification level, hence lower spalling or plowing from the wear out surface, and higher wear resistance. Whereas the composites reinforced with coarse TiC elements, which have poor interfacial bonds between the matrix and the particles, may create swelling and separation of particles from the wear-out surface under shear stress and hence poor wear resistance property.

Moreover, the shape and size of reinforcing elements can play a greater role in the AM-manufactured composites' fracture toughness. For example, the toughening mechanism of Ti-TiB composites is best illustrated by the presence of 3DQCN structure as well as finer TiB reinforcement when fabricating with higher laser power. These mechanisms can be reviewed as: In TiB–Ti composite containing 3DQCN structure, the cracks may propagate along the Ti grain boundaries (intercrystalline, rich with rigid TiB) or Ti grains (transcrystalline, rich with ductile TiB), as presented in Figure 9.13. Thereby the growing cracks are confronted by ductile TiB region of 3DQCN structure, leading to blunting and deflection of the cracks. Therefore, the formation and propagation of cracks demand extra fracture energy, which enhances the toughness of the composites.

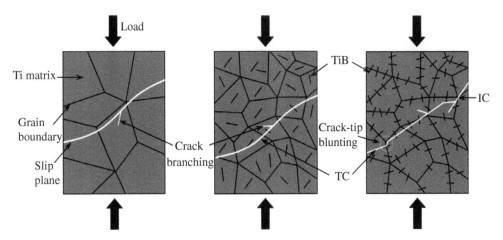

Figure 9.13 Demonstration of various failure approaches during compressive load. TC: Transcrystalline cracks, IC: Intercrystalline cracks. Source: Reprinted from [12] with permission from Elsevier.

9.11.3 Decomposition Temperature

In AM of MMCs, sometimes the reinforcing materials are directly mixed with the matrix powders prior to laser processing. If the reinforcing materials have a higher decomposition temperature and/or bigger particle size, there would be a possibility for the elements to be undissolved during laser processing. Therefore, a faulty interface may develop between the reinforcing elements and the matrix for the reasons of: (i) deficient wettability of the reinforcing materials, (ii) unequal thermal expansion coefficients of reinforcing and matrix materials, (iii) oxide layer formation on the undecomposed material surface. This makes in-situ reinforcements superior to the simply proposed ex-situ reinforcements. In addition, the reinforcing elements should be very fine, thermodynamically steady, and more homogeneously dispersed over the matrix.

9.11.4 Viscosity and Pore Formation

For a composite, the relative density varies with process input parameters such as laser power and scanning speed, which can be best supported by dynamic viscosity, flowability, as well as wettability of the liquid phases. During the LPBF process, the powder mixture is exposed to the laser beam and starts melting. At a lower melting temperature, matrix alloy particles melt initially, developing a stirring melted area. Afterward, the reinforcing phases dissipate partially in the matrix alloy melt, even having a higher melting temperature. Consequently, the dissolved reinforcing phases enhance the melt viscosity while limiting the flowability of the liquid melt.

For a composite melt, the dynamic viscosity (μ), as shown in Eq. (9.19) [8], clarifies the dependency of viscosity on temperature. At the lower laser input power or higher scan speed, the material receives inadequate energy to raise the melt pool temperature. This later increases the melt viscosity and lowers the wettability. Therefore, it restricts the proper dispersion of melt on the solidified layer, causing the appearance of uneven interlayer pores. As a result, densification of the printed parts is reduced. The higher energy input by constant laser scan speed with higher laser power and constant laser power with lower scan speed increase the melt temperature in several ways to: (i) lower the melt viscosity, (ii) improve flowability and wettability, and (iii) attain higher density. In another way, the lower scan speed allows a longer melt period; hence the melt gets sufficient time to merge to the subsequent layers and heal/minimize the porosity formation in the sample.

9.11.5   Volume of Reinforcing Elements and Pore Formation

From research, it is revealed that nanocomposites strengthened with higher contents of reinforcing elements show a greater tendency to form larger pores. The pore generation can be ascribed to the faster solidification of the liquid melts, which results in incomplete filling of voids inside the melted area. In addition, the melt pool dynamics offer asymmetric solidification because of having different regions with dissimilar melting and solidifying performances. With higher amounts of reinforcing elements, the surface tension and the Marangoni flow gradient become stronger from the middle to the border of the melt pool. This instability in the liquid melt results in several small droplets, which spatter from the liquid anterior as the liquid surface energy drops. The instability of the liquid, as well as the thermos-capillary forces, increase with higher amounts of reinforcing elements. During laser scanning, when the path of liquid flow changes from the border to the middle of the melt, the drastically inward liquid stream drives the liquid to spheroidize, as it comes closer to the laser beam. This afterward increases the possibility of forming bigger pores. In this circumstance, the surface tension, as well as the viscosity of the melt, becomes strong enough to disturb the melt stream, hereby resulting in the heterogeneous distribution of heat and mass flow. Consequently, the melting unsteadiness and the balling phenomena are greatly raised, directing to the reasonably lower densification in the solidified structure. At lower volume content, finer-sized TiC-reinforced ferrous composites possess minor porosity compared with the coarser TiC composites. It is known that the surface melting temperature drops with a reduction of particle size (i.e., melting point decline fact) [7]. Thus, a liquid phase may be created simply at low temperatures, as joining occurs within particles, resulting in greater density and lower porosity.

Furthermore, before laser melting, the reinforcing elements pass through the milling process, which makes them different in their dispersion and variation in sizes. Starting with finer TiC elements, a narrow down the particle size distribution results in variation in density. When the coarser particles with rough powder layer undertake laser melting, the solidification face of the running liquid melt is disrupted, resulting in intermittent and unevenly shaped pores. This decreases the wettability between the reinforcements and the matrix, as well as the consistency of the subsequent interfacial bonds because of the smaller surface area of the coarse TiC elements.

9.11.6   Buoyancy Effects and Surface Tension Forces

In laser AM, shape of the melt pool is influenced by fluid flow, which is also controlled by other factors of buoyancy effects as well as solid/liquid and liquid/gas surface tensions. Beyond the impact of buoyancy force, the flow shapes are strongly governed by the surface tensions, which depend on temperature and become lower with the increase of temperature. According to laser

Gaussian energy distribution, the laser beam delivers higher energy in the center rather than the periphery of the beam. Therefore, the difference of surface tension causes the melted liquid to pull from lesser surface tension to the greater surface tension area and thus, the subsequent flow would be external. Therefore, Marangoni convection generates thermal capillary force as the key force to rearrange particles. From the theory of Arafune and Hirata, Marangoni flow can be estimated using Eq. (9.24) [35].

$$M_a = \frac{\Delta\gamma L}{\mu_d D} \tag{9.24}$$

where $\Delta\gamma$ is the difference of surface tension of Marangoni flow (N.m), L is the length of the free surface (m), μ_d is the dynamic viscosity (Pa.s), and D is the diffusion constant of element causing the surface tension difference. Some researchers use the kinematic viscosity instead of D.

Moreover, as the distance between particles decreases, the Van der Waals force increases, which later intensifies the cluster of particles, as shown in Figure 9.14. At a higher energy density, the melt pool temperature obviously rises, thus boosting the Marangoni flow and consequently enhancing the capillary force and the wettability. Therefore, the capillary force combining the thermal energy becomes higher than the Van der Waals force. As a result, reinforcing elements become unable to overcome this energy obstacle to join each other, thereby influencing uniform distribution throughout the matrix.

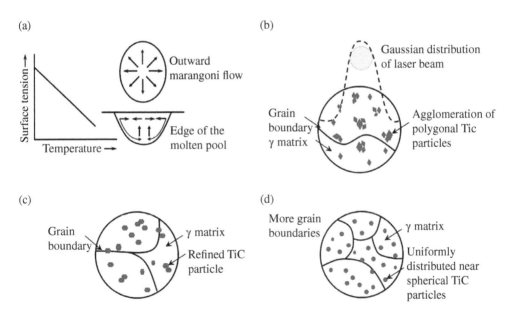

Figure 9.14 (a) The influence of Marangoni flow, (b) TiC particles under intensified energy density through LPBF process, (c) TiC particles are distributed in γ matrix, (d) more grains with uniformly distributed TiC particles. Source: Reprinted from [36] with permission from Elsevier.

9.12 Summary

Metal Matrix Composites

Composites are the combination of two or more fundamental materials having considerably different physical and chemical properties. Generally, composites are comprised of three parts: (i) matrix as an incessant phase, (ii) reinforcing elements as inhomogeneous or distributed phases, and (iii) an interfacial boundary or binder, if required.

Conventional Techniques of Metal Matrix Composites (MMCs)

The conventional techniques of MMCs are generalized into four as: (i) liquid phase, (ii) solid phase, (iii) solid/liquid dual phase, and (iv) deposition that includes in-situ fabrication too. These conventional methods are employed to fabricate simple geometrical MMC parts, which are not suitable for intricate parts.

Additive Manufacturing of MMCs

The possibility of fabricating composites in AM techniques creates the opportunity of their use in regions where lighter and lower-cost materials both have the preference, such as in the automobile industry as well as aerospace and biomedical sectors. Based on the nature of the reaction within phases throughout the fabrication of composites, the techniques can be generalized in two significant classes: ex-situ and in-situ methods. Compared with conventional casting, the laser-processed composite melt possesses nonequilibrium solidification due to faster heating and cooling on a localized area, which eventually causes finer structure and consistent distribution of strengthening materials.

Additive Manufacturing of Ferrous Matrix Composites

Generally, MMCs fabricate through AM techniques and are narrowed down to ferrous alloys such as 316 SS and H13 tool steel. For composites of ferrous alloys, the most used strengthening elements are the phases of carbides (SiC, TiC, WC), nitrides (TaN, TiN), borides (TiB, TiB_2, WB), metal oxides (Al_2O_3), and carbon fibers.

316 SS-TiC Composite

The reaction within Ti and C to form TiC is considered by the higher activation energy. This reaction takes place vigorously because of their exothermic nature. The strengthening element TiC can be produced in situ through direct reaction within elemental material Ti and graphite. Because of the reaction of Ti with other materials at higher temperatures, in-situ formed strengthening elements comprising Ti can be developed instantly.

316 SS-TiB_2 Composite

In 316–TiB_2 composites, the reinforcing materials are distributed homogeneously throughout the matrix, which also join each other and form a continuous ring-like pattern.

H13–TiB_2 Composite

The application of laser AM in the production of tools facilitates the accomplishment of intricate shaped parts using digital method sequence. The laser-processed H13–TiB_2 composite is distinguished with α-Fe and TiB_2 phases, except austenite.

Ferrous–WC Composites

WC is reflected as a high-quality ceramic reinforcing element having a higher melting temperature and excellent wettability with many ferrous alloys. Additionally, WC can keep the room temperature hardness as high as 1400 °C. In Fe–WC composite, a significant amount of WC reinforcing elements are dissolved in Fe-based matrix and release W and C atoms in the liquid melt. Some released W and C atoms react with ferrous alloy and form carbides distributing along the grain boundaries of the matrix.

Ferrous–VC Composites

The laser energy quickly warms up the mixture of 316L and VC, creating a molten area of fully dissolved V_8C_7/316L liquids. The V_8C_7 component can easily decompose and release V and C, because of their finer particle size, greater surface activity, and larger heat input from laser source. In addition, the microscopic pressure working on the ultrafine VC in the liquid melt accelerates the activity throughout the rapid melt process.

Additive Manufacturing of Titanium-Matrix Composites

The frequently used elements that offer efficient reinforcement for the production of TMCs include Cr_3C_2, TiC, TiN, TiO_2, $Si_3 N_4$, SiC, TiB_2, TiB, Al_2O_3, and Ti_5Si_3 [10]. In addition, boron particles and carbon as nanoparticles, nanotubes, and fibers are taken as effective additives to fabricate reinforcing compounds. Among the various reinforcing elements, TiC and TiB are regarded as in-situ types, while TiB_2 and SiC are considered as ex-situ compounds.

Ti–TiC Composite

Through the laser AM, TiC elements are formed by dissolution or precipitation mechanism including heterogeneous nucleation and subsequent grain development of TiC. Moreover, the ex-situ reaction of Ti-TiC and the dispersion of TiC reinforcement depend on processing conditions, such as the input energy density. At higher energy density, TiC reinforcing element disperses as network array with dendritic morphology, whereas at lower energy density, TiC becomes finer and distributes consistently.

Ti–TiB Composite

TiB is a steady element, even with a very small solubility of boron in titanium (<0.001 at.%), which is far low compared with carbon (22 at.%) and nitrogen (1.8 at.%) [11]. The laser-processed Ti-TiB$_2$ composites show two phases of α-Ti and TiB$_2$. The in-situ reaction occurs between TiB$_2$ and Ti through the diffusional removal of boron (B) from TiB$_2$ or by melting and successive solidification of TiB.

Ti–Hydroxyapatite (Ti–HA) Composites

HA is a calcium phosphate ($Ca_{10}(PO_4)_6(OH)_2$) compound having similar composition and morphology of human bone tissue. After decomposition of HA, Ca/P elements react with the α phase resulting in the formation of Ti_5P_3, $Ca_3(PO_4)_2$, $CaTiO_3$, and Ti_xO phases. Moreover, the remaining phases of Ti_5P_3 and $Ca_3(PO_4)_2$ disperse randomly throughout the α matrix. The formation of $CaTiO_3$ phase yields a small amount of $Ca_3(PO_4)_2$ as residue. The lower absorption rate of oxygen during the manufacturing process causes the occurrence of Ti_xO phase rather than TiO_2. However, the oxygen from HA acts as interstitial atom, which diffuses into Ti matrix and forms titanium oxides. A saturation then starts by faster solidification of LPBF and hence forms Ti_xO phase.

Ti-6Al-4V-Metallic Glass (MG) Composite

The incorporation of Zr-based MG reinforcement can create a new direction to modify the microstructure of TMCs, through stabilizing the β phase and refining both β and α′ phases; thereby can impart higher strength and better ductility in the TMC parts.

Through the fabrication mechanism of Ti-6Al-4V/MG composite, dendrite formation discards solute elements (Cu, Ni, and Al) through the solid/liquid interface to supercooled melt, which enriches the liquid melt with MG contents leading to the initiation of amorphous phases. In addition, the faster cooling process restricts the growth of the amorphous phases, resulting in the MG as nanobands.

Ti-6Al-4V + B4C Pre-alloyed Composite

The composite consists of a dual structure of α/β Widmanstätten Ti matrix, enclosing with non-uniformly distributed TiB whiskers. On the other hand, TiC is difficult to detect in the composite structure because of their finer size or higher solubility of carbon in α-Ti solid solution.

Ti-6Al-4V + Mo Composite

When Ti-6Al-4V powder is mixed with different percentages of Mo powder, a novel microstructure develops in the composite structure. Contrary to a complete α' martensitic structure of Ti-6Al-4V alloy, Ti-6Al-4V-Mo composite contains β-Ti matrix with nonuniformly distributed Mo particles.

Additive Manufacturing of Aluminum Matrix Composites

The lightweight AMCs are reinforced with hard ceramic particles and show better performance in aerospace, aircraft, and automotive industries, because of their higher stiffness, greater specific strength, and better wear property. The frequently used ceramic reinforcements are TiB_2, TiC, SiC, Al2O3, Si_3N_4, and B_4C.

Al-Fe_2O_3 Composite

Fe_2O_3 is another low-cost reinforcement, which minimizes energy required for the fabrication of AMC in the laser-processed AM technique. The stoichiometric reaction between Al and Fe_2O_3 generates additional heat and stimulates thermite interactions through the AM process. This excess heat of in-situ reaction lowers the energy required to melt powder mixture as well as solidification of the composite melt.

$AlSi_{10}Mg$–SiC Composite

The in-situ reaction between $AlSi_{10}Mg$ and SiC powder mixtures generates Al_4SiC_4 + SiC composite. The addition of SiC as a higher energy absorptive can enhance the energy absorption rate and thereby form adequate liquid with steady melt pools. Therefore, the intensified energy absorption rate, which raises the melt pool temperature, also improves the solid–liquid wettability.

2024Al–TiB_2 Composite

The addition of TiB_2 in 2024Al alloy causes the transformation of columnar to equiaxed grains. This is because TiB_2 acts as a grain growth inhibitor in Al matrix. In addition, TiB_2 creates heterogeneous nucleation sites in front of the solidification face to stimulate fine equiaxed structure.

$AlSi_{10}Mg$–TiB_2 Composite

The fabrication of in-situ nano-TiB_2-enhanced AlSi10Mg composite exhibits a very coherent interface with Al matrix, indicating good interfacial bonding. The microstructure development includes consequent solidification of both nonequilibrium and equilibrium (eutectic) phases of Al-Si through faster cooling of AM process.

AA7075–TiB_2 Composite

The laser-processed AA7075-TiB_2 composite forms $Mg(Zn,Cu,Al)_2$ and $Al_6(Cu,Fe)$ intermetallic phases along the grain boundaries and within the grains as well.

Additive Manufacturing of Nickel Matrix Composites

Ni and Ni-superalloys have extensive application in the aerospace industry (airplane jet engines, land-based turbines), petrochemical, and nuclear energy sections because of their outstanding properties. For further enhancement of mechanical performances, some reinforcing elements such as TiC, WC, CrC, and carbon nanotubes are added to reinforce nickel-based superalloys.

Inconel 625–TiC Composites

Ni shows a lower wetting angle with TiC, which results in major enhancements in interfacial bonding between TiC reinforcing elements and Ni matrix. Moreover, TiC-strengthened Ni matrix composites are taken as a better replacement of WC-Co-based wear-resistant alloy. This is because the graphite phase in Ni-TiC-C composite acts as an in-situ lubricant during the friction process and thereby improves wear performance.

Inconel 625–TiB$_2$ Composites

During Inconel 625–TiB$_2$ composite processing, the complete thermal gradient from melt pool bottom to top surface moves to an isolated area near TiB$_2$ particles. As a result, randomly oriented dendritic grains are formed in Ni matrix after solidification.

Factors Affecting Composite Property

At the starting of the composite manufacturing steps, both the matrix powder and the reinforcing elements are prepared by either mixing or ball milling with different volume percentages of reinforcing materials. The size and volume of the reinforcing elements have a significant impact on the microstructure, densification, hardness, toughness, and wear property of the composites. When the reinforcing materials have a higher decomposition temperature and/or larger particle size, there would be a possibility for the elements to be undissolved during laser processing. At lower laser input power or higher scan speed, the material receives inadequate energy to raise melt pool temperature. The pore generation can be ascribed to the faster solidification of the liquid melts, which results in incomplete filling of voids inside the molten area. In laser AM, the shape of the molten pool is influenced by fluid flow, which is also controlled by other factors of buoyancy effects and surface tension. All these factors depend on temperature and become lower with the increase of temperature.

References

1. B. Almangour, Additive manufacturing of emerging materials, Springer, 2018.
2. M. Abdellaoui and E. Gaffet, "The physics of mechanical alloying in a planetary ball mill: Mathematical treatment," *Acta Metall. Mater.*, vol. 43, pp. 1087–1098, 1995, doi: 10.1016/0956-7151(95)92625-7.
3. C. Suryanarayana, "Mechanical alloying and milling," *Prog. Mater. Sci.*, vol. 46, no. 1–2. pp. 1–184, 2001, doi: 10.1016/S0079-6425(99)00010-9.
4. B. S. Murty and S. Ranganathan, "Novel materials synthesis by mechanical alloying/milling," *Int. Mater. Rev.*, vol. 43, pp. 101–141, 2014, doi: 10.1179/imr.1998.43.3.101.
5. S. Jiang et al., "Discrete element simulation of particle motion in ball mills based on similarity," *Powder Technol.*, vol. 335, pp. 91–102, 2018, doi: 10.1016/j.powtec.2018.05.012.
6. H. Fayazfar et al., "A critical review of powder-based additive manufacturing of ferrous alloys: Process parameters, microstructure and mechanical properties," *Mater. Des.*, vol. 144, pp. 98–128, 2018, doi: 10.1016/j.matdes.2018.02.018.

7. B. AlMangour, D. Grzesiak, and J. M. Yang, "In-situ formation of novel TiC-particle-reinforced 316L stainless steel bulk-form composites by selective laser melting," *J. Alloys Compd.*, vol. 706, pp. 409–418, 2017, doi: 10.1016/j.jallcom.2017.01.149.

8. D. Gu, J. Ma, H. Chen, K. Lin, and L. Xi, "Laser additive manufactured WC reinforced Fe-based composites with gradient reinforcement/matrix interface and enhanced performance," *Compos. Struct.*, vol. 192, pp. 387–396, 2018, doi: 10.1016/j.compstruct.2018.03.008.

9. B. Li, B. Qian, Y. Xu, Z. Liu, J. Zhang, and F. Xuan, "Additive manufacturing of ultrafine-grained austenitic stainless steel matrix composite via vanadium carbide reinforcement addition and selective laser melting: Formation mechanism and strengthening effect," *Mater. Sci. Eng. A*, vol. 745, pp. 495–508, 2019, doi: 10.1016/j.msea.2019.01.008.

10. H. Attar, S. Ehtemam-Haghighi, D. Kent, and M. S. Dargusch, "Recent developments and opportunities in additive manufacturing of titanium-based matrix composites: A review," *Int. J. Machine Tools Manuf.*, vol. 133. pp. 85–102, 2018, doi: 10.1016/j.ijmachtools.2018.06.003.

11. A. Sabahi Namini and M. Azadbeh, "Microstructural characterisation and mechanical properties of spark plasma-sintered TiB2-reinforced titanium matrix composite," *Powder Metall.*, vol. 60, pp. 22–32, 2017, doi: 10.1080/00325899.2016.1265805.

12. Y. Hu, W. Cong, X. Wang, Y. Li, F. Ning, and H. Wang, "Laser deposition-additive manufacturing of TiB-Ti composites with novel three-dimensional quasi-continuous network microstructure: Effects on strengthening and toughening," *Compos. Part B Eng.*, vol. 133, pp. 91–100, 2018, doi: 10.1016/j.compositesb.2017.09.019.

13. R. Banerjee, A. Genç, P. C. Collins, and H. L. Fraser, "Comparison of microstructural evolution in laser-deposited and arc-melted in-situ Ti-TiB composites," *Metall. Mater. Trans. A Phys. Metall. Mater. Sci.*, vol. 35 A, no. 7, pp. 2143–2152, 2004, doi: 10.1007/s11661-004-0162-0.

14. C. Ning and Y. Zhou, "Correlations between the in vitro and in vivo bioactivity of the Ti/HA composites fabricated by a powder metallurgy method," *Acta Biomater.*, vol. 4, pp. 1944–1952, 2008, doi: 10.1016/j.actbio.2008.04.015.

15. S. Nath, R. Tripathi, and B. Basu, "Understanding phase stability, microstructure development and biocompatibility in calcium phosphate-titania composites, synthesized from hydroxyapatite and titanium powder mix," *Mater. Sci. Eng. C*, vol. 29, pp. 97–107, 2009, doi: 10.1016/j.msec.2008.05.019.

16. C. Han et al., "Microstructure and property evolutions of titanium/nano-hydroxyapatite composites in-situ prepared by selective laser melting," *J. Mech. Behav. Biomed. Mater.*, vol. 71, pp. 85–94, 2017, doi: 10.1016/j.jmbbm.2017.02.021.

17. B. Vrancken, L. Thijs, J. P. Kruth, and J. Van Humbeeck, "Microstructure and mechanical properties of a novel β titanium metallic composite by selective laser melting," *Acta Mater.*, vol. 68, pp. 150–158, 2014, doi: 10.1016/j.actamat.2014.01.018.

18. D. Gu et al., "Densification behavior, microstructure evolution, and wear performance of selective laser melting processed commercially pure titanium," *Acta Mater.*, vol. 60, pp. 3849–3860, 2012, doi: 10.1016/j.actamat.2012.04.006.

19. X. J. Shen, C. Zhang, Y. G. Yang, and L. Liu, "On the microstructure, mechanical properties and wear resistance of an additively manufactured Ti64/metallic glass composite," *Addit. Manuf.*, vol. 25, pp. 499–510, 2019, doi: 10.1016/j.addma.2018.12.006.

20. S. Pouzet et al., "Additive layer manufacturing of titanium matrix composites using the direct metal deposition laser process," *Mater. Sci. Eng. A*, 2016, vol. 677, pp. 171–181, doi: 10.1016/j.msea.2016.09.002.

21. W. Liu and J. N. DuPont, "Fabrication of functionally graded TiC/Ti composites by laser engineered net shaping," *Scr. Mater.*, vol. 48, pp. 1337–1342, 2003, doi: 10.1016/S1359-6462(03)00020-4.

22. H. Attar, M. Calin, L. C. Zhang, S. Scudino, and J. Eckert, "Manufacture by selective laser melting and mechanical behavior of commercially pure titanium," *Mater. Sci. Eng. A*, vol. 593, pp. 170–177, 2014, doi: 10.1016/j.msea.2013.11.038.

23. C. Kun, H. Beibei, W. Wenheng, and Z. Cailin, "The formation mechanism of TiC reinforcement and improved tensile strength in additive manufactured Ti matrix nanocomposite," *Vacuum*, vol. 143, pp. 23–27, 2017, doi: 10.1016/j.vacuum.2017.05.029.

24. D. Gu, G. Meng, C. Li, W. Meiners, and R. Poprawe, "Selective laser melting of TiC/Ti bulk nanocomposites: Influence of nanoscale reinforcement," *Scr. Mater.*, vol. 67, pp. 185–188, 2012, doi: 10.1016/j.scriptamat.2012.04.013.

25. D. Gu, Y. C. Hagedorn, W. Meiners, K. Wissenbach, and R. Poprawe, "Nanocrystalline TiC reinforced Ti matrix bulk-form nanocomposites by Selective Laser Melting (SLM): Densification, growth mechanism and wear behavior," *Compos. Sci. Technol.*, vol. 71, pp. 1612–1620, 2011, doi: 10.1016/j.compscitech.2011.07.010.

26. N. Kang, P. Coddet, Q. Liu, H. L. Liao, and C. Coddet, "In-situ TiB/near α Ti matrix composites manufactured by selective laser melting," *Addit. Manuf.*, vol. 11, pp. 1–6, 2016, doi: 10.1016/j.addma.2016.04.001.

27. W. Li et al., "Enhanced nanohardness and new insights into texture evolution and phase transformation of TiAl/TiB2 in-situ metal matrix composites prepared via selective laser melting," *Acta Mater.*, vol. 136, pp. 90–104, 2017, doi: 10.1016/j.actamat.2017.07.003.

28. Y. Yang, S. Wen, Q. Wei, W. Li, J. Liu, and Y. Shi, "Effect of scan line spacing on texture, phase and nanohardness of TiAl/TiB2 metal matrix composites fabricated by selective laser melting," *J. Alloys Compd.*, vol. 728, pp. 803–814, 2017, doi: 10.1016/j.jallcom.2017.09.053.

29. M. Xia, A. Liu, Z. Hou, N. Li, Z. Chen, and H. Ding, "Microstructure growth behavior and its evolution mechanism during laser additive manufacture of in-situ reinforced (TiB+TiC)/Ti composite," *J. Alloys Compd.*, vol. 728, pp. 436–444, 2017, doi: 10.1016/j.jallcom.2017.09.033.

30. D. Gu, C. Hong, and G. Meng, "Densification, microstructure, and wear property of in situ titanium nitride-reinforced titanium silicide matrix composites prepared by a novel selective laser melting process," *Metall. Mater. Trans. A Phys. Metall. Mater. Sci.*, vol. 43A, pp. 697–708, 2012, doi: 10.1007/s11661-011-0876-8.

31. D. Gu, Y. C. Hagedorn, W. Meiners, K. Wissenbach, and R. Poprawe, "Selective Laser Melting of in-situ TiC/Ti5Si3 composites with novel reinforcement architecture and elevated performance," *Surf. Coatings Technol.*, vol. 205, pp. 3285–3292, 2011, doi: 10.1016/j.surfcoat.2010.11.051.

32. T. Marcu, M. Todea, L. Maines, D. Leordean, P. Berce, and C. Popa, "Metallurgical and mechanical characterisation of titanium based materials for endosseous applications obtained by selective laser melting," *Powder Metall.*, vol. 55, pp. 309–314, 2012, doi: 10.1179/1743290112Y.0000000007.

33. M.-Y. Shen, X.-J. Tian, D. Liu, H.-B. Tang, and X. Cheng, "Microstructure and fracture behavior of TiC particles reinforced Inconel 625 composites prepared by laser additive manufacturing," *J. Alloys Compd.*, vol. 734, pp. 188–195, 2018, doi: 10.1016/J.JALLCOM.2017.10.280.

34. B. Zhang, G. Bi, S. Nai, C. N. Sun, and J. Wei, "Microhardness and microstructure evolution of TiB2 reinforced Inconel 625/TiB2 composite produced by selective laser melting," *Opt. Laser Technol.*, vol. 80, pp. 186–195, 2016, doi: 10.1016/j.optlastec.2016.01.010.

35. K. Arafune and A. Hirata, "Thermal and solutal marangoni convection in In-Ga-Sb system," *J. Cryst. Growth*, vol. 197, pp. 811–817, 1999, doi: 10.1016/S0022-0248(98)01071-9.

36. D. Gu, H. Zhang, D. Dai, M. Xia, C. Hong, and R. Poprawe, "Laser additive manufacturing of nano-TiC reinforced Ni-based nanocomposites with tailored microstructure and performance," *Compos. Part B Eng.*, vol. 163, pp. 585–597, 2019, doi: 10.1016/j.compositesb.2018.12.146.

10

Design for Metal Additive Manufacturing

Learning Objectives

At the end of this chapter, it is expected that you:

- Know and understand new design frameworks tailored to additive manufacturing (AM).
- Understand the importance of design rules and guidelines and how they differ between AM processes.
- Understand the theoretical framework supporting topology optimization, efforts toward including AM constraints in topology optimization models, and typical workflow of basic topology optimization in AM.
- Learn important terminologies that define lattice structures and understand some practical lattice design methodologies.
- Learn the significance and design strategies of support structures for success of the printing process and product quality.
- Understand design workflows for AM through case studies

10.1 Design Frameworks for Additive Manufacturing

The primary idea behind Design for Manufacturing (DfM) is considering manufacturing throughout the design process to ensure the design is manufacturable and product quality can be standardized/certified at a reasonable cost. When the complexity of a design increases,

Metal Additive Manufacturing, First Edition. Ehsan Toyserkani, Dyuti Sarker, Osezua Obehi Ibhadode,
Farzad Liravi, Paola Russo, and Katayoon Taherkhani.
© 2022 John Wiley & Sons Ltd. Published 2022 by John Wiley & Sons Ltd.

many traditional manufacturing processes tend to fail; however, in many instances, Additive Manufacturing (AM) performs excellently forasmuch as constraints are satisfied. Inclusive of its capability to build complex geometries, it produces hierarchically structured components, functionally integrated parts and develops multi-materials required of a part in a single build. Conventional design for manufacturing workflows do not take into consideration these added functionalities and so lack the capabilities to guide successful design processes. This is the rationale behind developing new design frameworks specifically suited for AM. Although these new frameworks have their differences from conventional design frameworks, they take their foundation from VDI 2221.

Design for additive manufacturing (DfAM) has been defined as "maximizing product performance through the synthesis of shapes, sizes, hierarchical structures, and material compositions, subject to the capabilities of AM technologies" [1]. DfAM is important for engineers who want to achieve a failure-free build process, standardized product quality, best product performance, or efficiency without exceeding a proposed cost limit. As we will discover further down this chapter, topology optimization and lattice structure design form the heart of the structural design component of the DfAM workflow because they are known to produce design solutions suitable for this manufacturing technology. This section will explore some design frameworks proposed, so far, by researchers for AM:

10.1.1   Integrated Topological and Functional Optimization DfAM

In this framework developed by Tang, Hascoet, and Zhao in 2014, as shown in Figure 10.1, multilevel and multidiscipline design methods are integrated into the design process where the initial design space is obtained from a conceptual model or scanned from an existing part. Two major aspects of this framework are generating an initial design space and performing structural optimization (topology optimization), which can be for single or multi-objective application(s). In the case of a multi-objective topology optimization, several objectives can be simultaneously carried out by optimizing a new consolidated function based on a weighted sum of all individual objectives. The weight assigned to each objective is usually chosen arbitrarily, therefore inherently a heuristic decision based on the experience of the user. However, there are now more systematic ways of arriving at optimal weight factors. A popular approach is to derive a *p-norm* expression of the consolidated objective function. Figure 10.2 shows an example of a multiphysics optimization flow chart for a Multifunctional Thermal Protection System (MTPS). This design framework ensures that design solutions are tailored within design requirements and manufacturing constraints; therefore, the optimization stages are guided by these constraints. This framework also considers the use of lattice patterns as the last step in optimization, where a link is established between the relative density of hexahedral mesh elements from topology optimization and lattice structure.

10.1.2   Additive Manufacturing-Enabled Design Framework

In 2015, Yang and Zhao proposed a simplistic AM design framework comprised of two steps: functional integration and structural optimization in the design process. The first step involves part consolidation from an initial CAD model based on its application while considering performance and functional requirements. In other words, the initial CAD model can be modified in such a way that it meets the working condition intended by the user while ensuring that

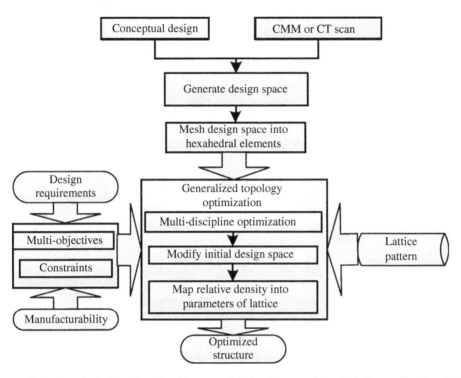

Figure 10.1 Topological and functional integrated design framework for AM. Source: Reprinted from [2] with permission from ASME.

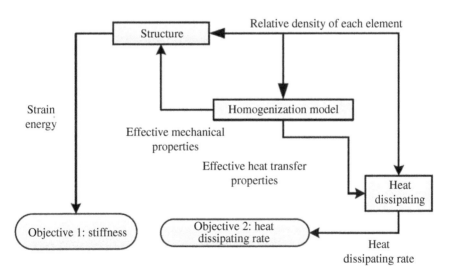

Figure 10.2 Multidiscipline optimization (MDO) for a multifunctional thermal protection system (MTPS). Source: Reprinted from [2] with permission from ASME.

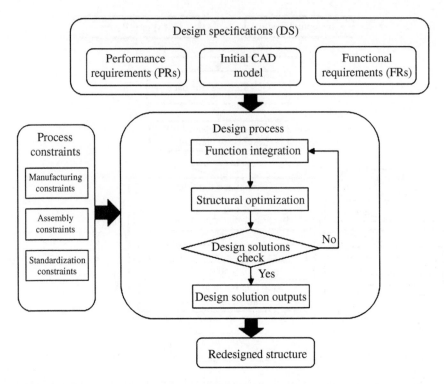

Figure 10.3 AM-enable design framework. Source: Redrawn and adapted from [3].

certain performance and functional specifications are considered. Structural optimization, taken as the second major step, is performed to transform the conceptual model obtained in Step 1 into one or series of designs to optimize specific functionalities such as stiffness maximization, better thermal conductivity, improved cooling abilities, etc. It is important to note that for both design steps, process constraints such as manufacturing and assembly factors as well as standardization constraints should be integrated. Structural optimization results in a design option that should be checked with desired requirements whereby the whole design process is reinitiated when there are shortfalls. A design solution can be arrived upon after several attempts through the design cycle. This framework is presented in Figure 10.3.

10.1.3 Product Design Framework for AM with Integration of Topology Optimization

Primo et al. [4] proposed a methodology for product design in AM that accommodates two alternative processes to generate a resultant hybrid solution: (i) topology optimization and (ii) lattice structure design. In general, the workflow follows the typical design for manufacturing, where specifications are outlined initially, and finite element (FE) analysis is performed during the final stages to ensure that the design solution checks out with the physics objectives.

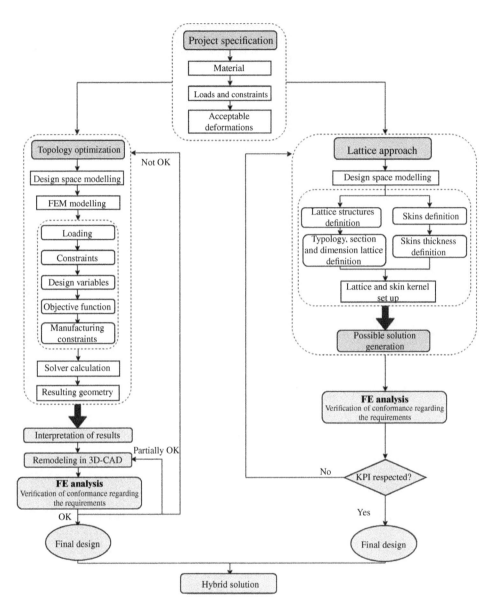

Figure 10.4 Design workflow for hybrid design solutions by topology optimization and lattice structure design. Source: Redrawn and adapted from [4].

This framework's unique aspect is the inclusion of a workflow for lattice structure design, which allows the user play around definitions of lattice such as typology, section type, and dimensions. In addition, the solution of the lattice design approach is integrated with that from topology optimization to give a final hybrid design. This workflow is shown in Figure 10.4.

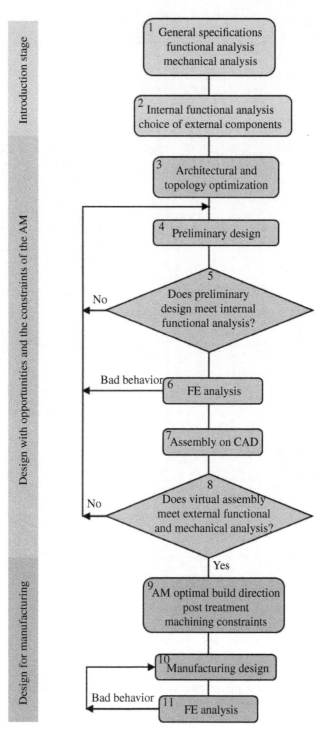

Figure 10.5 Multifunctional design methodology. Source: reprinted from [6] with permission from Elsevier.

10.1.4   Multifunctional Optimization Methodology for DfAM

Optimizing several important functional parameters in the design process ensures that the design's best performance is achieved. Orquera et al. opined that the AM potential could not be fully actualized except the conventional approach to product design is modified. This led to the development of a multifunctional design framework that consists of three major stages and 11 steps. The first stage, like conventional design methods, lists out functional requirements for the whole system and parts of the system. The second stage involves the actual design process that considers opportunities and constraints of the AM technology for production. Conceptual architectural modeling, topology optimization, and FE analysis form the critical aspects of this stage. In the final stage, manufacturing concerns such as build direction, overhanging features, and other constraints posttreatment (such as machining) are designed for following the structural solutions obtained in the previous stage. This stage ends with FE analysis to ensure that functional requirements are maintained. Figure 10.5 shows the design framework.

10.1.5   AM Process Model for Product Family Design

Lei et al. [7] developed a simplistic design framework with similar stages to previous frameworks. The model starts by specifying the requirements and constraints that define the product family. Topology optimization is introduced at the next step, specifically tailored for AM, generating several conceptual designs subjected to tests and analysis in the third step. FE analysis and cost analysis form the basis for assessing the designs' performances with the established requirements and constraints. A design that fulfills the requirements becomes a customized product family, which is additively manufactured and mechanically tested. This workflow is illustrated in Figure 10.6

10.2   Design Rules and Guidelines

As an evolving technology, AM has experienced contextual inconsistencies in establishing standard design rules and guidelines, which is not far-fetched from the inability to represent design criteria and compare it with AM processes and system performance [8]. Some researchers have used the terms rules and guidelines interchangeably in the literature, while others introduce terms such as design principles and fundamentals having similar meanings with few differences. Design rules stem from certain constraints in a design due to limitations in material or machine processes while the design guidelines form restrictions due to technicalities inherent in the specific AM process. There have been significant studies by researchers, research groups, and other collaborative efforts to outline some rules and guidelines for several AM processes. The following section will focus on three major metal AM processes: Laser Powder Bed Fusion (LPBF), Electron Beam Powder Bed Fusion (EB-PBF), and Binder Jetting (BJ).

10.2.1   Laser Powder Bed Fusion (LPBF)

For the class of PBF processes such as laser PBF (LPBF), important rules and guidelines to consider include: part orientation, support structure shape/distribution, tolerances, allowable feature sizes, and anisotropic properties.

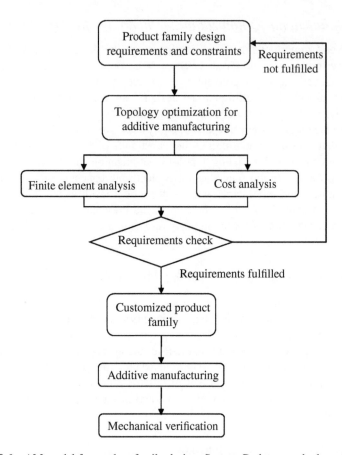

Figure 10.6   AM model for product family design. Source: Redrawn and adapted from [7].

a. **Part orientation**   The success of printed parts made by LPBF significantly depends on the models' orientation, location in the build plate, and arrangement on the substrate. It is important that the part is positioned such that a layer being melted has as little material area continuity as possible to avoid the melting and fusion of large solid layer portions. The larger the area to cool, the more the part is susceptible to residual stress-induced warping (popularly called the "curl effect"). This is a result of the induction of large thermal gradients due to insufficient cooling and high cooling rates during the process. Therefore, optimizing the part's orientation can reduce this curl effect, consequently improving part quality. Figure 10.7 shows the poor and good positioning of a part.

b. **Support structures**   During the building process, the movement of the roller (recoater) to spread a new powder layer may cause already built part sections to leave the position gently or entirely. Support structures are primarily necessary to hold the part firmly in place as any shift or movement can lead to print inhibition, part failure, or worse, cause damage to the printing machine. There are several other reasons why support structures are inevitable in LPBF, some of which are [9]:

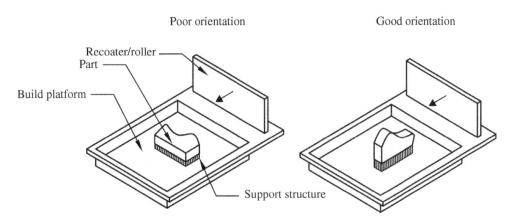

Figure 10.7 Poor and good part orientation for the avoidance of the "curl effect." Source: Redrawn and adapted from [9].

- To support overhanging features against self-weight, usually at angles < 45°.
- To ensure heat dissipation and minimization of residual stresses. Support structures play an important role in heat dissipation as well. Accumulative heat would cause hot zones that will result in excessive residual stresses, cracks, and porosity.
- To compensate for residual stress-induced warping.
  Figure 10.8 shows types of support structures; fill, lattice, offset, and gusset, while Figure 10.9 shows how support structures can be properly applied.
c. **Tolerances**   LPBF achieves accuracies relatively better than other metal AM technologies in the order of about ±0.1 mm. An effective approach to achieving a higher accuracy in a design involves an intentional geometric offset in the model at locations requiring high tolerances followed by a finishing process such as machining.
d. **Allowable feature size**   Features can either be positive (e.g. walls, ribs, drafts, vertices) or negative (hollow features), where minimum or maximum sizes achievable must be known. For positive features, the minimum size usually depends on the laser melting and cooling process. LPBF produces a minimum size of ~0.1 mm when maximum sizes are dependent on the machine fabrication envelope size. Wall thickness with the size of the process melt pool width can be achievable in LPBF [9]. In any case, the success and quality of printing

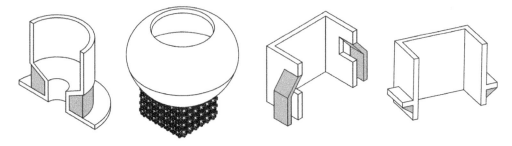

Figure 10.8 Types of support structures, from left to right: fill, lattice, offset, and gusset. Source: Redrawn and adapted from [10].

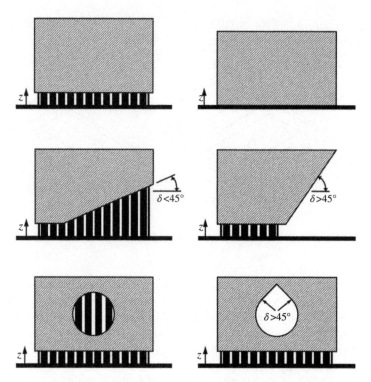

Figure 10.9   Guidance on the use of support structures. Source: Redrawn and adapted from [9].

these minimal sizes depend on the build orientation, support structure, aspect ratio (height to thickness), process parameters and the material used [9]. In many cases, hole resolution and quality depend on the wall thickness and orientation of the hollow feature. Thinner walls can resolve smaller holes, and as wall thickness increases, hole diameter has to increase to be resolvable [11]. A hole diameter of 1 mm is resolvable with a wall thickness of about 1 mm, but larger wall thicknesses (above 5 mm) can only resolve hole diameters above about 3 mm. It should be noted that hole diameters of 2 and 3 mm with large wall thicknesses present difficulties in clearing off adhering powders during post-processing. Hollow structures should normally have the axis of the circle parallel to the building direction. When the axis is orthogonal, the circular topology experiences a "dropping effect" if it prints successfully; and if the axis must be orthogonal, the shape of the circle might be modified as shown in Figure 10.10 to eliminate support structures during build.

Figure 10.11 shows several negative/hollow features printed with the hollow extrusion perpendicular to the build direction in (a), (b) and parallel to the build direction in (c), all without supports. The features in Figure 10.11(a) and (b) are poorly resolved with rough top surfaces seen because there were no supports to them. However, in Figure 10.11(c), correctly resolved holes are observed as there are no surfaces needing support due to their orientation.

It is noteworthy that in the design of a complex part with intricate internal cavities, loose powder removal mechanisms should be planned for by including open-ended features, reducing intricacy, or employing some post-processing method.

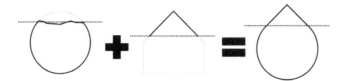

Figure 10.10   Modification of circular profile to avoid "dropping effect." Source: Redrawn and adapted from [9].

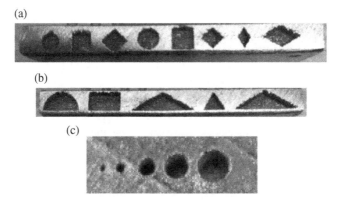

Figure 10.11   Hollow features printed with the hollow extrusion (a, b) perpendicular to the build direction and (c) parallel to the build direction.

e. **Anisotropic material properties**   Mechanical properties of printed parts usually vary in the building direction for a typical range of 5–15% [9]. This effect is more pronounced in lattice-like structures relative to more solid components. Therefore, during the design process, this should be factored in. It is recommended that the part's major load axis be aligned perpendicular to the building direction (parallel to the build substrate/platform). Post-processing techniques can be introduced to reduce anisotropic effects in parts.

10.2.2   Electron Beam Powder Bed Fusion (EB-PBF)

a. **Part orientation**   The orientation of components on the substrate affects the print quality but not as much as it does for LPBF because heat dissipation is relatively better controlled in EPBF. Thermal gradients are less, resulting in lower residual stress concentrations. Notwithstanding, it is recommended that part orientation, positioning, and location be optimized for best results as with LBPF.

b. **Support structures**   The need for support structures in EB-PBF is not paramount because there is less likelihood for prints to fail under subsidence. This results from the preheating that occurs before the actual print causing a "pre-sintered" body. Support structures become important, though, when areas under high thermal shocks are expected. There is the need to support such areas to ensure proper heat dissipation, consequently reducing residual stresses. A study by Vayre [12] reported that planar features at angles close to perpendicularity to the build direction or parallel to the build plate are more susceptible to the edge curling effect. In addition, parts built with heights close to the build plate experienced no significant deformation. These considerations can be made in the design stage for the reduction or elimination of support structures. In summary, compared with LPBF, thinner support structures are enough, thereby reducing material volume usage. In addition, optimization or intelligent manipulations can be performed on process parameters to further reduce residual stresses.

c. **Tolerances**   Accuracies are poorer with EB-PBF, and parts produced will usually require further machining for better precision. Current state-of-the-art EB-PBF machines give accuracies of about 500 μm. Surface roughness is not as good in EB-PBF-made parts compared to counterparts made by LPBF. In addition, geometric considerations can be made for this tolerance in the design process for parts that have high finishing requirements.

d. **Allowable feature size**   The minimum for positive features in EB-PBF is poorer than LPBF in the neighborhood of 0.5 mm as with the tolerance. It should be noted that robust walls should have dimensions in multiples of this allowable feature size for any AM process. The size of the build chamber determines the maximum feature size. Gaps between walls should have dimensions more than the melt pool width to prevent coalescence between walls during printing. The gap dimensions also depend on the type of material and process parameters.

e. **Powder removal**   During the build process, the temperature of the build platform is usually elevated to ensure good part integrity. This preheating step causes the unmelted powder to partially sinter, stick to the formed part, and cause difficulty during powder removal postprocessing step. This difficulty arises from the fact that the powder loses its flowability, and so a forced action such as blowing by compressed air must be adopted. Therefore, features that allow free flow passage of compressed air must be considered in the design. Figure 10.12 shows how a feature type can affect the passage of compressed air, while Figure 10.13 compares a feature that will enhance powder flow in (b) and one that will not in (a).

f. **Anisotropic properties**   Studies have shown that there is columnar or epitaxial grain growth microstructure in the build direction of parts manufactured by EB-PBF [14]. Therefore, a microstructure that is directionally dependent results in anisotropic mechanical properties, and this should be considered in the design and/or build setup similar to LPBF.

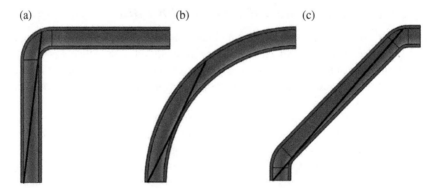

Figure 10.12   Line-of-sight powder removal: (a) small radius, (b) large radius, (c) double radii at exit ends. Source: Redrawn and adapted from [13].

(a)   (b)

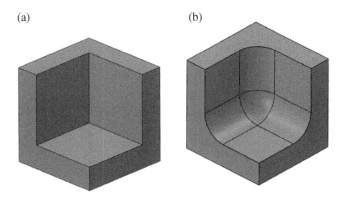

Figure 10.13 A feature with (a) sharp corners and (b) smooth corners. Source: Redrawn and adapted from [13].

10.2.3 Binder Jetting

a. **Part orientation** Parts produced by the binder-jet AM process require no support structures, experience no curl or warpage issues because there are no residual stresses induced during printing. In addition, the part being printed is supported by surrounding powder; however, unsupported features risk breaking when the part is handled in its green or uncured state. Therefore, it is recommended that unsupported features are made as short as possible. The fact that no support structure is needed makes the process significantly cheaper and less complicated compared to laser or electron beam powder bed technologies. Doyle et al. [15] carried out experiments to investigate the effects of layer thickness and part orientation on binder-jetted AM parts and discovered a 25–30% increase in the part's strength when layer thickness was decreased from 200 to 50 μm. It was also discovered that part orientation had negligible effect on the mechanical properties.

b. **Tolerances** Due to secondary processes such as infiltration, debinding, and/or sintering to make functional parts, thermal shrinkages occur and are largely unpredictable and nonuniform. These shrinkages are usually catered for by scaling up the original design by a percentage determined from experiment. Generally, a tolerance of about 0.2 mm is achievable depending on the material, powder sphericity, binder volume injected on loose powder, and debinding/sintering protocols.

c. **Allowable feature size** Feature sizes achievable are also poorer than laser powder bed fusion processes with a minimum wall thickness of about 1.5 mm. Although maximum sizes are restricted by the machine size, bigger parts are possible with binder jetting compared with powder bed systems. 3DHubs [16], a popular manufacturing service company, recommends a minimum value of 2 mm for wall thickness, 3 mm for unsupported walls, 0.5 mm for intricate features, 2 mm for unsupported edges, 1 mm for fillets, 1.5 mm for hole diameters, 5 mm for escape holes, and 2 mm for overall minimal feature size. Figure 10.14 shows a move from an unoptimized to an optimized design for binder-jetting AM.

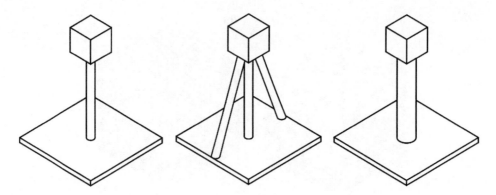

Figure 10.14 Design variations showing some thin sections that can encourage breakage in its green state. Source: Redrawn and adapted from 3Dhubs [16].

d. **Anisotropic properties** Studies show that sintered binder-jetting products have micro-structures with equiaxed grains that result in isotropic material properties in disparate printing directions. See the associated binder-jet chapter for further information.

10.2.4 Technologies Compared

Table 10.1 Metal additive technologies compared with key design considerations.

	Laser powder bed fusion	Electron beam powder bed fusion	Binder jetting
Part orientation	Key for print success	Affects part quality	Affects part quality
Support structure	Very important	Sometimes required for heat dissipation	Not required
Tolerance	~0.1 mm	~0.5 mm	~0.2 mm
Allowable feature size	Minimum: ~0.1 mm Maximum: Size of build chamber of machine	Minimum: ~0.5 mm Maximum: Size of build chamber of machine	Minimum: ~2 mm Maximum: Size of build chamber of machine
Anisotropy	Yes	Yes	No

10.3   Topology Optimization for Additive Manufacturing

In the design frameworks for AM earlier discussed, topology optimization and lattice structure design were identified to be the most common and effective tools for structural modeling in the design process. Topology optimization has gained much popularity because of the design free-dom it offers while achieving optimal functional designs. A major concern is that due to the structural complexities that the topology optimization algorithms generate, manufacturability might not be possible even with the capabilties of AM possesses. Therefore, engineers attempt to "interpret" these complex designs in the detailed design phase by fine-tuning the structure, which will, more often than not, translate to a loss in the previously achieved functional quality. In light of this, the narrative has moved on to formulating robust topology optimization models inclusive of AM constraints. In this section, the basic concepts of structural optimization will first be addressed, and then the focus will be drawn to the mathematical workflow of topology optimization for mechanical and thermomechanical problems and finally discussing AM-integrated topology optimization methodologies.

10.3.1 Structural Optimization

Structural optimization attempts to obtain the most efficient structural layout of a component or system. This form of optimization consists of three broad aspects: (i) sizing optimization, (ii) shape optimization, and (iii) structural optimization. A typical sizing optimization solution finds the optimal thickness of member areas in a truss structure [5]. The members' thicknesses become the design variable for the problem formulation and minimizing the structure's compliance, deflection, or peak stress is done as dictated by performance requirements. The challenge inherent in sizing optimization is the fact that there must be an already established structural layout which restricts design freedom. In shape optimization, the goal is to find the best shape for a prescribed domain; consequently, the shape of a solid or empty domain becomes the design variable that optimizes the structure's performance within a designable volume. Just as in sizing optimization, there is still restriction with design freedom because an a priori shape must be initially established at the commencement of the optimization. Topology optimization, which will be elaborated in the later sections, basically entails determining features such as the number and location of squares and the connectivity of the domain [5]. Figure 10.15 shows the different structural optimization methods.

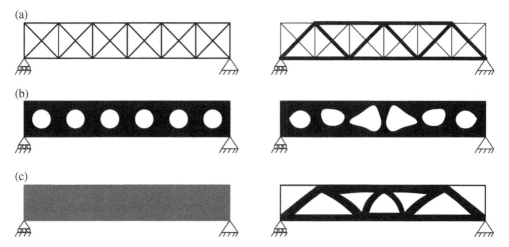

Figure 10.15 The three classifications of structural optimization. (a) Sizing optimization, (b) shape optimization, (c) topology optimization. Sizing and shape optimization initially start with a predefined structural layout. Source: Redrawn and adapted from [5].

The definition of a general structural optimization (SO) problem follows [17]:

$$(SO) \begin{cases} \text{minimize } f(x, y) \text{ with respect to } x \text{ and } y \\ \\ \text{subject to} \begin{cases} \text{behavioral constraints on } y \\ \\ \text{design constraints on } x \\ \\ \text{equilibrium constraint} \end{cases} \end{cases} \tag{10.1}$$

In the formulation in (10.1), x and y are design and behavioral variables. Design constraints are constraints imposed on the design variable x while behavioral constraints are constraints on the state variable y and the combination of these constraints is possible. The equilibrium constraint for a structure under static condition is

$$K(x)u = F(x) \tag{10.2}$$

where $K(x)$ is the stiffness matrix of the structure being a function of the design variable, u is the system's displacement vector, and $F(x)$ is the force vector resulting from external or internal load conditions (or combination of both), which may or may not be a function of the design variable. For a frequent case where the stiffness is invertible,

$$u = u(x) = K(x)^{-1}F(x) \tag{10.3}$$

The displacement vector in Eq. (10.3) can be treated as a state-space variable and substituted as a variable in the objective function

$$(SO) \begin{cases} \min\limits_{x} & f(x, u(x)) \\ \text{s.t.} & g(x, u(x)) \leq 0 \end{cases} \tag{10.4}$$

10.3.2   Topology Optimization

The aim of topology optimization is to obtain the best structural layout within a design space and usually does this by optimizing an objective function under constraints for a designable volume. The first known paper on topology optimization was published by Michell, an Australian inventor, in 1904. He derived an optimality criterion for the least weight layout of trusses.

Since then, there have been several efforts that have made topology optimization an attractive and efficient tool in the Design for Additive Manufacturing (DfAM) framework.

Currently, and amongst available classes, there are three popular methods for topology optimization: density-based, hard-kill, and level set in no order of popularity. Solid Isotropic Material with Penalization (SIMP) and Rational Approximation of Material Properties (RAMP) are density-based methods, with SIMP being the more popular scheme. In simple terms, SIMP applies a power-law penalization to relate stiffness and density where the density variables for mesh elements have values between 0 and 1; 0 meaning void element and 1 full dense/solid element. RAMP differs from SIMP by only the penalization or interpolation function, which is a ratio of the density variable and a penalized density function. Hard-kill methods consist of Evolutionary Structural Optimization (ESO) and Bidirectional Evolutionary Structural Optimization (BESO), with BESO generally more popular. Both methods are similar because they iteratively add a finite amount of material within a given design space; however, BESO has the added advantage of removing material when necessary. In these methods, heuristic criteria are adopted and may or may not have strict dependence on the structure's sensitivity information. In the Level Set Method (LSM), Level Set Functions (LSFs) are generated to determine the boundaries of a design, and topologies change with change in the LSF. Figure 10.16 shows a broad classification of topology optimization methodologies. A distinction between a conventional bracket design and a topology optimized version is shown in Figure 10.17. In this section, the theory behind density-based topology optimization for compliance minimization will be closely looked at and should help the reader better understand the science, mathematics, and importantly the workflow behind most commercially-available software on topology optimization.

The theory of topology optimization for compliance minimization is founded on the principle of virtual work. If we assume u is the displacement field that defines the equilibrium of a structure and v is taken as the kinematically admissible virtual displacement field, for an elastic structure with a fixed boundary Γ_d, we have [18]

$$\int_\Omega \epsilon^T(v)(D\epsilon(u))d\Omega = \int_\Omega f^T v d\Omega + \int_{\Gamma_t} t^T v d\Gamma \qquad (10.5)$$

The left-hand side of Eq. (10.5) is the system's strain energy in the elastic domain, which is a combination of the work done by forces f, applied on Ω and surface traction forces t, on Γ_t. By using notations from functional analysis and energy bilinear form for internal work and load linear form for external work,

$$a(u, v) = l(v), \qquad \forall v \in V \qquad (10.6)$$

where

$$a(u, v) = \int_\Omega \epsilon^T(v)(D\epsilon(u))d\Omega \qquad (10.7)$$

and

$$l(v) = \int_\Omega f^T v d\Omega + \int_{\Gamma_t} t^T v d\Gamma$$

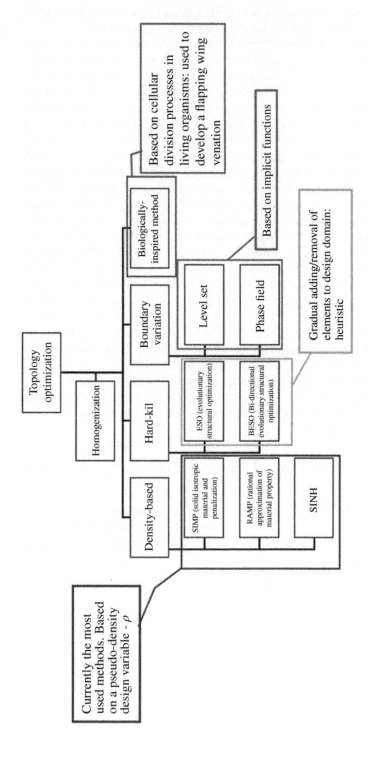

Figure 10.16 Classification of topology optimization methods.

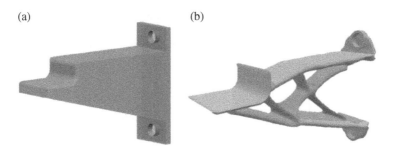

(a)   (b)

Figure 10.17 How topology optimization transforms the structural form of (a) a metal bracket to its (b) optimized design.

Topology optimization based on obtaining the stiffest structural layout means minimizing the compliance or strain energy. It aims at ensuring that the structural layout gives the least response possible to external loads and surface traction forces. Therefore, the structural problem by minimizing compliance $l(u)$ can be formulated thus:

$$\text{Minimize } l(u) \quad \text{subject to } a(u, v) = l(v), \quad \forall v \in V \text{ and design constraints} \tag{10.8}$$

Equation (10.8) is the comparison of internal work, $a(u, v)$ and external work, $l(v)$ on the system and which can be broken down to the FE equation for equilibrium condition:

$$K(\rho)U(\rho) = F \tag{10.9}$$

where K and U are the stiffness matrix and displacement vector respectively due to the external load vector F. With a compliance objective $l(u) = F^T U$ and volume constraint, the optimization problem in Eq. (10.8) transforms to:

$$\text{Minimize}: \quad C(\rho) = F^T U(\rho)$$
$$\text{subject to}: \quad K(\rho)U(\rho) = F$$
$$\frac{V(\rho)}{f_v V_0} \leq 1 \tag{10.10}$$
$$0 \leq \rho \leq 1$$

where f_v is the volume fraction constraint on the design. There are several methods to solve this nonlinear problem such as Optimality Criterial Method (OCM), Method of Moving Asymptotes (MMA), Convex Linearization (COLIN), Sequential Linear Programming (SLP), Sequential Quadratic Programming (SQP), etc. This text will briefly elaborate on OCM and MMA, which are popular optimizers in the literature.

a. To solve this nonlinear optimization problem by OCM, the following functions are implemented:

OCM updates the design variables depending on the optimality conditions being met as follows [19]:

$$
\rho_e^{new} = \begin{cases} \max\left(0, \rho_e - m\right) & \text{if } \rho_e B_e^{\eta} \leq \max\left(0, \rho_e - m\right) \\ \min\left(1, \rho_e + m\right) & \text{if } \rho_e B_e^{\eta} \geq \min\left(1, \rho_e - m\right) \\ \rho_e B_e^{\eta} & \text{otherwise} \end{cases} \tag{10.11}
$$

where m and η are move limits and numerical damping coefficient respectively. The numerical damping coefficient is commonly given a value of ½ and B_e is obtained as:

$$
B_e = \frac{-\frac{\partial c}{\partial \rho_e}}{\lambda \frac{\partial V}{\partial \rho_e}} \tag{10.12}
$$

λ is a Langrangian multiplier whose value must be chosen so that the volume constraint is met. Usually, the appropriate value is found using the numerical bisection method [19].

b. To solve this nonlinear optimization problem by the Method of Moving Asymptotes (MMA), the problem statement in (10.10) can be compared with a general optimization problem [20]:

$$
\text{Minimize}: \quad f_0(x) + a_0 z + \sum_{i=1}^{m}\left(c_i y_i + \frac{1}{2} d_i y_i^2\right)
$$

$$
\text{subject to}: \quad f_i(x) - a_i z - y_i \leq 0, \quad i = 1\ldots, m \tag{10.13}
$$

$$
x \in X, y \geq 0, z \geq 0.
$$

where x is the independent design variable and y, z are dependent state variables; also in addition, the objective and constraint functions f_0, f_1, ..., f_m must be differentiable. For Eqs. (10.10) and (10.13) to be equivalent, (10.13) must be written as

$$
\text{Minimize} \quad f_0(x)
$$

$$
\text{subject to } f_i(x) \leq 0, \quad i = 1\ldots, m \tag{10.14}
$$

$$
x \in X.
$$

where $a_0 = 1$, $a_i = 0$ for all $i > 0$, $z = 0$, $d_i = 1$ and $c_i =$ "a large number." Subproblems are obtained for $f_i(x)$, $i = 0, 1, \ldots, m$ and solved by either of two methods: dual approach or primal-dual interior-point approach. The subproblem for MMA is given as [20]:

$$
\tilde{f}_i^{(k)} = \sum_{j=1}^{n}\left(\frac{p_{ij}^{(k)}}{u_j^{(k)} - x_j} + \frac{q_{ij}^{(k)}}{x_j - l_j^{(k)}}\right) + r_i^{(k)}, \quad i = 0, 1, \ldots, m \tag{10.15}
$$

where p, q are coefficients of the convex subproblem, which are a function of the sensitivity information during the kth iteration. r is the residual of the value of subproblem at the "k^{th}" iteration subtracted from the actual function's value. u and l are the upper and lower asymptotes respectively. For an optimal solution of these subproblems, the Karush–Kuhn–Tucker(KKT) optimality conditions are both necessary and sufficient.

When a subproblem has been developed, an iterative solution step is taken to arrive at appropriate values of the design variables that give an optimum structure (minimum compliance in this case):

i. Step 0: Choose a starting point $x^{(0)}$, and let the iteration index be $k = 0$.
ii. Step 1: Given an iteration point $x^{(k)}$, calculate $f_i(x^{(k)})$ and the gradients $\nabla f_i(x^{(k)})$ for $i = 0, 1, \ldots, m$.
iii. Step 2: Generate a subproblem according to 10.15 and update the asymptotes u_j and l_j.
iv. Step 3: Solve the subproblems in step 2 and let the optimal solution of this subproblem be the next iteration point $x^{(k+1)}$. Let $k = k + 1$ and go to step 1.

MMA is done such that each $f_i^{(k)}$ is obtained by a linearization of f_i in variables of the type $\frac{1}{x_i - L_i}$ or $\frac{1}{U_j - x_j}$, which are dependent on the signs of the derivatives of f_i at $x^{(k)}$. The values of L_i and U_j are normally changed between the iterations and are popularly referred to as "moving asymptotes." The article on MMA and GCMMA (Globally Convergent version) by Svanberg [20] is recommended to the reader for more information on MMA.

10.3.2.1 Material Interpolation Functions

Density-based approaches for topology optimization require functions that assign values of material properties to elements depending on their pseudo-density variables (design variables). There have been many functions developed suitable for design-independent and design-dependent topology optimization problems. Very popular interpolation models are SIMP, RAMP, and SINH (coined after the sine hyperbolic function). Pioneers of SIMP are Kikuchi and Bendsoe in 1989, however, more researchers such as Zhou, Rozvany, and Sigmund further developed this method that has become widely accepted as a standard for density-based topology optimization approaches. SIMP being a power-law expression suffers much in performance when dealing with design-dependent loads (which will be discussed in subsequent sections) due to the undesirable effect it has on low-density elements as indicated by Bruyneel and Duysinx [21]. This is caused by its zero gradients when density values are near-zero or zero, resulting in 'vanishing' sensitivity values for design-dependent loads. Converse to this is the RAMP model that reserves a gradient value that depending on the penalization factor, for zero or near-zero density elements. Therefore, several researches [21–24] have concluded that SIMP is not an appropriate method for design-dependent topology optimization. A related function is SINH and is named so because of the hyperbolic sine function (sinh). In a similar fashion as SIMP, it ensures that intermediate density elements have less effect than solid or void elements. Unlike SIMP, the SINH interpolation operates such that intermediate elements are given more relevance and similar to RAMP in a way that the function does not varnish when the density of elements tends to zero. It should be noted that the density (pseudo-density design variable) of an element is likened to the volume fraction of that element with values ranging from 0 to 1 and these terms can be used interchangeably. Young's modulus functions based on SIMP and RAMP interpolation methods for classic and modified density-based methods are given in Eqs. (10.16)–(10.19) and graphically shown in Figure 10.18:

Classic model:

$$\text{SIMP}: \quad E(\rho_e) = \eta_{SIMP}E_0 = \rho_e^p E_0 \tag{10.16}$$

$$\text{RAMP}: \quad E(\rho_e) = \eta_{RAMP}E_0 = \frac{\rho_e}{1 + p(1 - \rho_e)}E_0 \tag{10.17}$$

$$\text{SINH}: \quad E(\rho_e) = \eta_{SINH}E_0 \tag{10.18}$$

$$\text{where } \eta_{SINH} = \begin{cases} 1 & \text{if } p = 1 \\ 1 - \dfrac{\sinh\left(p(1 - x_e)\right)}{\sinh(p)} & \text{if } p > 1 \end{cases}$$

$$0 \leq \rho_{min} \leq \rho_e \leq 1$$

Modified model:

$$\text{SIMP}: \quad E(\rho_e) = E_{min} + \rho_e^p(E_0 - E_{min}) \tag{10.19}$$

$$0 \leq \rho_e \leq 1$$

$$0 \leq E_{min} \leq E_e \leq E_0$$

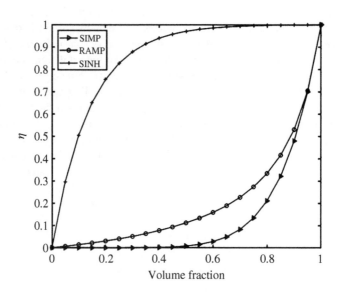

Figure 10.18 Influence of volume fraction on interpolation function for SIMP, RAMP, and SINH where penalty, $p = 7$.

10.3.2.2 Sensitivity Analysis and Filtering

For all gradient-based methods, the optimization process inherently depends on how changes in design variables affect the values of the objective function(s) and constraint functions: the process of ascertaining these effects is referred to as sensitivity analysis. There are two major ways to arrive at the sensitivities: the direct method and the adjunct method. For a minimum compliance problem with volume fraction as constraint, the sensitivities of compliance and volume with respect to the density design variable can be obtained with the direct method as follows:

$$\frac{\partial C}{\partial \rho} = -\frac{\partial E}{\partial \rho} U K_0 U \tag{10.20}$$

$$= -\frac{\partial \eta}{\partial \rho} E_0 U K_0 U \tag{10.21}$$

We can differentiate the equation for system equilibrium in (10.9) as:

$$\frac{\partial K}{\partial \rho} U + K \frac{\partial U}{\partial \rho} = \frac{\partial F}{\partial \rho} \tag{10.22}$$

Combining Eqs. (10.22) and (10.21) and for a design-independent load $\left(\frac{\partial F}{\partial \rho} = 0 \right)$, we have

$$\frac{\partial C}{\partial \rho} = -U \frac{\partial K}{\partial \rho} U \tag{10.23}$$

The global material stiffness is the product of the material's Young modulus and stiffness matrix of the primitive mesh element, therefore, (10.23) becomes

$$\frac{\partial C}{\partial \rho} = -\frac{\partial E}{\partial \rho} U K_0 U$$

$$= -\frac{\partial \eta}{\partial \rho} E_0 U K_0 U \tag{10.24}$$

Assuming each element has a unit volume,

$$\frac{\partial V}{\partial \rho} = 1$$

Checker-boarding features are normally formed in topology-optimized results due to the binary values density variables are forced to take. To resolve this problem and obtain realistic solutions, filtering becomes important. Several filtering methods exist, among which sensitivity, density, and Heaviside function filtering are most popular. A sensitivity filter is expressed as [19]:

$$\frac{\widehat{\partial c}}{\partial \rho_e} = \frac{1}{\max(\gamma, \rho_e) \sum\limits_{i \in N_e} H_{ei}} \sum\limits_{i \in N_e} H_{ei} \rho_i \frac{\partial c}{\partial \rho_i} \tag{10.25}$$

Density filter is expressed as:

$$\check{\rho}_e = \frac{1}{\sum\limits_{i \in N_e} H_{ei}} \sum\limits_{i \in N_e} H_{ei} \rho_i \qquad (10.26)$$

where N_e is the set of neighboring elements i, which have center-to-center distance $d(e, i)$ to element e less than set filter radius r_{min}. In order to avoid divisions by zero in the fraction in Eq. (10.26), γ is assigned a small value as 10^{-3}. H_{ei} is a weight factor expressed as:

$$H_{ei} = \max\left(0, r_{min} - d(e, i)\right) \qquad (10.27)$$

A linear filter is described graphically in Figure 10.19

10.3.3   Design-Dependent Topology Optimization

Many studies on structural optimization, in particular topology optimization, have focused on point loads while some extensions have been made to uniformly distributed loads and pressure on non-designable regions. A challenging aspect of topology optimization is the presence of design-dependent loads in an application. In such a case, the loads depend on the material volume and/or layout over the design domain and are subject to changes during the optimization process [22]. As illustrated in Figure 10.20, design-dependent loads can be classified into:

a. **Transmissible or sliding force:** This force maintains its magnitude but consistently changes its line of action as the topology changes during optimization.
b. **Body load**: This load is a result of the inertia load of the design. As material distribution changes during the optimization process, the magnitude and centroid of this load change. Centrifugal loads can come under this class because the force exerted on a rotating body largely depends on its weight, the position of the center of rotation, and the rotational speed.

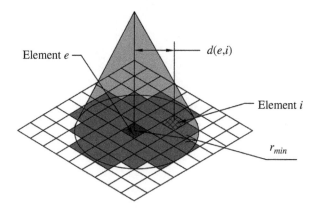

Figure 10.19   Linear filter for a 2D mesh.

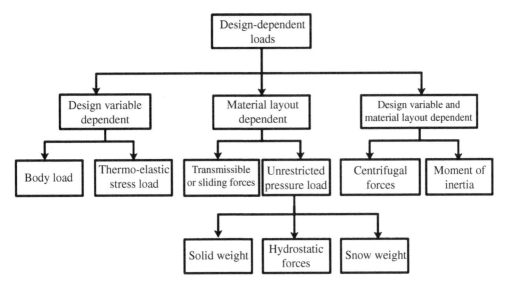

Figure 10.20 Classification of design-dependent loads for topology optimization.

c. **Surface pressure load**: When the surface that carries the pressure load is designable, this load can be described as design-dependent. Gao and Zhang [22] further classified this load into solid weight, hydrostatic, and snow weight pressure.

d. **Thermoelastic stress load**: Thermal stress has gained considerable interest in recent times because of the diverse practical applications and numerical complications that can arise from thermal stresses which depend not only on the changing topology but also on Young's modulus and coefficient of thermal expansion. For this kind of load, a term called Thermal Stress Coefficient (TSC) is usually introduced within the thermo-structural topology optimization formulation to ensure simultaneous material interpolation and penalization of the Young's Modulus and the coefficient of thermal expansion [22].

10.3.3.1 Review on Efforts to Tackle Multiobjective Thermomechanical Topology Optimization Problems

There have been several interesting studies on design-dependent topology optimization, especially related to thermomechanical problems. As early as 1995, Rodrigues and Fernandes [25] formulated a computational model for topology optimization based on the material distribution for a 2-D linear-elastic solid subjected to thermal loads. In their work, the temperature distribution was a uniform steady-state temperature independent of the design. Their study showed that even for simple models and at low-temperature changes, optimal topologies obtained strongly depend on these temperature changes. In 2006, Li et al. [26] proposed an evolutionary optimization procedure by thermoelasticity to address varying temperature distribution throughout the design domain. Their model's iterative loop includes transient heat analysis by conduction, FE thermoelastic analysis, and design modifications. Studies conducted separately by Hou et al. [27] and Deaton and Grandhi [28] addressed thermomechanical topology optimization by stress-based criteria while maximizing stiffness as in [27]. This they achieved by including a constraint of a global stress measure based on a p-norm function so that the number of stress constraints is reduced in the optimization. Compliance minimization was imposed as the objective function. Again, a stress constraint in the form of a p-norm function was added to the regular volume constraint. It is worthy to note that Gao and Zhang [22] and Deaton and Grandhi [28] both concluded that the classical SIMP material interpolation function encountered numerical difficulties for thermomechanical topology optimization which might be largely due to the existence of null values in the stiffness matrix for elements with zero density design values. As this can cause singularity issues, they suggested that the RAMP interpolation function might be more efficient for thermomechanical problems. This conclusion might be extensible to most design-dependent topology optimization formulations. SIMP and RAMP interpolation functions are defined by the functions:

$$\text{SIMP}: \quad E(x_e) = \rho_e^p E_0 \tag{10.28}$$

$$\text{RAMP}: \quad E(x_e) = \frac{\rho_e}{1 + p(1 - \rho_e)} E_0 \tag{10.29}$$

For density values close to zero ($x_e \cong 0$), the interpolation schemes in the sensitivity function become

$$\text{SIMP}: \quad \frac{dE(\rho_e)}{d\rho_e} = p\rho_e^{p-1} E_0 \xrightarrow[\rho_e \to 0]{} 0 \tag{10.30}$$

$$\text{RAMP}: \quad \frac{dE(\rho_e)}{d\rho_e} = \frac{1 + p(1 - \rho_e) + p\rho_e^2}{(1 + p(1 - \rho_e))^2} E_0 \xrightarrow[\rho_e \to 0]{} \frac{1}{p+1} E_0 \tag{10.31}$$

Equation (10.30) shows that SIMP neglects the significance of elements with design values close to zero but RAMP gives zero density elements a factor of $\frac{1}{p+1}$ of the material property (Young's modulus).

From studies, it is common for thermomechanical optimization to produce results with significant gray/intermediate elements, more so when lower penalty values are used. Since the structural interpretation of these gray elements is difficult, several research efforts have focused on using multi-material models to handle this problem as in the works by Vantyghem et al. [29], Takezawa and Kobashi [30]. The application of thermomechanical topology optimization was extended to compliant actuators by Du et al. [31], where they used mesh-free methods; thermal actuated compliant and electrothermal complaint mechanisms were done by Ansola et al. [32, 33]. Jahan et al. [34] developed a numerical model for topology optimization for redesigning traditional injection molding tools. They aimed at minimizing the weight of the design (without losing stiffness) and maximizing heat conduction, and this led to a multi-objective formulation based on compliance and heat conduction.

The following sections will introduce the theoretical background of thermomechanical topology optimization.

10.3.3.2 Thermomechanical Topology Optimization Based on Compliance Minimization

Structures under combined thermal and mechanical loads experience significant topological changes during optimization compared to when the are under only mechanical loads. The reason is that a body under elevated temperature deforms according to its coefficient of thermal expansion in addition to its Young's modulus. As will be shown in later sections, the thermal stress load on an element is dependent on its amount of material. Therefore, as the topology is being updated, the magnitude and direction of this load on every element change accordingly, the reason why they are referred to as design-dependent. In this section, topology optimization for design-dependent thermoelastic loads using FE analysis, and material interpolation functions are formulated. Figure 10.21 shows a structure subjected to constant, elevated temperature in (a) and steady temperature distribution in (b) while under mechanical loads and other boundary conditions.

10.3.3.3 Thermal Stress Coefficient and Load

In a thermoelastic problem, the cell elastic strain energy of an element in a discretized material space depends on its temperature difference with ambience, material coefficient of thermal

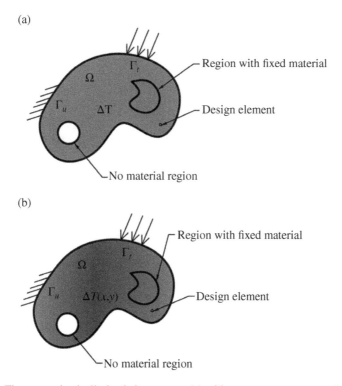

Figure 10.21 Thermomechanically loaded structures (a) with constant temperature change (b) with a temperature distribution.

expansion, and Young's modulus. The cell's elastic strain energy can be defined as the potential mechanical energy in the elastic body [27]. If we assume that the cell is under combined mechanical and thermal loads, while within the elastic limit of the material, its total strain is a summation of mechanical and thermal strains in Eq. (10.32):

$$\varepsilon_e = \varepsilon_e^m + \varepsilon_e^{th} \tag{10.32}$$

In addition, suppose we assume that the material properties are temperature-independent and the temperature distribution remains constant through any given period (steady state), we can express the cell's mechanical strain energy as:

$$\Psi_e^m = \frac{1}{2} \int_\Omega \sigma_e^m \varepsilon_e^m d\Omega \tag{10.33}$$

For a solid element,

$$\begin{aligned}
\Psi_e^m &= \frac{1}{2} \int\int_v \sigma_e^m \varepsilon_e^m dv \\
&= \frac{1}{2} \int\int_A \left(\varepsilon_e - \varepsilon_e^{th}\right)^T D_e \left(\varepsilon_e - \varepsilon_e^{th}\right) h dx dy \\
&= \frac{1}{2} \int\int_A \left(\varepsilon_e^T D_e \varepsilon_e - 2\varepsilon_e^T D_e \varepsilon_e^{th} + \varepsilon_e^{th,T} D_e \varepsilon_e^{th}\right) h dx dy
\end{aligned}$$

Thermal strain vector

$$\varepsilon_e^{th} = \alpha_e (T_e - T_{amb}) \phi^T = \alpha_e \Delta T \phi^T \tag{10.34}$$

$$\Psi_e^m = \frac{1}{2} \int\int_A \left(u_e^T B^T D_e B u_e - 2u_e^T B^T D_e \alpha_e \Delta T \phi^T + \alpha_e \Delta T^T D_e \alpha_e \Delta T \phi^T \phi\right) h dx dy \tag{10.35}$$

For a rigid and static body in equilibrium and under elasticity, the differential of strain energy w.r.t displacement should be zero; therefore, Eq. (10.35) becomes

$$\frac{\partial \Psi_e^m}{\partial u_e} = \frac{1}{2} \int\int_A \left(2B^T D_e B u_e - 2B^T D_e \alpha_e \Delta T \phi^T\right) h dx dy = 0 \tag{10.36}$$

The first term in (10.36) is the equivalent load, which deforms the element and is attributed to the effect of thermal expansion from the second term; therefore, the thermal load for an element with a unit thickness can be expressed as

$$F_e^{th} = \int\int_A B^T D_e \alpha_e \Delta T \phi^T dx dy \tag{10.37}$$

where

$$B_i = \begin{bmatrix} \dfrac{\partial N_i}{\partial x} & 0 \\ 0 & \dfrac{\partial N_i}{\partial y} \\ \dfrac{\partial N_i}{\partial y} & \dfrac{\partial N_i}{\partial x} \end{bmatrix} \tag{10.38}$$

$$D_e = \frac{E_e}{1-v^2} \begin{bmatrix} 1 & v & 0 \\ v & 1 & 0 \\ 0 & 0 & \dfrac{1-v}{2} \end{bmatrix}$$

$$\varepsilon_e^m = \varepsilon_e - \varepsilon_e^{th} \tag{10.39}$$

where B and D_e are differential shape function matrix and material matrix for plane stress respectively, α is a coefficient of linear thermal expansivity, E_e is Young's modulus, u_e is displacement vector of nodes of an element, v is Poisson's ratio, and ΔT is change in temperature with ambience constant throughout the design domain in this case. It should also be noted that ΔT is independent of time. $\phi = [1 \ 1 \ 0]$ is the thermal strain unit displacement vector for 2D problems.

D_e and α_e are material properties that both depend on the design variable (x_e). It should be noted that variables with subscript "e" are for an element and depend on its density design variable. The material matrix is expressed as:

$$D_e = E_e D_0 \tag{10.40}$$

Substituting Eqs. (10.40) into (10.37) gives

$$F_e^{th} = \int \int_A B^T E_e D_0 \alpha_e \Delta T \phi^T dx dy$$

$$= E_e \alpha_e \Delta T \int \int_A B^T D \phi^T dx dy \tag{10.41}$$

E_e and α_e can be combined into one variable called TSC and denoted by β_e and can be expressed by the RAMP function in Eq. (10.42)

$$\beta_e = E_e \alpha_e$$

$$= \frac{\rho_e}{1 + p_2(1-\rho_e)} E_0 \alpha_0 \tag{10.42}$$

Therefore, thermal stress load can be written as

$$F_e^{th} = \beta_e \Delta T \int \int_A B^T D_0 \phi^T dx dy \tag{10.43}$$

10.3.3.4 Thermomechanical Topology Optimization Problem Definition

In summary, the theory of steady-state thermomechanical topology optimization is a combination of equations from thermal and mechanical disciplines. We can, therefore, introduce weighted compliance functions for thermal and mechanical loads required for sensitivities to be comparable and convergence possible.

Problem statement:

$$\min c = w_1 F_m^T U_m + w_2 F_{th}^T U_{th} = w_1 U_m^T K U_m + w_2 U_{th}^T K U_{th}$$

$$\text{subject to} : K U_m = F_m$$

$$K U_{th} = F_{th}$$

$$(K_c + K_h)T = R_Q + R_q + R_h$$

$$\frac{\sum_{e=1}^{n} V(\rho)}{f V_0} \leq 1 \tag{10.44}$$

$$0 \leq \rho \leq 1$$

$$0 \leq E_{min} \leq E \leq E_0$$

Material function based on RAMP:

$$E = \frac{\rho}{1 + p_1(1-\rho)} E_0 \tag{10.45}$$

$$\beta = \frac{\rho}{1 + p_2(1-\rho)} E_0 \alpha_0$$

Sensitivity analysis for each element:

$$\frac{\partial c_e}{\partial \rho_e} = -w_1 \frac{\partial E_e}{\partial \rho_e} \left(U_e^m\right)^T k_0 U_e^m + w_2 \left(2\left(U_e^{th}\right)^T \frac{\partial F_e^{th}}{\partial \rho_e} - \frac{\partial E_e}{\partial \rho_e} \left(U_e^{th}\right)^T k_0 U_e^{th}\right) \tag{10.46}$$

Assuming each element has a unit volume,

$$\frac{\partial V}{\partial \rho_e} = 1$$

In Eq. (10.44), $[K_c]$ is the conductivity matrix, $[K_h]$ is the convective matrix, and $\{R_Q\}$, $\{R_q\}$, $\{R_h\}$ are heat loads by volume conduction, surface conduction, and convection respectively. A workflow summary of major aspects of topology optimization is shown in Figure 10.22.

Table 10.2 and Figure 10.23 show some simple 2D design problems and topology-optimized solutions.

10.3.4 Efforts in AM-Constrained Topology Optimization

With design freedom as a major advantage of AM, many industries and researchers have adopted topology optimization as a critical tool in designing additively manufactured parts.

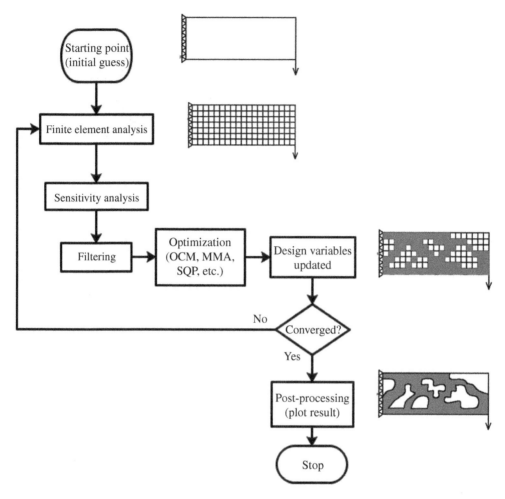

Figure 10.22 Workflow for density-based topology optimization methods.

For topology optimization to be successfully integrated into the design for AM, the constraints posed by this manufacturing technology must be captured in the optimization algorithms. This section will examine several efforts by researchers to incorporate AM constraints within topology optimization. Several AM constraints that have been captured are overhang minimization or elimination, minimum feature thickness, self-supporting features, void elimination, support structure minimization or elimination, residual stress or deformation reduction, etc. Some of these constraints will be discussed.

10.3.4.1 Overhang Minimization and Elimination

Overhanging features of parts built by LPBF AM cause a significant increase in cost and printing time because more sacrificial material is needed to support them. Overhangs have been the

Table 10.2   Topology optimization of point loaded structures.

Design problem	Optimized topology

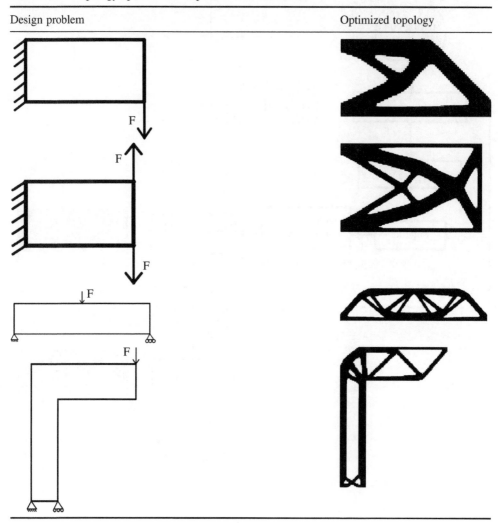

most studied AM constraint in topology optimization. An overhanging feature is one that lacks sufficient supporting material beneath in such a way that it becomes perpendicular or at an angle subtended to the building direction; a well-supported element in a discretized finite element domain is shown in Figure 10.24. Although several researchers have suggested 45° as the minimum self-supporting angle, it is important that the angle be governed by details of the process [35]. Gaynor and Guest [35] proposed three minimum overhang angle constraints which depend on the feature thickness shown in Figure 10.25.

Typically, in defining or building up the overhang constraint, evaluating the contour of resulting topologies on every iteration of the optimization should be done for edge detection [36].

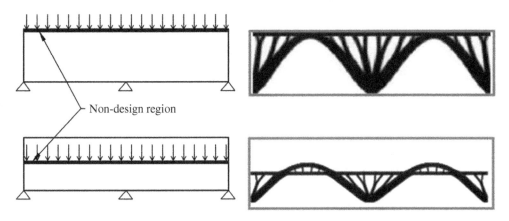

Figure 10.23   Initial design domain and optimized bridge-like designs with initial model space shown in the gray rectangle.

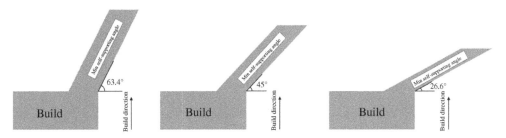

Figure 10.24   Different allowable minimum self-supporting angles for satisfying overhang constraints. Source: Redrawn and adapted from [35].

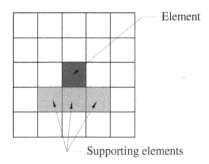

Figure 10.25   Element density with supporting elements in a 2D FE mesh.

(a)

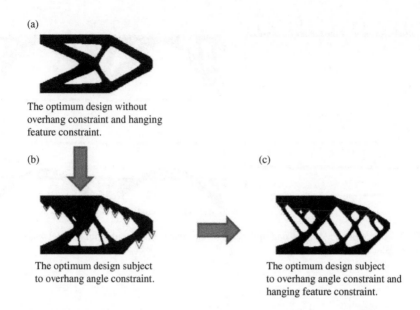

The optimum design without
overhang constraint and hanging
feature constraint.

(b)   (c)

The optimum design subject   The optimum design subject
to overhang angle constraint.   to overhang angle constraint and
                                        hanging feature constraint.

Figure 10.26 Optimum designs showing difference in the topologies resulting from an overhanging constraint alone against combined overhanging and hanging constraints. Source: Reprinted from [38] with permission from Elsevier.

Several researchers have deployed different methodologies to identify edges and formulate an overhang expression as a constraint within the optimization problem. Two common aspects in resolving overhanging features are predetermining a build direction or orientation and formulating an element overhang expression coupled to the objective function or included as a constraint in the optimization problem. Overhang formulations solved as constraints can be found in [36, 37], Zhang et al. [38] considered hanging features alongside overhang features as shown in Figure 10.26, Wang et al. [39] studied overhang reduction by coupling normalized compliance and overhang expressions serving as a multiobjective function for minimization.

Van de Ven et al. [40] made use of a "printable" density variable, which could either be the objective or a constraint function. This required obtaining the sensitivity function using the chain rule for the objective and volume constraint functions. RAMP material interpolation was used in conjunction with the Method of Moving Asymptotes optimizer by Svanberg [20].

10.3.4.2 Self-Supporting Constraint

AM technologies such as LPBF need sacrificial structures for mechanical support and proper heat dissipation, without which the part quality cannot be guaranteed. Support structures are therefore inevitable to support the base of the part but can be minimized or eliminated altogether when printing regions within the part. The self-supporting constraint is very similar to an overhang and/or hanging constraint but slightly differs in that it usually deals with the whole part while overhang constraint deals with features or regions in the part. In general, the definitions of these constraints have either been interchanged or taken as the same based on several research efforts investigated.

Guo et al. [41] proposed two methods to ensure optimized topologies comply with certain self-supporting angles. The methods are known as Moving Morphable Components (MMC) and Moving Morphable Voids (MMV). The first method takes the inclination angle of features or components in the topology as the design variable, while the second introduces printable features (voids) such that their interfaces are represented by B-spline curves in the problem formulation. The problem is then solved as a shape optimization approach rather than topology optimization. Langelaar [42] developed an AM filter for density-based topology optimization taking its foundation from the typical layer-wise AM process. In this way, any feature inclined at an unacceptable angle will be banned from the design space. The drawback to several of these constraints is that they only serve to uphold the structure's geometric printability without regard to its effects on the developed stresses or distortion. Figure 10.27 shows Langelaar's AM filter for an optimized 2D half MBB beam built using different baseplate positions.

Other efforts to factor-in self-supporting constraints in topology optimization can be seen in the work by Mezzadri et al. [43], focusing on volume and compliance constraint-based topology optimization for self-supporting support structures. Zhang and Zhou [44] introduced polygon-featured holes as basic primitive designs whose movements, deformations, and intersections can control the outcome of the structural topology. They made use of a finite cell method (FCM), which has finite element method (FEM) incorporated in it. Other innovative methods that capture self-supporting and overhang constraints can be seen in [45–48].

10.3.4.3 Void Filling Constraint

Powder bed AM technologies build a part in a layer-wise manner so that unmelted powders fill holes and cavities within the part. Therefore, in most cases, these unmelted powders must be removed during post-processing so a usual design criterion is to ensure that there are channels through which these powders can be taken out. During topology optimization, there are the possibilities of obtaining topologies with completely or partially enclosed cavities that are unwanted. Research has resulted in the formulation of void-filling constraints within topology optimization. An innovative methodology was formulated by Liu et al. [49], which they referred to as the Virtual Temperature Method (VTM). The first step was to establish the structure's connectivity by classifying structures without voids as simply-connected and with voids as multiply-connected. Simply-connected structures should have holes or voids with channels leading to the structure's surrounding while multiply connected structures have completely closed voids. Once these definitions are established, the voids are assigned as "virtual" heat sources while solid elements are assigned as thermally insulating materials. A heat conduction problem is formulated and solved to obtain temperature values of all elements within the design domain. A constraint is then set up within the optimization algorithm such that the maximum temperature in the domain during the process should be less than or equal to a threshold temperature set prior to the optimization's commencement. This will ensure that voids are avoided or minimized during topology formation. A graphic comparison between two topologically optimized brackets designed with and without a void constraint is shown in Figure 10.28.

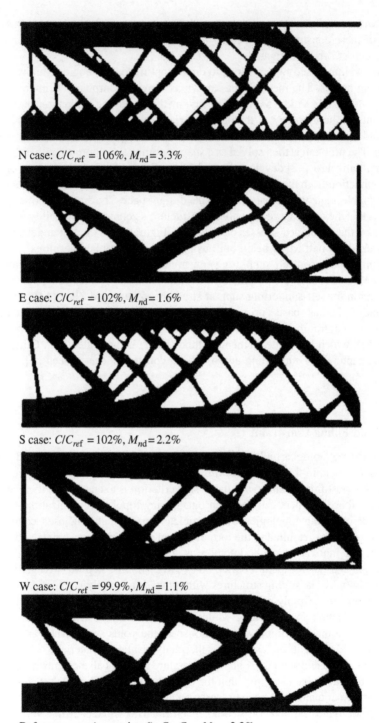

N case: $C/C_{ref} = 106\%$, $M_{nd} = 3.3\%$

E case: $C/C_{ref} = 102\%$, $M_{nd} = 1.6\%$

S case: $C/C_{ref} = 102\%$, $M_{nd} = 2.2\%$

W case: $C/C_{ref} = 99.9\%$, $M_{nd} = 1.1\%$

Reference case (unrestricted): $C = C_{ref}$, $M_{nd} = 2.2\%$

Figure 10.27 Topology-optimized half MBB beam with AM filter and Heaviside projection with baseplate indicated by a black straight line. Source: Reprinted from [42] with permission from Springer.

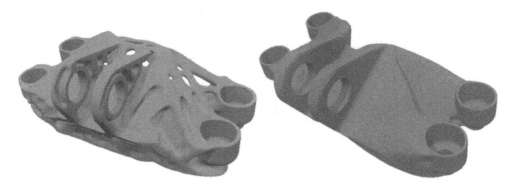

Figure 10.28   Optimized designs of an aerospace bracket. The right bracket was optimized with a void-filling constraint, while the left bracket was done without the constraint.

10.3.4.4   Minimum Feature Thickness

Very thin features are undesirable due to the minimum sizes allowable for different AM technologies. The minimum wall thickness is usually a function of the melt pool width in the case of LPBF. In most topology optimization algorithms, the minimum feature size is controlled by filtering methods, where sensitivity and density filters are the most popular. Osanov and Guest [50] addressed minimum feature size by adopting the layer-wise nature of AM process through a modified Heaviside Projection Method (HPM), which is common in setting geometric constraints in topology optimization. Together with a filtering method, the HPM is a popular density thresholding and projection method for improving the topological boundary crispness and solving mesh dependency and checker-boarding problems in topology optimization. In [50], the modification of the HPM was in the search volume of neighboring elements for filtering. The standard HPM uses a spherical search volume, while Osanov and Guest used a cylindrical volume to mimic the layer-by-layer AM process.

The minimization of thin features and support structures was carried out by Mhapsekar et al. [48]. They implemented the density filter in SIMP for feature size control by using a cylindrical search volume as Osanov and Guest. Figures 10.29 and 10.30 show the size feature control by Mhapsekar et al.

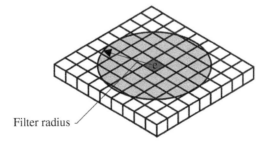

Filter radius

Figure 10.29   Neighboring elements to element *e* for a single layer. Source: Redrawn and adapted from [48].

(a)   (b)

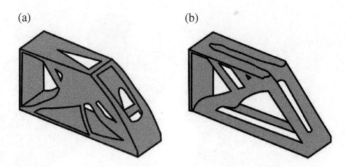

Figure 10.30 (a) Unconstrained and (b) constrained topology-optimized cantilever topologies. It is observable that the right optimized cantilever has controlled feature sizes unlike that on the left. Source: Redrawn and adapted from [48].

10.3.4.5 Other AM Constraints

There have been many attempts at capturing several other AM-based constraints within topology optimization such as the combined part, support structure, and build orientation optimization by Langelaar [51], support structure optimization by Kuo [52], stress-constrained topology optimization under uniform manufacturing uncertainties by da Silva et al. [53], residual thermal stresses in topology optimization for AM by Allaire and Jakabcin [54].

10.4 Lattice Structure Design

To fully understand the process and benefits of integrating lattice structures in design for AM, critical terminologies must be established. By way of definition, a lattice consists of interconnected struts or walls uniformly or randomly patterned, resulting in a cellular-like structural configuration. For most design applications, the structure is patterned so that material is distributed to critical areas in the design domain. The repeatability and adaptability of these "cell" structures give engineers the power to develop highly optimized lattice structures for different applications. Some benefits of lattice design, are a high strength to weight ratio, negative Poisson's ratio, enhancement of ossesointegration of bio-implants in orthopedic applications, high cooling rates due to better heat dissipation, low thermal expansion coefficient, high acoustic insulation, etc. [55, 56]. There are several terminologies that have been adopted by different authors regarding this subject area, but some unifying contextual definitions are given in the following sections.

10.4.1   Unit Cell

This can be described as the smallest structural arrangement captured in the lattice and usually possesses some repeatability from location to location within the domain concerned. Commonly used unit cells in 2D and 3D are shown in Figures 10.31 and 10.32.

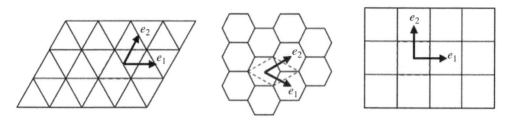

Figure 10.31   Some examples of 2D unit cells. Source: Adapted from [55].

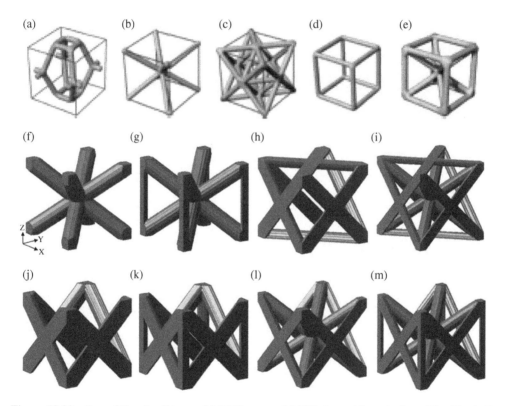

Figure 10.32   Some 3D unit cell types. (a) 3d Hexagon, (b) "X" shape, (c) octahedron, (d) cubic, (e,f) BCC, (g) BCCz, (h) FCC, (i) FBCC, (j) S-FCC, (k) S-FCCz, (l) S-FBCC, and (m) S-FBCCz. Source: Reprinted from [57] with permission from Elsevier.

10.4.2   Lattice Framework

The skeletal arrangement of a lattice is simply referred to as its framework. Technically, a framework is the structural arrangement of the base wireframe, which dictates how the material is distributed throughout the design domain. Functional lattice structures can be optimized by determining the best thickness values of the unit cell for every frame through resolving an

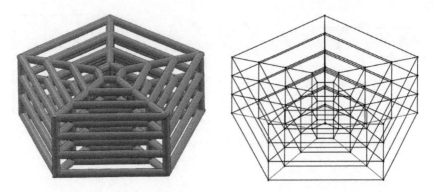

Figure 10.33 Relationship between a lattice structure and its framework.

objective function such as displacement or stress minimization. An example of a lattice structure and its framework is shown in Figure 10.33.

10.4.3 Uniform Lattice

These are commonly used lattices where a unit cell is replicated throughout a uniform grid. They can also be referred to as Orthogonal Translational Lattices (OTLs). Since the lattice is a translational pattern of a primitive or unit cell, the centroid of every unit cell can be referred to as the kernel point, and it represents in space, the position of the unique cell unit. When a lattice is assigned a global coordinate system as a means of reference, there also exists a local coordinate system for every unit cell that is developed by translating base vectors. If the centroid of a unit cell is considered the origin of the local coordinate system, we can derive the equation for a lattice node in terms of this system as well as the global coordinate system. Therefore, for a lattice node $p = (x, y, z)$ defined by its local coordinate system for a unit cell, its position $p = (x', y', z')$ defined in its global coordinate system is given by Eq. (10.47):

$$\begin{bmatrix} x' \\ y' \\ z' \end{bmatrix} = T + L \begin{bmatrix} x \\ y \\ z \end{bmatrix} = \begin{bmatrix} t_x \\ t_y \\ t_z \end{bmatrix} + \begin{bmatrix} l_{11} & l_{12} & l_{13} \\ l_{21} & l_{22} & l_{23} \\ l_{31} & l_{32} & l_{33} \end{bmatrix} \begin{bmatrix} x \\ y \\ z \end{bmatrix} \tag{10.47}$$

Here, (t_x, t_y, t_z) is the centroid of the lattice unit cell in terms of its global coordinate system, l_{ij} is a component of the transformation matrix L, and the components can be calculated in terms of Euler angles (α, β, γ), which define the orientation of a lattice unit cell, these angles are shown in Figure 10.34. Thus, L can be further defined by Eq. (10.48):

$$L = \begin{bmatrix} \cos\alpha.\cos\beta.\cos\gamma - \sin\alpha.\sin\gamma & \sin\alpha.\cos\beta.\cos\gamma + \cos\alpha.\sin\gamma & -\sin\beta.\cos\gamma \\ -\cos\alpha.\cos\beta.\sin\gamma - \sin\alpha.\cos\gamma & -\sin\alpha.\cos\beta.\sin\gamma & \sin\beta.\sin\gamma \\ \cos\alpha.\sin\beta & \sin\alpha.\sin\beta & \cos\beta \end{bmatrix}$$

$$\tag{10.48}$$

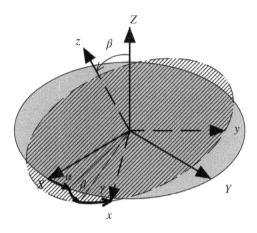

Figure 10.34 Lattice orientation showing Euler angles α, β, γ.

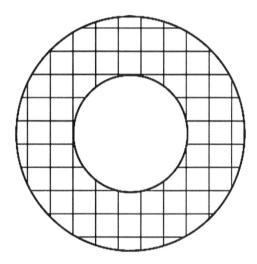

Figure 10.35 Uniform lattice structure. Source: Redrawn and adapted from [55].

Equation (10.47) can be used to resolve both cuboid and non-cuboid unit structures into a global coordinate system from a local one, but the transformation matrix L in Eq. (10.48) cannot be used for non-cuboid unit structures, therefore, has to be redefined. The scope of this text is limited to introducing basic mathematical concepts guiding the definition of unit cell structures, for a greater knowledge depth, the reader is referred to [55, 58]. A uniform lattice is shown in Figure 10.35.

10.4.4   Conformal Lattices

This lattice-type varies from the uniform lattice since on any given plane along the axis of lattice formation, the edges or vertices of the unit cells are aligned with the evolution of boundary iso-parametric surface curves; this is shown in Figure 10.36. The alignment can also be done on a volumetric level such that unit cell edges or vertices always align to boundary iso-parametric curves irrespective of the cutting plane's orientation. Unit cells are described differently for conformal lattices than their uniform counterparts because of variance in cell sizes; therefore, a cell containing the smallest number of lattice nodes stipulates the unit cell for a conformal lattice. For many conformal lattices, the framework can be defined by a volumetric mesh such as a tetrahedral or hexahedral mesh. In addition, the lattice framework can be defined by distance fields or implicit functions. Conformal lattices defer from random or heterogeneous lattices because their cell topologies remain the same, although cell shapes may vary based on location in the framework. Cell topology essentially relates to the basic structural layout of its framework, while cell shape typifies configuration in terms of frame angles and lengths.

10.4.5   Irregular/Randomized Lattices

Going by the name, these lattices do not have a regular pattern, thereby elevating the difficulty to characterize and effectively control them. The location and orientation of every unit cell are randomly distributed and can take several forms. The first step in an attempt to control these lattices is to identify primitive unit cells within the design domain in order to generate a lattice frame. Since the structure is randomized, the framework cannot be directly obtained from the primitive cells but can certainly be developed from their edges or vertices. A common type of

(a)

(b)

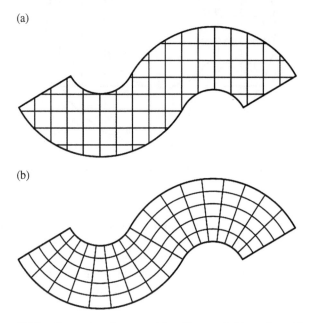

Figure 10.36   Comparison between (a) uniform and (b) conformal lattices.

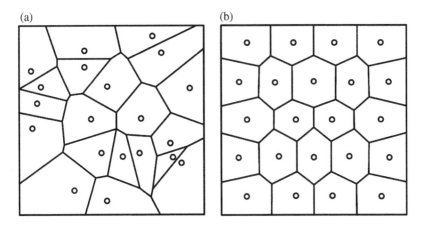

Figure 10.37　Voronoi-based lattice structures. (a) Normal Voronoi structure, (b) centroidal Voronoi structure (licensed under CC).

irregular lattice is the tetrahedron-based structure developed based on tetrahedron elements from the meshed design. For lattice generation of these structures, the Delaunay triangulation technique has been identified to be very suitable because the cell size distribution can be easily controlled by designers depending on their design's functionality and specifications. Additionally, this technique has proven to generate tetrahedron cells that are boundary controlled, leading to conformal lattice-type of structures. Another randomized structure is the Voronoi-based lattice structure shown in Figure 10.37.

The major difference between Voronoi cells and tetrahedron cells is that the former can combine several polyhedron shapes, unlike the latter, which mixes different tetrahedral sizes. It is, therefore, obvious that Voronoi lattices are more randomized like foam structures. The unit cell definition of both random lattice structures can be expressed in terms of their polyhedron shapes.

10.4.6　Design Workflows for Lattice Structures

There are several workflows proposed to integrate lattice structures in design for AM. Three methods: Hybrid geometric modeling, combined Topology Optimization and Lattice Design, Relative Density Mapping (RDM), will be discussed further.

10.4.6.1　Hybrid Geometric Model

This workflow for generating lattice structures is undertaken in three major steps [55]: lattice frame generation, development of geometric functions, and voxelization. The general workflow is shown in Figure 10.38, and each step of the model is explained in the preceding paragraphs.

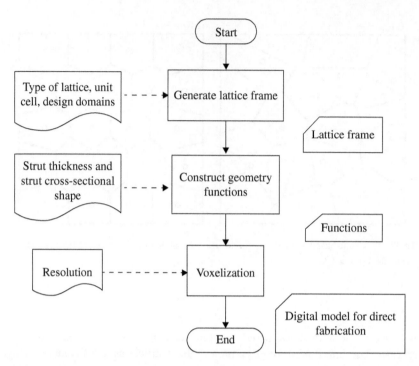

Figure 10.38   Design workflow of multiscale geometric modeling of lattice structures. Source: Redrawn and adapted from [55].

a. **Frame generation**   Steps in generating the lattice framework for uniform, conformal, and random lattices are explained:

Frame generation of uniform lattices can be achieved in six steps:

    i. The primitive cells of the lattice structure are first obtained.

    ii. The kernel points are generated by a kernel algorithm. A kernel is simply the center of a primitive cell. The kernel points can be of three types: internal kernel, external kernel, and boundary kernel.

    iii. Based on internal and boundary kernel points, nodes and struts can be generated for each lattice cell.

    iv. The frame formed from boundary kernel points has to be trimmed such that struts do not outstep design boundaries.

    v. Frames from internal and boundary kernel points are combined.

    vi. Duplicate struts are removed.

Frame generation of conformal lattice structures can be achieved in three steps:

    i. An auxiliary domain is generated based on the designer's conformal surface. The auxiliary domain should essentially contain the original domain and share the same original domain's conformal surfaces as its boundary surfaces. It should either be a mapped shape or be sliced into a mapped shape.

ii. From the auxiliary design domain obtained in (i), a 3D mesh generation is made using the "sweep" method to develop the hexahedron primitives. The sweep method replicates a 2D mesh through a 3D model's length from a source to a target surface.

iii. The final step entails trimming cells situated at nonconformal surfaces.

For the Voronoi-based irregular lattice structure, there are three major steps in generating its framework [55].

i. In the first step, it is required that points are randomly distributed within the bounding box of the design domain. These points randomly distributed are considered the centers of eventual Voronoi cells.

ii. From these points, Voronoi polyhedrons are generated within the design domain based on the Voronoi tessellation algorithm.

iii. In the final step, polyhedrons that outstep the boundaries of the design domain are trimmed.

b. **Development of geometric functions**   Next step in the hybrid geometric method is the development of geometric functions. These are implicitly expressed functions to determine the geometry of every strut [55]. There are several ways to define this function, but we will look at that proposed by Tang et al. [55], which is based on the generated lattice frame. In its general form, the implicit function to determine all the points in every solid strut in the lattice is expressed as:

$$S_t = \{P(x, y, z) \mid f(x, y, z) > 0, P \in \mathbb{R}^3\} \tag{10.49}$$

Here, f is the implicit function for a specific lattice frame and P is a p-norm function defined in the 3D Euclidean space. A piecewise polynomial is used to define this implicit function for every strut ei thus:

$$f_{ei}(r) = \begin{cases} a\left(1 - \dfrac{3r^2}{b^2}\right) - c & 0 \le r < \dfrac{b}{3} \\[2mm] \dfrac{3a}{2}\left(1 - \dfrac{r}{b}\right)^2 - c & \dfrac{b}{3} \le r \le 3 \\[2mm] -c & r \ge b \end{cases} \tag{10.50}$$

where r is defined as

$$r = min \, \|P - P_e\|_p, P_e \in e_i \subset \mathbb{R}^3, P \in \mathbb{R}^3 \tag{10.51}$$

r is the minimum distance between a point and strut while a and c are predefined constants but b is a function defined as:

$$b = \begin{cases} R\sqrt{\dfrac{3a}{a-c}} & \dfrac{c}{a} > \dfrac{2}{3} \\[4mm] \dfrac{R}{1 - \sqrt{\dfrac{2c}{3a}}} & \dfrac{c}{a} \le \dfrac{2}{3} \end{cases} \tag{10.52}$$

In Eq. (10.52), R is the strut size and to control the strut cross-sectional shape, P can be varied from 1 to infinity. When P is 1, the strut's cross section assumes a rhombic geometry, when P is 2, it transforms to a circle and increasing this value toward infinity, the cross section ends up in a rectangular shape with each side having a length of $2R$.

c. **Voxelization**   This final step of the hybrid geometric model converts the previous step's obtained geometric functions into a voxel representation. A voxel is the value of a specific point within the grid defined in the design domain. This value can be represented by a float, integer, Boolean, or vectors, however, this methodology focuses on using floating numbers. The voxelization process is divided into two substeps; the first is obtaining the bounding box of the design domain. The resolution of the selected AM process and struts' diameter determine the vectors of grid spacing. The second step involves evaluating the value of each point within the grid based on already established geometric functions. A lot more information and graphical illustrations on this methodology can be found in [55].

10.4.6.2   Combined Topology Optimization and Lattice Design for Constructing Functionally Graded Lattices

This methodology was proposed by Panesar et al. [57], and it combines topology optimization with lattice design to develop functionally graded lattices. The methodology consists of the following major steps: carrying out topology optimization that enables the realization of certain lattice strategies, using some proposed strategies, and developing functionally graded lattices.

a. **Obtaining optimal topologies**   Topology optimization has been extensively discussed in Section 10.3. This proposed method employs SIMP to obtain the optimum material density distribution on the basis of minimizing compliance under the volume constraint. Since this is a density-based approach, an interpolation function which is a simple power-law expression is required to assign a material property between void and solid to every density design variable. The power-law function contains a penalty value, which, when increased from 1 to 3, enhances the distinction between solid and void regions. This methodology assigns 1 as penalty for gray topologies and 3 for more unique solid and void distributions as shown in Figure 10.39. Once the topology is obtained either in discrete solid-void or gray forms, it is intersected with the design domain-filled lattice structure. The discrete solid-void regions are obtained by assigning a threshold to the density results with an iso-value of 0.5 [57].
b. **Lattice design strategies**   There are three different strategies with which this methodology introduces a lattice design within an optimized topology solution:
   i. **Intersected lattice:** This is obtained from the intersection of a uniform material density lattice structure within a design domain and the topologically optimized solid-void solution of the same design domain. The uniform material density lattice structure is such that every lattice cell within the domain has the same volume fraction as the intended topology-optimized solution. This type of lattice is shown in Figure 10.40b; it is noticed that for a volume fraction of 0.5, the square root 0.7071 is applied as the volume fraction of the lattice domain and the optimized topology domain before intersection such that the resulting solution reverts back to 0.5 volume fraction after the intersection.

(a)   (b)

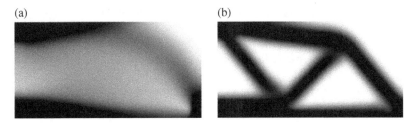

Figure 10.39   SIMP results for 0.5 volume fraction with penalty value (a) set to 1 and (b) set to 3. Source: Reprinted from [57] with permission from Elsevier.

(a)   (b)

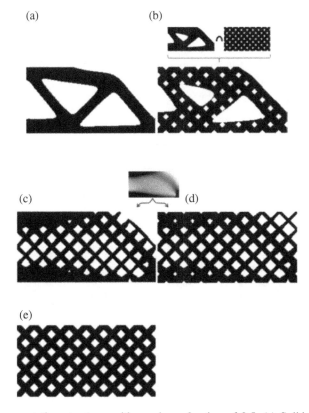

(c)   (d)

(e)

Figure 10.40   Representative structures with a volume fraction of 0.5. (a) Solid SIMP solution, (b) Intersected Lattice, (c) Graded Lattice, (d) Scaled Lattice, (e) Uniform Lattice. Source: Reprinted from [57] with permission from Elsevier.

ii. **Graded lattice:** This involves mapping a lattice with varying cell-to-cell material density onto a grayscale topology-optimized solution such that cells with density values below a minimum threshold are limited to 0 material density (void). Conversely, cells with material density above a maximum threshold are assigned 1 (solid), this type of lattice is shown in Figure 10.40c

iii. **Scaled lattice:** Here, a lattice with varying material density is mapped onto a rescaled grayscale topology-optimized solution such that the density is bounded between two limits slightly similar to graded lattices. The difference between graded and scaled lattices exists in the grayscale topology-optimized solution and application of minimum and maximum threshold densities. Depending on the threshold, densities out of bound are either driven to 0 or 1 in the case of graded lattices, but these values are driven to the exact threshold values for scaled lattices. Figure 10.40d shows a scaled lattice.

c. **Generation of functionally graded lattices** Generally, lattice structures can be developed using strut-based members as discussed in Section 10.4.6.1 or using surface-based representation, a popular type being Triply Periodic Minimal Surfaces (TPMSs). TPMSs are governed by implicit functions as given in Eq. (10.53)

$$f(x, y, z) = t \tag{10.53}$$

where t is an iso-value that dictates the offset from the level sets, that is, when the function value equates 0 [57]. In practice, nonetheless, surface representations of TPMS for solid structures are possible by inequality conditions expressed as

$$f(x, y, z) \le t \tag{10.54}$$

There are several TPMS structures, where their detailed geometric description and analysis are beyond the scope of this text, but they are presented in Figure 10.41.

(a) (b) (c) (d)

(e) (f) (g) (h)

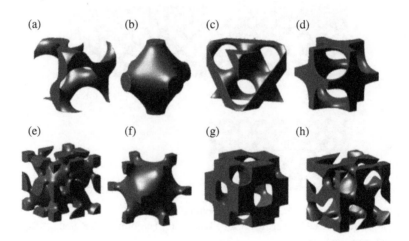

Figure 10.41 Library of surface-based unit cells: (a) G (Schoen's Gyroid), (b) P (Schwarz's Primitive), (c) D (Schwarz's Diamond), (d) W (Schoen's iWP), (e) Lidinoid (by Sven Lidin), (f) Neovius (by Schoen's student Neovius), (g) Octo (by Schoen), and (h) Split P. Source: Reprinted from [57] with permission from Elsevier.

The Schwarz's P surface in Figure 10.42 is an application of the inequality, when $t = 0.37$ in Figure 10.42a, -0.37 in Figure 10.42b, and the functional surface in Figure 10.42c is between -0.37 and 0.37. The surface function for Schwarz's P is defined as

$$f_P(x,y,z) = cos\left(\lambda_x x\right) + cos\left(\lambda_y y\right) + cos\left(\lambda_z z\right) \quad (10.55)$$

In Eq. (10.55), λ_i is the function periodicity expressed as $\lambda_i = 2\pi \times \frac{n_i}{L_i}$, where $i = x, y, z$; also, n_i is the number of times the cells are repeated while L_i is the lattice domain's global dimension.

To obtain a graded material as opposed to a homogenously dense material for a 3D model, a 4D formulation can be implemented by making the iso-value of the TPMS surface function (t) dependent on spatial coordinates (x, y, and z). The dependence of t on spatial coordinates controls the variation of material densities at different locations of the lattice structure. Therefore, Eq. (10.55) becomes

$$f(x,y,z) \leq t(x,y,z) \quad (10.56)$$

Figure 10.43 shows examples of linear grading for strut-based BCC and surface-based D–P unit cells where the volume fraction varies from 0.1 to 0.5, starting at the bottom to the top of each lattice structure.

(a)   (b)   (c)

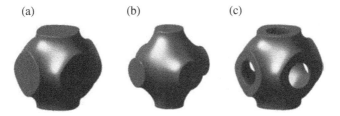

Figure 10.42 Schwarz's P surfaces (for value of $t = 0.37$); (a) f_P $(x,y,z) \leq t$, (b) f_P $(x,y,z) \leq -t$, and (c) $-t \leq f_P$ $(x,y,z) \leq t$. Source: Reprinted from [57] with permission from Elsevier.

(a)   (b)

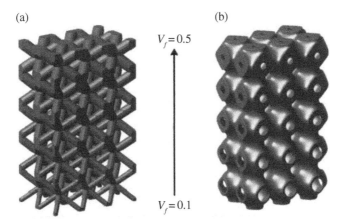

$V_f = 0.5$

$V_f = 0.1$

Figure 10.43 Example of a linear material grading for (a) strut-based BCC lattice and (b) surface-based D–P lattice. Source: Reprinted from [57] with permission from Elsevier.

10.4.6.3 Relative Density Mapping (RDM) Method

This method, proposed by Alzahrani et al. [56], uses the relative densities of the result from topology optimization to develop a lattice structure design. In a sense, it is likened to the previously discussed hybrid topology optimization and lattice design methodology, but there exist inherent differences. The major difference between RDM and the hybrid method is that in RDM, when a uniform lattice is superimposed on a topology-optimized solution, the cross-sectional area of struts overlay on denser elements will receive higher values compared to struts on less dense elements.

The workflow for RDM is given in the following paragraph and illustrated in Figure 10.44. It essentially begins with two inputs: a topology-optimized solution and a coarse FE mesh of the design domain.

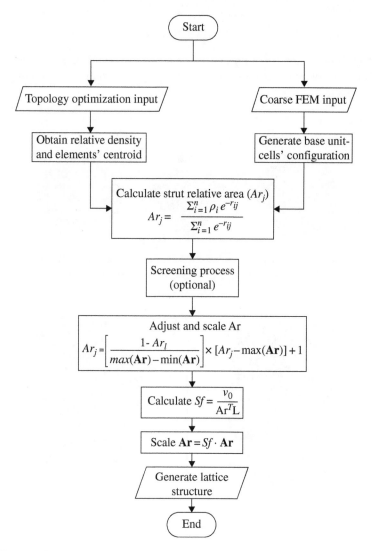

Figure 10.44　Workflow for RDM method. Source: Reprinted from [56] with permission from Elsevier.

The optimized topology solution is expected to contain the relative density of every finite element, the FE mesh nodal coordinates, and connectivity information. The second input should contain the nodal coordinates of the mesh as well as their connectivity information. This second input introduces a base lattice structure, from which the final lattice design will be built after it is interfaced with the topology-optimized solution. Several primitive unit cell-types can be utilized for the base lattice depending on the designer's specifications or requirements.

As aforementioned, this method begins with two inputs: a topology-optimized solution (density values of a FE mesh) and a base lattice structure. The elements' (obtained from the optimized solution) centroids are computed and stored in a vector, E_c, which is in turn used to calculate the distance of each element centroid r_{ij}, from strut j, as shown in Figure 10.45. Thereafter, the relative cross-sectional area of each strut can be obtained from ρ and r_j in:

$$Ar_j = \frac{\sum_{i=1}^{n} \rho_i e^{-r_{ij}}}{\sum_{i=1}^{n} e^{-r_{ij}}} \tag{10.57}$$

The equation obtains the cross-sectional area for every strut within the lattice. Since FEM analysis is done using beam elements, numerical instabilities will arise for extremely small cross-sectional areas; therefore, a minimum threshold is applied to Ar

$$min\,(Ar) = Ar_l \tag{10.58}$$

$$max\,(Ar) = 1 \tag{10.59}$$

In Eqn. (10.58), Ar_l, is a minimum value assigned to the relative cross-sectional area. In addition, the ratio between two relative areas must be left the same after this threshold has been applied, therefore, a scaling and adjustment are applied thus:

$$Ar = \left[\frac{1 - Ar_l}{max\,(Ar) - min\,(Ar)}\right] \times [Ar - max\,(Ar)] + 1 \tag{10.60}$$

In Eqn. (10.60), the term on the left side of the multiplication sign is the scaling portion while that on the right is the adjustment portion. A screening process can be carried out after the relative areas are calculated. This stage is optional because it is only meant to remove struts that do not contribute significantly to the design's functional performance. The screening process is shown in Figure 10.46 but will not be discussed further.

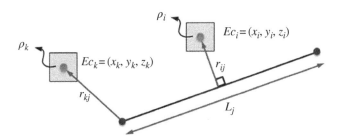

Figure 10.45 Elements' centroids from strut j. Source: Reprinted from [56] with permission from Elsevier.

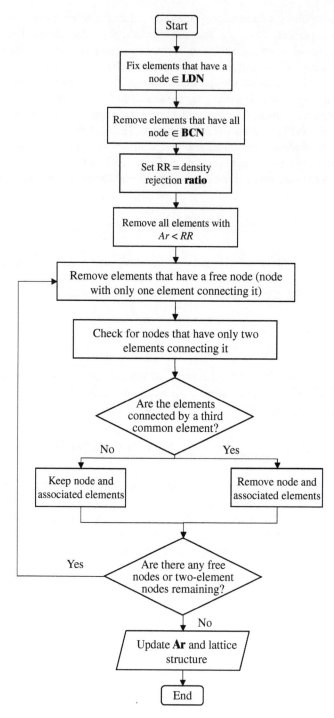

Figure 10.46 Screening process for RDM. Source: Reprinted from [56] with permission from Elsevier.

The final step of the RDM methodology is the calculation of the exact cross-sectional area of every strut. To obtain this, a scaling factor SF is computed and V_0, the volume constraint, is applied to ensure that the total volume of the resultant lattice structure adheres to this constraint. The new **Ar** values are then computed as follows:

$$SF = \frac{V_0}{\sum_{j=1}^{N} Ar_j \times L_j} \tag{10.61}$$

$$Ar = SF \times Ar \tag{10.62}$$

The steps in the previous paragraph outline the processes involved in generating a lattice by the RDM method. Detailed information, especially regarding the screening phase, can be obtained from the journal paper by Alzahrani et al. [56].

10.5 Design for Support Structures

In metal AM, the need for support structures cannot be overemphasized; therefore attention has to be paid to designing efficient supports specially used in PBF processes such as LPBF and EB-PBF. Although the narrative has moved to optimize supports to reduce material, printing time, and cost, there are core reasons why they cannot be completely eliminated yet in thermal-based additive technologies.

a. Thermal-based AM technologies are characterized by high thermal gradients induced during the build process; therefore, part deformation and residual stresses are usually inevitable due to significant heat accumulation. Support structures, therefore, help to conduct accumulated heat from the part to the build substrate. The effective thermal conductivity of the powder bed is significantly low compared to the built part and is insufficient to perform this heat transfer task.
b. As the part being built solidifies and is formed, support structures ensure the part does not fail under its own load, especially for features that are not self-supporting such as overhangs and holes. Supports act as fixtures that ensure the part remains rigidly fixed to the substrate as the printing progresses.
c. If we imagine a part built directly on a substrate, there will be significant difficulties attempting to detach the part from the substrate when the printing is done. Therefore, support structures are required for ease of removal as they are usually printed with process parameters inferior to that of the part.

Although supports offer huge benefits, as presented in the aforementioned points, there are several challenges they pose:

a. The most obvious challenge is an increase in material usage, which in turn contributes to an extended printing time, energy for removal, and cost.
b. The removal of supports might be challenging if there are tool access difficulties within the part.
c. The part surface finish may be seriously affected as support structures are being taken off.
d. To ensure the part can accommodate supports, extra time has to be dedicated to designing the part such that supports can be included optimally and easily removed eventually.

e. More often than not, supports end up as wasted materials because they might not be recyclable or reusable.

In order to utilize the benefits and get around the challenges supports pose, several methods for support structure generation have been proposed, and a number of them are already in use. Two major approaches for developing optimal support structures in metal AM are build orientation optimization and support structure optimization. Before going into the approaches for developing optimal support structures, we will look at principles that should guide the design process for support structures.

10.5.1 Principles that Should Guide Support Structure Design

There are general principles that should be considered when designing supports. Some of these principles are:

a. The support must prevent features from collapsing, curling, warping, or undergoing any other failure type unique to metal additive methods. Since most defects arise from thermal effects, it is advisable to carry out thermo-mechanical simulation of any setup to understand what the stress and deformation maps might be during printing. Some popular commercial simulation tools for this are Autodesk Netfabb, Simufact Additive, AdditiveLab.
b. Large holes should be oriented such that they are either parallel to print direction or subtended at small angles ($< 30°$). This is done to eliminate supports within these holes such that the smoothness and fit of the curved surfaces are preserved after general support removal.
c. Overhanging features with large areas that need support should be avoided. As much as it is possible, overhanging features should be aligned to or close to the build direction. This will reduce downward facing surfaces that will require support.
d. Ensure that features needing support have easy accessibility for support removal after printing. If there must be intricate surfaces, then it is advisable they are printed without support structures.
e. For easy support removal, support regions at contact surfaces of the part should be made as weak as possible. They should be strong enough to hold down the part but weak enough for easy removal.
f. Contact areas for support should be reduced as much as possible to preserve smoothness of surfaces.
g. To reduce material, printing time, and cost, build and/or support structure optimization should be considered.

10.5.2 Build Orientation Optimization

Build orientation optimization entails obtaining the best print or build layout for a part to reduce support material volume, deformation or residual stresses, printing time, or a combination of any. Several researchers have adopted this approach in efforts to optimize the printing process. Figure 10.47 shows how the build orientation of a dogbone sample can affect the support material volume used and residual stress formation.

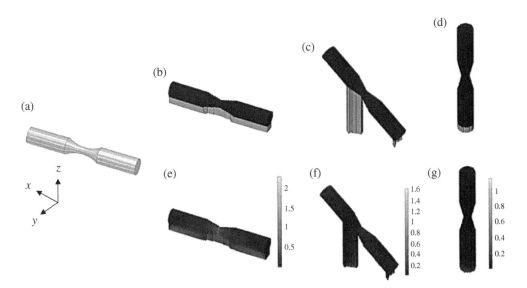

Figure 10.47 Effect of build orientation on a dogbone sample. (a) CAD model (b) dogbone oriented at 0° along the x–y plane, (c) dogbone oriented at 45° about the y-axis, (d) dogbone oriented parallel to the z-axis, (e) normalized stress distribution for (b) (f) normal. Source: Reprinted from [59] with permission from Elsevier.

Figure 10.50b–d shows that change in orientation significantly affects support material volume with 90° about the y-axis giving the least volume. When the dogbone is inclined at 45°, we observe that some support is introduced at the middle to ensure the part is securely fixed to the substrate or baseplate during printing. To observe the effects of these orientations on residual stress formation, a fast predictive model called the inherent strain method is used. The inherent strain method is an elastic FE procedure carried out layer by layer on the part and uses an experimentally or numerically validated inherent strain value as a boundary condition on every layer. In the dogbone example, this model is used to obtain the normalized residual stresses by the Hill's stress measure in the three different orientations and 2.3, 1.62, 1.14 were observed as maximum stresses for 0°, 45°, and 90° about the y-axis, respectively [59]. This indicates that the build orientation has considerable contributions in determining residual stress distributions, support material volume, and invariably overall print success.

A simple build orientation optimization of the framework of an arbor press assembly will be presented here.

10.5.2.1 Build Orientation Optimization of an Arbor Press Framework

Objectives of orientation optimization: Residual stress, deformation, and support structure volume minimization.

To explain the concept more effectively, an example is presented for which the following steps were taken to achieve the objectives:

a. The LPBF process simulation was done based on a mechanical inherent model using Simufact Additive 3.0.1. The inherent strain model is based on the elastic FE method, which accounts for combined thermal, plastic, and phase change strains during the build process of LPBF [60]. In this case, the inherent strains are obtained as estimated values generated by Simufact Additive and based on process parameters such as laser power, speed, beam diameter, and absorptivity (which is represented as an efficiency variable in Simufact Additive). Inherent strain values used in the simulation are

$$\varepsilon_x = -0.00121$$

$$\varepsilon_y = -0.000542$$

$$\varepsilon_z = -0.00505$$

b. The design variables for optimization are the rotational angles about x and y-axis on the build platform as shown in Figure 10.48.
c. A range of angles from 0 (horizontal) to 30° (about x and y axes) in steps of 7.5° can be selected as the design variables. Special preference is given to the cylindrical feature that houses the sleeve and spur gear, and it is the reason for not going beyond 30° rotation to avoid using support structures within the feature. The axis of the cylindrical feature is kept less than or equal to 30° to ensure the inner surfaces have reduced surface roughness; this is shown in Figure 10.49.
d. Results of residual stress, deformation, and support structure volume for each orientation are shown in the plots in Figures 10.50–10.52.

An increase in orientation angle increases the maximum deformation on the framework for x rotation while it fairly maintains the same value for y rotation. Maximum residual stress values are comparable for both x and y rotation except for 15° in y seen in while showing that support structures used considerably increases as the angle increases in x. Based on minimizing the objectives (residual stresses, displacement, and support structure volume), 22.5° and 30° subtended to the y-axis are comparatively the better orientations. Peak stresses and deformation are experienced in sharp features or at corners shown in Figure 10.53, and these features can be modified to obtain lower residual stress distributions in these areas.A more robust and holistic approach to this optimization will be to adopt the Euler angles of the part as the design variables, run the inherent strain simulations, and utilize a heuristic or mathematical-based optimizer to arrive at the global optimum of Euler angles.

10.5.3 Support Structure Optimization

The essence of optimizing supports is to obtain the best structural layout while keeping material and printing times at minimal levels. Magics Materialize software which is popular for preprocessing LPBF prints has utilized struts and box or "honeycomb-like" structures as support. More recently, though, innovative support structure optimization strategies have been developed and implemented: a few are "tree-like", lattice designed-, or topology optimized-structures. These three strategies will be discussed further:

a. **"Tree-like" supports:** Tree and inverted tree structures have been investigated and found to be of practical use as supports. Gan and Wong [61] in 2016 proposed an inverted Y support as they found out that parts printed with this support type performed best compared with Y and pin supports. The low performance of the Y support used could be attributed to the uneven spacing of the Y branches on the contact surface of the parts printed. The inverted Y and Y supports are shown in Figure 10.54.

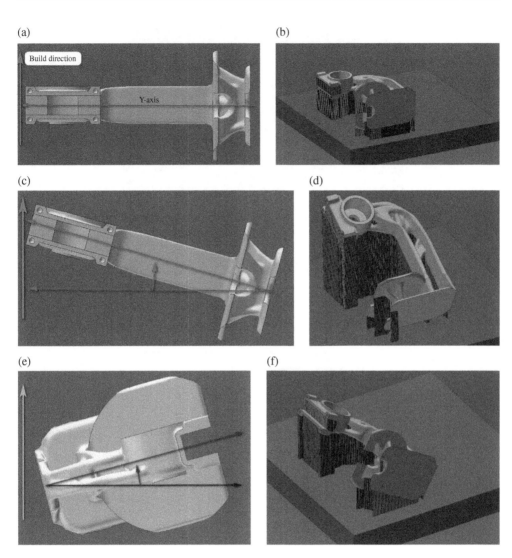

Figure 10.48 Orientation angles for framework build showing support structures required. (a) horizontal orientation, (b) support structures for horizontal orientation, (c) rotation about *x*-axis, (d) support structures for rotation about *x*-axis, (e) rotation about *y*-axis, and (f) support structures for rotation about *y*-axis.

Opposed to what Gan and Wong discovered, interestingly, Magics Materialize has reported using Y or tree-like structures the right way up while recording excellent print success. Their structure slightly differed from Gan and Wong's with each unitary support structure having more branches that are shorter and closer to the contact surface.

b. **Lattice and Cellular Support Structures:** Lattice structures were discussed extensively in Section 10.4 mainly with regard to part design. They can also be applied as supports because

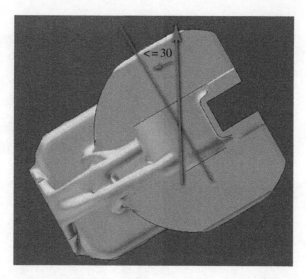

Figure 10.49 Angle threshold of ≤30° of cylindrical axis to building direction.

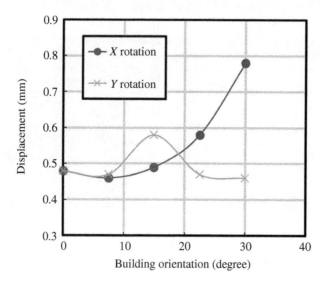

Figure 10.50 Maximum displacement in the framework.

of their proven structural strength and excellent material savings with volume fractions as low as 10%. Hussein et al. [62] discovered lattice support structures could be made using a gyroid unit cell with 8% volume fraction consequently removing and recycling 92% of unused and loose powders. Therefore, with low volume fractions of lattice structures, considerable material and energy savings can be made. Notwithstanding these immense benefits, caution has to be taken that lattice struts are not too thin as a result of very low volume fractions which will compromise the supports' structural integrity. In addition,

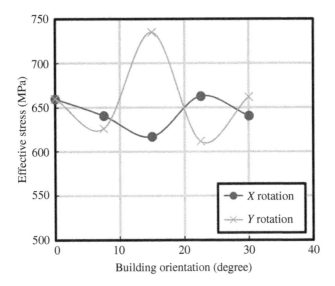

Figure 10.51   Maximum residual stress in the framework.

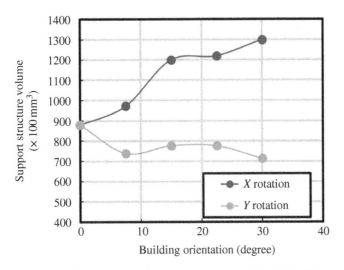

Figure 10.52   Volume of support structure used for the build.

when the lattice cell sizes are too large, there will be large unsupported areas at the contact between the support structure and part, and this can result in distortions due to thermal stresses. Figure 10.55 shows a cantilever structure using a gyroid and diamond lattice cell type as support structure.

In a similar fashion to lattice structures, minimal cellular support can be implemented, as demonstrated by Vaidya and Anand [63]. Dijkstra's shortest path algorithm can be used to

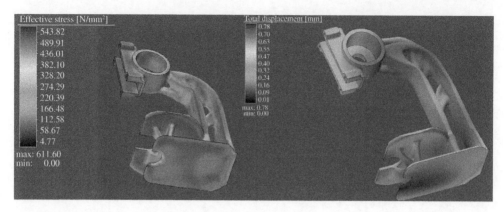

Figure 10.53　Residual stress and deformation plots for 22.5° rotation about the *y*-axis. Peak stresses and deformation can be seen on edges and corners.

(a)

(b)

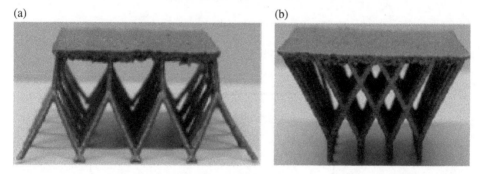

Figure 10.54　Inverted Y (a) and Y "tree-like" (b) structures supporting a thin plate. Source: Reprinted from [61] with permission from Elsevier.

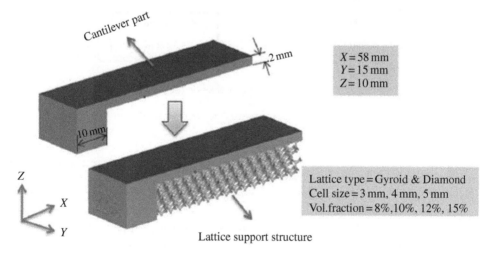

Figure 10.55　Cantilever CAD geometry supported by lattice structures. Source: Reprinted from [62] with permission from Elsevier.

generate minimal cell supports as opposed to filling up the support domain with lattice cells. The Dijkstra's shortest path algorithm generates the minimum cellular path between the interface unit cell and part unit cell voxels. The interface unit cell voxels are unit cells close to or touching the part's contact surfaces, as shown in Figure 10.56.

This method of support generation is especially suitable for parts with internal geometrical intricacies as shown in Figure 10.57.

c. **Topology optimization:** This is a powerful tool for obtaining the optimum structural layout in a design domain while achieving an objective or set of objectives under certain constraints. It can, therefore, be excellently applied to designing support structures. To incorporate topology optimization in support structure design, the build orientation must be determined first, then the support domain should be obtained from overhanging, unsupported, and hole features; this is shown in Figure 10.58. Before topology optimization

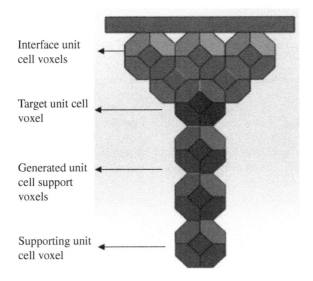

Interface unit
cell voxels

Target unit cell
voxel

Generated unit
cell support
voxels

Supporting unit
cell voxel

Figure 10.56   Unit cell voxel types. Source: Reprinted from [63] with permission from Elsevier.

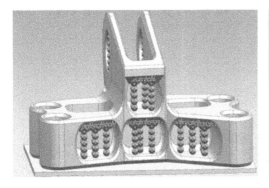

Figure 10.57   Designs with intricate features supported by minimal cellular supports. Source: Reprinted from [63] with permission from Elsevier.

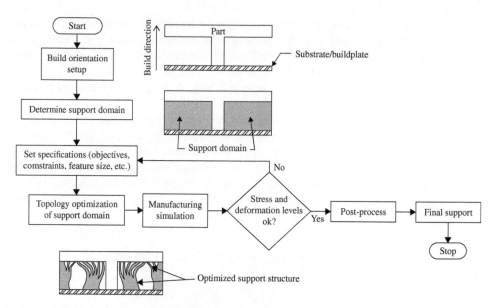

Figure 10.58 Workflow for integrated topology optimization and support structure design. Source: Adapted from [64].

can be carried out, objectives, constraints, feature size restrictions, and other desired specifications have to be set depending on the requirements or demands of the manufacturing method or designer. Apart from compliance minimization as an objective for support structure topology optimization, heat transfer conditions can also be optimized such as minimizing the structure's overall thermal compliance. After topology optimization is done, manufacturing simulation is recommended to check that the support and part have well-distributed, low residual stresses and deformation. Features with unusually high-stress concentrations should be noted and either eliminated or redesigned. In addition, for overall undesirable stress and deformation results, some specifications such as volume fraction, minimum feature thickness can be changed, and topology optimization carried out again. When desirable results are obtained, post-processing can be done and the final support structure arrived at.

A study by Zhou et al. [65] expressed how topology optimization can be integrated into the design process for support structures. They minimized the temperature gradient between selected points in the part design and ambience. They developed support structures that attempted to effectively dissipate heat generated during the printing process. Some of their results are presented in Figure 10.59.

The workflow shown in Figure 10.58 can be enhanced when manufacturing simulation is integrated into topology optimization. This eliminates the need to rerun topology optimization every time there is a disparity between simulation results and designer's or manufacturing requirements. In this case, there will be the need for a fast-predictive manufacturing simulation model, which the inherent strain model is popular for. A new and enhanced workflow for topology optimization integrated support structure design can be developed, shown in Figure 10.60.

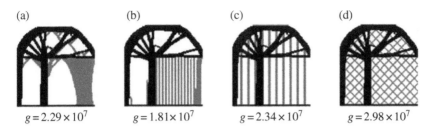

Figure 10.59 Different support types compared (Part in black and support in gray). (a) Optimized support with overhang constraint, (b) optimized support without overhang constraint, (c) and (d) are referenced supports. The variable g is the objective function. Source: Reprinted from [65] with permission from Elsevier.

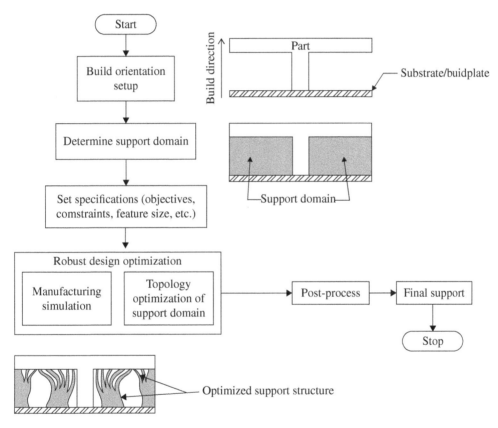

Figure 10.60 Enhanced workflow of a topology optimization integrated support structure design. Source: Adapted from [64].

10.6 Design Case Studies

In this section, several design case studies are summarized and presented for metal AM. This will enable the reader to understand, step-by-step, DfAM procedures.

10.6.1 Redesign of an Aerospace Bracket to be Made by LPBF

A redesign of an existing aerospace bracket was done by topology optimization. The workflow is shown in Figure 10.61 and the steps taken are outlined further:

a. Specifications for the bracket are established: Original bracket was made from aluminum alloy 7050-T7441, but the optimized bracket was targeted for titanium alloy Ti6Al4V. The boundary and load conditions were determined with three load cases at three different bearing sites. The original CAD domain is shown in Figure 10.62.
b. Topology optimization is done iteratively until the desired functional specifications are met with regard to volume fraction and stress levels.
c. The final topology-optimized result is reconstructed to smoothen features and aid Finite Element (FE) analysis. A comparison between the raw topology optimized result and the reconstructed CAD is shown in Figure 10.63.
d. FE analysis is done to compare the stress distribution of original and optimized designs under static loading for the third load case. It is observed that peak equivalent Von Mises stress is lower in the optimized design compared with the original model as in Figure 10.64.
e. The optimized design is printed taking into account the process accuracy, build orientation, support structures, surface roughness, and geometrical feasibilities. The printed part was subjected to Hot Isostatic Pressing (HIP) as a post-processing step to reduce pores and release residual stresses. Thereafter, the printed part is tested for metrological and mechanical conformities. A maximum deviation of 0.8 mm was observed when the scanned printed part was superimposed on the reconstructed optimized CAD model, this can be attributed to residual stresses or support removal. Mechanical tests showed that the factor of safety increased by a factor of 2 and can be a result of redesign and change of material from aluminum alloy to titanium alloy. In addition, an overall volume reduction of 54% and weight reduction of 28% were achieved.

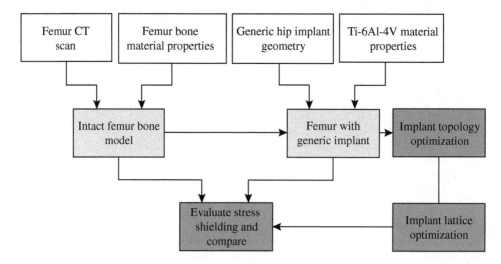

Figure 10.61 Workflow for redesign of aerospace bracket. Source: Redrawn and adapted from [64].

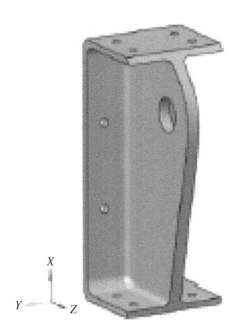

Figure 10.62 Original CAD. Source: Reprinted from [64] with permission from Elsevier.

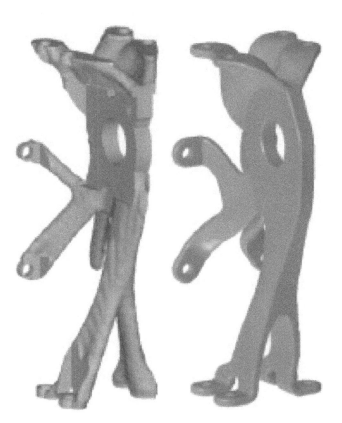

Figure 10.63 Topology-optimized solution and reconstructed design. Source: Reprinted from [64] with permission from Elsevier.

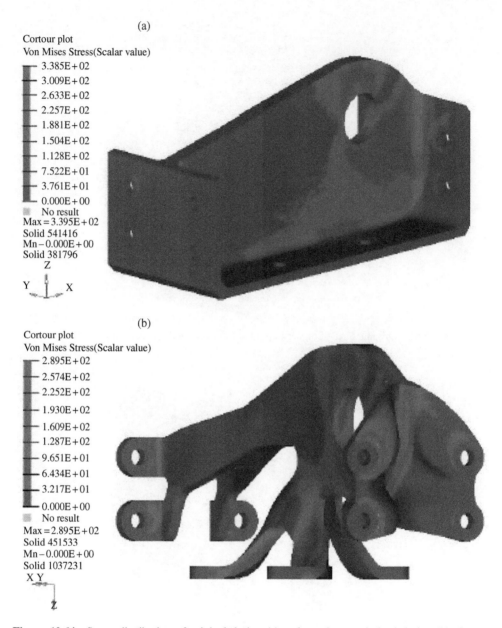

Figure 10.64 Stress distribution of original design (a) and topology-optimized design (b). Source: Reprinted from [64] with permission from Elsevier.

10.6.2 *Design and Development of a Structural Member in a Suspension Assembly Using EB Powder Bed Fusion*

This study was drawn from a project by the University of Warwick's Formula Student Team [66], and they followed the following steps to achieve their design:

a. Specifications were laid out: one was to develop a lightweight structure as an alternative to the original design another was to change the material from aluminum alloy to Ti6Al4V using EB-PBF, also, the component needed an exactly identical twin part.
b. The design domain, load, and boundary conditions are shown in Figure 10.65. It is important to note that the component's functional conditions must be well known, its interaction with other components, if in a system, must be well understood and properly represented as a boundary or load condition.
c. Other preprocessing actions such as applying shape control (e.g. implementing an overhang or a draw constraint), setting objectives and constraints were implemented. Altair Hyper-Mesh and Optistruct were used as pre-processor and solvers for this study. Some geometrical constraints placed on the optimization due to EB-PBF process and HyperMesh restriction were:
   i. Minimum member size of 6 mm to ensure connectivity of elements in the design and to prevent the printing process from geometrically distorting any feature within the component.
   ii. Maximum member size of 12 mm because HyperMesh requires this size to be at least twice that of the minimum size.

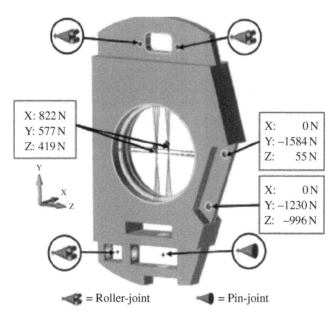

Figure 10.65 Design domain, boundary, and load conditions. Source: Reprinted from [66] with permission from Elsevier.

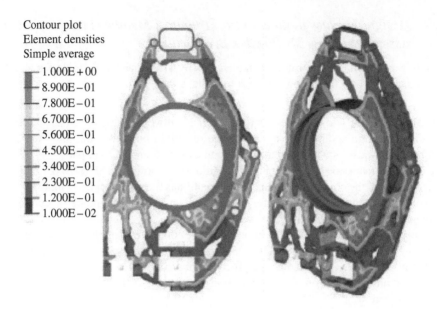

Figure 10.66 Topology-optimized result from Altair HyperMesh. Source: Reprinted from [66] with permission from Elsevier.

   iii. A symmetry constraint introduced along the y–z plane (as shown in Figure 10.53) so the
        result is exactly replicated.
  d. Topology optimization was done as shown in Figure 10.66 where a value of 1 means solid
     and 0 void, values in between are intermediate or gray areas. Again, reconstruction was
     done, and slight modifications such as edge filleting were made for neatness and easy tool
     access for support removal. The final design in Figure 10.54 achieved a 36% mass reduction
     with a 1.7 factor of safety. A maximum displacement of 0.32 mm was noticed from static
     FEA which was a 28% increase from the previous design.
  e. The final designs were printed by an ARCAM A2X EB-PBF machine using Ti6Al4V on a
     single build. For fits and smoothness, the bearing faces were machined by a CNC, but there
     were no other finishing or heat treatment given. The final components are shown in
     Figure 10.67.

10.6.3   Binder Jetting of the Framework of a Partial Metal Denture

This case study is an example of the design process for binder jetting applied in the medical
industry and how the consolidation of a binder-jetted part by infiltration and sintering can result
in a relatively high density of 99% while controlling the shrinkage. The steps followed to
achieve the objectives are outlined further:

  a. A 3D model of the denture was generated. The framework of the partial denture was made
     available and scanned using a micro Computer Tomography (μCT) to obtain an STL model.
     The initial and μCT-scanned framework are shown in Figure 10.68.
  b. The part was printed and the printing process is briefly described: The scanned denture is
     printed beginning with a thin layer of alloy 625 metal powder spread across the powder bed,

Figure 10.67 Printed and machined structural members. Source: Reprinted from [66] with permission from Elsevier.

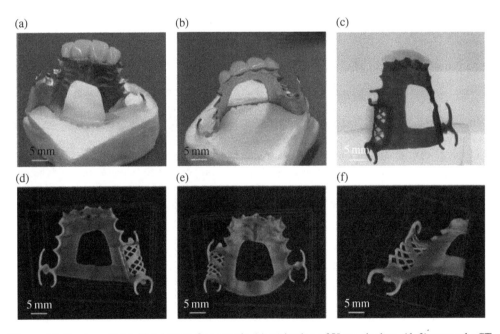

Figure 10.68 (a and b) Initial denture framework, (c) projection of X-ray shadow, (d–f) scanned µCT micrographs of the initial denture framework. Source: Reprinted from [67] with permission from Elsevier.

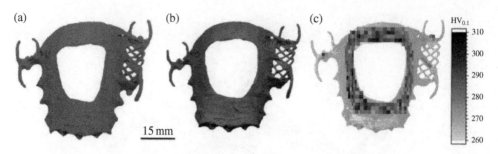

Figure 10.69 Optical images of dental framework's (a) green state (b) sintered state and (c) micro-hardness measurements. Source: Reprinted from [67] with permission from Elsevier.

then an ExOne M-Flex binder joins the loose powder according to STL information. The process of powder spreading and binder application happens cyclically until the part is fully printed.

c. After printing, the green part (as-printed denture framework) was cured in a box furnace at 175 °C to remove the binder.

d. Infiltration was done by embedding the part in alumina powder, and sintering was done in a Lindberg tube furnace at a constant temperature of 1285 °C for four hours.

e. To compensate for shrinkages, the scanned denture framework was scaled up to about 17% in the x and y direction, 19% in the Z direction before printing. The green part was therefore larger than the normal size and it shrinked during sintering and sandblasting. Figure 10.69 shows optical images of the denture framework's green state, sintered state, and micro-hardness measurements.

10.6.4   Redesign of a Crank and Connecting Rod

This case study shows why there might be a need to reconstruct optimized components in order to modify irregular features and/or carry out more complex analysis needed before AM production. A simple one-cylinder internal combustion engine is shown in Figure 10.70, with the components of interest captured within the rectangle.

Although compliance minimization in the static condition is the objective function for this optimization problem, it is important to analyze the components for stresses under dynamic response, which is their actual working state. This analysis can only be properly achieved when the components are reconstructed following the topology optimized results. A brief workflow for this problem is outlined and also shown in Figure 10.71:

a. **Objective:** To optimize (by compliance minimization) the crank and connecting rod in a simple one-cylinder IC engine and compare optimized design's dynamic structural response with that of initial design.

b. **Methodology:**
   i. Adopt the original design as design space for optimization (this is at the discretion of whoever is carrying out the optimization).
   ii. Apply load and boundary conditions within the optimization software.
   iii. Include manufacturing constraints such as overhang elimination, void filling depending on software capabilities.
   iv. Ensure optimized result meets static structural requirements. (Note: Steps c and d are iterative, and multiple results are usually obtained from which one is selected)

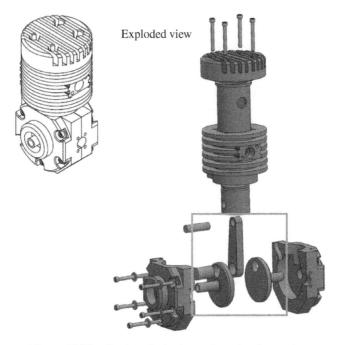

Figure 10.70   Single-cylinder internal combustion engine.

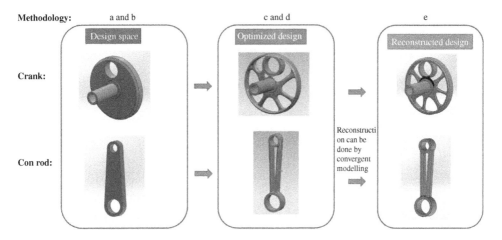

Figure 10.71   Topology optimization of the crank and connecting rod with reconstruction.

   v. Reconstruct results to allow use for further analysis (dynamic structural requirements) and/or to eliminate unwanted features.

c. **Results:** Minimum factor of safety in dynamic response through the time duration of two seconds for the original and optimized models are 175 and 91 respectively shown in Figure 10.72; with these values, further decisions and actions can be taken to update the design or move forward with a working design based on the results.

10.6.5   Redesign of a Mechanical Assembly

This case study looks at a complete redesign of an arbor press assembly shown in Figure 10.73.

a. Objectives
   a. Reduce the overall weight by at least 50% while maintaining a safe minimum safety factor of 1.5.
   b. Obtain optimum build orientation by process simulation for the column and base.

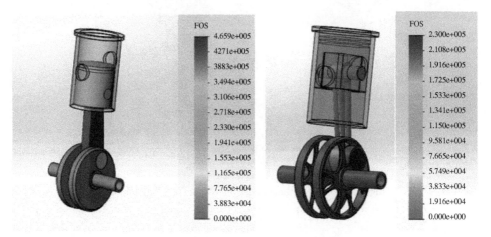

Figure 10.72   Dynamic response of original (left) and topology-optimized crank and connecting rod.

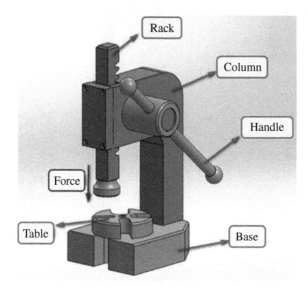

Figure 10.73   Arbor press assembly.

b. **Design Specifications**
   a. Press force capacity: 1/2 tonne force
   b. Material: Stainless Steel 316L
   c. Bounding box dimension: $160 \times 90 \times 80$ mm
   d. Minimum Factor of safety: 1.5

c. **AM Technology:** LPBF, ElectroOptical Systems (EOS) M290 with a build volume of $250 \times 250 \times 325$ mm.

d. **Topology Optimizer:** TrueSOLID by Frustum in NX 12

e. **Requirements and Manufacturing Constraints:** Functional and performance requirements were specified and serve as benchmarks during the design process. Manufacturing constraints were also considered during topology optimization and process simulation to ensure the LPBF process was successfully executed and unhindered by unwanted design features. These requirements and constraints are outlined in Table 10.3.

f. **Mass Analysis of components** A simple mass analysis was performed for every component of the assembly to investigate the percentage of each of the total. Table 10.4 shows that the significantly heavy components are the column and base amounting to 2.179 kg or 76.5% of the total mass; therefore, these components are given high importance during optimization.

Table 10.3 Objectives and constraints for topology optimization of arbor press.

No	Functional requirements	Performance requirement	Manufacturing constraints
1	Press capacity of 1/2 tonne force	Minimum factor of safety factor of 1.5 for peak stresses on any component in comparison to the material's proof or yield stress	Overhang and void elimination
2	All parts must hold together on and off operation	—	Residual stress and deformation minimization by orientation optimization
3	—	—	Support structure minimization

Table 10.4 Arbor press components and mass composition.

Part no.	Mass (kg)	Percentage of total (%)
1 Column	1.163	40.83
2 Cover	0.0955	3.35
3 Table	0.1181	4.15
4 Base	1.016	35.67
5 Rack	0.0906	3.18
6 Gear	0.1150	4.03
7 Gear Sleeve	0.1231	4.32
8 Ball Ends	0.0087	0.31
9 Handle	0.0689	2.42
10 Rack Pad	0.0357	1.25
11 Gear Cap	0.0084	0.30
12 Pin	0.0057	0.20
Total:	**2.8487**	**100**

g. **Topology optimization and LPBF manufacture of components**   The components considered for are the column, base, table, cover, rack, and gear. The design space for each component was their original design. Table 10.5 shows the load and boundary conditions of the components and optimized results obtained after a number of design iterations.

The optimized parts were printed using an EOS M290, which is a LPBF machine. The following process parameters were used for printing components:

- Skin Power: 140 W
- Skin Speed: 1250 mm/s
- Core Power: 160 W
- Core Speed: 1300 mm/s
- Layer Thickness: 40 μm
- Hatching distance: 90 μm
- Scanning Strategy: x,y with 67° rotation

Some of the printed parts are shown in Figure 10.74.

h. **Why the column failed during support removal**   Due to time constraints on the machine for this project, process simulation was skipped, and although other parts were printed successfully, the column was not as lucky. From investigations on the crack formed during failure and real-time images from the machine's monitoring system, failure occurred because:
  i. A thin feature in the column had a corresponding thin and lengthy support structure during print setup.
  ii. Due to this support structure's slender nature, the movement of the machine's recoater was sufficient to flex this structure and shift its tip from the position.
  iii. This resulted in a ditch in the powder bed, causing a poor powder distribution in the region below the thin feature and discontinuation of the support structure build shown in Figure 10.75a.
  iv. A combination of residual stresses in the feature and poor print quality in and around the feature resulted in a huge crack in the column's boss during support structure removal shown in Figure 10.75b.
i. **Redesigning the column and base – Assembly coupling**   In an attempt to redesign and reprint the column, a decision to make the assembly simpler by combining the column and base to form a unitary framework was made. This combination was possible because there is no relative motion between these two components and no cogent need to have them apart. This framework was topologically optimized with an attempt to get the least weight possible while adhering to the minimum safety factor benchmark based on observable maximum Von Mises stress during FEA. The design domain is shown in Figure 10.76.
j. **Topology optimization of framework**   Several optimizations were done on the framework taking into account important manufacturing constraints such as overhang restriction in the direction of the cylindrical boss axis, which houses the spur gear and sleeve. The justification

Table 10.5 Design space and optimized result of arbor press components.

Part no.	Design space, load and boundary conditions	Optimized result
1 Column		
2 Base		
3 Gear		

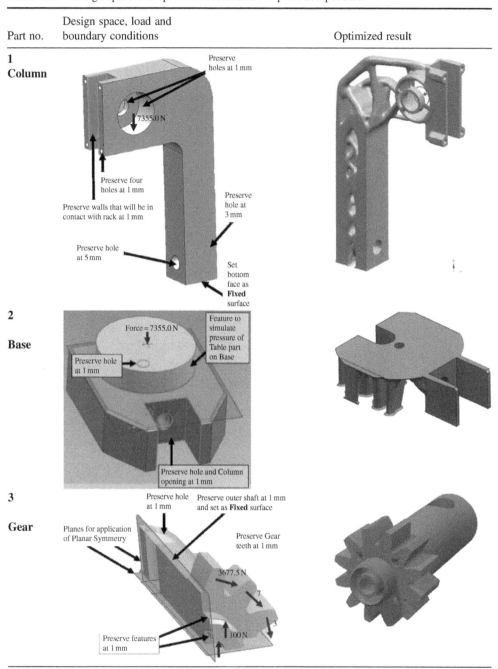

(*continued overleaf*)

Table 10.5 *(continued)*

Part no.	Design space, load and boundary conditions	Optimized result

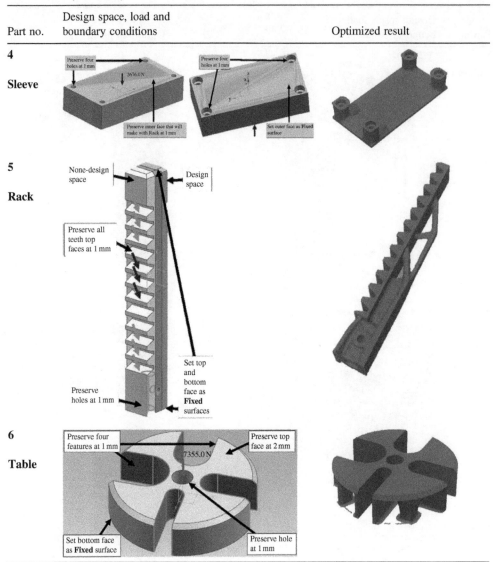

4 Sleeve		
5 Rack		
6 Table		

for choosing this overhang direction was to ensure the framework was printed with the boss axis either parallel or near parallel to the building direction while ensuring features are not overhanging. The essence was to have the surface roughness in this cylindrical feature low enough to allow smooth relative motion between the rotating sleeve and stationary framework.

(a)

(b)

(c)

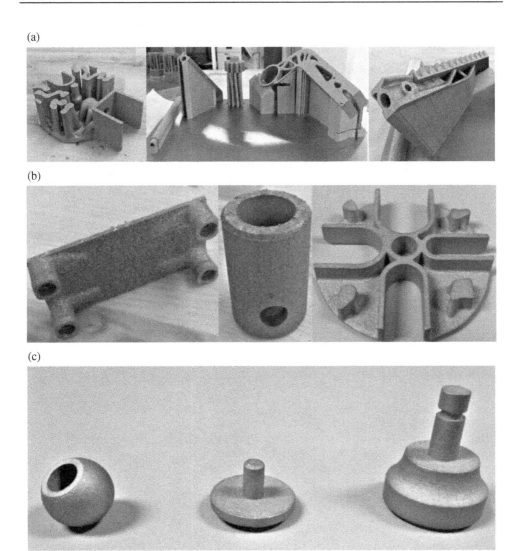

Figure 10.74   Images of printed components (a) base, rack, gear, handle, column still on the substrate (b) cover, sleeve, and table, (c) ball end, pin, and rack pad.

Wang et al. [68] and Hitzler et al. [69] report the yield stress of Stainless Steel 316 L LPBF-manufactured parts between 400 and 650 MPa. In Table 10.6, designs 1–4 satisfy the minimum safety factor of 1.5, while design 5 is much lower. Design 2 was adopted as the final framework because it was within a safe region considering maximum stress and displacement and the factor of safety.

(a)   (b)

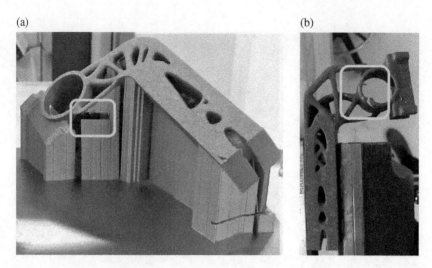

Figure 10.75 (a) Printed column before cutting from substrate, (b) printed column after cutting and support removal. The highlighted regions show where the column failed during printing (a) and after support removal (b).

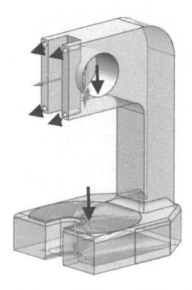

Figure 10.76 Design space for framework. The arrows indicate the loads on the framework.

10.6.6 Solid-Lattice Hip Prosthesis Design

This study was carried out to develop a hip prosthesis that causes reduced stress shielding and meets some fatigue requirements compared with the existing hip prosthesis. To achieve this, a generic implant model was generated, then topology optimization was carried out to have an

Table 10.6 Design iterations for framework.

No	Design 1	Design 2	Design 3	Design 4	Design 5
Volume Fraction	0.36	0.3	0.3	0.23	0.14
Max. stress (MPa)	103	142	149	151	1225
Max disp. (mm)	0.29	0.40	0.44	0.51	5.55
FOS (ys: 400 MPa)	3.9	2.8	2.7	2.6	0.32

idea of the optimum material structural layout. When this was determined, a lattice structure was designed to replace this layout. Major steps are explained further:

a. A workflow was first established and it began with getting a CT scanned STL file of the femur bone. Femur bone material properties were determined and this was meant for comparisons between an intact bone model and the eventual optimized model.

b. The CT scanned femur bone was used to develop a generic implant model; a reconstructed model was made to closely mimic the scanned bone. The implant model was to be obtained by sectioning the reconstructed bone model close to its greater trochanter. A freeze contact was used to bond the implant head and stem, also to bond the surface of the implant stem and the femur bone. A freeze contact is one that fixes a node on one part with a master surface on another part.

c. Material properties of all parts were assumed isotropic except the cortical bone which was assigned orthotropic properties. For material type, the implant head was assigned Cobalt Chrome Alloy, Implant stem Ti6Al4V; the bone properties were obtained from literature. Load conditions were characterized for the femur standing-up, standing, climbing stairs, and jogging. The surfaces of the bottom portion of the femur bone were assumed to be securely fixed.

d. Topology optimization. Four major aspects are considered in this stage

   i. **Design space:** The stem (without the neck region) of the implant served as the design domain for topology optimization.

   ii. **Constraints:** Limitations were placed on the compliance function, which was the objective function of optimization. It was observed from literature that the head region of the femur bone suffers the most from stress shielding. Therefore, compliance computation was done regionally. A constraint to keep the compliance of each section within 5% of the original femur model was placed. Volume fractions of about 10%, 20%, and 30% were also placed as constraints. In addition, draw and extrusion manufacturing constraints were applied to the optimization model.

   iii. **Objective:** a global compliance minimization. The optimization was carried out in Altair's OptiStruct.

e. **Lattice Optimization:** This stage consisted of four major considerations.

   i. **Construction of lattice-solid model:** To represent the topology-optimized result in a solid-lattice form, full dense elements were assigned shell and tetra elements while semi-dense elements were assigned face and body-centered cubic unit cells having vertical struts (FBCCZ). Since the implant was to be printed standing up, this unit cell was suitable because no strut was oriented less than 45° to the build-plate. Each unit cell was assigned a size between 4 and 7 mm, which varied throughout the implant. The lattice optimization was also implemented in Altair's OptiStruct.

   ii. **Design variables:** Since a standard FBCCZ unit cell was used, the variable subject to change was the diameter of struts co-joined at every node. The diameter on a strut could be varied, but every strut at a node had to possess the same diameter. Specification for this diameter was a range of 0.5–2 mm while the skin shell thickness varied between 0.5 and 5 mm.

   iii. **Constraints:** Fatigue requirement was set at a standard ISO7206-4 load for lattice optimization, which limited the implant's von Mises stress to no more than 600 MPa.

   iv. **Objective:** The overall goal for the lattice structure optimization was to minimize total material volume to reduce printing time and cost.

     The final part was printed by LPBF technology using an EOS M290 printer. The full design process can be accessed in [70].

10.7 Summary

The following paragraphs summarize the learning objectives for this chapter:

Design frameworks for AM
DfAM is important for engineers who want to achieve a failure-free build process, standardized product quality, optimal product performance, or efficiency without exceeding a proposed cost limit. Several frameworks specially tailored for AM have been established to ensure the most is obtained from the technology. Topology optimization forms the heart of the design workflow for AM because it is known to produce design solutions suitable for this manufacturing technology.

Design rules and guidelines
Design rules refer to certain constraints in a design domain due to limitations in material or machine processes, while design guidelines form restrictions that stem from technicalities inherent in the manufacturing technology.

Topology optimization for AM
Topology optimization obtains the best structural layout for a structural design problem and usually does this by optimizing an objective function under constraints for a designable volume. For topology optimization to be successfully integrated into the design for AM, the constraints posed by this manufacturing technology must be captured in the optimization algorithms. Amongst several AM constraints that have been captured are overhang minimization or elimination, feature size control, self-supporting features, void elimination, support structure minimization or elimination, residual stress or deformation reduction.

Lattice structure design
A lattice consists of interconnected struts or walls uniformly or randomly patterned, resulting in a cellular-like structural configuration. Many lattice structure design methodologies typically begin with topology optimization of the structural design problem, then implement strategies that infuse lattice designs within the topological results. As mentioned severally in the text, topology optimization forms an important aspect of the general structural design process for many AM processes.

Design for support structures
Two major approaches to arrive at optimal support structures in metal AM are build orientation optimization and support structure optimization, and they can be combined to achieve better results. Build orientation optimization entails obtaining the best layout for printing a part (usually obtaining the optimal Euler angles of the part) to reduce support material volume, residual stresses, deformation, printing time, or a combination of any. Support structure optimization obtains the best structural layout of the support that achieves its set objectives while keeping material and printing times at minimal levels.

References

1. D. W. Rosen, "Computer-aided design for additive manufacturing of cellular structures," Comput. Aided. Des. Appl., vol. 4, no. 5, pp. 585–594, 2007, doi: https://doi.org/10.1080/16864360.2007.10738493.
2. Y. Tang and Y. F. Zhao, "Esda2014-20381 integration of topological and functional optimization in design for additive manufacturing," Proc. ASME 2014 12th Bienn. Conf. Eng. Syst. Des. Anal." pp. 1–8, 2015.

3. S. Yang and Y. F. Zhao, "Additive manufacturing-enabled design theory and methodology: a critical review," Int. J. Adv. Manuf. Technol., vol. 80, no. 1–4, pp. 327–342, 2015, doi: https://doi.org/10.1007/s00170-015-6994-5.

4. T. Primo, M. Calabrese, A. Del Prete, and A. Anglani, "Additive manufacturing integration with topology optimization methodology for innovative product design," Int. J. Adv. Manuf. Technol., vol. 93, no. 1–4, pp. 467–479, 2017, doi: https://doi.org/10.1007/s00170-017-0112-9.

5. M. P. Bendsøe and O. Sigmund, Topology optimization: theory, methods, and applications, 2nd edition, no. 724, 2003.

6. M. Orquéra, S. Campocasso, and D. Millet, "Design for additive manufacturing method for a mechanical system downsizing," Procedia CIRP, vol. 60, pp. 223–228, 2017, doi: https://doi.org/10.1016/j.procir.2017.02.011.

7. N. Lei, X. Yao, S. K. Moon, and G. Bi, "An additive manufacturing process model for product family design," J. Eng. Des., vol. 27, no. 11, pp. 751–767, 2016, doi: https://doi.org/10.1080/09544828.2016.1228101.

8. M. Mani, P. Witherell, and J. Jee, "Design rules for additive manufacturing: a categorisation," ASME 2017 Int. Des. Eng. Tech. Conf. Comput. Inf. Eng. Conf., vol. 65, no. 7–8, pp. 77–82, 2017, doi: https://doi.org/10.1115/DETC2017-68446.

9. T. Schnabel, M. Oettel, and B. Mueller, "Design for Additive Manufacturing Guidelines and Case Studies for Metal Applications," no. May 2017, pp. 1–110, 2017. [Online]. Available: http://canadamakes.ca/wp-content/uploads/2017/05/2017-05-15_Industry-Canada_Design4AM_141283.pdf.

10. V. On, "Design guide: direct metal laser sintering (DMLS)," Xometry, pp. 1–16, 2017. [Online]. Available: https://cdn2.hubspot.net/hubfs/340051/Design_Guides/Xometry_DesignGuide_DMLS.pdf%0Ahttps://cdn2.hubspot.net/hubfs/340051/Design_Guides/design-guide-dmls-2016.pdf?submissionGuid=b32-de25b-1089-4288-b601-c251ca887c3e.

11. J. Allison, C. Sharpe, and C. Seepersad, "Powder bed fusion metrology for additive manufacturing design guidance," Addit. Manuf., vol. 25, no. June, pp. 239–251, 2019, doi: https://doi.org/10.1016/j.addma.2018.10.035.

12. B. Vayre, F. Vignat, and F. Villeneuve, "Identification on some design key parameters for additive manufacturing: application on electron beam melting identification on some design key parameters for additive manufacturing: application on electron beam melting," Procedia CIRP, vol. 7, no. December, pp. 264–269, 2013, doi: https://doi.org/10.1016/j.procir.2013.05.045.

13. C. Oswald, "Additive manufacturing design considerations for production in aerospace additive manufacturing design: considerations for production," LAI International. http://app.emarketeer.com/resources/12517/My_Documents/AddManufacturingDesignPart1-LAI-FINAL.pdf (accessed 1 October 2019).

14. P. Edwards, A. O'Conner, and M. Ramulu, "Electron beam additive manufacturing of titanium components: properties and performance," J. Manuf. Sci. Eng., vol. 135, no. 6, p. 061016, 2013, doi: https://doi.org/10.1115/1.4025773.

15. M. Doyle, K. Agarwal, W. Sealy, and K. Schull, "Effect of layer thickness and orientation on mechanical behavior of binder jet stainless steel 420 + bronze parts," Procedia Manuf., vol. 1, pp. 251–262, 2015, doi: https://doi.org/10.1016/j.promfg.2015.09.016.

16. "3Dhubs," "Metal 3D Printing," 3Dhubs. https://www.3dhubs.com/guides/metal-3d-printing/ (accessed 1 October 2019).

17. P. W. Christensen and A. Klarbring, An introduction to structural optimization, vol. 153, Springer, 2009, ISSN: 0925-0042 Series: www.springer.com/series/6557.

18. B. Hassani and E. Hinton, Homogenization and structural topology optimization, Springer, 1999. ISBN-13: 978-1-4471-1229-7 e-ISBN-13:978-1-4471-0891-7

19. E. Andreassen, A. Clausen, M. Schevenels, B. S. Lazarov, and O. Sigmund, "Efficient topology optimization in MATLAB using 88 lines of code," Struct. Multidiscip. Optim., vol. 43, no. 1, pp. 1–16, 2011, doi: https://doi.org/10.1007/s00158-010-0594-7.

20. K. Svanberg, "MMA and GCMMA – two methods for nonlinear optimization," Kth, vol. 1, pp. 1–15, 2007.

21. M. Bruyneel and P. Duysinx, "Note on topology optimization of continuum structures including self-weight," Struct. Multidiscip. Optim., vol. 29, no. 4, pp. 245–256, 2005, doi: https://doi.org/10.1007/s00158-004-0484-y.

22. T. Gao and W. Zhang, "Topology optimization involving thermo-elastic stress loads," Struct. Multidiscip. Optim., vol. 42, no. 5, pp. 725–738, 2010, doi: https://doi.org/10.1007/s00158-010-0527-5.

23. J. D. Deaton and R. V. Grandhi, "Stress-based design of thermal structures via topology optimization," Struct. Multidiscip. Optim., vol. 53, no. 2, pp. 253–270, 2016, doi: https://doi.org/10.1007/s00158-015-1331-z.

24. E. Hooijkamp and F. Van Keulen, "Topology optimization for transient thermo-mechanical problems," In 24th International Congress of Theoretical and Applied Mechanics, no. August, pp. 2–3, 2016.

25. H. Rodrigues and P. Fernandes, "A material based model for topology optimization of thermoelastic structures," Int. J. Numer. Methods Eng., vol. 38, no. 12, pp. 1951–1965, 1995, doi: https://doi.org/10.1002/nme.1620381202.

26. Q. Li, G. P. Steven, and Y. M. Xie, "Thermoelastic topology optimization for problems with varying temperature fields," J. Therm. Stress., vol. 24, no. 4, pp. 347–366, 2001, doi: https://doi.org/10.1080/01495730151078153.

27. J. Hou, J. H. Zhu, and Q. Li, "On the topology optimization of elastic supporting structures under thermo-mechanical loads," Int. J. Aerosp. Eng., vol. 2016, 2016, doi: https://doi.org/10.1155/2016/7372603.

28. J. D. Deaton and R. V Grandhi, "Stress-based topology optimization of thermal structures," 10th World Congr. Struct. Multidiscip. Optim., pp. 1–10, 2013, doi: https://doi.org/10.2514/6.2013-1466.

29. G. Vantyghem, V. Boel, M. Steeman, and W. De Corte, "Multi-material topology optimization involving simultaneous structural and thermal analyses," Struct. Multidiscip. Optim., vol. 59, no. 3, pp. 731–743, 2019, doi: https://doi.org/10.1007/s00158-018-2095-z.

30. A. Takezawa and M. Kobashi, "Design methodology for porous composites with tunable thermal expansion produced by multi-material topology optimization and additive manufacturing," Compos. Part B Eng., vol. 131, pp. 21–29, 2017, doi: https://doi.org/10.1016/j.compositesb.2017.07.054.

31. Y. Du, Z. Luo, Q. Tian, and L. Chen, "Topology optimization for thermo-mechanical compliant actuators using mesh-free methods," Eng. Optim., vol. 41, no. 8, pp. 753–772, 2009, doi: https://doi.org/10.1080/03052150902834989.

32. R. Ansola, E. Vegueria, and J. Canales, "An element addition strategy for thermally actuated compliant mechanism topology optimization," Eng. Comput. (Swansea, Wales), vol. 27, no. 6, pp. 694–711, 2010, doi: https://doi.org/10.1108/02644401011062090.

33. R. Ansola, E. Veguería, J. Canales, and C. Alonso, "Electro-thermal compliant mechanisms design by an evolutionary topology optimization method," Eng. Comput. (Swansea, Wales), vol. 30, no. 7, pp. 961–981, 2013, doi: https://doi.org/10.1108/EC-12-2011-0150.

34. S. A. Jahan, T. Z. Wu, E. M. Yi, T. Hazim, Z. Andres, A. Jing, N. Douglas, G. Razi, L. Xingye, and H. Weng, "Implementation of conformal cooling & topology optimization in 3D printed stainless steel porous structure injection molds," Procedia Manuf., vol. 5, pp. 901–915, 2016, doi: https://doi.org/10.1016/j.promfg.2016.08.077.

35. A. T. Gaynor and J. K. Guest, "Topology optimization considering overhang constraints: eliminating sacrificial support material in additive manufacturing through design," Struct. Multidiscip. Optim., vol. 54, no. 5, pp. 1157–1172, 2016, doi: https://doi.org/10.1007/s00158-016-1551-x.

36. A. Garaigordobil, R. Ansola, J. Santamaría, and I. Fernández de Bustos, "A new overhang constraint for topology optimization of self-supporting structures in additive manufacturing," Struct. Multidiscip. Optim., pp. 1–15, 2018, doi: https://doi.org/10.1007/s00158-018-2010-7.

37. Y. Mass and O. Amir, "Topology optimization for additive manufacturing: accounting for overhang limitations using a virtual skeleton," Addit. Manuf., vol. 18, pp. 58–73, 2017, doi: https://doi.org/10.1016/j. addma.2017.08.001.

38. K. Zhang, G. Cheng, and L. Xu, "Topology optimization considering overhang constraint in additive manufacturing," Comput. Struct., vol. 212, pp. 86–100, 2019, doi: https://doi.org/10.1016/j. compstruc.2018.10.011.

39. X. Wang, C. Zhang, and T. Liu, "A topology optimization algorithm based on the overhang sensitivity analysis for additive manufacturing," IOP Conf. Ser. Mater. Sci. Eng., vol. 382, p. 032036, 2018, doi: https:// doi.org/10.1088/1757-899X/382/3/032036.

40. E. van de Ven, R. Maas, C. Ayas, M. Langelaar, and F. van Keulen, "Continuous front propagation-based overhang control for topology optimization with additive manufacturing," Struct. Multidiscip. Optim., vol. 57, no. 5, pp. 2075–2091, 2018, doi: https://doi.org/10.1007/s00158-017-1880-4.

41. X. Guo, J. Zhou, W. Zhang, Z. Du, C. Liu, and Y. Liu, "Self-supporting structure design in additive manufacturing through explicit topology optimization," Comput. Methods Appl. Mech. Eng., vol. 323, pp. 27–63, 2017, doi: https://doi.org/10.1016/j.cma.2017.05.003.

42. M. Langelaar, "An additive manufacturing filter for topology optimization of print-ready designs," Struct. Multidiscip. Optim., vol. 55, no. 3, pp. 871–883, 2017, doi: https://doi.org/10.1007/s00158-016-1522-2.

43. F. Mezzadri, V. Bouriakov, and X. Qian, "Topology optimization of self-supporting support structures for additive manufacturing," Addit. Manuf., vol. 21, no. March, pp. 666–682, 2018, doi: https://doi.org/ 10.1016/j.addma.2018.04.016.

44. W. Zhang and L. Zhou, "Topology optimization of self-supporting structures with polygon features for additive manufacturing," Comput. Methods Appl. Mech. Eng., vol. 334, pp. 56–78, 2018, doi: https:// doi.org/10.1016/j.cma.2018.01.037.

45. X. Qian, "Undercut and overhang angle control in topology optimization: A density gradient based integral approach," Int. J. Numer. Methods Eng., vol. 111, no. 3, pp. 247–272, 2017, doi: https://doi.org/10.1002/ nme.5461.

46. M. Langelaar, "Topology optimization of 3D self-supporting structures for additive manufacturing," Addit. Manuf., vol. 12, pp. 60–70, 2016, doi: https://doi.org/10.1016/j.addma.2016.06.010.

47. Z. Li, D. Z. Zhang, P. Dong, and I. Kucukkoc, "A lightweight and support-free design method for selective laser melting," Int. J. Adv. Manuf. Technol., vol. 90, no. 9–12, pp. 2943–2953, 2017, doi: https://doi.org/ 10.1007/s00170-016-9509-0.

48. K. Mhapsekar, M. McConaha, and S. Anand, "Additive manufacturing constraints in topology optimization for improved manufacturability," J. Manuf. Sci. Eng., vol. 140, no. May, pp. 1–16, 2018, doi: https://doi. org/10.1115/1.4039198.

49. S. Liu, Q. Li, W. Chen, L. Tong, and G. Cheng, "An identification method for enclosed voids restriction in manufacturability design for additive manufacturing structures," Front. Mech. Eng., vol. 10, no. 2, pp. 126–137, 2015, doi: https://doi.org/10.1007/s11465-015-0340-3.

50. M. Osanov and J. K. Guest, "Topology optimization for additive manufacturing considering layer-based minimum feature sizes," Proc. ASME 2017 Int. Des. Eng. Tech. Conf. Comput. Inf. Eng. Conf., pp. 1–8, 2018.

51. M. Langelaar, "Combined optimization of part topology, support structure layout and build orientation for additive manufacturing," Struct. Multidiscip. Optim., vol. 57, pp. 1985–2004, 2018.

52. Y. Kuo, C. Cheng, Y. Lin, and C. San, "Support structure design in additive manufacturing based on topology optimization," Struct. Multidiscip. Optim., vol. 57, no. 168, pp. 183–195, 2018, doi: https://doi.org/10.1007/s00158-017-1743-z.

53. G. A. da Silva, A. T. Beck, and O. Sigmund, "Stress-constrained topology optimization considering uniform manufacturing uncertainties," Comput. Methods Appl. Mech. Eng., vol. 344, pp. 512–537, 2019, doi: https://doi.org/10.1016/j.cma.2018.10.020.

54. G. Allaire and L. Jakabčin, "Taking into account thermal residual stresses in topology optimization of structures built by additive manufacturing," Math. Model. Methods Appl. Sci., vol. 28, no. 12, pp. 1–45, 2017.

55. Y. Tang, G. Dong, and Y. F. Zhao, "A hybrid geometric modeling method for lattice structures fabricated by additive manufacturing," Int. J. Adv. Manuf. Technol., pp. 1–20, 2019, doi: https://doi.org/10.1007/s00170-019-03308-x.

56. M. Alzahrani, S. K. Choi, and D. W. Rosen, "Design of truss-like cellular structures using relative density mapping method," Mater. Des., vol. 85, pp. 349–360, 2015, doi: https://doi.org/10.1016/j.matdes.2015.06.180.

57. A. Panesar, M. Abdi, D. Hickman, and I. Ashcroft, "Strategies for functionally graded lattice structures derived using topology optimisation for additive manufacturing," Addit. Manuf., vol. 19, pp. 81–94, 2018, doi: https://doi.org/10.1016/j.addma.2017.11.008.

58. A. Gupta, K. Kurzeja, J. Rossignac, G. Allen, P. Srinivas, and S. Musuvathy, "Designing and processing parametric models of steady lattices," Georgia Tech Library, 2018. http://hdl.handle.net/1853/60058 (accessed 1 October 2019).

59. L. Cheng and A. To, "Part-scale build orientation optimization for minimizing residual stress and support volume for metal additive manufacturing : theory and experimental validation," Comput. Des., 2019, doi: https://doi.org/10.1016/j.cad.2019.03.004.

60. C. Paper, "A modified inherent strain method for fast prediction of residual deformation in additive manufacturing of metal parts," Solid Free. Fabr. Symp., no. August, pp. 2539–2545, 2017.

61. M. X. Gan and C. H. Wong, "Journal of materials processing technology practical support structures for selective laser melting," J. Mater. Process. Tech., vol. 238, pp. 474–484, 2016, doi: https://doi.org/10.1016/j.jmatprotec.2016.08.006.

62. A. Hussein, L. Hao, C. Yan, R. Everson, and P. Young, "Advanced lattice support structures for metal additive manufacturing," J. Mater. Process. Tech., vol. 213, no. 7, pp. 1019–1026, 2013, doi: https://doi.org/10.1016/j.jmatprotec.2013.01.020.

63. R. Vaidya and S. Anand, "Optimum support structure generation for additive manufacturing using unit cell structures and support removal constraint," Procedia Manuf., vol. 5, pp. 1043–1059, 2016, doi: https://doi.org/10.1016/j.promfg.2016.08.072.

64. M. Seabra Azevedo, José Araújo, Aurélio Reis, Luís Pinto, Elodie Alves, Nuno Santos, Rui Pedro, Mortágua, João, "Selective laser melting (SLM) and topology optimization for lighter aerospace componentes," Procedia Struct. Integr., vol. 1, pp. 289–296, 2016, doi: https://doi.org/10.1016/j.prostr.2016.02.039.

65. M. Zhou, Y. Liu, and Z. Lin, "ScienceDirect topology optimization of thermal conductive support structures for laser additive manufacturing," Comput. Methods Appl. Mech. Eng., vol. 353, pp. 24–43, 2019, doi: https://doi.org/10.1016/j.cma.2019.03.054.

66. D. Walton and H. Moztarzadeh, "Design and development of an additive manufactured component by topology optimisation," Procedia CIRP, vol. 60, pp. 205–210, 2017, doi: https://doi.org/10.1016/j.procir.2017.03.027.

67. A. Mostafaei, E. L. Stevens, J. J. Ference, D. E. Schmidt, and M. Chmielus, "Binder jetting of a complex-shaped metal partial denture framework," Addit. Manuf., vol. 21, no. January, pp. 63–68, 2018, doi: https://doi.org/10.1016/j.addma.2018.02.014.

68. Y. M. Wang Voisin, Thomas, Mckeown, Joseph T, Ye, Jianchao, Calta, Nicholas P, Li, Zan, Zeng, Zhi, Zhang, Yin, Chen, Wen, Roehling, Tien Tran, Ott, Ryan T, Santala, Melissa K, Depond, Philip J, Matthews, Manyalibo J, Hamza, Alex V, Zhu, Ting, "Additively manufactured hierarchical stainless steels with high strength and ductility," Nat. Mater., vol. 17, no. January, pp. 63–69, 2018, doi: https://doi.org/10.1038/NMAT5021.

69. L. Hitzler, J. Hirsch, B. Heine, M. Merkel, W. Hall, and A. Öchsner, "On the anisotropic mechanical properties of selective laser-melted stainless steel," Materials (Basel)., vol. 10, no. 1136, pp. 1–19, 2017, doi: https://doi.org/10.3390/ma10101136.

70. Y. He, D. Burkhalter, D. Durocher, and M. J. Gilbert, "Solid-lattice hip prosthesis design: applying topology and lattice optimization to reduce stress shielding from hip implants," in Proceedings of the 2018 Design of Medical Devices Conference, 2018, pp. 1–21, 2018.

11

Monitoring and Quality Assurance for Metal Additive Manufacturing

---

Learning Objectives

After reading this chapter, you should learn

- Different classes of in-situ sensing devices used for metal AM processes.
- Theories of some sensors used in metal AM.
- Various applications of statistical approaches as quality assurance paradigms.
- Fundamental and applications of machine learning techniques in AM.

11.1 Why are Closed-Loop and Quality Assurance Platforms Essential?[1][1]

AM has changed the entire manufacturing enterprise by offering unique features to fabricate complex shapes with superior mechanical properties. In the last decades, through an exponential advancement, AM has been promoted from a prototyping to a series and mass production platform. Like all conventional techniques, quality assurance procedures/tools are of the utmost importance in aiding manufacturers in quality management and certification. For this purpose,

---

[1] This section is partially reproduced from the previous work of one of the authors [1] with permission from the publisher conveyed through Copyright Clearance Center, Inc.

Metal Additive Manufacturing, First Edition. Ehsan Toyserkani, Dyuti Sarker, Osezua Obehi Ibhadode, Farzad Liravi, Paola Russo, and Katayoon Taherkhani.
© 2022 John Wiley & Sons Ltd. Published 2022 by John Wiley & Sons Ltd.

in-line melt pool monitoring (MPM) devices, installed in laser-based AM systems, provide vital real-time information about process characteristics implicitly or explicitly leading toward understanding the quality of printed parts.

Besides, many AM challenges remain unaddressed such as dimensional errors, undesired porosity, delamination of layers, and undesired material properties. The high sensitivity of AM processes to disturbances is one of the main challenges. A small change in absorbed laser power, for example, can cause a large change in the melt pool size in laser powder bed fusion (LPBF). Also, changes in the mass flow of the powder in powder-fed directed energy deposition (DED) can produce significant variations in the overall geometry and microstructure of the deposited track. The metal AM processes are governed by many process parameters (such as laser power, scan strategy, hatch spacing, scan speed, gas flow, and recoater speed). These factors profoundly affect the process, causing the formation of defects [2, 3].

To make the process more stable and less susceptible to disturbances, it is important to understand the effects of involved process parameters on the printed part quality.

Typically, the AM process parameters can be categorized into two groups:

- Intrinsic
- Extrinsic

Intrinsic parameters are those related to the substrate and powder properties.

Some of these parameters include absorptivity, thermal conductivity, heat capacity, thermal diffusivity, and substrate geometry. Extrinsic parameters are related to the hardware used in the process, such as laser, powder feeder, recoater, and positioning system. Some of these parameters include laser average energy, laser focal point, powder feeder mass flow, nozzle position and orientation, and the laser and substrate's relative position and velocity.

In general, there is no direct control on intrinsic parameters; however, the effects of changes in intrinsic parameters can be compensated by controlling the extrinsic parameters. For example, a change in the absorptivity can be adjusted by controlling the average laser power.

The control of extrinsic parameters, on the other hand, is relatively easy for most of the parameters mentioned above. Commercial lasers, powder feeders, and positioning systems are equipped with built-in controllers that allow the user to set the desired values. It is, in fact, the approach that is used in many applications. After much trial and error, the parameters that result in a good part quality for a specific application are obtained. These parameters, such as average laser power, focal point, speed, and powder feed rate, are then used in open-loop process control.

In Figure 11.1, the block diagram of a typical closed-loop system is shown. When for the open-loop process, the controllable parameters are set to predefined values, and there is no feedback from the process output to adjust the parameters, closed-loop control of the process will monitor one of the process parameters (e.g. deposition temperature) in real-time in order to compare it with the desired value. Then, a controller takes the difference between the measured and desired values and sends a signal to an actuator(s) to tune one or more process parameters.

Some output parameters can be measured directly such as deposition dimensions and surface roughness, where they can be measured using a camera or a laser scanner. Indirect parameters include the melt pool temperature and size that indirectly indicate the part quality. The microstructure and rate of solidification are other indirect parameters that can be used to determine the state of the solidified parts.

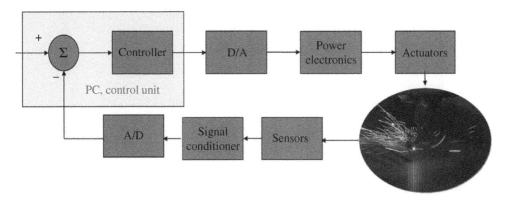

Figure 11.1 Concept of closed-loop control for a typical AM process.

Having a measure of the deposition condition and comparing this condition with the desired one, a controller can be used to change the controllable parameters on the fly to compensate for any disturbance or change in the process. It is important to note that all the controllable parameters are not independent, and only some of them are needed to achieve closed-loop control. For example, laser power and velocity are correlated, and only one of them is needed to change the effective power injected to a volume of the substrate.

In general, open-loop control of AM processes is useful when the application is fixed and is repeated over and over. Finding the right parameters is very time-consuming; however, as long as there is no change in the application, this control method is effective and efficient. However, the need for quality assurance has pushed industry to substantially look at sensors that predict the part quality as it builds. It is very important for the certification purpose to understand a printed part has any defects or not. Thus, the implementation of closed-loop control and/or quality assurance platforms to AM processes is important for the widespread adoption of AM in industry. Since quality assurance and repeatability are critical problems, researchers and commercial vendors have installed in-situ sensing equipment to capture information from the process to understand better and control the process. Types and setup strategies of sensors are discussed in the following section.

11.2   In-Situ Sensing Devices and Setups

There are a variety of sensors used in metal AM processes. Measurable parameters related to the melt pool, such as temperature, size, rate of solidification, and the composition of the plume are measured with different devices when the measurable signals are correlated with the quality of printed parts. The sensors that are used for these measurements can be categorized into different classes as elaborated in the following section.

11.2.1   Types of Sensors Used in Metal AM

In-situ sensors can be categorized into two major groups: radiative and non-radiative. Radiative sensors measure the radiation emission emitted from the melt pool; however, non-radiative sensors may measure other features of the process, such as wrinkles on the powder bed. Both sensors are discussed in the following:

11.2.1.1　Radiative Sensors

Radiative sensors are required to measure radiative rays being emitted from the melt pool and surrounding zones. Radiative sensors are widely used in in-situ monitoring because they normally provide fast responses, high reliability, high resistance to dust, and temperature variations. Besides, the data collected from radiative sensors provide vital information about the AM process. The radiative sensors, based on their working principles and outputs, can be classified into cameras, inline coherent imaging (ICI), X-ray imaging, photodiode, and pyrometer.

Cameras

Cameras are one of the popular radiative sensors in AM. They are used to capture images and videos from the process. In general, cameras are divided into analog and digital. Analog cameras transmit an electronic analog signal, whereas digital cameras transmit binary data in the electronic signal by an A/D converter. Analog cameras are less expensive and less complicated than digital ones; however, analog cameras suffer from a few limitations compared to digital cameras. Digital cameras provide images with higher resolution at higher frame rates with less noise. Different cameras have been widely used to monitor the metal AM processes to provide information about surface quality, dimensional accuracy of printed parts, porosity, and deformation. The in-situ cameras for metal AM are usually categorized into two groups based on the wavelength spectrum, as shown in Figure 11.2: (i) from visible to near-infrared (NIR) and (ii) from NIR to long wavelength.

Visible to NIR

The visible to NIR images have a wavelength between 400 and 1000 nm that have been used to monitor surface temperature [5–7], dimensional accuracy [8–10], and deformation [11, 12]. Charged-coupled device (CCD) and the complementary metal–oxide–semiconductor (CMOS) are two popular devices used in AM monitoring. To choose an appropriate camera, the pixel, frame rates, and dynamic ranges should be considered to take high-quality images. For example, a Photron FASTCAM SA5 model 1300K C2 high-speed camera is placed above the powder bed surface in LPBF to record video at a frame rate of 10 000 fps from the consolidation of metal to investigate the effect of layer thickness [5]. CCD cameras with an

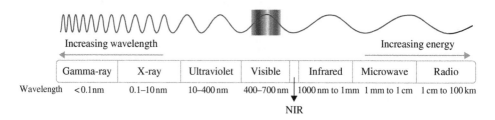

Figure 11.2　Light spectrum. Source: Redrawn and adapted from [4].

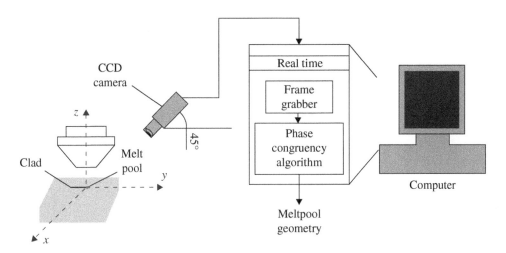

Figure 11.3 Schematic diagram of sensor installation for detection melt pool in DED. Source: Republished with permission of Elsevier, from [13].

8-bit-depth grayscale and resolution of 648×488 pixels are used to monitor the melt pool geometry, as shown in Figure 11.3 [13]. A CMOS camera can also be used to detect the effect of process parameters on the melt pool geometry [14]. Figure 11.4 shows the types of radiations emitted from LPBF and captured by a CMOS camera.

One challenge encountered using a camera in the visible range is image acquisition in the presence of a heat source, spatter, and small dynamic molten pool. As a remedy, coupled optical

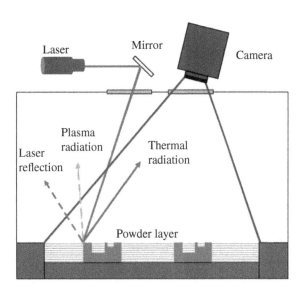

Figure 11.4 Types of radiations emitted from LPBF and captured by CMOS camera. Source: Redrawn and adapted from [15].

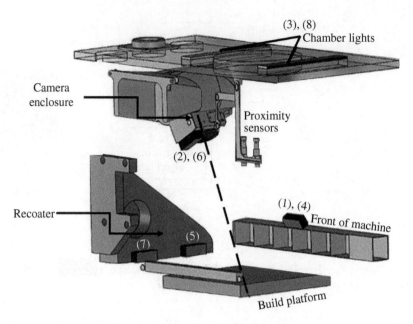

Figure 11.5 Location of camera and light sources in the build chamber. Source: Republished with permission of Elsevier, from [19].

filters such as interference filters can be used. Choosing proper technical parameters with a proper calibration and using illumination devices will resolve these issues. For example, an SVCam-hr29050, SVS-VISTEK monochrome CCD camera can be mounted on the LPBF setup when an image calibration technique can be used to improve the quality of captured images [16]. The tilt and shift lenses can be implemented to minimize the perspective distortion. With these remedies, in-process instabilities (e.g. insufficient powder, poor support structure, and coater damage) can be detected [16]. A camera coupled with a visible wavelength filter can be used with a 430-nm band-pass filter on LDED [17], and a high-speed camera with a visible waveband cutoff filter (350–800 nm) can be used for the same process [18]. High-resolution digital single-lens reflex (DSLR) camera (Nikon D800E) along with several flashing lights can be used to generate eight different lightening conditions, as shown in Figure 11.5 [19, 20].

NIR to LWIR

NIR to long-wavelength infrared (LWIR) cameras capture images between 1000 and 1 mm. Although infrared spectrums can eliminate noises and disturbance, several factors should be considered while working with this type of camera [21]. For example, infrared cameras with short wavelengths need an interference filter to avoid interference from the laser, since many laser sources' wavelength is in the range of the camera's wavelength. While infrared cameras with long wavelengths have a strong resistance to high temperatures, such cameras are too expensive. The thermal or infrared camera is installed in AM to monitor the process's thermal behavior, such as maximum temperature and cooling rates. Thermal images are then analyzed to identify process quality, such as dimensional accuracy [22–24], porosity [25–27], and cooling rate [28–30].

For example, the thermographic camera PYROVIEW 640G/50 Hz/25° × 19°/compact + (DIAS Infrared GmbH, Dresden, Germany) can be placed on an SLM 280HL machine (SLM Solutions AG, Lübeck, Germany) to take images with 640 × 480 optical resolution and with 4.8–5.2 μm spectrum range [22]. Images can then be used to extract melt pool geometry, delamination, and splatter. A micro-bolometer IR camera with 8-ms response time, 50-Hz maximum frame rate, and 250-μm pixel resolution can be used with EOS M270 to monitor temperature gradients and thermal inhomogeneity [31]. Besides, CCD and CMOS cameras equipped with an infrared filter are sometimes used to take images in the infrared range. For example, a CCD camera coupled with an infrared filter on the laser head can be used to monitor the melt pool [32].

Inline Coherent Imaging (ICI)

ICI is a type of optical coherence tomography (OCT) imaging, a noninvasive test that is suitable for taking low-coherence long wavelength to capture high resolution images. This technique was developed to take cross-sectional images of the anterior eye in 1991 [33]. A typical OCT setup consists of a fiber-based Michelson interferometer, a low-coherence light source (e.g. a superluminescent diode [SLD]), and a high-speed spectrum, as shown in Figure 11.6 [34].

A low-coherence light source generates the fiber-coupled broadband light that passes through an optical isolator to protect against back reflection. After that, the light is split by the 50/50 coupler into sample and reference arms. The input beam of each arm travels to a mirror, causing the beam to reflect back. The reflected signals are recombined by the splitter and transmitted to the spectrometer. The spectrometer is used to resolve the fringe pattern of an electric phase difference between arms. For each sample point, the reference mirror is scanned in the z-direction, and the light intensity is recorded on the photodetector [33–35]. Many versions of OCT have been developed over the last decades in the time domain [33, 36–38] and the Fourier domain [39, 40]. In general, the Fourier domain-based OCT has better sensitivity than the time domain OCT technique, since the Fourier domain OCT is less sensitive to noise [41]. ICI works based on the inline SLD-OCT technique.

ICI is commonly used in biomedical applications [35]; however, this technique is used for the LPBF process scarcely to show the melt pool characterization and morphology [34, 42].

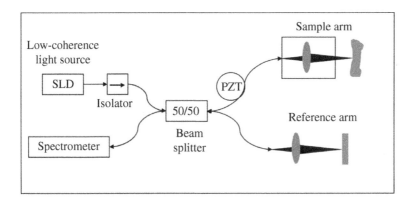

Figure 11.6 The inline coherent imaging system. Source: Republished with permission of Elsevier, from [34].

For example, the Precitec IDM sensor, a commercial ICI setup, can be installed on a typical LPBF setup to extract keyhole depth, topology information, and reconstruction for the molten material and loose powder. The sensor can capture the surface topology data sequentially with the scanning unit taken in a range of milliseconds to several seconds corresponding to the area size. The captured ICI data can then be analyzed to highlight the keyhole depth, melted surface characterization, and distortion [42].

X-Ray Imaging

X-ray imaging is commonly used in the medical sector in which an X-ray source produces a high-energy beam that is filtered to reduce the amount of scattered or low-energy rays, and then the beam is passed through the objects. Common types of X-ray imaging in engineering are computed tomography (CT), where the transmitted beam from an object is monitored by a detector (see Figure 11.7), and X-ray diffraction (XRD), where the reflected X-Ray is detected. The CT is used to analyze the internal microstructure of materials based on their X-ray absorption, which is used as a post-processing evaluation of printed parts. XRD is a technique to analyze the material composition and measure residual stresses.

XRD technique is used to monitor the LPBF process [43]. Polychromatic X-rays has penetrated the sample from the side, and the diffracted signal was detected by imaging and diffraction detector. The sensors have been reported to be installed 300 mm away from the Ti-6Al-4V sample. The X-ray imaging provides information about laser power's effect on the melt pool dimension, keyhole mode, particle motions, cooling rate, and phase transformation rate [43]. An X-ray sensor can also be installed in the LDED process to evaluate phase transformation by monitoring the heating and cooling cycle of Ti-6Al-4V [44].

The X-ray beam can be passed through the chamber's observation windows to monitor the melt pool and consequently penetrated through the region of interest. In the next step, the beam can be converted to visible light by the scintillator to be detected by a high-speed camera, as demonstrated in Figure 11.8. The detected image resolution can be 1024×1024 pixels with a 1.97-μm pixel size to shed some light on the scatter, interactions, the relation of cooling rate, and keyhole formation [45].

Figure 11.7   A computed tomography (CT) setup, showing the X-Ray source, object, and detector.

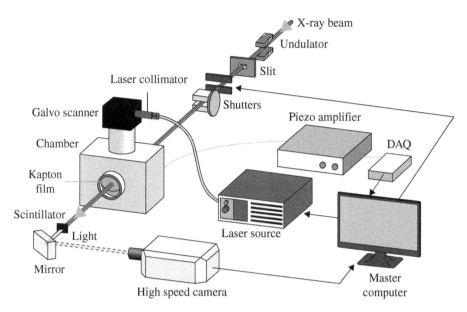

Figure 11.8 The schematic of the beamline setup within the chamber [45]. Published under an open access CC BY 4.0 license.

Photodiode

Photodiodes are widely used in metal AM as they are inexpensive, where they provide vital information about the process. They are used to sense thermal radiation and light emission. A photodiode is a semiconductor device with a P–N junction that converts photons into an electrical current. The P-layer junction has positive charges, and the N-layer junction contains electrons or negative charges. Semiconductor materials (e.g. silicon and germanium) have a band gap (energy gap) between their conduction and depletion bands. Enough energy (more than the energy gap) is required to stimulate electrons to move and transfer from the conduction band to the depletion band [46, 47]. When the light intensity greater than the gap energy hits the photodiode, it results in electrons' movement. The electrons' movement toward the cathode and holes' movement toward the anode create the electrical current, as shown in Figure 11.9. The generated electrical current is a function of applied voltage, electron charge, Boltzmann constant, and absolute temperature.

For instance, silicon has an average of 1.12 eV energy gap between its bands [48]. It means silicon requires more than 1.12 eV energy to transfer the electrons. 1.12 eV energy is corresponding to a wavelength shorter than 1100 nm based on Planck's law.

Thorlabs PDA36A photodiode (wavelength range of 350–1100 nm) can be used to sense light intensity signal and two CMOS cameras (wavelength range of 700–1000 nm) to capture images from the melt pool. The data can be used to determine melt pool size, intensity, and shape to provide information about the overheating and balling effect [49]. Two photodiodes' complex arrangements with 100 kHz sampling time might be able to monitor thermal radiation and plasma emission. The first photodiode can be installed to detect plasma emission with a wavelength of 700–1050 nm. The second photodiode may be used to detect thermal radiation

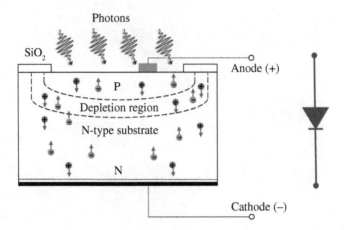

Figure 11.9 Schematic cross section of P–N photodiode. Source: redrawn and adapted from [46].

with a wavelength of 1100–1700 nm [50]. In addition to thermal radiation and plasma emission, a laser beam-reflected signal may also be measured with this setup.

For example, three photodiodes, a power meter, and a CCD camera on the laser head are represented in Figure 11.10. Photodiodes 1, 2, and 3 may be used to detect the laser beam-reflected signal, thermal radiation, and plume emission, respectively. Also, a power meter can be used to measure laser power, and a CCD camera can be used to observe the melt pool in LDED. The information can be used to find the relation between the deposited width and process parameters [51].

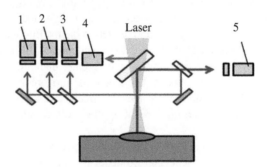

1. Photodiode (wavelength range: 1070 nm)
2. Photodiode (wavelength range: 1100–1800 nm)
3. Photodiode (wavelength range < 600 nm)
4. Power meter
5. CCD camera

Figure 11.10 Schematic of in situ sensing equipment integrated on LDED. (Source: Republished with permission of AIP Publishing, from [51]; permission conveyed through Copyright Clearance Center, Inc.).

Pyrometer

A pyrometer is a noncontact sensor to estimate the temperature from the thermal electromagnetic radiation received from the targeted object in two spectrums: visible and infrared ranges. The pyrometer has two basic components: an optical system and a detector. The optical system is used to capture the emitted light from the objects. The light is then transferred to the detector, which is sensitive to the radiation wave. As the thermal radiation is related to the temperature by Stefan–Boltzmann law, the temperature is calculated as an output of the pyrometer. Pyrometer is used in the metal AM processes to detect temperature since it fuses the material by thermal energy. Besides, in some studies, the calculated temperature is analyzed to identify geometry distortion and overheating.

The melt pool temperature can be indirectly calculated using a pyrometer [52–54]. Pyrometer with a temperature range of 900–2600°C, a wavelength of 900–1200 nm, and a sampling time of 50 ms can be used to capture melt pool temperature, as demonstrated in Figure 11.11 [52, 53]. A dual-wavelength pyrometer and a thermal camera can be used in an LDED machine (e.g. Optomec system) to capture the temperature distribution from the top and side views [26]. A single-color infrared pyrometer can also be placed to provide evidence of how temperature can affect the part's geometry [54].

A combination of the pyrometer and other sensors may be used to identify the overhang and overheating structure [55]. To identify overheating, a pyrometer and an IR camera can be mounted on an LPBF system to generate a spatial map of melt pool temperature, and data then can be analyzed to identify overheating [56]. Moreover, a calibrated pyrometer and a CCD camera can be fixed to monitor the melting temperature to shed some light on the effect of temperature on the overheating and overhang layer [57].

11.2.1.2 Non-Radiative

Non-radiative sensors detect the physical properties without the use of radiation light emitted from the process. Acoustic sensors, thermocouple, and displacement sensors are common types of non-radiative sensors used in the metal AM processes. This class of sensors can interpret the dynamics and physical nature of the process to the electrical signal, as described in the following sections.

Acoustic Sensors

The expansion of medium and physical compression with a specific frequency is known as a sound wave associated with the human hearing ranges (20–20 000 Hz). The wave below 20 Hz are called infrasound, and above 20 000 Hz is known as ultrasound [58]. Many sensors are used to sense the acoustic signal, such as microphones, solid-state acoustic detectors, and ultrasonic detectors [58]. A microphone is a sensor or transducer to convert sound to an electrical signal. Different microphones can convert sounds to the electrical signal differently; however, their main component consists of an extremely thin and light diaphragm. When the sound wave enters the microphone, the diaphragm starts to vibrate. Then, the vibration is transmitted to a piezoelectric or a coil part to generate an electrical current. For instance, in the dynamic microphone, vibration waves are transmitted to a coil, transferring the vibration to the electrical wires (Figure 11.12a). In the condenser microphone, a parallel plate was used instead of a coil,

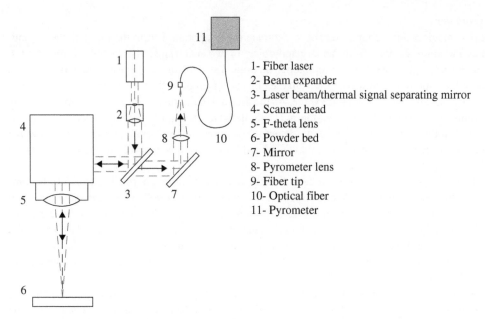

1- Fiber laser
2- Beam expander
3- Laser beam/thermal signal separating mirror
4- Scanner head
5- F-theta lens
6- Powder bed
7- Mirror
8- Pyrometer lens
9- Fiber tip
10- Optical fiber
11- Pyrometer

Figure 11.11 Schematic of the optical system using a pyrometer. Source: Republished with permission of Elsevier, from [52].

and the diaphragm vibration leads to alter the distance between two plates (Figure 11.12b). Then, the distance of the plates is digitalized as an electrical signal [59].

Another type of acoustic sensor works based on the surface acoustic wave (SAW). These sensors sense the SAWs of physical phenomena and then convert them to an electrical signals. The sensor consists of a piezoelectric substrate with input and output interdigitated transducers (IDTs) located on either side of the substrate. The input IDT is connected to the voltage source, and the output IDT is connected to the signal detection circuits. The input IDT converts the electrical voltage to an acoustic Rayleigh wave that propagates on the material's surface. Any characterization change of material can lead to altering the wave velocity or amplitude.

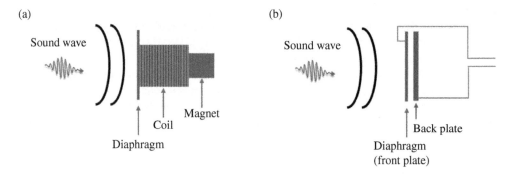

Figure 11.12 Schematic of (a) dynamic and (b) condenser microphones.

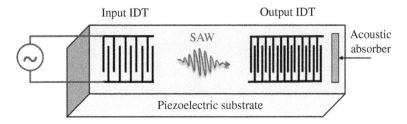

Figure 11.13 The schematic of the surface acoustic wave (SAW) sensor. Source: redrawn and adapted from [55].

Such changes can be detected by the output of IDT [60]. The basic schematic of the SAW sensor is demonstrated in Figure 11.13.

Acoustic sensors are connected to the processing and analysis units to amplify the sensor and filter the noise. The emitted signal's characteristic depends on the voltage source. As a result, different sources can emit different signals [61]. Major issues with acoustics sensors are the noise/signal ratio that might be high due to environmental disturbances.

Acoustic sensors have the potential to monitor metal AM because surface defects, residual stresses, feedstock/powder motions, and cracks may affect the signature of sound and might be detectable by acoustic sensors. A Printed Circuit Board (PCB) microphone fixed at the angle of 30° over the LPBF platform with a frequency response from 0 Hz to 100 kHz can be used to detect sound. Then, the signals can be converted to the frequency domain by power spectral density (PSD). Data were then can be used to detect the porosity formation [62]. A Kistler 8152B211 acoustic emission sensor can be placed on the LDED process to capture a 100–1000 kHz bandwidth signal. A (Digital Wave B1025-MAE broadband) transducer can also be used to acquire a 300 Hz signal, where the signal can be digitalized at a 5-MHz sampling frequency. The recorded data can then be processed to identify crack and porosity [63]. With the wavelength ranges of 400 kHz to 30 MHz, an ultrasound sensor can also be used to understand the complex dynamics of the LPBF process. The four-channel ultrasonic transmitter and receiver can be located and sealed at the lower side of the platform. The sensor could acquire the data layer by layer with 250 mega-samples per second and 14-bit resolution. The data can be stored for further offline analysis to detect residual stresses [64].

Thermocouple

A thermocouple is a type of contact-based temperature sensor measuring the temperature based on the produced voltage through a media. The thermocouple consists of two metal conductors welded together, creating a junction, as shown in Figure 11.14. One junction is connected to the body of unknown temperature (hot junction), and another junction is attached to another known temperature body. It is used as a reference temperature (cold junction). Due to temperature differences, electric and magnetic fields (EMFs) are generated. The fields can be measured based on the temperature of junctions and metal properties [58]. Then, the voltage can be interpreted using reference tables to measure temperature. The thermocouple is classified based on their characteristics, and each type has its reference table. Thermocouple has a broad range of temperatures (−200 to +2000 °C); however, one critical point that should be considered is that temperature rising could result from irradiation.

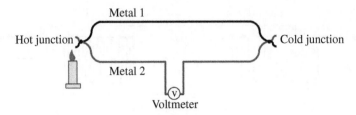

Figure 11.14   The basic schematic of thermocouple. Source: Redrawn and adapted from [65].

Thermocouple has been used in the metal AM processes to measure the temperature distribution and verify the simulation accuracy [66, 67]. For example, a J-type thermocouple can be mounted on the LDED process to monitor the nozzle temperature. A controller can also be designed to maintain temperature to prevent powder clogging [66]. A baseplate temperature in LPBF and DED can be measured by a K-type thermocouple to study the effect of different scan strategies on the build distortion [68]. A K-type thermocouple and strain gauges can be positioned at the center of the support structure, as shown in Figure 11.15, to record the temperature evolution and strain variation, respectively [69, 70].

Displacement and Stress–Strain Sensors
Displacement sensors are devices to measure the distance between objects or part elongation. A displacement sensor is also known as a distance or proximity sensor. Stain gauges may also fall under this category. Two different types of sensors can be considered to measure distance: contact and noncontact methods. The contact method uses a dial gauge, strain gauge, or differential transformer. A noncontact sensor consists of two parts: a transmitter and a receiver. The transmitter sends the signal, which is generated by a laser diode or magnetism. Then, the transmitted signal hits the monitored surface and returns to the receiver. The time that takes the

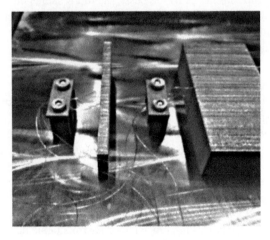

Figure 11.15   Thermocouple sensors placement in LPBF. Source: Republished with permission of Elsevier, from [69].

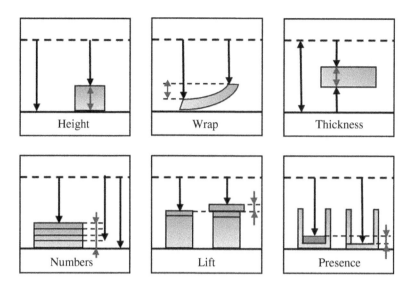

Figure 11.16 Example of measurement by displacement sensor. Source: Redrawn and adapted from [71].

laser diode signal to be sent, hit the receiver's surface, and captured by the receiver is translated into a distance between the sensor and the surface [71].

The displacement sensor could be generally used to detect height, wrap, thickness, numbers, lift, and object presence, as shown in Figure 11.16. In AM, displacement sensors are implemented to detect beam deflection [72], deformation [11, 12], and surface quality [73]. A differential variable reluctance transducer (DVRT) displacement sensor and K-type thermocouple can be installed on an LPBF machine to detect distortion [68]. Strain gauges can be installed to measure residual stresses.

11.2.2 Mounting Strategies for In-line Monitoring Sensors in Metal AM Setups

During the parts qualification process, AM monitoring systems must provide vital information from the build process to enable the users to document the process quality. Two types of mounting strategies are normally used:

1. Co-axial
2. Off-axial

The co-axial sensor is placed in the power source's optical path, and data are transferred through the f-θ lens, mirror, and beam splitter to the sensor, as demonstrated in Figure 11.17. However, the off-axial sensor is installed outside the optical path with a given

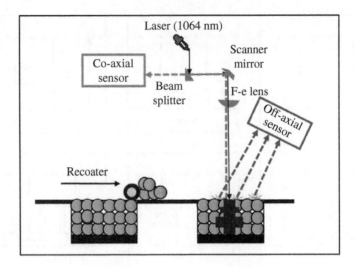

Figure 11.17   Schematic of co-axial and off-axial setup in LPBF.

angle of view, as demonstrated in Figure 11.17. Tables 11.1 and 11.2 categorize different co-axial and off-axial sensor arrangements used in the literature based on their types used in the LPBF and LDED process.

11.3   Commercially Available Sensors

Although many vendors have installed sensor(s) on the LPBF and LDED machines, most of them are mainly designed to gather in-situ data that need offline processing. As a result, to automatically generate an alarm or control the process, further developments are needed. Some of the available commercial sensors installed in the LPBF and LDED systems will be discussed in the following sections.

11.3.1   LPBF Commercial Sensors

Different developers of LPBF systems have offered process monitoring toolkits. Table 11.3 summarizes information of some of the commercial sensors mounted on the LPBF machine.

EOS: EOS offers two modules to capture and analyze the data:

- EOSTATE MeltPool Monitoring, which includes co-axial and off-axial photodiodes. Both sensors measure the light intensity emitted from the melt pool with 60-kHz sampling time. After the process, the data can be analyzed using the EOSTATE monitoring software to find the intensity signal disturbance. MPM data are very sensitive to deviations resulting from power fluctuations.
- EOSTATE Exposure Optical Tomography includes an integrated camera to capture powder beds' images. The camera uniformly captures images in the NIR spectral range and with a resolution of 2560 × 2160 pixels. Then, the images are correlated with thermal radiation. Analysis of the correlated data is provided in the EOSTATE Exposure OT software to

Table 11.1 Co-axial mounted based sensors used in LPBF and LDED.

Types	Equipment	LPBF	LDED
Radiative			Dimension measurement [10, 14, 51, 72, 73, 79–85]
	Camera	Abnormal process [74]	Residual stress [67]
	Visible to NIR	Dimension measurement [8, 9, 75]	Temperature measurement [67, 82]
		Deformation [76]	Cooling rate [86, 87]
		Spatter [16, 77]	Powder stream monitoring [88]
		Plume [18]	Surface quality [89, 90]
		Balling [9, 54, 77]	
		Delamination [49]	
	Camera	Overheating [56]	Feedstock motion/wrinkles [32]
	NIR to LWIR		Dimension measurement [24, 91]
	Optical inline coherence imaging	Dimension accuracy [34]	
		Inhomogeneity [42]	
		Distortion [42]	
		Overhang structure [92]	
	Photodiode	Dimension accuracy [8, 9]	Plume instability [51]
		Overheating [49, 93, 94]	Surface quality [89]
		Balling [49]	
	Pyrometer	Overhang layer [57]	Dimension measurement [54]
		Surface temperature [52, 95]	Temperature measurement [72]
		Overheating [56]	
Non-radiative	Acoustic sensor	—	Feedstock motion [98]
	Thermocouple	—	Temperature measurement to measure thermal stresses [66, 67]

identify any thermal deviation. Exposure OT is more sensitive to scanning speed and hatching distance deviations [119].

SLM Solution: Selective Laser Melting (SLM) Solution offers two toolkits to monitor the melt pool and laser power:

- MPM includes a co-axial pyrometer to measure thermal radiation from the melt pool at a rate of up to 100 000 times per second [120].
- Laser power monitoring (LPM) is recorded as the actual emitted laser output at a rate of 100 000 times per second. The laser output is measured by uncoupling the laser beam in the optical path and reflecting it to a sensor [121].

Table 11.2 Off-axial mounted based sensors used in LPBF and LDED.

Types	Equipment	LPBF	LDED
Radiative	Camera Visible to NIR	Lack of fusion [19, 20, 78, 99] Keyhole [99] Delamination [100] Overhang [16, 100, 101] Overheating [19, 20] Recoater blade [100] Surface temperature [5, 6] Balling [102]	Dimension measurement [13, 103, 104] Deformation [11, 12] Plume [17, 105] Surface temperature [7]
	Camera NIR to LWIR	Delamination and splatter [22] Dimension measurement [106] Surface temperature [95] Lack of fusion [25, 107] Keyhole [25] Inhomogeneity [108] Spatter [109] Plume [110] Deformation [31]	Dimension measurement [23, 111, 112] Cooling rate [23, 28–30] Solidification rate [112]
	Pyrometer	Surface temperature [6, 95, 102, 113]	Temperature measurement [26, 73, 111, 114]
Non- radiative	Acoustic sensor	Overheating and balling [62] Lack of fusion [115, 116] Residual stress [115]	Crack [68] Porosity [63] Process parameters [117, 118]
	Displacement	Distortion [68]	Surface quality [73] Beam deflection [72] Deformation [11, 12]
	Thermocouple	Residual stress [70]	—

Table 11.3 Some commercial sensors mounted on the LPBF machines.

Developer	Module name	In situ sensor	Monitored items
EOS	EOSTATE MeltPool	Two photo- diodes	Melt pool
	EOSTATE Exposure OT	Camera	Powder bed thermal map
SLM Solution	Melt pool monitoring	Pyrometer	Melt pool
	Layer control system	Camera	Powder bed
GE additive (Concept Laser)	QM melt pool 3D	Photodiode	Melt pool (area and intensity)
Renishaw	InfiniAM	Photodiode	Melt pool
Trumpf	Truprint monitoring	Camera	Powder bed and part geometry
B6 Sigma	PrintRite3D	photodiode pyrometer	Different monitoring possibilities

GE Additive (Concept Laser): The company offers two separate modules to collect data from the melt pool:

- QM melt pool 3D module is a monitoring tool with a co-axial monitoring setup of the photodiode to estimate the melt pool area and intensity. This system provides two kinds of data. The first dataset is the average intensity of the component after completing the print, and another one generates a 3D dataset of the part and its structure. This information is accessible after the process, and there is not currently used for defect correction and feedback control [122].
- QM coating focuses on the dose factor, that is, the control amount of the powder released by the powder hopper before the recoating system. This factor depends on the sufficient powder released before the recoating operation [106].

Renishaw: InfiniAM has been offered by Renishaw to monitor the energy input and melt pool emission. This module provides information about melt pool characteristics in high-temporal resolution to provide a 2D and 3D view of the build [123].

Trumpf: Truprint consists of a high-resolution camera to monitor melt pool, powder bed, and part geometry. The integrated camera captures powder bed images layer by layer. Image analysis modules are also provided, and comprehensive analysis of each layer can be performed and compared with the Computer-Aided Design (CAD) model.

B6 Sigma, Inc.: B6 Sigma has developed the PrintRite3D that includes Sensorpak that is a set of co-axial and off-axial sensors. A software called "Inspect" is developed to determine quality metrics and to identify suspicious patterns layer by layer. A software called "Contour" is developed for real-time monitoring and reconstruction of the part geometry and a module to compare the result with the CAD model.

11.3.2   LDED Commercial Sensors

Different vendors of LDED systems have introduced process monitoring toolkits. Table 11.4 summarizes information about some commercial sensors mounted on the LDED machines.

DM3D technology: It offers a DMD closed-loop feedback system that monitors the melt pool by photodiodes, dual-color pyrometer, or three high-speed CCD cameras to measure build height.

Table 11.4   Some commercial sensors mounted on LDED machines.

Developer	In situ sensors	Monitored items
DM3D technology	Dual-color pyrometer and three CCD cameras	Melt pool (height and temperature)
Stratonics	Infrared cameras	Melt pool (height and temperature), heating, and cooling rates
DMG MORI	Dual-color pyrometer	Melt pool temperature
Optomec	Thermal camera, CCD, dual-color pyrometer	Melt pool temperature, microstructure, and size

Stratonics: It does not make LDED systems but has developed two modules to capture and analyze the online data. These modules can be installed on commercially available systems. Their setup includes a co-axial CMOS imaging detector with 25 frames per second. The data acquisition system is synchronized to the external hardware. Displays profiles, time histories, and histograms of thermal and dimensional data are exported after the process for post-processing. Their system also has co-axial CMOS imaging technology with 10 000 frames per second. The video is recorded by the sensor and archived for post-analysis. In the post-processing software, melt pool temperature, size, heating, and cooling rates are measured.

DMG MORI: The DMG MORI LASERTEC 3D can continuously analyze the homogeneous component quality via the melt pool temperature data of dual-color pyrometer.

OPTOMEC: This company offers the LENS Melt-Pool Sensor that provides a closed-loop option to control heat flux and microstructure of their LDED process. This sensor monitors the melt pool's size while continuously adjusting the laser power to keep it fixed. They also offer a thermal imager (Stratonic's ThermaViz system) that uses a two-wavelength imaging pyrometer to provide real-time, true temperature images and measurements. It provides high-resolution, high-accuracy temperature information. ThermaViz is coaxially integrated into their setup.

11.4 Signal/Data Conditioning, Methodologies, and Classic Controllers for Monitoring, Control, and Quality Assurance in Metal AM Processes

Achieving excellent quality of metallic AM parts is a challenging task due to the high complexity of the physical phenomena and the lack of high fidelity and high-speed mathematical and statistical models. To address some of these challenges, much emphasis has been recently placed on the monitoring of AM processes [43] to shed some light on the process dynamics and signatures. From a statistical viewpoint, monitoring is used to indicate both data collection and signal conditioning. After collecting in-situ data, the sensor's data should be processed/conditioned and analyzed to extract the desired information. Different signal conditioning and statistical methods are discussed in the literature to monitor the AM processes effectively. However, these methods are discussed here based on the type of flaws that are classified under geometry, temperature, porosity, cracks, and material ejection. Further details will be discussed in the following sections.

11.4.1 Signal/Data Conditioning and Controllers for Melt Pool Geometrical Analysis

Melt pool/clad geometry is one of the output quality that can affect the final part dimensional and physical properties. For the geometry measurement, cameras and photodiodes are mostly used to capture in situ data. The collected data are then analyzed to correlate the geometrical features.

For example, feedback control for LPBF or DED processes can be used to capture the data from a photodiode and a high-speed CMOS camera. Both sensors may calculate the melt pool geometry area and the area measurement repeatedly compared with the desired point. A proportional–integral–derivative (PID) controller may be designed and tuned to stabilize

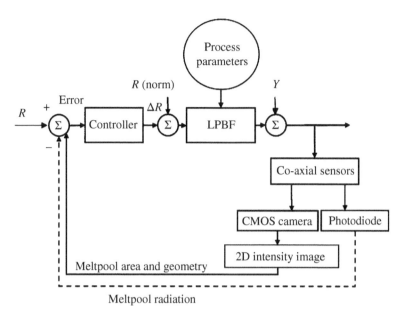

Figure 11.18   Schematic overview of the feedback loop proposed for LPBF but can be used for other DED processes. Source: Redrawn and adapted from [8].

the melt pool dynamics by controlling the laser power (Figure 11.18) [8]. However, the implementation of this closed-loop control for LPBF is challenging due to the high speed of the LPBF process.

The above-mentioned methods can be constructed out of four steps:

1. Collecting data by the installed sensors,
2. Data processing unit to identify the melt pool width, length, area, and emitted light intensity.
3. The reference data and measured data should be compared.
4. The quality of parts can be estimated by mapping algorithms or a conventional controller to measure a position domain from the time domain and then send the adjustment signal to the laser power [104].

In one case, a proportional–integral (PI) controller based on the dynamic between laser power and photodiode signals can be used where the light sensitivity is in the range of 400—900 nm for LPBF. During the scanning, the photodiode may capture a signal with a 20-kHz sampling time. The photodiode signal is then analyzed by the average Fourier transform and fitted to the second-order model. Then as shown in Figure 11.19, three PI controllers can be used with bandwidths of 3600 rad/s (very fast), 660 rad/s (intermediate), and 95 rad/s (slow), respectively, to change the laser power.

According to Figure 11.19a, the photodiode signal response results in an overshoot because of sparks emitted from the melt pool. The third controller reacts too slowly to reach 1 (Figure 11.19c). So, the intermediate controller may be chosen for further study. It is claimed that the geometrical accuracy of overhang regions and surface roughness may improve using the intermediate PI controller [94].

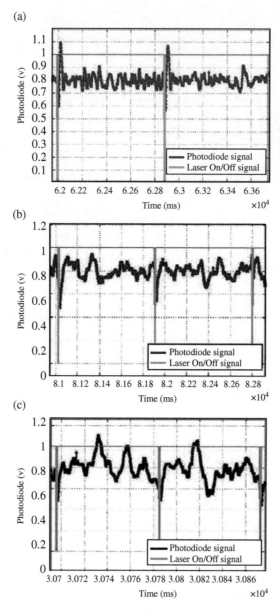

Figure 11.19 Photodiode signal used in LPBF during the scanning using (a) very fast, (b) intermediate, and (c) slow PI controllers. Source: Republished with permission of Elsevier, from [94].

In LDED, three photodiodes (1070 nm, 1100–1800 nm, and <600 nm) can be installed as well as a CCD camera to determine the correlation of their signals with the melt pool size under various laser power, scanning speed, and defocusing distance conditions. Figure 11.20 shows the correlation between the average intensity signal and laser power, scanning speed, and

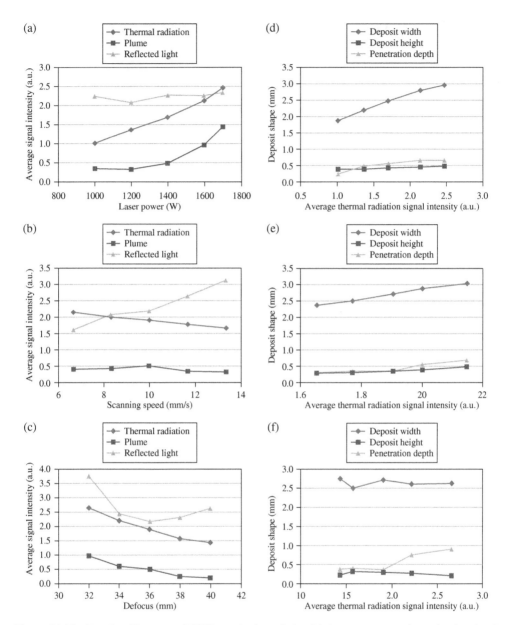

Figure 11.20   Results of in-process LDED monitoring relationship between average intensity signal and (a) laser power, (b) scanning speed, (c) defocus distance and relationship between thermal radiation signal and deposit change under the variation of (d) laser power, (e) scanning speed, and (f) defocus distance (Source: Republished with permission of AIP Publishing, from [51]; permission conveyed through Copyright Clearance Center, Inc.).

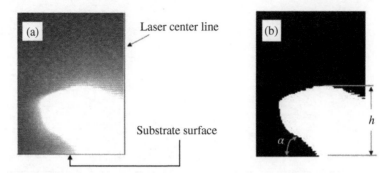

Figure 11.21 Typical process zone in LDED and image processing: (a) original image and (b) processed image (Source: Republished with permission of Taylor & Francis Group, from [1]; permission conveyed through Copyright Clearance Center, Inc.).

defocusing distance based on the photodiode signals. By comparing the average thermal signal and the considered parameters (Figure 11.20d–f), variation in the laser power results in the significant change of the deposit width, thus leading to the development of an adaptive PID controller. The controller may maintain the melt pool size and reduced the width variation from 63.6% to 12.5% [51].

One of the challenges encountered when using cameras and photodiodes is the contamination of data with plasma and injected spatters, so the image conditioning techniques are essential to be used.

For example, for LDED, a CCD camera and an interference filter (bandwidth of 500–700) can be used to provide images from the process zone. In pre-processing, images' brightness and contrast can be decreased and increased by 10%, respectively. Then, images will be fed into an image processing and pattern recognition algorithm to identify dimension, roughness, and the solidification rate. The image processing algorithm consists of three steps to reach the goal:

1. Images to be converted from RGB to the gray level (Figure 11.21a).
2. The image threshold level using Otsu's method to be applied to every 10th frame.
3. The dimensions and the interface angle are calculated by finding the white object's border. As shown in Figure 11.21b, the track (known as clad) height (h) and the angle (α) are obtained based on the number of bright pixels and the angle between the border of the bright area and a reference line, respectively.

The results can then be compared with real-time measurements and demonstrated the performance of the proposed algorithm to find the clad height [1].

The clad roughness and surface quality can be obtained by analyzing the fluctuation of clad height. Classical and fuzzy logic controllers can be implemented to the LDED process to adjust the laser energy and the processing speed. The PID controller can be developed to control the clad height in LDED based on the quadratic Hessian optimization function. For example, a PID controller can be considered with the identified model (see Chapter 4) and a threshold limiting the upper and lower ranges of the laser pulse energy as shown in Figure 11.22.

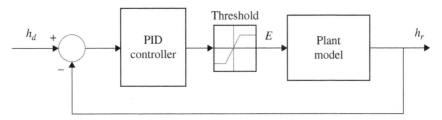

Figure 11.22 PID controller integrated with a threshold and the identified plant model used in LDED (Source: Republished with permission of Taylor & Francis Group, from [1]; permission conveyed through Copyright Clearance Center, Inc.).

In this figure, h_d is the desired and h_r is the actual generated clad height during LDED. The main reason for using the lower and upper thresholds is to address the deposited track quality and the laser system limitations, respectively. For optimizing the PID gains, an optimization algorithm can be used based on the minimization of the quadratic Hessian function. The algorithm optimizes the PID gains based on the desired overshoot, settling time, and rising time. For a typical LDED process, the desired situation is when the overshoot, settling time, and the oscillations around the setpoint become minimum, so that a bead with the desired height can be produced rapidly and precisely [1].

11.4.2 Signal/Data Conditioning and Methodologies for Temperature Monitoring and Analysis

Temporal and maximum temperatures in metal AM are other factors that can be used to predict the quality of the printed part. For example, K-type thermocouple with −200 to 1250 °C to measure temperature in LDED can be used when a protective sheet solution can be included to decrease the thermocouple failure at high temperatures [94].

A CCD camera and two-wavelength pyrometer setup can be used to find temperature distribution for LPBF [53, 124]. The camera and pyrometer measure the temperature distribution and the maximum surface temperature, respectively. However, the main challenge is that the sensory data may not provide accurate temperature distribution, so the IR camera can be off-axially used instead of the pyrometer to achieve better results [106]. A thermal camera (with a temporal resolution of 10 μs) and a pyrometer can be co-axially mounted to monitor the melt pool's temperature. The scanner position is also acquired, and the pyrometer signal should be quantized with an AD convertor at 100 kHz. Then, scanner position, pyrometer signals, and camera images are transferred into Field Programmable Gate Array (FPGA) to synchronize the signals' timing. The data are then sent to PC memory for recording and later processing, as represented in Figure 11.23. Even using FPGA for data synchronization, the monochromatic aberrations can be observed, so the prefocus system is needed to resolve the problem. The spatial map obtained through this setup can be used to detect overheating regions [56].

Additional actions may be needed to better visualize the melt pool dynamic by adding a coaxial CMOS camera and external illuminations to capture melt pool shape and surface structure at a high sampling rate. The image processing technique can be performed to analyze the

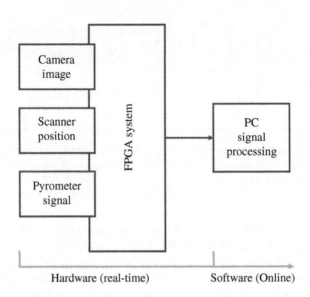

Figure 11.23 Data acquisition for temperature signal recording and processing used in LBPF and LDED [56]. Published under the open access CC BY 4.0 license.

captured images. Spot diagram and fast Fourier transformation (FFT) can be conducted to measure melt pool dynamics and temperature [125].

11.4.3 Signal/Data Conditioning and Methodologies for the Detection of Porosity

The formation of porosity seriously affects the quality of the printed part. Porosity is commonly generated in LPBF due to the "lack of fusion (LoF)" and keyhole phenomena. The formation of porosity in LDED is limited to inherent physics that minimizes the formation of pores. In LPBF, image grabbing from the process and image segmentation method can help to detect porosity. High-speed cameras and several flash modules can be mounted on the LPBF build chamber to provide high image contrast. Another approach is to cut and normalize images and compare them using a level-set segment method as represented in Figure 11.24 [126]. To calibrate the method, intentional defects can be printed and then correlated by a CT scan using a confusion matrix. It should be noted that although the accuracy of the matrix may show a high level of the true-positive rate (true detection of pores), the false-positive rate (wrong detection) may also be too high, which limits the feasibility of controlling and predicting the process. Thus, these methods must carefully be calibrated and assessed before becoming a reliable quality assurance platform for LPBF.

Another method to detect porosity in LPBF and/or DED is based on the ultrasonic wave measurement. A transducer can generate and receive ultrasonic waves with a 30-MHz bandwidth while the received signals can be transferred into an oscilloscope to measure the time between the signal peaks. The technique may be implemented to detect absolute changes of 5% and larger, which has shown its linear relationship with porosity [116].

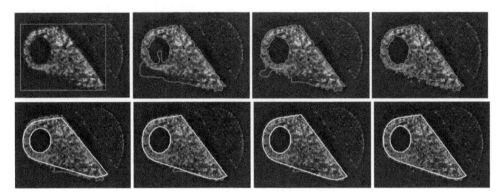

Figure 11.24 Image segmentation method from initialization (first row) to the final map (second row) used in LPBF (Source: Republished with permission of Elsevier, from [126]).

One of the emerging technologies for pore detection in LPBF is a commercialized system that was developed by the MTU Aero Engines team [15, 127, 128]. They use an sCMOS camera system with a 5 megapixels detector. The integrated sensor captures 10 frames per second. Then, all 10 capture images are combined into one single image. In the next step, the final image is correlated with the light emitted from the process. However, the emitted light is a combination of three sources (laser radiation, thermal emission, and plasma emission), as shown in Figure 11.25, so a band-pass filter was also used to filter out the laser reflection and plasma emissions from thermal radiation [49]. Then, the derived thermal signal can be used to identify its deviation. Any thermal deviation is interpreted as a potential defect in the final printed part. The final datum is a layer-by-layer exposure image in which cold and hot spots correspond to the nominal and abnormal areas with a resolution of 0.1 mm × 0.1 mm [15], as shown in Figure 11.25. This technique is used in the EOS Exposure OT [119].

This technique is correlated to micro-computed tomography [128] and digital radiography [15] to identify a LoF.

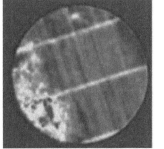 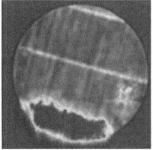 

Figure 11.25 Three layerwise optical tomography (OT) images used in LPBF (Source: Republished with permission of AIP Publishing, from [127]; permission conveyed through Copyright Clearance Center, Inc.).

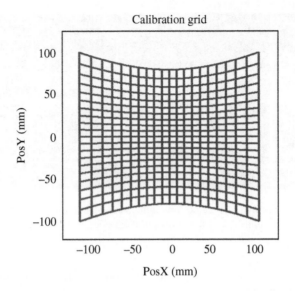

Figure 11.26 Geometry distortion due to the use of a Galvo scanner.

In another method for LPBF, the melt pool emitted light can be monitored by on-axial photodiode, and signal fluctuation may be correlated to the pore formation in the printed parts. Photodiode raw signals must be corrected for geometry and light intensity. The reasons for these corrections are explained below:

Geometry correction: The scanner would not get the absolute coordinate on the build considerably, causing image distortion, as shown in Figure 11.26 [129]. To calculate the absolute position, the mask file, the angle, center, offset, and resolution of geometry are needed to correct distortion. The following equations on x and y datasets are usually used to correct the geometry:

$$data.x = \frac{data.x}{geometry\ x\ correction\ center} \tag{11.1}$$

$$data.y = \frac{data.y}{geometry\ y\ correction\ center} \tag{11.2}$$

Then, rotation correction is applied by:

$$x = data.x - data.y*(angle\ correction\ factor) \tag{11.3}$$

$$y = data.y - data.x*(angle\ correction\ factor) \tag{11.4}$$

In the last step, the offset influence requires to be implemented by the offset deduction provided by the vendor:

$$x = x - x.offset \tag{11.5}$$

$$y = y - y.offset \tag{11.6}$$

Light intensity correction: The on-axis light intensity correction is necessary for eliminating the dispersion and angle-dependent transmission of the optical system, mainly due to the f-θ lens. The scanner mirrors and the f-θ lens are optimized to have a homogenous reflection/transmission at the wavelength of the laser (1064 nm). Nevertheless, since the wavelengths of light recorded by the on-axis photodiode are shorter than 1064 nm, this light will be affected by the "chromatic aberration," resulting in an inhomogeneous intensity distribution across the platform that does not correctly reflect the actual process emissions depending on the position [119]. The use of an achromatic doublet can correct chromatic aberration. However, it can also be corrected through a mask file, which is identified by scaling or subtracting the fringed colors (e.g. R, G, B). All spatially overlapped colors are corrected in the final image. By applying the mask file of the intensity signal, the correction can be obtained:

$$intensity\ signal = intensity\ signal * mask\ file \tag{11.7}$$

After correcting the raw data, three detection algorithms have been suggested by vendors like EOS [129] as follows:

1. **Absolute limits (AL) algorithms**: The algorithm has been developed to apply the moving average on the signal, as represented in [130]. The user can set the filter length of the moving average. The algorithm is applied to the build by a threshold approach in which an exit from the range means a disturbance in the signal as shown in Figure 11.27 [119]. The algorithm may be used to detect defects created due to sudden laser power variation:

$$moving\ average\ of\ intesity\ signal = \frac{1}{n}\sum_{i=0}^{n-1} intensity\ signal(i) \tag{11.8}$$

2. **Signal dynamics (SD)**: The algorithm can calculate the signal disturbance in which each measurement point will be compared to the mean value. This characteristic is also filtered *by the moving average and can* be *parameterized by the user. The threshold algorithm can*

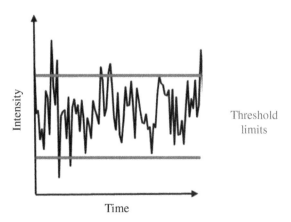

Figure 11.27   Concept of absolute limits algorithm.

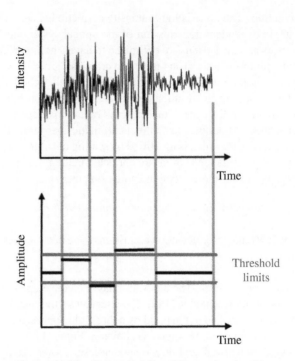

Figure 11.28   Concept of Signal Dynamics algorithm.

identify the disturbance in the signal, as shown in Figure 11.28. High-dynamic process changes like gas flow issues may be identified by the algorithm:

$$Signal\ Dynamics(intensity) = moving\ average(intesity) - mean\ value(intensity) \quad (11.9)$$

3. **Short-term fluctuations (STF):** Moving average with longer filter length is calculated, and the tolerance band is calculated around the moving average signal, as shown in Figure 11.29.

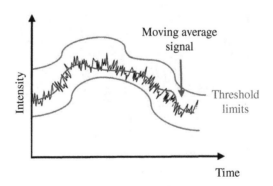

Figure 11.29   Concept of short-term fluctuations algorithm.

Band ranges are defined in percentage of limits window graph (e.g. upper level 120% and lower level 80%). The algorithm may be used to detect short instabilities like noisy processes, pores, and ejects.

A case study based on the above-mentioned method will be discussed at the end of this chapter.

11.4.4   Signal/Data Conditioning and Methodologies for Detection of Crack and Delamination

Cracks and delamination influence the part quality substantially. Rapid solidification under a moving laser energy input results in a large residual thermal stress induced all through the printed parts [131]. The high residual stress normally leads to the initiation and propagation of the cracks in a fabricated part.

To detect these defects, a regression model can be used to identify crack and porosity. A camera (e.g. a single-lens reflex, SLR, and camera) can take surface images at periodic intervals. The color of images normally shows the visible change of RGB color due to the temperature variation, as demonstrated in Table 11.5.

The RGB and the corresponding temperature values can be used to identify the linear regression model to determine surface temperature. The regression equation is given by [7]:

$$T = 1930 + 0.603R + 0.706G - 4.98B \qquad (11.10)$$

A moving average filter can be applied to the regression curve to minimize the effect of noise. In one study, intentional cracks were created by joining two stainless steel 316L. The study shows that the presence of a flaw changed the conduction of heat and can lead to a deviation in the temperature curve [7]. However, a mismatch between camera frame rate (7 frames/s) and the software processing time (1 s) has been reported, which could influence real-time image processing feasibility.

It should be noted that the approaches mentioned under Sections 11.4.1 to 11.4.4 may be applicable to identify cracks and delimitation. To this end, a proper calibration lookup table needs to be developed.

Table 11.5   Calibration of RGB values with temperature. For colorful spectrum, see the original reference.

R	G	B	Temperature (°C)	Colour
2	1	251	650	
0	2	254	700	
0	145	254	750	
0	223	255	800	
0	255	255	850	
15	255	244	900	
212	255	252	950	
167	254	250	1000	

Source: Redrawn and adapted from [7].

11.4.5 Signal/Data Conditioning and Methodologies for Detection of Plasma Plume and Spatters

The effect of plasma plume and hot particle ejection from the melt zone is critical to the printability and quality of parts. Camera calibration and image resolution by knife-edge detection measurement can improve the resolution and reduce the blurring level, thus enabling to detect the plumes and spatters. The thermal video can be converted to temperature value by the National Institute of Standards and Technology (NIST) calibration equation given by [132]:

$$S_{meas} = \varepsilon_t.F(T_{true}) + (1 - \varepsilon_t).F(T_{amb}) \tag{11.11}$$

where S_{meas} is the measured signal, ε_t is the surface emissivity (assuming $\varepsilon_t = 0.5$), T_{true} is the true object temperature, and T_{amb} is the temperature of the ambient environment or source contributing to reflections. F is the Sakuma-Hattori calibration function, a nonlinear regression fit, that is $F(T) = C / (exp\ (c^2 / A.T + B) - 1)$, where A, B, C and c^2 are constants. Based on the calculated temperature gradient and cooling rate, the spatter ejection can be identified [109].

The plume formation detection can be formulated for pure zinc. Low melting and vaporization points of pure zinc make it sensitive to generate plume. An off-axial IR camera can be installed with a spectral range of 8–9 μm and a resolution of 320 × 240 pixels to capture in-situ data [110]. The thermal images can be normalized into a grayscale pixel intensity ranging from black (cold) to white (hot). In the next step, different image segmentation methods (such as IsoData, Otsu, Li, Huang, and K-means) can be compared to extract features, as shown in Figure 11.30. The Otsu method may provide better results in the presence of a small plume and avoid the risk of local minimum convergence. Then, the bi-level thresholding method can be applied to the extracted feature to distinguish between plume and spatter [110].

Random and systematic flaws can be fabricated under the change of print parameters (laser power, powder flow rate, and hatching pattern) during LDED of Ti-6Al–4V. During the print, a

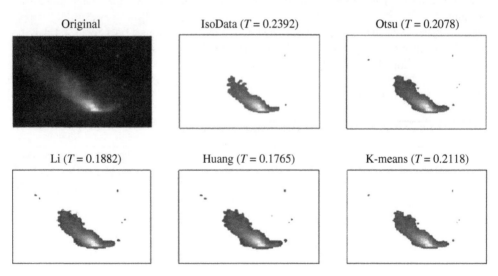

Figure 11.30 Spatter detection: segmentation results based on IsoData, Otsu's, Li's, Huang's, and K-means method (Source: Republished with permission of Elsevier, from [110]).

spectrometer collected uncalibrated spectra of laser interaction zone and a CCD camera, coupled with a band-pass (430 nm) filter, capture plume images. Then, spectra and images are analyzed to calculate the line-to-continuum ratio and plume area, respectively. The line-to-continuum ratio is calculated in [133], and the plume area is identified by applying the threshold method on the grayscale images. The result can be compared with the CT scan dataset to validate the detection. Although this study may identify plume, a low data acquisition rate and low SNR are some of its limitations [17].

11.5 Machine Learning for Data Analytics and Quality Assurance in Metal AM

Machine learning is an application of artificial intelligence (AI), providing the opportunity to learn from processes and then predict the processes through data patterns. The learning process consists of data acquisition and pattern recognition. The machine learning methods can predict the formation of flaws in real-time from metal AM due to their fast processing. Machine learning algorithms can be classified into three main groups:

1. **Supervised learning**
    a. *Classification methods*: Support vector machine (SVM), neural network (NN), deep learning (DL), adaptive neuro-fuzzy inference system (ANFIS), and K-nearest neighbors (KNN)
    b. *Regression methods*: Gaussian process (GP) and neural network (NN)
2. **Unsupervised learning**
    a. *Clustering methods*: K-means and self-organizing map (SOM)
    b. *Dimensionality reduction methods*: Principal component analysis (PCA) and singular value decomposition (SVD)
3. **Reinforcement learning** These methods include several subtechniques and some are used in metal AM quality assurance platforms. Table 11.6 lists some of these techniques used in metal AM for both LPBF and LDED processes to predict different defects and flaws. These techniques can be used for other metal AM processes. It should be noted that reinforcement learning has not been applied in metal AM yet. In the following, each method and its classifications will be explained.

In the following, the basics of these methods are explained.

11.5.1 Supervised Learning

The supervised learning algorithm is trained with a labeled dataset in which each input feature X is labeled with an output Y. Knowing the input and output datasets, the model is trained to identify the relationship between input and output variables. The classification and regression problems can be solved by supervised learning. Classification maps the function from the input to discrete output; however, the regression model is a predictive model to estimate continuous output. In the following sections, the applications of classification and regression models in AM will be discussed.

Table 11.6 Machine learning methods used in LPBF and LDED.

Method	LPBF	LDED
SVM	Overheating [19, 20] Porosity [19, 20, 134, 135] Spatter [18, 135] Plume [18] Balling [135] Powder classification [136]	Porosity [26, 137] Deposition height [138]
MLP	Overheating [20, 62] Balling [62] Porosity [20, 134]	Deposition height [138] Porosity [139, 140]
CNN	Porosity [97, 99] Track width continuity [141] Abnormal process [74] Spatter [18] Plume [18] Thin wall quality [75]	Process parameters classification [142] Porosity [143] Crack [143]
RNN	—	Thermal history [144]
SNN	Porosity [97]	—
DBN	Overheating and balling [62]	—
Depth wise-separable neural deep network	Delamination and splatter [22]	—
ANFIS	Fatigue life [145] Surface roughness [146]	Catchment efficiency [147] Clad height [147–150] Clad width [147, 150]
KNN	Porosity [134]	Porosity [26, 139]
GP-unsupervised	Melt pool geometry [151–153] Porosity [154]	Clad geometry [155]
NN-unsupervised	Porosity [156]	Clad geometry [157–159]
K-means	Delamination and overhang structure, part failure, and recoater blade [100, 135, 160] Plume [110] Single-track depth [152]	Crack [63] Porosity [63] Process parameters classification [117]
SOM	—	Porosity [27, 139]
PCA	Data reduction [18, 74, 97, 100, 160]	Data reduction [26, 63, 137, 161]
SVD	Feature extraction [50]	—

11.5.1.1 Classification

In the classification method, the ultimate goal is to predict output labels based on the dataset's previous observation. Various classification methods are proposed to evaluate prediction performance; however, the most common way is to calculate the accuracy. The accuracy is the percentage of correct classification out of all predictions. Table 11.6 lists the recent classification applications in metal AM. According to Table 11.6, SVM, neural network (NN),

ANFIS, and KNN are applied in LPBF and LDED to classify different types of flaws. In the following sections, each algorithm and its application in AM will be explained.

Support Vector Machine (SVM)

The SVM is used to find the hyperplane in N-dimensional space to classify the dataset. There are many possible hyperplanes to classify two sets of features, as shown in Figure 11.31a. SVM's objective is to find the maximum distance between data points, as demonstrated in Figure 11.31b. The maximum margin can provide more confidence for the classification of future and unseen data points not included in the training. The optimum hyperplane is the decision boundary, which is measured by the training dataset. The hyperplanes are then used to classify test data by which each data point falling on either side of the line can be attributed to different classes. The number of features can determine the number of optimum hyperlines. For example, three classes of input features result in two optimum hyperlines.

The linear data can be classified by:

$$f_{w,b}(x) = sgn\left(w^T x + b\right) = \begin{cases} w^T x_i + b \geq 1 & x \geq 0 \\ w^T x_i + b \leq 1 & x < 0 \end{cases} \qquad (11.12)$$

where w is a weight vector and b is a bias. SVM can easily classify two linear datasets, but different classes are not linearly separable for more complex datasets. One possible solution is to transfer input data into higher dimensional space by adding kernel function to the SVM technique. Kernel function can be categorized into linear, nonlinear, polynomial, radial basis, and sigmoid [162, 163].

SVM is applied in the AM process to classify nominal and abnormal regions. For example, a linear and Gaussian kernel SVM can be used when the coaxial thermal camera captures the input data during the fabrication of a thin wall, and the output data (anomalies) can be labeled based on the anomalies identified by the X-ray CT scan. After dimensional reduction and

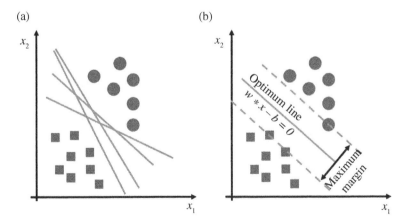

Figure 11.31 Support vector machine method: (a) possible hyperplane and (b) maximum margin (Source: Redrawn and adapted from [162]).

feature extraction, the SVM models can be trained, and the accuracy of predictions can be calculated using recall, precision, and F-score, resulting in the ability of the algorithm to predict anomalies [137]. The same approach can be used to classify normal and abnormal melt pool [26].

Linear kernel SVM is used to identify porosity [19, 20]. This can be done through a DSLR camera mounted on the LPBF setup and five flashing lights to generate eight different lighting conditions. Three images after recoating and five images after scanning can be captured from the melt pool. The step cylindrical sample can be designed to create discontinuity. Post-build CT scans can be obtained from the part, and the anomaly detection of CT scan can be labeled manually to generate a ground-truth table. A linear SVM classifier is trained using ground-truth labels to classify voxels into anomalous (porosity) or nominal. Fourfold cross-validation is implemented to validate the effectiveness of the linear SVM method. In this approach, it is reported that the results show 65% accuracy using an individual flash module and 85% accuracy using all eight modules. The algorithm can detect porosity larger than 47 μm; however, the detection near the edges of parts was not considered in the study [18].

Neural Network (NN) and Deep Learning (DL)

An artificial neural network (ANN) is a computational model inspired by the structure of the brain network. Each network has one input layer, hidden layers (at least one), and one output layer, as shown in Figure 11.32a. The input and output layers may have multiple inputs/outputs. Each layer consists of computational devices (neuron/node) that are connected (weights). Neurons are a mathematical function that feeds the input signal into an activation function and weights, which carried values between neurons as represented in Figure 11.32b. With a forward pass, the output is calculated, and an error between labeled and calculated output is estimated. To minimize errors, weights are updated with a backward pass [164].

Various types of NN are proposed based on their connection, the number of hidden layers, and applications. These include but not limited to:

1. Feedforward neural network (FF-NN): In this NN class, the data passes only through one direction. The FF-NN is used to face recognition and computer vision [165].

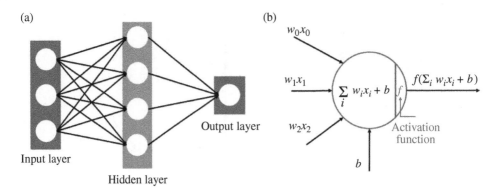

Figure 11.32 Neural network: (a) 3 input–1 output neural network and (b) schematic of one neuron (Source: redrawn and adapted from [164]).

2. Multilayer perceptron neural network (MLP-NN): It has three or more layers with nonlinear activation function (hyperbolic tangent or logistic) to classify nonlinear data. MLP-NN is commonly used in machine translation technologies [166].
3. Deep learning (DL): More complex and realistic models with complex computational principles are known as deep learning. Deep learning is associated with the transformation and extraction of features to establish a relation between the labeled input and output dataset. Deep learning network could be classified into various types based on their connections, the number of neurons, hidden layers, and applications; however, the most common types are convolutional neural networks (CNN), recurrent neural networks (RNN), deep belief network (DBN), and auto-encoder (AE) [167].

Various NNs can be used to predict track geometry [20, 138, 141] (see Table 11.6). The CNN model can be developed, trained, and evaluated to predict LPBF track widths, width standard deviations, and track continuity from in-situ video data captured from the melt pool by applying a variety of laser power and scan speed settings [141].

Defect detection is another application of deep learning neural networks in AM [140, 168]. For example, the off-axial videos captured from LPBF under different process parameters conditions can be analyzed to extract two images per layer: (i) image which is taken after powder deposition and (ii) image which is captured after laser scanning. The extracted images can be used to train the CNN model. The model can be used to detect anomalies in the solidified zones [99]. Images can also be captured from the melt pool area and then pass through a series of transformations to crop and add the Gaussian noise and blur to expand the dataset. Many images can be considered to feed into the CNN models, and L2 regularization can also avoid overfitting. After training the dataset, the model can be validated by evaluation metrics to predict crack, keyhole, good quality (nominal), and LoF [143]. Although the study shows improvement to classify flaws, some of the pores cannot be classified correctly in the nominal group, mainly due to the process's uncertainties and complexities.

In general, two types of information from the melt pool (the "keyhole" information and the melt pool temperature information) can be collected by the sensors installed in LPBF, while five features from the plume (the plume area, intensity, direction, major axial length, and minor axial length), and six spatter-related features (the spatter number and average of spatter area, gray value, orientation, and velocity) can be extracted from images. The RBF kernel SVM and CNN algorithms can classify the input into low-dimensional feature spaces. The AlexNet can be used for the kernel size and pooling strategy, and the layer depth and kernel number can be selected based on a trade-off between the accuracy and training time in the CNN algorithm [18].

A more complex model can be proposed when the temperature data (e.g. extracted from a dual-wavelength pyrometer) is used [169]. The model is called Convolutional and Artificial Neural Network for Additive Manufacturing Prediction using Big Data (CAMP-BD) [169]. The dataset can be used to train the CAMP-BD deep learning model using CNN and MLP as inputs. CNN and MLP networks can be applied to analyze thermal images and relevant process/design parameters. The output of the CNN and ANN are joined to pass through CAMP-BD as shown in Figure 11.33. The CAMP-BD is propagated to optimize the final pointwise distortion prediction by Adam optimizer [169].

Adaptive Neuro-Fuzzy Inference System (ANFIS)
An ANFIS works based on the Takagi–Sugeno fuzzy inference system (FIS) [170]. Since the principle of the algorithm is based on a combination of an ANN and a FIS, it has the potential to

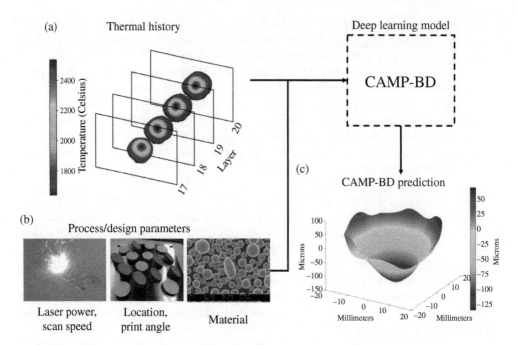

Figure 11.33 A deep learning approach to LPBF based on "Convolutional and Artificial Neural Network for Additive Manufacturing Prediction using Big Data": (a) CNN model analysis, (b) ANN model analysis, and (c) a prediction example (Source: Republished with permission of Elsevier, from [169]).

estimate nonlinear functions. ANFIS uses the labeled input–output data during training to obtain IF-THEN fuzzy rules. The architecture of ANFIS consists of five layers, as shown in Figure 11.34.

The first layer is known as the fuzzification layer in which input data are used to calculate membership functions, which can be stated as [171]:

$$O_i^1 = \mu_{A_i}(x_1) = gbellmf(x; a, b, c) = \frac{1}{1 + \left|\frac{x-c}{a}\right|^{2b}} \tag{11.13}$$

where a, b, and c are called the premise parameter. In the second layer (rule layer), the membership function is used to generate firing strength (w_i):

$$O_i^2 = w_i = \mu_{A_i}(x_1).\mu_B(x_2) \quad i = 1, 2 \tag{11.14}$$

The third layer is called the normalization layer, in which the normalized firing strength is calculated:

$$O_i^3 = \overline{w_i} = \frac{w_i}{w_1 + w_2 + w_3 + w_4} \quad i \in \{1, 2, 3, 4\} \tag{11.15}$$

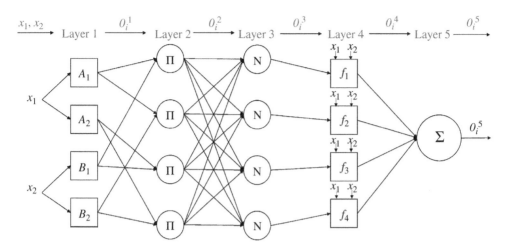

Figure 11.34 ANFIS structure with two inputs and one output (Source: redrawn and adapted from [171]).

In the fourth layer, which defuzzification, weighted values of each node are calculated using the first-order polynomial:

$$O_i^4 = \overline{w}_i f_i = \overline{w}_i (p_i x_1 + q_i x_2 + r_i) \tag{11.16}$$

where p_i, q_i, and r_i are the consequence parameters. In the last layer, known as the summation layer, the actual output is calculated by summing the outputs of each rule gained from layer 4.

$$O_i^5 = \sum_i \overline{w}_i f_i = \frac{\sum_i w_i f_i}{\sum_i w_i} \tag{11.17}$$

In the training phase, premise and consequence parameters are updated to achieve effective results [171]. Many optimization algorithms are suggested to train ANFIS parameters such as backpropagation (BP) [170], gradient descent (GD) [172], Levenberg–Marquardt (LM) [173], recursive least square (RLS) [174], Kafman filter (KF) [175], and extended Kafman filter (EKF) [176].

ANFIS is used in the AM process to predict fatigue life cycle [145], surface roughness [146], and clad geometry [147–150]. Another artificial intelligence-based algorithm that can be implemented for DED for controlling the geometry of the deposited bead is a fuzzy logic controller. The basic paradigm for fuzzy logic control is a linguistic or rule-based control strategy, which maps the given physical system's observable inputs into its controllable outputs by applying a set of linguistic implication rules [1]. Figure 11.35 shows a typical fuzzy control strategy for an LDED process, where a fuzzy logic controller and fuzzy logic gain scheduler are incorporated into the controller decision-making part. At the heart of this control, the scheme is a fuzzy logic control algorithm that maps the normalized error e_k and rate of error d_k to the change in the control output or δu_k. As an example, a typical logistic rule in the fuzzy logic controller is

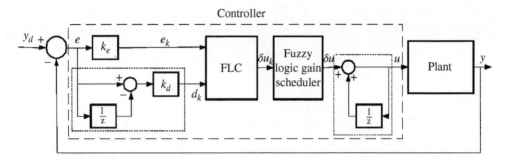

Figure 11.35 Proposed fuzzy logic controller for LDED to control the bead height by adjusting process speed (Source: Republished with permission of Taylor & Francis Group, from [1]; permission conveyed through Copyright Clearance Center, Inc.).

if e_k is large Positive (LP) or small Positive (SP) and d_k is Large Negative (LN) or Small Negative (SN), then δu_k is Zero

Figure 11.36 shows fuzzy membership functions over the normalized domains of definition of the relevant variables and the mathematical meanings of LP, SP, LN, SN, and Z in a normalized range. For a system with a large settling time such as LDED, the process command can also be amplified at the beginning of the process. The basics of the developed fuzzy logic control

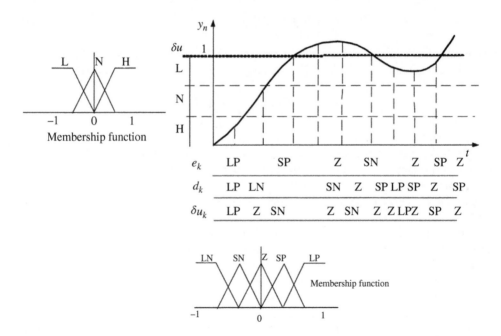

Figure 11.36 Typical membership function in fuzzy logic controller used in LDED (Source: Republished with permission of Taylor & Francis Group, from [1]; permission conveyed through Copyright Clearance Center, Inc.).

paradigm for LDED is to choose the value of normalization factor in the output of the standard fuzzy controller by another set of fuzzy rules, which is a so-called fuzzy logic gain scheduler. To keep control over the response, it is proposed that the area below the steady-state line be divided into several appropriate subregions of L, N, and H, as shown in Figure 11.36. During the startup, the value of gain is switched to a higher value to accelerate the system response. Conversely, it is decreased when the response approaches the steady-state response. Implementation of the designed fuzzy logic controller to the LDED process is subjected to the nonlinear terms identified in the Hammerstein–Wiener model (see Chapter 4, Section 4.7.5). As a result, the inverse functions of f and g are incorporated in the fuzzy controller [1].

K-Nearest Neighbors (KNN)

KNN is a pattern recognition method [177]. The method is used to estimate the relationship between input and output in a multidimensional feature space. The algorithm trains the dataset in two steps. In the first step, the similarity between the two data points (x and x') is calculated. The most popular method to calculate similarity is Euclidean distance given by [178]:

$$d(x, x') = \sqrt{\left(x_1 - x_1'\right)^2 + \left(x_2 - x_2'\right)^2 + \ldots + \left(x_n - x_n'\right)^2} = \sqrt{\sum_{i=1}^{n} (x_i - x_i')^2} \qquad (11.18)$$

After computing distance (d) between x and each training set, in the second step, K (defined by the user) closest points to x are selected in one class called $\mathcal{A}$. Then, the conditional probability of each class is calculated by [179]:

$$P(y = j | X = x) = \frac{1}{K} \sum_{i \in \mathcal{A}} I\left(y^{(i)} = j\right) \qquad (11.19)$$

where $I(x)$ is the indicator function. Finally, the input data are assigned to the class with the largest probability [179]. Although the KNN algorithm is easy to implement, finding the optimum k is one challenge. Besides, the accuracy of the algorithm depends on the scale, dimension, and quality of data. KNN classification method is slightly used in the AM process to compare its result with other classification methods [26, 134, 139].

SVM, KNN, and Feed Forward Neural Network (FFNN) have been used to classify a LoF. F-score can be used to evaluate the results [134].

11.5.1.2   Regression Model

The regression model predicts the mapping function, a mathematical equation, from the input (x) variable to the output (y) variable. The regression model can solve many types of problems with discrete input, multi-input, and time-series input variables. The performance of the regression model is evaluated by mean squared error (MSE) and root mean square error (RMSE). Many types of regression models are proposed according to the number of input data, the shape of the regression line, and the input variable type. According to Table 11.6, the Gaussian process (GP) and neural network (NN) are regression models applied in LPBF and LDED. In the following section, each algorithm and its application in AM are discussed.

Gaussian Process (GP)

The Gaussian process is a random process to assign input variable x to a random function $f(x)$ with a Gaussian distribution. Gaussian function defines by mean (m) and covariance (σ) given by [180]:

$$p(f \mid X) = \mathcal{N}\left(f \mid \mu, \sum\right) \tag{11.20}$$

where $f = (f(x_1), f(x_2), \ldots, f(x_n))$, $\mu = (m(x_1), m(x_2), \ldots, m(x_n))$, and $\sum = \sigma(x_i, x_j)$. m is the mean function and commonly is zero ($m(x)=0$). σ is a positive covariance function. A GP prior $p(f \mid X)$ can be converted to GP posterior $p(f \mid X, y)$ by [180]:

$$p(f_* \mid X_*, X, y) = \int p(f_* \mid X_*, f) p(f \mid X, y) df = \mathcal{N}\left(f_* \mid \mu_*, \sum\nolimits_*\right) \tag{11.21}$$

The posterior function is then used to make predictions of f_* by a given new input (X_*) [180].

Four kernel hyperparameters are involved in the GP model, including the variances of the two Matern 5/2 kernel functions and the length scales of the two input dimensions (the laser scan speed and the laser power). These parameters can be updated once a new observation was made to reduce the mean absolute prediction error (MAPE). Since the number of observation points is limited, the regression surface is generated using the n-fold cross-validation method to avoid overfitting issues and maximize the utilization of observation points.

Neural Network (NN)

The neural network also can use as a regression model to estimate continuous output. The NN regression model can be used to predict the flaws happening in the LPBF process. To this end, spatially resolved acoustic spectroscopy (SRAS) can be used to monitor the frequency of acoustic signals from the building of 10 samples. The SRAS characteristics can then be extracted using the Fourier transform. The DCB-MIR network, which is a type of fully convolutional block densely connected network and builds upon ResNet [181] and DenseNet [182] models, can be trained offline. The cosine similarity can be considered to identify the similarity between the proposed model and the optical micrograph. This criterion can be notified of the progress in detecting size and location of defects, especially the LoF and scratches [156].

A NN regression model can be used to correlate the LDED process parameters and geometrical features. Then process and geometrical parameters can be utilized to feed into NN trained using the Levenberg–Marquardt algorithm. The RMSE can be considered to compare the ANN prediction and experimental results, showing the best performance belongs to the nine hidden layer nodes [157].

11.5.2 Unsupervised Learning

Unsupervised learning is applied to find groups or clusters in a feature space without any labeled data. The most common task in unsupervised learning is clustering, in which the pattern among input data can be identified. Another task is data reduction by PCA and SVD.

11.5.2.1 Clustering

Clustering analysis is a task to group the dataset based on their similarity. The algorithm had been introduced in anthropology [183], but it has recently been applied in other fields. Clustering algorithms are also used to pre-process the dataset by feature extraction and data reduction. Table 11.6 demonstrates the recent application of the clustering algorithm in LPBF and LDED to group similarity among data and reduce the data dimension. According to Table 11.6, K-means and self-organized map (SOM) algorithms are applied in LPBF and LDED to predict abnormal processes. In the following sections, these popular clustering algorithms are explained.

K-Means
K-means is a simple type of clustering algorithm. To apply K-means, only four iterative steps are required.

1. The K centroids (center of the cluster) are randomly selected.
2. Each data point is assigned to its nearest centroids called expectation step.
3. The mean of all points for each cluster is calculated, and a new centroid is set called as maximization step.
4. Then, the algorithm is repeated until the centroid positions do not change.

Although K-means has some weaknesses in clustering data with complex shapes and different densities, it is simple to implement and guarantee convergence. The algorithm is commonly used in AM to classify delamination and an overhang structure [77, 100, 160], crack [63], plume [110], and process parameters [117].

The K-means algorithm is used to classify consequential anomalies related to the powder spreading process in LPBF [77]. The camera is used to capture images from the melt pool, and lightening condition is also implemented to enhance the quality of images. Then, many image patches should be convolved with six types of filters (Gaussian, difference of Gaussian (DoG), oriented edge detectors, oriented line detectors, streak detectors, and debris), as shown in Figure 11.37 [77, 135]. Similar features can be categorized into one cluster by the K-means algorithm (Figure 11.37c). The methodology to find the optimum number of the cluster is the cluster validity. Cluster validity is an optimization technique in which the requested number of clusters is increased until no noticeable improvement in the results, leading to 100 groups or more. Besides, the cluster seeding is repeated 100 times to avoid trapping in a local minimum. In the next step, the fingerprint of each clustered group is calculated by histogram and stored in a table for future use (Figure 11.37g). By generating and comparing test-patched fingerprints with the table, the top three matches are chosen and weighted to identify the type of anomaly (Figure 11.37i and Figure 11.37j) [77].

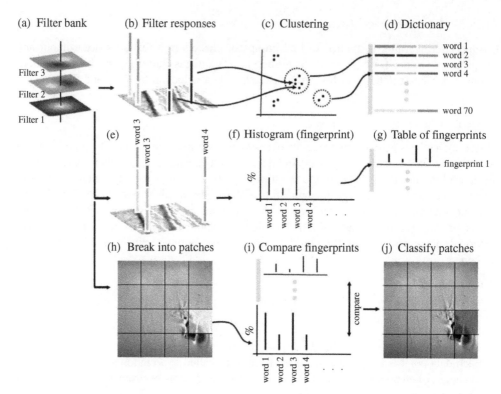

Figure 11.37   Use of K-means in LPBF: flowchart of the proposed process (Source: Republished with permission of Elsevier, from [77]).

K-means can apply to detect and localize the existence of hot spots and spatters. The cluster validity can be used to optimize the cluster number. For example, for an LPBF, the normal process may have only two clusters: one refers to a laser scan and another corresponds to the background. As soon as the K-means algorithm finds more than two clusters known as a hot spot, and the alarm is signaled, as demonstrated in Figure 11.38 [100].

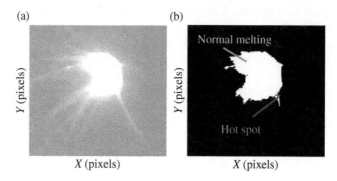

Figure 11.38   (a) An original image and (b) using K-means to identify hot spot and spatters based on clustering the data (Source: Redrawn and adapted from [100]).

Self-Organized Map (SOM)

The SOM algorithm is a type of ANN unsupervised learning. The SOM maps the high-dimensional input into the two-dimensional output. The algorithm is working based on a neighborhood function and mapping the dataset by multiple steps [184], as shown in Figure 11.39.

The initialization is the first step in which weights and biases are chosen randomly. The second step is known as competition in which discriminant function is given by [186]:

$$d_j(x) = \sum_{i=1}^{D} (x_i - w_{ji})^2 \tag{11.22}$$

where x is a D-dimensional input space and w_{ji} is the connection weight between neurons i and j. At the end of this step, input space is mapped to the discrete output space by a simple process of competition between the neurons. In the third step, called the cooperative, winner neurons affect their neighborhood neurons, which is the same as the real neural network in the human brain. So, lateral interaction or topological neighborhood effect is defined by [186]:

$$T_{j,I(x)} = exp\left(-S^2_{j,I(x)}/2\sigma^2\right) \tag{11.23}$$

where S_{ij} is the lateral distance between the winner neuron and its neighbor (j) and $I(x)$ is the index of the winner neuron. The topological neighborhood ($T_{j,\ I(x)}$) exponentially shrinks with time by σ. In the fourth step, known as the adaption, weights are updated by [186]:

$$\Delta w_{ji} = \eta(t).T_{j,I(x)}(t).(x_i - w_{ji}) \tag{11.24}$$

where t is the iteration number and η is a learning rate. In the last step called continuation, the mapping process keeps returning to step 2 until the map stops changing.

The SOM is a popular technique in fused filament fabrication (FFF) additive manufacturing to detect geometry deviation [187] and defect [188]. Recently, some researches were carried out to identify porosity in LDED [27, 139].

The SOM based on the temperature distribution of the top surface of the melt pool can be applied to LDED. The thermal history is recorded and stored in many images. The melt pool

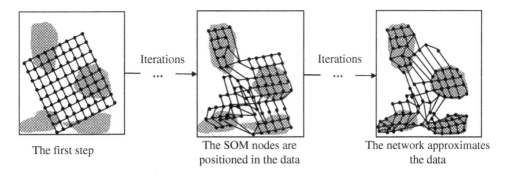

The first step	The SOM nodes are positioned in the data	The network approximates the data

Figure 11.39 SOM concept and iterations to map nodes to the input space (Source: Redrawn and adapted from [185]).

Table 11.7 Performance of different SOM dimensions based on the SOEDNN.

SOM dimension	Average MC	Best MC	Average accuracy	Best accuracy
5×5	25.86	12.5	0.9668	0.9871
10×10	23.75	5.93	0.9672	0.9839
15×15	20.32	3	0.9676	0.9807
20×20	26.53	2	0.9501	0.9871

Source: Republished with permission of Elsevier, from [139].

images are converted to the continuous thermal distribution model to feed into the SOM algorithm. The algorithm is applied to predict normal and abnormal melt pool with four cluster sizes (4 × 4, 5 × 5, 6 × 6, and 7 × 7). Also, X-ray CT can be used to find the actual positions of defects [27].

The porosity identification in the single track Ti-6Al-4V thin wall based on the in-situ melt pool images is investigated [139]. Images are captured by an IR camera and then used for porosity prediction and characterization. X-ray CT is conducted to provide microstructure information to enable supervised learning. The SOM algorithm is selected with different dimensions from 5 × 5 to 25 × 25. The best accuracy and the average and best misclassification cost of the SOM can be analyzed by Self-Organizing Error-Driven Neural Networks (SOEDNN) [189], as represented in Table 11.7.

The 15 × 15 dimensions is chosen because of its lower average on the misclassification cost and the highest accuracy than other dimensions [139].

11.5.2.2   Data Reduction

Data reduction is used as a data pre-processing method to reduce the data dimension and highlight the most significant features. Various techniques are defined to reduce the high-dimensional data. However, in AM, PCA and SVD are mainly used, as listed in Table 11.6. Although the SVD technique is originally a type of clustering algorithm, it is used to reduce data and extract features in the AM process. PCA and SVD techniques are discussed in the following sections:

Principal Component Analysis (PCA)
PCA is a technique to find patterns, similarities, and differences in high-dimensional data. Although the algorithm has a principle based on the statistical concept, the PCA can be used to implement the data reduction automatically. To implement the algorithm, the covariance matrix of data is calculated by [190]:

$$\Sigma = \frac{1}{m}\sum_{i=1}^{n}\left(x^i\right)\left(x^i\right)^T \tag{11.25}$$

where x is the high-dimensional input data and Σ is the covariance matrix. In the next step, the eigenvalue and eigenvector of the covariance matrix are calculated. The eigenvalue is sorted from the highest to the lowest value. The eigenvector with the highest eigenvalue is the principal component of the data and data with lower eigenvalues can be ignored. Then, the principal component of the data is used as the reduced dataset [190].

PCA is used to extract significant features in LPBF [18, 74, 97, 100, 160] and LDED [26, 63, 137, 161]. It should be noted that ignoring many input features can adversely affect the accuracy of the performance. It is shown that how a different number of input features can affect the performance of the SVM. The PCA improves the accuracy of the SVM from 89.7 to 90.1% using 33 input features; however, it reduces the SVM accuracy from 89.2 to 88.3% using 17 input features [18]. The study revealed that PCA could also have a negative influence on performance if too many features are eliminated.

More complex types of PCA can also be proposed. For example, spatially weighted PCA (ST-PCA) and T-mode PCA can be compared to classify local hot spots. The simulation analysis shows that ST-PCA is more effective and faster than the basic T-mode PCA in detecting the onset of a local defect. The recursive and the moving window updating methods, coupled with the ST-PCA, provide almost similar performances [100].

Singular Value Decomposition (SVD)
The SVD is a technique to factorize a matrix into three matrices given by [191]:

$$A = USV^T \tag{11.26}$$

where A is $m \times n$, U is an $m \times n$ orthogonal, S is an $n \times n$ diagonal, and V is an $n \times n$ orthogonal matrices. S matrix is called a singular value and normally arranged from the largest to the smallest value. The singular value matrix can approximate the A matrix. This feature makes the SVD appropriate to reduce data dimension and feature extraction [191]. The technique is widely used in the design [192] and process optimization [193] of the AM process; however, only one study is conducted in the quality assurance field [50]. This study suggests using two photodiodes coaxially mounted on an LPBF setup to detect plasma emission (700–1050 nm) and thermal radiation (1100–1700 nm). A combination of supervised and unsupervised learning can be used to identify the tensile test bars fault. In the unsupervised learning part, the SVD is applied to extract the key features. Although some information is lost during the feature extraction, which could decrease the algorithm's accuracy, this methodology reduced the need for costly certification experiments like CT scan of all samples to provide labeled data.

11.6 Case Study

The case study is based on the previous study of authors [199]. The focus of this part is to highlight the results of in-situ monitoring for in-process identification of porosity formed due to the LoF phenomenon. This study aims to apply a systematic approach in which the defects are intentionally designed and embedded in the coupon samples to resemble the effect of pores due to LoF. Then through the model developed for the detection of artificial defects, randomized defects are detected. The effectiveness of an algorithm developed in the EOS M290 MeltPool Monitoring commercial software is investigated. The algorithm is applied to understand the impact of artificial defects on the light intensity collected by an on-axis photodiode to detect an LoF porosity. The impact of variation in print parameters on the detection of defects is also discussed. The results are validated through micro-CT scanning datasets.

11.6.1 Design of Experiments

Two sets of cubical samples are devised and labeled as R-series and T-series. All samples, composed of Hastelloy-X and size of 8 × 8 × 10 (W × L × H) mm, are fabricated by an EOS M290. In R-series, the effect of LoF is mimicked by embedding artificial voids in samples R2, R3, R4, and R5, as shown in Figure 11.40a–d. One control sample (R1) is also printed without any artificial void. The print parameters, listed in Table 11.8, are selected according to the print parameters used to obtaining high-quality Hastelloy-X (EOS Nickel Alloy HX, Krailling, Germany) parts [194]. Additionally, three samples (R6, R7, and R8) with a similar design to R2 (Figure 11.40a) are included in the first set of design of experiments (DoEs). Still, they are printed with different process parameters, as highlighted in Table 11.8. Each sample was printed eight times at different locations of the build plate with respect to the direction of the gas flow as well as the recoater (Figure 11.41a).

In T-series, parts are designed and printed by only varying print parameters to create randomized voids due to the LoF. The geometry of these samples is similar to R1, where the print

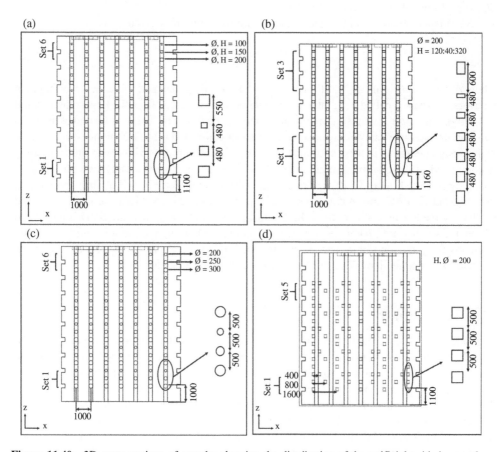

Figure 11.40 2D cross sections of samples showing the distribution of the artificial voids in samples: (a) R2, (b) R3, (c) R4, and (d) R5. Ø (μm) is diameter and H (μm) is height [199]. Published under an open access CC BY 4.0 license

Table 11.8 Types and print parameters in print 1.

Sample no.	Type of artificial voids	Power (W)	Hatching distance (µm)	Speed (mm/s)
R1	Standard (no artificial voids)	200	90	1000
R2	Cylindrical voids	200	90	1000
R3	Cylindrical voids	200	90	1000
R4	Spherical voids	200	90	1000
R5	Cylindrical voids	200	90	1000
R6	Cylindrical voids at the different laser power	100	90	1000
R7	Cylindrical voids with the different hatching distance	200	150	1000
R8	Cylindrical voids with the different process speed	200	90	1500

parameters for each one are listed in Table 11.9. Six samples from each design are labeled and arranged on the build plate (Figure 11.41b).

11.6.2 In-Situ Sensors and Quality Assurance Algorithm

The integrated sensors of EOS M290 are used to capture light intensity signals from the melt pool. The co-axial and off-axial photodiodes collect intensity signals in the visible–NIR range with a 60-kHz sampling frequency. The intensity signal is collected during the print, and it is accessible afterward for analysis using commercial EOSTATE MeltPool Monitoring software (EOS GmbH, Krailling, Germany). Three algorithms are incorporated into the software: AL, SD, and STF [119]. These three algorithms are designed to detect specific process phenomena that might affect the quality of the printed part by applying the threshold on the collected photodiode(s) signal of each layer of the build. The schematic of signals and detection method is shown in Figure 11.42. After applying the algorithm and threshold ranges (Figure 11.42c), some signals might exist within the threshold ranges (Figure 11.42d). This signal perturbation/fluctuation may be corresponding to the location of defects in the last tiff images (Figure 11.42e light gray pixels). A tiff image of each layer is exported for offline analysis.

A data screening procedure is pursued when the following ranges of threshold and the different number of window lengths for the moving average window are selected:

1. AL: threshold 10 000–22 000 and window length 5–30;
2. SD: threshold of 20–1000 and window length 5–30, and;
3. STF: threshold 65–140 and window length 5–20.

From observing the results of this step, each threshold is narrowed down to a smaller band to highlight the artificial defect's effect. Besides, the results from this step demonstrate a high signal fluctuation in layer 1 printed on the consecutive of artificial defect zones (see light gray corresponding to the artificial defects in Table 11.10).

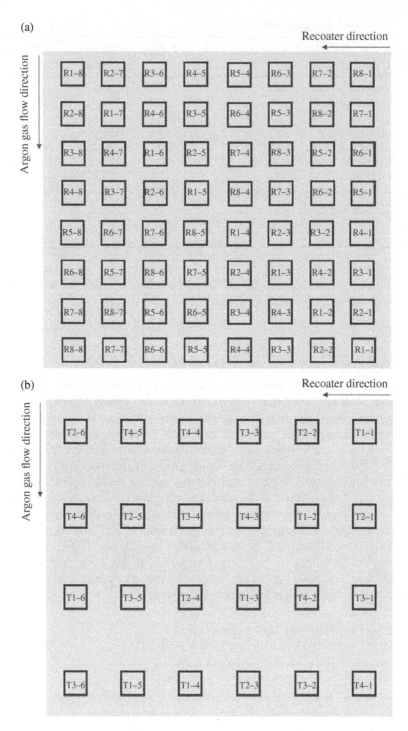

Figure 11.41 Samples layout on the build plate: (a) R-series and (b) T-series [199]. Published under an open access CC BY 4.0 license

Table 11.9 Types and print parameters in print 2.

Sample no.	Type of variation	Power (W)	Hatching distance (µm)	Speed (mm/s)
T1	Standard	200	90	1000
T2	Power	100	90	1000
T3	Hatching distance	200	150	1000
T4	Speed	200	90	1500

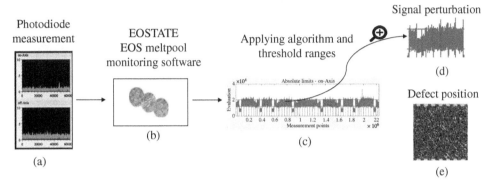

Figure 11.42 Schematic of the defect detection process in EOSTATE MeltPool Monitoring software [199]. Published under an open access CC BY 4.0 license

Table 11.10 Applying the detection algorithms on the five consecutive layers after the defect (sample R3, defect radius = 200 µm, and defect height = 280 µm) [199]. Published under an open access CC BY 4.0 license

	Algorithm	Layer 0[1]	Layer 1	Layer 2	Layer 3	Layer 4	Layer 5
On-axial photodiode	Absolute limits						
	Signal dynamics						
	Short-term fluctuations						
Off-axial photodiode	Absolute limits						
	Signal dynamics						
	Short-term fluctuations						

"1" means the layer which is included defects.

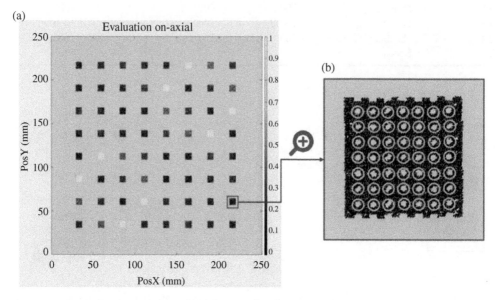

Figure 11.43 (a) An example of an image generated by the EOSTATE software for one layer of the print and (b) selection of 49 regions of interest around each defect in sample R2 – Location 1 [199]. Published under an open access CC BY 4.0 license

The matching result between the location of artificial defects and light gray pixels demonstrates that the AL algorithm can effectively identify the location of defects. However, by analyzing photodiodes' data, the result shows that applying the AL algorithm on the off-axial photodiode signal leads to fewer light gray indicators than the co-axial photodiode signal. After applying the AL algorithm, the images generated by the software are segmented into areas of interest around each artificial defect using MATLAB (Figure 11.43). The experiments' results are compared through the determination of the average and standard deviation of indicators' (light gray pixels) percentage. Besides, to investigate the statistical significance of each input, a single factor (size and distribution separately) analysis of variance (ANOVA) with a significance level (α) of 0.05 is performed. The ANOVA is conducted on the number of indicators to study if the size/distribution of artificial defects and their location on the build plate affect the light intensity signal (p-value $< \alpha$).

All samples are analyzed for micro-CT scan (μCT) to validate the actual pores' positions and distribution. The coupon samples are subjected to the X-ray μCT on Zeiss Xradia Versa 520 system (Carl Zeiss Microscopy GmbH, Germany) with ~10-W and 160-kV source voltage.

The results are presented for the spatial distribution of the light gray pixels/indicators for each sample of R-series on selected layers (Figure 11.44). The average number of light gray pixels/indicators and their standard deviations in the regions of interest for R5 is shown in Figure 11.45. When the following discussion is valid for most printed parts, we focus the discussion on R5.

The effect of the distribution in sample R5 establishes that approximately $17.44 \pm 6.5\%$, $16.24 \pm 6.33\%$, and $15.09 \pm 4.17\%$ of the selected areas are covered by indicators. The same range of results proved that the distribution could not significantly affect the light intensity signal (Figure 11.45).

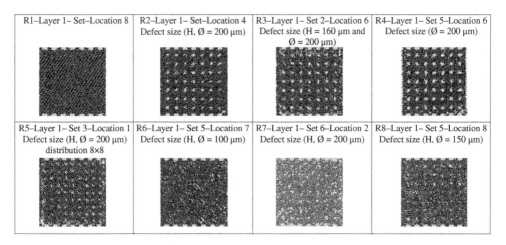

Figure 11.44 The population and distribution of indications for AL algorithm – R-series (see Figure 11.41a) [199]. Published under an open access CC BY 4.0 license

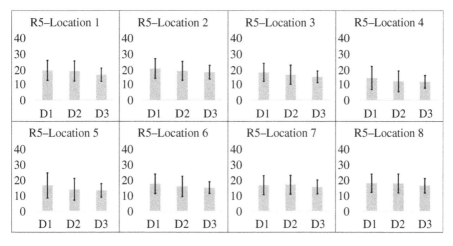

Figure 11.45 The percentage of average and standard deviation of indications in the region of interests for the selected layers and after applying the AL algorithm (sample R5). D1, D2, and D3 mean three different void distribution embedded in sample R5 (D1 = 5 × 5, D2 = 8 × 8, and D3 = 9 × 9) [199]. Published under an open access CC BY 4.0 license

The CT result of samples R2 and R3 show 120 µm is the smallest size of detection by EOS-TATE software. The average number of indicators in 100-µm defects (sample R2) is half of the average number of indicators in 120-µm defects (sample R3) and for defects larger than 120 µm, the gap (the difference between indicators' number) between the sizes is insignificant. It is also absorbed that the locations of samples, in the build, are realized that the direction of gas flow and recoater could affect the LoF porosity, as discussed before [195]. Besides, the gas flow distribution is not uniform in the build plate [196], resulting in different numbers of light gray pixels/indicators when printing one design in different positions.

Table 11.11 The range of *p*-value from ANOVA single-factor analysis of the variance.

Sample	Defect type	Defect size (μm)	Effect of different locations	Size of defects	Distribution of defects
R2	Cylindrical artificial defect	Ø, H = 200	<0.05	< 0.05	–
		Ø, H = 150	<0.05		
		Ø, H = 100	<0.05		
R3	Cylindrical artificial defect	200*200*320	<0.05	< 0.05	–
		200*200*280	<0.05		
		200*200*240	<0.05		
		200*200*200	<0.05		
		200*200*160	<0.05		
		200*200*120	<0.05		
R4	Spherical artificial defect	Ø = 300	<0.05	< 0.05	–
		Ø = 250	<0.05		
		Ø = 200	<0.05		
R5	Cylindrical artificial defect	Distribution: 5*5	<0.05	-	>0.05
		Distribution: 9*9	<0.05		
		Distribution: 8*8	<0.05		
R6	Cylindrical artificial defect	Ø, H = 200	<0.05		–
	and randomized defect	Ø, H = 150	<0.05	< 0.05	
		Ø, H = 100	<0.05		
R7	Cylindrical artificial defect	Ø, H = 200	<0.05	> 0.05	–
	and randomized defect	Ø, H = 150	<0.05		
		Ø, H = 100	>0.05		
R8	Cylindrical artificial defect	Ø, H = 200	<0.05	< 0.05	–
	and randomized defect	Ø, H = 150	<0.05		
		Ø, H = 100	<0.05		

After evaluating the data for all samples by average and standard deviation, the ANOVA is conducted for exploring the effect of sample positions in the build plate, as well as the effect of size (samples R2, R3, R4, R6, R7, and R8) and distribution (sample R5) of artificial defects (Table 11.11).

The ANOVA analysis endorses the effect of samples' positions on light gray pixels/indicators in which most of the samples showed a *p*-value < 0.05 (Table 11.11). Additionally, the effect of artificial defects' size is discussed above. A similar conclusion is drawn when the ANOVA is conducted (Table 11.11), resulting in a significant statistical difference between different sizes of defects on light intensity signal and, consequently, on the light gray pixels/indicators. However, the ANOVA analysis demonstrates an insignificant difference between scenarios of distribution (sample R5).

11.6.3 *Correlation Between CT Scan and Analyzed Data*

The CT scan analysis of samples R1–R5 is conducted for Location 4 on the build plate. However, some pores are healed in the process based on the CT scan result. By removing those

Table 11.12 The percentage of identification for each size of the defect and density of each sample.

Sample	Defect type (µm)	Percentage of identification by EOSTATE	Density
	Ø,H = 200	100	99.89
R2 – Location 4	Ø,H = 150	100	
	Ø,H = 100	100	
	200*200*320	99.31	99.28
	200*200*280	100	
R3 – Location 4	200*200*240	100	
	200*200*200	100	
	200*200*160	100	
	200*200*120	100	
	Ø = 300	100	99.69
R4 – Location 4	Ø = 250	100	
	Ø = 200	100	
	Distribution: 5*5	76	98.08
R5 – Location 4	Distribution: 9*9	77.78	
	Distribution: 8*8	72.5	

pores, the detection of MPM analysis shows most of the defects are identified (Table 11.12). The CT scan analysis of R2, R3, R4, and R5 demonstrates the dimension deviation of defects (Figure 11.46). The deviation is limited to 150 µm in samples R2 and R3 (Figure 11.46a and b). The sample R4 defects have more deviation due to the spherical shape and size of artificial defects (Figure 11.46c). However, the deviation result of sample R5 cannot result in the rational conclusion due to this sample's position in the build plate (Figure 11.46d). Also, the CT scan result of samples R5–4 (Figure 11.46d) confirms that defects in the gas flow direction were partially detectable. The same analysis can be done for others.

Based on the knowledge gained from the artificial defects (samples R2, R3, R4, and R5), the rest of the samples in R-series (R6, R7, and R8) and T-series are investigated through a new image segmentation approach as explained in the following sentences. In each sample, 82 segmental batches are assumed (Figure 11.47a), each of which includes three layers of the print. Then each batch is subdivided into 4356 voxels (Figure 11.47b). The size of each voxel is 120 (µm) × 120 (µm) × 120 (µm) which selected based on the smallest size of the artificial defects detected by analyzing the data of artificial defects. Also, a 60-µm overlap is considered between adjacent voxels.

The stack of three layerwise images collected is labeled to create 120-µm height. At the end, the result of 357 192 cells for each sample is compared with the corresponding CT data to establish the matching matrix [197]. The matching matrix is used a ground-truth table in which each voxel is labeled based on the presence of porosity detected by the CT scan and AL algorithm. Then, each 3D neighborhood is labeled as either pore or nominal (Table 11.13) [19, 20].

The matching matrix can evaluate the algorithm by four criteria:

- TP (true-positive): actual anomaly voxel that is correctly predicted,
- FP (false-positive): actual anomaly voxel that is wrongly predicted,

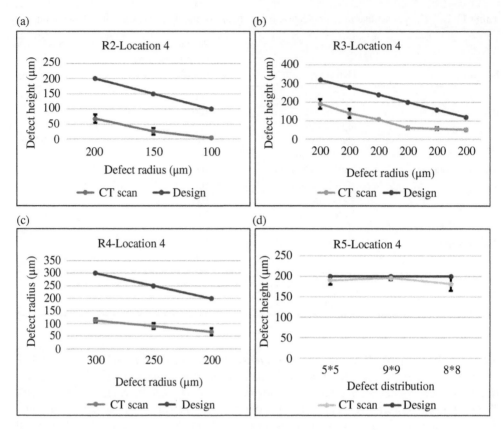

Figure 11.46 The amount of shrinkage in the sample: (a) R2 – Location 4, (b) R3 – Location 4, (c) R4 – Location 4, and (d) R5 – Location 4 for each type of defects [199]. Published under an open access CC BY 4.0 license

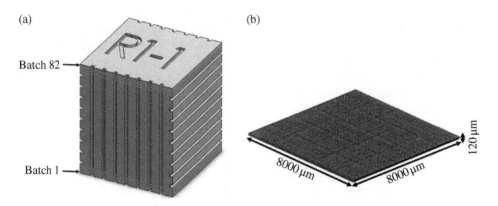

Figure 11.47 Schematic of (a) a sample that is segmented in 82 batches and (b) a batch which is divided into 4356 voxels. It includes three layers [199]. Published under an open access CC BY 4.0 license

Table 11.13   Confusion matrix to compare the prediction of the algorithm (AL) and the actual defect (CT scan).

		Prediction (AL algorithm)	
		Anomaly	Nominal
Actual defect (CT scan)	Anomaly	TP	FN
	Nominal	FP	TN

- FN (false-negative): actual nominal voxel that is wrongly predicted, and
- TN (true-negative): actual nominal voxel that is correctly predicted [198].

Besides, the accuracy and TN and TP rates of the prediction are considered given by [198]:

$$Accuracy = \frac{TP + TN}{TP + FP + FN + TN} \tag{11.27}$$

$$TP\ rate = \frac{TP}{TP + FN}, \ TN\ rate = \frac{TN}{TN + FP} \tag{11.28}$$

A high-level schematic of the overall process is represented in Figure 11.48.

Correlation between CT scan and EOSTATE MPM result for samples incorporated randomized defect (T-series) is conducted. The evaluation metrics for samples T3 and T4 are represented in Tables 11.14 and 11.15.

The result represents that the TP rate of the algorithm to find the true-positive rate for T3 (high hatching distance) and T4 (high speed) is more than 73% and 75%, respectively, which confirm that the algorithm could detect the pore's location. Still, the true negative rate limits the specificity of the algorithm. The AL algorithm's accuracy can be improved by reducing the number of false-positive cells, which may be provided by changing the pixel resolution and threshold method. Nevertheless, the algorithm should be improved to provide better defect detection, which is the first step toward online control.

11.7   Summary

The following paragraphs recap the learning objectives:

In-situ sensors: In-situ sensors can be categorized into two groups: radiative and non-radiative. The radiative sensors used in AM are camera, photodiode, and pyrometer, and the non-radiative sensors are acoustic, thermocouple, and displacement. These sensors are used to monitor different behavior of the process. Besides, two types of installation are used to mount sensors on the system: co-axial and off-axial. The co-axial sensors are placed in the energy source path, and the off-axial sensors are fixed and mounted outside the optical path.

Quality assurance algorithms: In-situ sensors are installed to capture information during the fabrication of parts. Afterward, the sensor's data are analyzed to identify the formation of

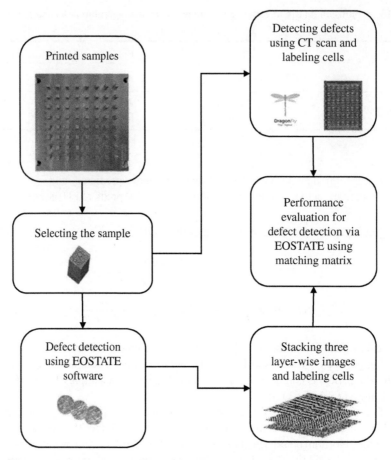

Figure 11.48 Schematic to compare the CT scan and EOSTATE software results to establish a matching matrix [199]. Published under an open access CC BY 4.0 license

Table 11.14 Evaluation metrics of AL algorithm based on the CT scan result for sample T3.

Performance	Location 1	Location 2	Location 3	Location 4	Location 5	Location 6
TP (%)	73.83	74.42	73.52	74.26	74.47	74.05
TN (%)	51.56	57.46	50.78	51.76	54.42	53.81
Accuracy (%)	70.28	73.15	71.97	66.79	68.94	69.74

Table 11.15 Evaluation metrics of AL algorithm based on the CT scan result for sample T4.

Performance	Location 1	Location 2	Location 3	Location 4	Location 5	Location 6
TP (%)	75.18	75.30	76.37	75.20	75.15	75.47
TN (%)	70.44	73.38	70.41	63.82	64.28	65.19
Accuracy (%)	72.21	74.71	71.78	73.41	73.84	71.01

defects. The statistical and machine learning algorithms are applied to detect different kinds of flaws created in printed parts. The statistical method is discussed based on the defect type. The machine learning algorithms are categorized based on their type, and it could be classified into supervised and unsupervised learning. Supervised learning is applied to continuous sensor data to train the labeled dataset and develop the prediction model to identify the relationship between the input and output variables. Supervised learning can solve classification and regression problems. Supervised learning is mostly used in AM to predict melt pool/track geometry, porosity, balling, spatter, and crack in the manufactured part. On the other hand, the unsupervised learning method is used to cluster input dataset without any labeled data and reduce the high-dimensional dataset into low-dimension features. The unsupervised learning techniques are applied to metal AM to identify potential delamination, plume, crack, and porosity in the printed parts.

References

1. E. Toyserkani, A. Khajepour, and S. Corbin, Laser Cladding. CRC Press LLC, 2004.
2. M. J. Matthews, G. Guss, S. A. Khairallah, A. M. Rubenchik, P. J. Depond, and W. E. King, "Denudation of metal powder layers in laser powder bed fusion processes," *Acta Mater.*, vol. 114, pp. 33–42, 2016.
3. W. Shifeng, L. Shuai, W. Qingsong, C. Yan, Z. Sheng, and S. Yusheng, "Effect of molten pool boundaries on the mechanical properties of selective laser melting parts," *J. Mater. Process. Technol.*, vol. 214, no. 11, pp. 2660–2667, 2014.
4. D. W. Ball, The basics of spectroscopy, vol. 49, SPIE Press, 2001.
5. T. Furumoto, M. R. Alkahari, T. Ueda, M. S. A. Aziz, and A. Hosokawa, "Monitoring of laser consolidation process of metal powder with high speed video camera," *Phys. Procedia*, vol. 39, pp. 760–766, 2012.
6. T. Furumoto, T. Ueda, M. R. Alkahari, and A. Hosokawa, "Investigation of laser consolidation process for metal powder by two-color pyrometer and high-speed video camera," *CIRP Ann. - Manuf. Technol.*, vol. 62, no. 1, pp. 223–226, 2013.
7. S. Barua, F. Liou, J. Newkirk, and T. Sparks, "Vision-based defect detection in laser metal deposition process," *Rapid Prototyp. J.*, vol. 20, no. 1, pp. 77–85, 2014.
8. J. Kruth, P. Mercelis, J. Van Vaerenbergh, and T. Craeghs, "Feedback control of Selective Laser Melting," in Proceedings of 3rd International Conference on Advanced Research in Virtual and Rapid Prototyping, pp. 1–7, 2007.
9. S. Berumen, F. Bechmann, S. Lindner, J. P. Kruth, and T. Craeghs, "Quality control of laser- and powder bed-based Additive Manufacturing (AM) technologies," *Phys. Procedia*, vol. 5, no. PART 2, pp. 617–622, 2010.
10. D. O. Griffith, M. L. Hofmeister, W. H. Knorovsky, G. A. MacCallum and J. E. Schlienger, and M. E. Smugeresky, "(12) United States Patent," US Pat. No. 6,459,951, vol. 1, no. 12, 2017.
11. M. Biegler, A. Marko, B. Graf, and M. Rethmeier, "Finite element analysis of in-situ distortion and bulging for an arbitrarily curved additive manufacturing directed energy deposition geometry," *Addit. Manuf.*, vol. 24, pp. 264–272, 2018.
12. M. Biegler, B. Graf, and M. Rethmeier, "Assessing the predictive capability of numerical additive manufacturing simulations via in-situ distortion measurements on a LMD component during build-up," in Procedia CIRP, pp. 158–162, 2018.
13. L. Song, F. Wang, S. Li, and X. Han, "Phase congruency melt pool edge extraction for laser additive manufacturing," *J. Mater. Process. Technol.*, vol. 250, pp. 261–269, 2017.

14. S. Ocylok, E. Alexeev, S. Mann, A. Weisheit, K. Wissenbach, and I. Kelbassa, "Correlations of melt pool geometry and process parameters during laser metal deposition by coaxial process monitoring," *Phys. Procedia*, vol. 56, pp. 228–238, 2014.
15. A. Gögelein, A. Ladewig, G. Zenzinger, and J. Bamberg, "Process monitoring of additive manufacturing by using optical tomography," 14th Quantitative InfraRed Thermography Conference, pp. 266–272, 2018.
16. S. Kleszczynski, J. Zur Jacobsmühlen, J. T. Sehrt, and G. Witt, "Error detection in laser beam melting systems by high resolution imaging," in International Solid Freeform Fabrication Symposium, 2012.
17. C. B. Stutzman, A. R. Nassar, and E. W. Reutzel, "Multi-sensor investigations of optical emissions and their relations to directed energy deposition processes and quality," *Addit. Manuf.*, vol. 21, pp. 333–339, 2018.
18. Y. Zhang, G. S. Hong, D. Ye, K. Zhu, and J. Y. H. Fuh, "Extraction and evaluation of melt pool, plume and spatter information for powder bed fusion AM process monitoring," *Mater. Des.*, vol. 156, pp. 458–469, 2018.
19. C. Gobert, E. W. Reutzel, J. Petrich, A. R. Nassar, and S. Phoha, "Application of supervised machine learning for defect detection during metallic powder bed fusion additive manufacturing using high resolution imaging.," *Addit. Manuf.*, vol. 21, pp. 517–528, 2018.
20. J. Petrich, C. Gobert, S. Phoha, A. R. Nassar, and E. W. Reutzel, "Machine learning for defect detection for PBFAm using high resolution layerwise imaging coupled with post-build CT scans," in Solid Freeform Fabrication 2017: Proceedings of the 28th Annual International Solid Freeform Fabrication Symposium – An Additive Manufacturing Conference, SFF 2017, 2020.
21. Z. Y. Chua, I. H. Ahn, and S. K. Moon, "Process monitoring and inspection systems in metal additive manufacturing: Status and applications," *Int. J. Precis. Eng. Manuf. - Green Technol.*, vol. 4, no. 2, pp. 235–245, 2017.
22. H. Baumgartl, J. Tomas, R. Buettner, and M. Merkel, "A deep learning-based model for defect detection in laser-powder bed fusion using in-situ thermographic monitoring," *Prog. Addit. Manuf.*, vol. 5, pp. 277–285, 2020.
23. S. J. Wolff et al., "Experimentally validated predictions of thermal history and microhardness in laser-deposited Inconel 718 on carbon steel," *Addit. Manuf.*, vol. 27, pp. 540–551, 2019.
24. D. Hu and R. Kovacevic, "Sensing, modeling and control for laser-based additive manufacturing," *Int. J. Mach. Tools Manuf.*, vol. 43, no. 1, pp. 51–60, 2003.
25. S. Moylan, E. Whitenton, B. Lane, and J. Slotwinski, "Infrared thermography for laser-based powder bed fusion additive manufacturing processes," in AIP Conference Proceedings, 2014.
26. M. Khanzadeh, S. Chowdhury, M. Marufuzzaman, M. A. Tschopp, and L. Bian, "Porosity prediction: Supervised-learning of thermal history for direct laser deposition," *J. Manuf. Syst.*, vol. 47, pp. 69–82, 2018.
27. M. Khanzadeh, S. Chowdhury, M. A. Tschopp, H. R. Doude, M. Marufuzzaman, and L. Bian, "In-situ monitoring of melt pool images for porosity prediction in directed energy deposition processes," *IISE Trans.*, vol. 51, no. 5, pp. 437–455, 2019.
28. M. H. Farshidianfar, A. Khajepour, and A. P. Gerlich, "Effect of real-time cooling rate on microstructure in laser additive manufacturing," *J. Mater. Process. Technol.*, vol. 231, pp. 468–478, 2016.
29. M. H. Farshidianfar, A. Khajepour, and A. Gerlich, "Real-time control of microstructure in laser additive manufacturing," *Int. J. Adv. Manuf. Technol.*, vol. 82, no. 5–8, pp. 1173–1186, 2016.
30. J. L. Bennett et al., "Cooling rate effect on tensile strength of laser deposited Inconel 718," in Procedia Manufacturing, 2018.
31. J. Schilp, C. Seidel, H. Krauss, and J. Weirather, "Investigations on temperature fields during laser beam melting by means of process monitoring and multiscale process modelling," *Adv. Mech. Eng.*, vol. 2014, pp. 217584–217590, 2014.

32. Y. Ding, J. Warton, and R. Kovacevic, "Development of sensing and control system for robotized laser-based direct metal addition system," *Addit. Manuf.*, vol. 10, pp. 24–35, 2016.
33. D. Huang et al., "Optical coherence tomography," *Science*, vol. 254, no. 5035, pp. 1178–1181, 1991.
34. J. A. Kanko, A. P. Sibley, and J. M. Fraser, "In situ morphology-based defect detection of selective laser melting through inline coherent imaging," *J. Mater. Process. Technol.*, vol. 231, pp. 488–500, 2016.
35. S. W. Hell and R. N. Weinreb, High resolution imaging in microscopy and ophthalmology: new frontiers in biomedical optics, Springer, 2019.
36. M. R. Hee et al., "Quantitative assessment of macular edema with optical coherence tomography," *Arch. Ophthalmol.*, vol. 113, no. 8, pp. 1019–1029, 1995.
37. M. R. Hee et al., "Topography of diabetic macular edema with optical coherence tomography," *Ophthalmology*, vol. 105, no. 2, pp. 360–370, 1998.
38. C. Hauger, M. Wörz, and T. Hellmuth, "Interferometer for optical coherence tomography," *Appl. Opt.*, vol. 42, no. 19, pp. 3896–3902, 2003.
39. S. R. Chinn, E. A. Swanson, and J. G. Fujimoto, "Optical coherence tomography using a frequency-tunable optical source," *Opt. Lett.*, vol. 22, no. 5, pp. 340–342, 1997.
40. B. Potsaid et al., "Ultrahigh speed spectral/fourier domain OCT ophthalmic imaging at 70,000 to 312,500 axial scans per second," *Opt. Express*, vol. 16, no. 19, pp. 15149–15169, 2008.
41. R. Leitgeb, C. Hitzenberger, and A. Fercher, "Performance of fourier domain vs time domain optical coherence tomography," *Opt. Express*, vol. 11, no. 8, pp. 889–894, 2003.
42. A. Neef, V. Seyda, D. Herzog, C. Emmelmann, M. Schönleber, and M. Kogel-Hollacher, "Low coherence interferometry in selective laser melting," *Phys. Procedia*, vol. 56, no. C, pp. 82–89, 2014.
43. C. Zhao et al., "Real-time monitoring of laser powder bed fusion process using high-speed X-ray imaging and diffraction," *Sci. Rep.*, vol. 7, no. 1, pp. 1–11, 2017.
44. C. Kenel et al., "In situ investigation of phase transformations in Ti-6Al-4V under additive manufacturing conditions combining laser melting and high-speed micro-X-ray diffraction," *Sci. Rep.*, vol. 7, no. 1, pp. 1–10, 2017.
45. S. J. Wolff et al., "In-situ high-speed X-ray imaging of piezo-driven directed energy deposition additive manufacturing," *Sci. Rep.*, vol. 5, pp. 523–531, 2019.
46. Osi Optoelectronics, "Photodiode characteristics," *Osi Optoelectron.*, 2009.
47. W. Electronics Inc, "AN-LD17: Photodiode Basics: Selection & Operation," no. 406, pp. 1–4, 2020.
48. J. Low, M. Kreider, D. Pulsifer, A. Jones, and T. Gilani, "Band gap energy in silicon," *Am. J. Undergrad. Res.*, vol. 7, no. 1, pp. 27–32, 2009.
49. C. Van Gestel, "Study of physical phenomena of selective laser melting towards increased productivity," *Nucl. Inst. Methods Phys. Res. A*, No. THESIS. EPFL, 2015.
50. I. A. Okaro, S. Jayasinghe, C. Sutcliffe, K. Black, P. Paoletti, and P. L. Green, "Automatic fault detection for laser powder bed fusion using semi-supervised machine learning," *Addit. Manuf.*, vol. 27, pp. 42–53, 2019.
51. M. Miyagi, T. Tsukamoto, and H. Kawanaka, "Adaptive shape control of laser-deposited metal structures by adjusting weld pool size," *J. Laser Appl.*, vol. 26, no. 3, p. 032003, 2014.
52. M. Pavlov, M. Doubenskaia, and I. Smurov, "Pyrometric analysis of thermal processes in SLM technology," *Phys. Procedia*, vol. 5, pp. 523–531, 2010.
53. M. Doubenskaia, M. Pavlov, S. Grigoriev, E. Tikhonova, and I. Smurov, "Comprehensive optical monitoring of selective laser melting," *J. Laser Micro Nanoeng.*, vol. 7, no. 3, pp. 236–243, 2012.
54. G. Bi, C. N. Sun, and A. Gasser, "Study on influential factors for process monitoring and control in laser aided additive manufacturing," *J. Mater. Process. Technol.*, vol. 213, no. 3, pp. 463–468, 2013.
55. T. Islam, S. C. Mukhopadhyay, and N. K. Suryadevara, "Smart sensors and internet of things: A postgraduate paper," *IEEE Sensors J.*, vol. 17, no. 3, pp. 577–584, 2017.
56. U. Thombansen, A. Gatej, and M. Pereira, "Process observation in fiber laser–based selective laser melting," *Opt. Eng.*, vol. 54, no. 1, p. 011008, 2014.

57. Y. Chivel, "Optical in-process temperature monitoring of selective laser melting," *Phys. Procedia*, vol. 41, pp. 904–910, 2013.

58. J. Fraden, Handbook of Modern Sensors, 4. Springer, 2004.

59. L. L. Beranek and T. Mellow, Acoustics: sound fields and transducers, Academic Press, 2012.

60. J. X. J. Zhang and K. Hoshino, "Mechanical transducers: Cantilevers, acoustic wave sensors, and thermal sensors," in Molecular Sensors and Nanodevices, 2019.

61. Q. Y. Lu and C. H. Wong, "Additive manufacturing process monitoring and control by non-destructive testing techniques: Challenges and in-process monitoring," *Virtual Phys. Prototy.*, 2018.

62. D. Ye, G. S. Hong, Y. Zhang, K. Zhu, J. Ying, and H. Fuh, "Defect detection in selective laser melting technology by acoustic signals with deep belief networks," *Int. J. Adv. Manuf. Technol.*, vol. 96, no. 5. pp. 2791–2801, 2018.

63. H. Gaja and F. Liou, "Defects monitoring of laser metal deposition using acoustic emission sensor," *Int. J. Adv. Manuf. Technol.*, vol. 90, pp. 561–574, 2017.

64. H. Rieder, A. Dillhöfer, M. Spies, J. Bamberg, and T. Hess, "Online monitoring of additive manufacturing processes using ultrasound," in Proceedings of 11th European Conference on Non-Destructive Testing, 2014.

65. A. J. Welch and M. J. C. Van Gemert, Optical-thermal response of laser-irradiated tissue, vol. 2. Springer, 2011.

66. J. K. Kelly, "Direct-metal-deposition (DMD) nozzle fault detection using temperature measurements," U.S. Patent 6,423,926, issued July 23, 2002.

67. M. L. Griffith et al., "Understanding thermal behavior in the LENS process," *Mater. Des.*, vol. 20, pp. 107–113, 1999.

68. A. J. Dunbar, "Analysis of the laser powder bed fusion additive manufacturing process through experimental measurement and finite element modeling," *Dep. Mech. Eng.*, No. Thesis, Pennsylvania State University, 2016.

69. M. Chiumenti et al., "Numerical modelling and experimental validation in Selective Laser Melting," *Addit. Manuf.*, vol. 18, pp. 171–185, 2017.

70. L. Van Belle, G. Vansteenkiste, and J. C. Boyer, "Investigation of residual stresses induced during the selective laser melting process," in Key Engineering Materials, vol. 554, pp. 1828–1834, Trans Tech Publications Ltd, 2013.

71. Omron Corporation, "Displacement sensor technical guide," 2017. [Online]. Available: "https://assets.omron.eu/downloads/manual/en/q257_displacement_sensor_technical_manual_en.pdf"

72. F. Liou, K. Slattery, M. Kinsella, J. Newkirk, H. N. Chou, and R. Landers, "Applications of a hybrid manufacturing process for fabrication of metallic structures," *Rapid Prototyp. J.*, vol. 13, no. 4, pp. 236–244, 2007.

73. M. R. Boddu, S. Musti, R. G. Landers, S. Agarwal, and F. Liou, "Empirical modeling and vision based control for laser aided metal deposition process," in Proceedings of Annual International Solid Freeform Fabrication Symposium, 2001.

74. O. Kwon et al., "A deep neural network for classification of melt-pool images in metal additive manufacturing," *J. Intell. Manuf.*, vol. 31, no. 2, pp. 375–386, 2020.

75. A. Gaikwad, F. Imani, P. Rao, H. Yang, and E. Reutzel, "Design rules and in-situ quality monitoring of thin-wall features made using laser powder bed fusion," in ASME 2019 14th International Manufacturing Science and Engineering Conference, MSEC 2019, 2019.

76. T. Craeghs, S. Clijsters, J. P. Kruth, F. Bechmann, and M. C. Ebert, "Detection of process failures in layer-wise laser melting with optical process monitoring," *Phys. Procedia*, vol. 39, pp. 753–759, 2012.

77. L. Scime and J. Beuth, "Anomaly detection and classification in a laser powder bed additive manufacturing process using a trained computer vision algorithm," *Addit. Manuf.*, vol. 19, pp. 114–126, 2018.

78. M. Aminzadeh and T. R. Kurfess, "Online quality inspection using Bayesian classification in powder bed additive manufacturing from high-resolution visual camera images," *J. Intell. Manuf.*, vol. 30, pp. 2505–2523, 2019.

79. D. Hu, H. Mei, and R. Kovacevic, "Improving solid freeform fabrication by laser-based additive manufacturing," *Proc. Inst. Mech. Eng. Part B J. Eng. Manuf.*, vol. 216, no, 9, pp. 1253–1264, 2002.

80. D. Hu and R. Kovacevic, "Modelling and measuring the thermal behaviour of the molten pool in closed-loop controlled laser-based additive manufacturing," *Proc. Inst. Mech. Eng. Part B J. Eng. Manuf.*, vol. 217, no, 4, pp. 441–452, 2003.

81. J. T. Hofman, B. Pathiraj, J. Van Dijk, D. F. De Lange, and J. Meijer, "A camera based feedback control strategy for the laser cladding process," *J. Mater. Process. Technol.*, vol. 212, no. 11, pp. 2455–2462, 2012.

82. D. A. Kriczky, J. Irwin, E. W. Reutzel, P. Michaleris, A. R. Nassar, and J. Craig, "3D spatial reconstruction of thermal characteristics in directed energy deposition through optical thermal imaging," *J. Mater. Process. Technol.*, vol. 221, pp. 172–186, 2015.

83. S. Moralejo et al., "A feedforward controller for tuning laser cladding melt pool geometry in real time," *Int. J. Adv. Manuf. Technol.*, vol. 89, pp. 821–831, 2017.

84. S. Donadello, M. Motta, A. G. Demir, and B. Previtali, "Monitoring of laser metal deposition height by means of coaxial laser triangulation," *Opt. Lasers Eng.*, vol. 112, pp. 136–144, 2019.

85. Y. Liu, L. Wang, and M. Brandt, "Model predictive control of laser metal deposition," *Int. J. Adv. Manuf. Technol.*, vol. 105, no. 1, pp. 1055–1067, 2019.

86. W. Hofmeister, M. Griffith, M. Ensz, and J. Smugeresky, "Solidification in direct metal deposition by LENS processing," *JOM*, vol. 53, no. 9, pp. 30–34, 2001.

87. Hofmeister, Knorovsky, and Maccallum, Video Monitoring and Control of the LENS Process, Conference: American Welding Society 9th International Conference of Computer Technology in Welding, Detroit, MI (US), 28 September 1999–30 September 1999, 1999.

88. H. K. Lee, "Effects of the cladding parameters on the deposition efficiency in pulsed Nd:YAG laser cladding," *J. Mater. Process. Technol.*, vol. 202, pp. 321–327, 2008.

89. G. Bi, B. Schürmann, A. Gasser, K. Wissenbach, and R. Poprawe, "Development and qualification of a novel laser-cladding head with integrated sensors," *Int. J. Mach. Tools Manuf.*, vol. 47, pp. 555–561, 2007.

90. M. Gharbi et al., "Influence of various process conditions on surface finishes induced by the direct metal deposition laser technique on a Ti-6Al-4V alloy," *J. Mater. Process. Technol.*, vol. 213, no. 5, pp. 791–800, 2013.

91. D. Hu, H. Mei, G. Tao, and R. Kovacevic, "Closed loop control of 3D laser cladding based on infrared sensing," Proc. Solid Free. Fabr. Symp., International Solid Freeform Fabrication Symposium 2001, pp. 129–137 2001.

92. P. J. DePond et al., "In situ measurements of layer roughness during laser powder bed fusion additive manufacturing using low coherence scanning interferometry," *Mater. Des.*, vol. 154, pp. 347–359, 2018.

93. S. Clijsters, T. Craeghs, S. Buls, K. Kempen, and J. P. Kruth, "In situ quality control of the selective laser melting process using a high-speed, real-time melt pool monitoring system," *Int. J. Adv. Manuf. Technol.*, vol. 75, no. 5–8, pp. 1089–1101, 2014.

94. T. Craeghs, F. Bechmann, S. Berumen, and J. P. Kruth, "Feedback control of layerwise laser melting using optical sensors," *Phys. Procedia*, vol. 5, no. PART 2, pp. 505–514, 2010.

95. F. Bayle and M. Doubenskaia, "Selective laser melting process monitoring with high speed infra-red camera and pyrometer," *Int. Soc. Opt. Photonics*, vol. 6985, no. January 2008, pp. 698505–698508, 2008.

96. R. J. Smith, M. Hirsch, R. Patel, W. Li, A. T. Clare, and S. D. Sharples, "Spatially resolved acoustic spectroscopy for selective laser melting," *J. Mater. Process. Technol.*, vol. 236, pp. 93–102, 2016.

97. S. A. Shevchik, C. Kenel, C. Leinenbach, and K. Wasmer, "Acoustic emission for in situ quality monitoring in additive manufacturing using spectral convolutional neural networks," *Addit. Manuf.*, vol. 21, pp. 598–604, 2018.

98. J. Whiting, A. Springer, and F. Sciammarella, "Real-time acoustic emission monitoring of powder mass flow rate for directed energy deposition," *Addit. Manuf.*, vol. 23, pp. 312–318, 2018.

99. A. Caggiano, J. Zhang, V. Alfieri, F. Caiazzo, R. Gao, and R. Teti, "Machine learning-based image processing for on-line defect recognition in additive manufacturing," *CIRP Ann.*, vol. 68, no.1, pp. 451–454, 2019.

100. B. M. Colosimo and M. Grasso, "Spatially weighted PCA for monitoring video image data with application to additive manufacturing," *J. Qual. Technol.*, vol. 50, no. 4, pp. 391–417, 2018.

101. J. zur Jacobsmühlen, S. Kleszczynski, G. Witt, and D. Merhof, "Elevated region area measurement for quantitative analysis of laser beam melting process stability," in Proceedings – 26th Annual International Solid Freeform Fabrication Symposium – An Additive Manufacturing Conference, SFF 2015, 2020.

102. M. Islam, T. Purtonen, H. Piili, A. Salminen, and O. Nyrhilä, "Temperature profile and imaging analysis of laser additive manufacturing of stainless steel," *Phys. Procedia*, vol. 41, pp. 835–842, 2013.

103. L. Song, V. Bagavath-Singh, B. Dutta, and J. Mazumder, "Control of melt pool temperature and deposition height during direct metal deposition process," *Int. J. Adv. Manuf. Technol.*, vol. 58, pp. 247–256, 2012.

104. H. W. Hsu, Y. L. Lo, and M. H. Lee, "Vision-based inspection system for cladding height measurement in direct energy deposition (DED)," *Addit. Manuf.*, vol. 27, pp. 372–378, 2019.

105. A. R. Nassar, B. Starr, and E. W. Reutzel, "Process monitoring of directed-energy deposition of Inconel-718 via plume imaging," in Proceedings – 26th Annual International Solid Freeform Fabrication Symposium – An Additive Manufacturing Conference, SFF 2015, 2020.

106. M. A. Doubenskaia, I. Y. Smurov, V. I. Teleshevskiy, P. Bertrand, and I. V. Zhirnov, "Determination of true temperature in selective laser melting of metal powder using infrared camera," *Mater. Sci. Forum*, vol. 834, pp. 93–102, 2015.

107. H. Krauss, C. Eschey, and M. F. Zaeh, "Thermography for monitoring the selective laser melting process," in 23rd Annual International Solid Freeform Fabrication Symposium - An Additive Manufacturing Conference, SFF 2012, 2012.

108. H. Krauss, T. Zeugner, and M. F. Zaeh, "Layerwise monitoring of the selective laser melting process by thermography," *Phys. Procedia*, vol. 56, no. C, pp. 64–71, 2014.

109. B. Lane, S. Moylan, E. P. Whitenton, and L. Ma, "Thermographic measurements of the commercial laser powder bed fusion process at NIST," *Rapid Prototyp. J.*, vol. 22, no. 5, pp. 778–787, 2016.

110. M. Grasso, A. G. Demir, B. Previtali, and B. M. Colosimo, "In situ monitoring of selective laser melting of zinc powder via infrared imaging of the process plume," *Robot. Comput. Integr. Manuf.*, vol. 49, pp. 229–239, 2018.

111. M. Doubenskaia, I. Smurov, S. Grigoriev, M. Pavlov, and E. Tikhonova, "Optical monitoring in elaboration of metal matrix composites by direct metal deposition," *Phys. Procedia*, vol. 39, pp. 767–775, 2012.

112. J. L. Bennett, S. J. Wolff, G. Hyatt, K. Ehmann, and J. Cao, "Thermal effect on clad dimension for laser deposited Inconel 718," *J. Manuf. Process.*, vol. 28, pp. 550–557, 2017.

113. T. Furumoto, T. Ueda, N. Kobayashi, A. Yassin, A. Hosokawa, and S. Abe, "Study on laser consolidation of metal powder with Yb:fiber laser-evaluation of line consolidation structure," *J. Mater. Process. Technol.*, vol. 209, pp. 5973–5980, 2009.

114. L. Tang and R. G. Landers, "Melt pool temperature control for laser metal deposition processes-part I: Online temperature control," *J. Manuf. Sci. Eng. Trans. ASME*, vol. 132, pp. 011011–011019, 2010.

115. H. Rieder, A. Dillhöfer, M. Spies, J. Bamberg, and T. Hess, "Online monitoring of additive manufacturing processes using ultrasound 2. Additive manufacturing and quality assurance considerations 3. Ultrasonic process monitoring," in Proceedings of 11th European Conference on Non-Destructive Testing, 2014.

116. J. A. Slotwinski, E. J. Garboczi, and K. M. Hebenstreit, "Porosity measurements and analysis for metal additive manufacturing process control," *J. Res. Natl. Inst. Stand. Technol.*, vol. 119, p. 494, 2014.

117. H. Taheri, L. W. Koester, T. A. Bigelow, E. J. Faierson, and L. J. Bond, "In situ additive manufacturing process monitoring with an acoustic technique: Clustering performance evaluation using K-means algorithm," *J. Manuf. Sci. Eng. Trans. ASME*, vol. 141, no. 4, pp. 041011–11, 2019.

118. L. J. Bond, L. W. Koester, and H. Taheri, "NDE in-process for metal parts fabricated using powder based additive manufacturing," in Smart Structures and NDE for Energy Systems and Industry 4.0, vol. 10973, p. 1097302, International Society for Optics and Photonics, 2019, p. 1.

119. L. Fuchs and C. Eischer, "In-process monitoring systems for metal additive manufacturing," EOS GmbH, Germany, p. 20, 2018. [Online]. Available: https://www.semanticscholar.org/paper/In-process-monitoring-systems-for-metal-additive-Fuchs-Eischer/0b4a7bbeff6c2faec5d67c0f0c23a47bf2f45991.

120. SLM Solution, "Melt pool monitoring," p. 1, [Online]. Available: https://pdf.aeroexpo.online/pdf/slm-solutions-gmbh/melt-pool-monitoring-mpm/170578-4425.html.

121. SLM Solution, "Laser Power Monitoring," Report, pp. 1–2.

122. T. Toeppel et al., "3D analysis in laser beam melting based on real-time process monitoring," *Mater. Sci. Technol. Conf. Exhib. 2016, MS T 2016*, vol. 1, pp. 123–132, 2016.

123. Renishaw PLC, "InfiniAM Spectral – Energy input and melt pool emissions monitoring for AM systems," Report, pp. 1–5, 2017.

124. Y. Chivel and I. Smurov, "On-line temperature monitoring in selective laser sintering/melting," *Phys. Procedia*, vol. 5, pp. 515–521, 2010.

125. P. Lott, H. Schleifenbaum, W. Meiners, K. Wissenbach, C. Hinke, and J. Bültmann, "Design of an optical system for the in situ process monitoring of Selective Laser Melting (SLM)," *Phys. Procedia*, vol. 12, pp. 683–690, 2011.

126. M. Abdelrahman, E. W. Reutzel, A. R. Nassar, and T. L. Starr, "Flaw detection in powder bed fusion using optical imaging," *Addit. Manuf.*, vol. 15, pp. 1–11, 2017.

127. G. Zenzinger, J. Bamberg, A. Ladewig, T. Hess, B. Henkel, and W. Satzger, "Process monitoring of additive manufacturing by using optical tomography," *AIP Conf. Proc.*, vol. 1650, no. 1, 2015.

128. J. Bamberg, G. Zenzinger, and A. Ladewig, "In-process control of selective laser melting by quantitative optical tomography," in 19th World Conference on Non-Destructive Testing, 2016.

129. L. Fuchs, "Expert Training @ University of Waterloo."

130. B. N. Bauer and Y. Chou, "Statistical analysis: With business and economic applications.," *J. Am. Stat. Assoc.*, vol. 66, no, 334, p. 427, 1971.

131. D. Gu et al., "Densification behavior, microstructure evolution, and wear performance of selective laser melting processed commercially pure titanium," *Acta Mater.*, vol. 60, no. 9, pp. 3849–3860, 2012.

132. B. Lane and E. P. Whitenton, "Calibration and measurement procedures for a high magnification thermal camera," *Natl. Inst. Stand. Technol.*, Report no. NISTIR8098, 2015.

133. A. R. Nassar, T. J. Spurgeon, and E. W. Reutzel, "Sensing defects during directed-energy additive manufacturing of metal parts using optical emissions spectroscopy," in 25th Annual International Solid Freeform Fabrication Symposium &#65533; An Additive Manufacturing Conference, SFF 2014, 2014.

134. F. Imani, A. Gaikwad, M. Montazeri, P. Rao, H. Yang, and E. Reutzel, "Layerwise in-process quality monitoring in laser powder bed fusion," in ASME 2018 13th International Manufacturing Science and Engineering Conference, MSEC 2018, 2018.

135. L. Scime and J. Beuth, "Using machine learning to identify in-situ melt pool signatures indicative of flaw formation in a laser powder bed fusion additive manufacturing process," *Addit. Manuf.*, vol. 25, pp. 151–165, 2019.

136. B. L. DeCost, H. Jain, A. D. Rollett, and E. A. Holm, "Computer vision and machine learning for autonomous characterization of AM powder feedstocks," *JOM*, vol. 69, no. 3, pp. 456–465, 2017.

137. S. H. Seifi, W. Tian, H. Doude, M. A. Tschopp, and L. Bian, "Layer-wise modeling and anomaly detection for laser-based additive manufacturing," *J. Manuf. Sci. Eng. Trans. ASME*, vol. 141, no. 8, pp. 081013–12, 2019.

138. Z. L. Lu, D. C. Li, B. H. Lu, A. F. Zhang, G. X. Zhu, and G. Pi, "The prediction of the building precision in the laser engineered net shaping process using advanced networks," *Opt. Lasers Eng.*, vol. 48, no. 5, pp. 519–525, 2010.

139. R. Jafari-Marandi, M. Khanzadeh, W. Tian, B. Smith, and L. Bian, "From in-situ monitoring toward high-throughput process control: cost-driven decision-making framework for laser-based additive manufacturing," *J. Manuf. Syst.*, vol. 51, pp. 29–41, 2019.

140. H. Gaja and F. Liou, "Defect classification of laser metal deposition using logistic regression and artificial neural networks for pattern recognition," *Int. J. Adv. Manuf. Technol.*, vol. 94, pp. 315–326, 2018.

141. B. Yuan et al., "Machine-learning-based monitoring of laser powder bed fusion," *Adv. Mater. Technol.*, vol. 3, no. 12, pp. 1800136–6, 2018.

142. X. Li, S. Siahpour, J. Lee, Y. Wang, and J. Shi, "Deep learning-based intelligent process monitoring of directed energy deposition in additive manufacturing with thermal images," *Procedia Manuf.*, vol. 48, pp. 643–649, 2020.

143. W. Cui, Y. Zhang, X. Zhang, L. Li, and F. Liou, "Metal additive manufacturing parts inspection using convolutional neural network," *Appl. Sci.*, vol. 10, p. 545, 2020.

144. M. Mozaffar et al., "Data-driven prediction of the high-dimensional thermal history in directed energy deposition processes via recurrent neural networks," *Manuf. Lett.*, vol. 18, pp. 35–39, 2018.

145. M. Zhang et al., "High cycle fatigue life prediction of laser additive manufactured stainless steel: A machine learning approach," *Int. J. Fatigue*, vol. 128, p. 105194–13, 2019.

146. D. N. Aqilah, A. K. Mohd Sayuti, Y. Farazila, D. Y. Suleiman, M. A. Nor Amirah, and W. B. Wan Nur Izzati, "Effects of process parameters on the surface roughness of stainless steel 316L parts produced by selective laser melting," *J. Test. Eval.*, vol. 46, no. 4, pp. 1673–1683, 2018.

147. H. Sohrabpoor, "Analysis of laser powder deposition parameters: ANFIS modeling and ICA optimization," *Optik (Stuttg).*, vol. 127, no. 8, pp. 4031–4038, 2016.

148. M. Alimardani and E. Toyserkani, "Prediction of laser solid freeform fabrication using neuro-fuzzy method," *Appl. Soft Comput. J.*, vol. 8, pp. 316–323, 2008.

149. M. H. Farshidianfar, A. Khajepour, M. Zeinali, and A. Gelrich, "System identification and height control of laser cladding using adaptive neuro-fuzzy inference systems," in ICALEO 2013 – 32nd International Congress on Applications of Lasers and Electro-Optics, vol. 2013, pp. 615–623, 2013.

150. A. Foorginejad, M. Azargoman, N. Mollayi, and M. Taheri, "Modeling of weld bead geometry using adaptive neuro-fuzzy inference system (ANFIS) in additive manufacturing," *J. Appl. Comput. Mech.*, vol. 6, pp. 160–170, 2020.

151. L. Meng and J. Zhang, "Process design of laser powder bed fusion of stainless steel using a Gaussian process-based machine learning model," *JOM*, vol. 72, pp. 420–428, 2020.

152. G. Tapia, S. Khairallah, M. Matthews, W. E. King, and A. Elwany, "Gaussian process-based surrogate modeling framework for process planning in laser powder bed fusion additive manufacturing of 316L stainless steel," *Int. J. Adv. Manuf. Technol.*, vol. 94, no. 9, pp. 3591–3603, 2018.

153. C. Kamath, "Data mining and statistical inference in selective laser melting," *Int. J. Adv. Manuf. Technol.*, vol. 86, no. 5, pp. 1659–1677, 2016.

154. G. Tapia, A. H. Elwany, and H. Sang, "Prediction of porosity in metal-based additive manufacturing using spatial Gaussian process models," *Addit. Manuf.*, vol. 12, pp. 282–290, 2016.

155. S. Wang, L. Zhu, J. Y. H. Fuh, H. Zhang, and W. Yan, "Multi-physics modeling and Gaussian process regression analysis of cladding track geometry for direct energy deposition," *Opt. Lasers Eng.*, vol. 127, p. 105950, 2020.

156. J. Williams, P. Dryburgh, A. Clare, P. Rao, and A. Samal, "Defect detection and monitoring in metal additive manufactured parts through deep learning of spatially resolved acoustic spectroscopy signals," *Smart Sustain. Manuf. Syst.*, vol. 2, pp. 204–226, 2018.

157. F. Caiazzo and A. Caggiano, "Laser direct metal deposition of 2024 al alloy: Trace geometry prediction via machine learning," *Materials (Basel).*, vol. 11, no. 3, p. 444, 2018.

158. S. Saqiba, R. J. Urbanica, and K. Aggarwal, "Analysis of laser cladding bead morphology for developing additive manufacturing travel paths," *Procedia CIRP*, vol. 17, pp. 824–829, 2014.

159. P. Sreeraj and T. Kannan, "Modelling and prediction of stainless steel clad bead geometry deposited by GMAW using regression and artificial neural network models," *Adv. Mech. Eng.*, vol. 4, p. 237379, 2012.

160. M. Grasso and B. M. Colosimo, "Process defects and in situ monitoring methods in metal powder bed fusion: A review," *Meas. Sci. Technol.*, vol. 28, no. 4, p. aa5c4f, 2017.

161. L. W. Koester, H. Taheri, T. A. Bigelow, L. J. Bond, and E. J. Faierson, "In-situ acoustic signature monitoring in additive manufacturing processes," in AIP Conference Proceedings, 2018.

162. S. B. Kotsiantis, "Supervised machine learning: A review of classification techniques," *Informatica (Ljubljana)*, vol. 160, pp. 3–24, 2007.

163. S. Shalev-Shwartz and S. Ben-David, Understanding Machine Learning: From Theory to Algorithms, Cambridge University Press, 2013.

164. C. Gallo, "Artificial neural networks tutorial," in Mehdi Khosrow-Pour Encyclopedia of information science and technology, Third edition, pp. 6369–6378, 2014.

165. M. Awais et al., "Real-time surveillance through face recognition using HOG and feedforward neural networks," *IEEE Access*, 2019, vol. 7, pp. 121236–121244.

166. M. P. Ghaemmaghami, H. Sametit, F. Razzazi, B. BabaAli, and S. Dabbaghchiarr, "Robust speech recognition using MLP neural network in log-spectral domain," in IEEE International Symposium on Signal Processing and Information Technology, ISSPIT 2009, 2009.

167. A. C. Ian Goodfellow, Y. Bengio, Deep Learning – Ian Goodfellow, Yoshua Bengio, Aaron Courville – Google Books, 2016.

168. T. Wang, Y. Chen, M. Qiao, and H. Snoussi, "A fast and robust convolutional neural network-based defect detection model in product quality control," *Int. J. Adv. Manuf. Technol.*, vol. 94, no. 9, pp. 3465–3471, 2018.

169. J. Francis and L. Bian, "Deep learning for distortion prediction in laser-based additive manufacturing using big data," *Manuf. Lett.*, vol. 20, pp. 10–14, 2019.

170. J. S. R. Jang, "ANFIS: Adaptive-network-based fuzzy inference system," *IEEE Trans. Syst. Man. Cybern.*, vol. 23, no. 3, pp. 665–685, 1993.

171. D. Karaboga and E. Kaya, "Adaptive network based fuzzy inference system (ANFIS) training approaches: a comprehensive survey," *Artif. Intell. Rev.*, vol. 52, no. 4, pp. 2263–2293, 2019.

172. S. Barada and H. Singh, "Generating optimal adaptive fuzzy-neural models of dynamical systems with applications to control," *IEEE Trans. Syst. Man Cybern. Part C Appl. Rev.*, vol. 28, no. 3, pp. 371–391, 1998.

173. J. S. R. Jang and E. Mizutani, "Levenberg-Marquardt method for ANFIS learning," in Biennial Conference of the North American Fuzzy Information Processing Society – NAFIPS, 1996.

174. K. Premkumar and B. V. Manikandan, "Fuzzy PID supervised online ANFIS based speed controller for brushless dc motor," *Neurocomputing*, vol. 157, p. 76–90, 2015.

175. X. Wang, H. Hu, and A. Zhang, "Concentration measurement of three-phase flow based on multi-sensor data fusion using adaptive fuzzy inference system," *Flow Meas. Instrum.*, vol. 39, pp. 1–8, 2014.

176. A. Dehghanian Serej and H. Mojallali, "Speed control of elliptec motor using adaptive neural-fuzzy controller with on-line learning simulated under MATLAB/SIMULINK," in IntelliSys 2015 – Proceedings of 2015 SAI Intelligent Systems Conference, 2015.

177. N. S. Altman, "An introduction to kernel and nearest-neighbor nonparametric regression," *Am. Stat.*, vol. 46, no. 3, pp. 175–185, 1992.

178. H. Anton and C. Rorres, Elementary linear algebra – applications version, John Wiley & Sons, 2008.

179. O. Kramer, "Dimensionality reduction with unsupervised nearest neighbors," *Intell. Syst. Ref. Libr.*, vol. 51, Springer, Berlin, 137pp, 2013.

180. K. P. Murphy, Machine learning: a probabilistic perspective (adaptive computation and machine learning series), MIT Press, 2012.

181. K. He, X. Zhang, S. Ren, and J. Sun, "Deep residual learning for image recognition," in Proceedings of the IEEE Computer Society Conference on Computer Vision and Pattern Recognition, 2016.

182. G. Huang, Z. Liu, L. Van Der Maaten, and K. Q. Weinberger, "Densely connected convolutional networks," in Proceedings – 30th IEEE Conference on Computer Vision and Pattern Recognition, CVPR 2017, 2017.

183. H. E. Driver and A. L. Kroeber, "Quantitative expression of cultural relationships," *Univ. Calif. Publ. Am. Archaeol. Ethnol.*, vol. 31, no. 4, pp. 211–256, 1932.

184. S. Haykin, "Self-organizing maps," Neural Networks A Compr. Found., second edition, 842 pp., Pearson Education, 1999.

185. V. Chaudhary, R. S. Bhatia, and A. K. Ahlawat, "A novel self-organizing map (SOM) learning algorithm with nearest and farthest neurons," *Alex. Eng. J.*, vol. 53, no. 4, pp. 827–831, 2014.

186. R. Kamimura, "Information-theoretic approach to interpret internal representations of self-organizing maps," in Josphat Igadwa Mwasiagi Self Organizing Maps – Applications and Novel Algorithm Design, bod–books, pp. 3–31, 2011.

187. M. Khanzadeh, P. Rao, R. Jafari-Marandi, B. K. Smith, M. A. Tschopp, and L. Bian, "Quantifying geometric accuracy with unsupervised machine learning: Using self-organizing map on fused filament fabrication additive manufacturing parts," *J. Manuf. Sci. Eng. Trans. ASME*, vol. 140, no. 3, pp. 031011, 2018.

188. H. Wu, Z. Yu, and Y. Wang, "Experimental study of the process failure diagnosis in additive manufacturing based on acoustic emission," *Meas. J. Int. Meas. Confed.*, vol. 136, pp. 445–453, 2019.

189. R. Jafari-Marandi, M. Khanzadeh, B. K. Smith, and L. Bian, "Self-organizing and error driven (SOED) artificial neural network for smarter classifications," *J. Comput. Des. Eng.*, vol. 4, no. 4, pp. 282–304, 2017.

190. L. I. Smith, "A tutorial on principal components analysis introduction," Statistics (Ber)., Technical Report, Department of Computer Science, University of Otago, 26pp., 2002.

191. J. Hopcroft and R. Kannan, "Computer science theory for the information age. Markov chains," in Introduction to queueing systems with telecommunication applications, Springer, pp. 93–177 2012.

192. Y. Shi, Y. Zhang, S. Baek, W. De Backer, and R. Harik, "Manufacturability analysis for additive manufacturing using a novel feature recognition technique," *Comput. Aided Des. Appl.*, vol. 15, no. 6, pp. 941–952, 2018.

193. P. Nath, Z. Hu, and S. Mahadevan, "Multi-level uncertainty quantification in additive manufacturing," in Solid Freeform Fabrication 2017: Proceedings of the 28th Annual International Solid Freeform Fabrication Symposium – An Additive Manufacturing Conference, SFF 2017, 2020.

194. R. Esmaeilizadeh et al., "Customizing mechanical properties of additively manufactured Hastelloy X parts by adjusting laser scanning speed," *J. Alloys Compd.*, vol. 812, p. 152097, 2020.

195. J. Reijonen, A. Revuelta, T. Riipinen, K. Ruusuvuori, and P. Puukko, "On the effect of shielding gas flow on porosity and melt pool geometry in laser powder bed fusion additive manufacturing," *Addit. Manuf.*, vol. 32, p. 101030, 2020.

196. B. Ferrar, L. Mullen, E. Jones, R. Stamp, and C. J. Sutcliffe, "Gas flow effects on selective laser melting (SLM) manufacturing performance," *J. Mater. Process. Technol.*, vol. 212, no. 2, pp. 355–364, 2012.

197. T. Fawcett, "An introduction to ROC analysis," *Pattern Recogn. Lett.*, vol. 27, no. 8, pp. 861–874, 2006.

198. D. Chicco and G. Jurman, "The advantages of the Matthews correlation coefficient (MCC) over F1 score and accuracy in binary classification evaluation," *BMC Genomics*, vol. 21, pp. 1–13, 2020.

199. K. Taherkhani, E. Sheydaeian, C. Eischer, M. Otto, E. Toyserkani, "Development of a defect-detection platform using photodiode signals collected from the meltpool of laser powder bed fusion," *Addit. Manuf.*, vol. 46, p. 102152, 2021.

12

Safety

Learning Objectives

After reading this chapter, you should learn about the following initial safety measures for AM:

- A basic understanding of AM process hazards.
- Insight on hazardous materials in AM.
- Associated safety matters in laser-based AM techniques.
- Associated safety matters in electron beam AM.
- Human health hazards in AM.
- Basic steps necessary for safety management.

12.1   Introduction

The application of additive manufacturing techniques can bring evolution in conventional manufacturing systems, with some additional precautions for industrialists and their workers. These precautions involve possible safety hazards connected with AM fabrication processes, equipment, and materials. The advancement of AM facilities can be complicated with current manufacturing capabilities with conventional safety management practices.

Metal Additive Manufacturing, First Edition. Ehsan Toyserkani, Dyuti Sarker, Osezua Obehi Ibhadode, Farzad Liravi, Paola Russo, and Katayoon Taherkhani.
© 2022 John Wiley & Sons Ltd. Published 2022 by John Wiley & Sons Ltd.

The Underwriters Laboratories (UL: a world leader in safety testing and certification) white papers and online materials explain safety matters in AM, emphasizing hazards connected with metal AM [1]. The UL's documents focuse on the origin of AM hazards and the appropriate approach for safety management or modification of existing practice in the overview of AM potentials.

Despite remarkable development in AM applications and acceptance, there is still a comparatively limited number of theoretical studies available on AM's possible propositions for employee security, safety, and health. Though several methods of AM present likenesses with current techniques, variations in resources, equipment, functions, and control institute can generate possible hazards that are adequately different as to declare revamped matter or are completely novel. These challenges may be similar to nanotechnology, where the combination of previous and recent techniques, innovative atmosphere, and characterization and evaluation of hazards are continuing trials.

12.2   Overview of Hazards

Though AM technology has a substantial impact on industrial advancement, its implication also concerns potential safety because of the associated AM hazards. The most common hazards related to metal AM can be generated from materials, AM processes, equipment, and other manufacturing facilities [14], as summarized in Figure 12.1.

12.3   AM Process Hazards

To focus on AM technology's possible hazards, it is necessary to develop a structure for hazard identification. This type of structure can be developed from the various AM processes. According to the international agreement, all AM processes can fall into the seven groups (see Chapter 1) as follows: material extrusion, powder bed fusion, vat photopolymerization, material jetting, binder jetting, sheet lamination, and directed energy deposition [2]. The individual technique is characterized through feedstock materials and their condition (phase or form, i.e. solid, liquid, or powder), techniques (mechanical forces or energies applied to join materials), and structural design of machines. A complete understanding of these features and other factors related to preprocessing, post-processing, manufacturing conditions, and application may lead to real hazard evaluations. A summary of the AM processes being used for metal printing and their possible hazards is presented in Table 12.1.

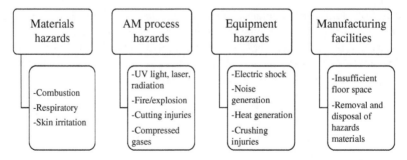

Figure 12.1   Overview of hazards in additive manufacturing.

Table 12.1 Related potential hazards by metal AM process category [3].

Category	Feedstock materials	Feedstock form	Binding/fusing	Most prominent potential hazards
Powder bed fusion	Metal, ceramic, or plastic	Powder	High-powered laser or electron beam heating	Inhalation/dermal exposure to powder, fume; explosion; laser/electron-beam radiation exposure
Binder jetting	Metal, ceramic, plastic, or sand	Powder	Adhesive	Inhalation/dermal exposure to powder; explosion; inhalation of volatile organic compound (VOCs), dermal exposure to binders
Sheet lamination	Metal, ceramic, or plastic	Rolled film or sheet	Adhesive or ultrasonic welding	Inhalation of fumes, VOCs; shock, laser/radiation exposure
Directed energy deposition	Metal	Powder or wire	Laser/electron beam heating	Inhalation/dermal exposure to powder, fume; explosion; laser/electron-beam radiation exposure

Source: Reprinted by permission of the publisher Taylor & Francis Ltd.

In the powder bed process, various heating sources of laser or electron beam can fuse metal powders. Both techniques use a heated chamber along with an inert atmosphere for the laser process and a high vacuum for the electron beam process. Afterward, post-processing is continued with the removal of unfused powder as well as powder re-coating.

Unlike others, in directed energy deposition, the feedstock material as a wire or powder can be added continuously rather than stepwise. Therefore, potential hazards in AM techniques depend on both feedstock materials type and the manufacturing mechanism. Metals may have toxic or exposing properties, and finer powders can easily breathe and skin contacts. Moreover, from airborne powder, fire or explosion may occur. The three significant manufacturing strategies using laser, electron beam, and plasma arc generate burn hazards, creating distinctive hazards like vision loss, contact with ionizing radiation, electric shock, etc.

12.4 Laser Safety in Additive Manufacturing[1]

12.4.1 Laser Categorization

All lasers are classified by the manufacturer and labeled with the appropriate warning labels. Any modification of an existing laser or an unclassified laser must be classified by a laser safety officer prior to use. The following criteria are used to classify lasers:

1. Wavelength: if the laser is designed to emit multiple wavelengths, the classification is based on the most hazardous wavelength.

---

[1] This section is entirely reprinted from [14] by permission of the publisher Taylor & Francis Ltd.

2. For continuous wave or repetitively pulsed lasers, the average power output and limiting exposure time inherent in the design are considered.
3. For pulsed lasers, the total energy per pulse, pulse duration, pulse repetition frequency, and emergent beam radiant exposure are considered.

12.4.1.1   Class 1 Lasers

These are lasers that are not hazardous for continuous viewing or are designed to prevent human access to laser radiation. These consist of low-power lasers or higher-power embedded lasers (e.g. laser printers).

12.4.1.2   Class 2 Visible Lasers (400–700 nm)

These are the lasers emitting visible light that, because of normal human aversion responses, do not normally present a hazard but would if viewed directly for extended periods (i.e. many conventional light sources).

12.4.1.3   Class 3A Lasers

These are the lasers that normally would not cause injury to the eye if viewed momentarily but would present a hazard if viewed using collecting optics (fiber optics loupe or telescope).

12.4.1.4   Class 3B Lasers

These are the lasers that present an eye and skin hazard if viewed directly. This includes both intrabeam viewing and specular reflections. Class 3B lasers do not produce a hazardous diffuse reflection except when viewed at close proximity.

12.4.1.5   Class 4 Lasers

Lasers that present an eye hazard from direct and specular reflections. Besides, such lasers may be fire hazards and produce skin burns.

In this class, some gas lasers such as CO_2 are much safer. The wavelength of a CO_2 beam is 10.6 µm, a wavelength that is strongly absorbed by water. Since more than 70% of the human body consists of water, an unfocused or reflected CO_2 beam does not penetrate beyond the first few layers of skin. More importantly, this wavelength does not penetrate the eye if a scattered beam is observed. Processing can usually be directly observed with minimal precautions. Laser safety eyewear is normally sufficient to meet the guidelines.

Several solid-state lasers, such as Yttrium Aluminum Garnet (YAG), have a 1.06-µm wavelength – just below the visible deep red. This wavelength deeply penetrates the body. If you get hit by a YAG beam, there is more damage than just a surface burn. Worse, the 1.06-µm wavelength is focused on the eye just like "normal" light. A beam scattered from a process will be focused to a point in the eye, probably destroying the spot where it was focused. As such, all YAG workstations must be sealed off and light-tight during processing.

12.4.2 Laser Hazards

12.4.2.1 Eye Hazards

The potential for injury to the different structures of the eye depends on which structure absorbs the energy. Laser radiation may damage the cornea, lens, or retina depending on the wavelength, intensity of the radiation, and the absorption characteristics of different eye tissues.

Wavelengths between 400 and 1400 nm are transmitted through the curved cornea and lens and focused on the retina. Intrabeam viewing of a point source of light produces a very small spot on the retina resulting in a greatly increased power density and an increased chance of damage. A large source of light, such as a diffuse reflection of a laser beam, produces light called an extended source that enters the eye at a large angle. An extended source produces a relatively large image on the retina; energy is not concentrated on a small retina area as in a point source.

Details of Irradiation Effects on Eyes
Cornea absorbs all UV light, which produces photokeratitis (weld).

Ultraviolet – B+C (100–315 nm)
The surface of the flash by a photochemical process causes the denaturation of proteins in the cornea. This is a temporary condition because the corneal tissues regenerate very quickly.

Ultraviolet –A (315–400 nm)
The cornea, lens, and aqueous humor allow ultraviolet radiation of 315–400 nm wavelengths, of which the principal absorber is the lens. Photochemical processes denature proteins in the lens resulting in the formation of cataracts.

Visible Light and Infrared-A (400–1400 nm)
The cornea, lens, and vitreous fluid are transparent to electromagnetic radiation of these wavelengths. Damage to the retinal tissue occurs by absorption of light and its conversion to heat by the melanin granules in the pigmented epithelium or photochemical action to the photoreceptor. The focusing effects of the cornea and lens will increase the irradiance on the retina by up to 100 000 times. For visible light (400–700 nm), the aversion reflex, which takes 0.25 seconds, may reduce exposure causing the subject to turn away from a bright light source. However, this will not occur if the intensity of the laser is great enough to produce damage in less than 0.25 seconds or when the light of 700–1400 nm (near-infrared) is used since the human eye is insensitive to these wavelengths.

Infrared-B (1400–3000 nm) and Infrared-C (3000–10000 nm)
Corneal tissue will absorb light with a wavelength longer than 1400 nm. Damage to the cornea results from the absorption of energy by tears and tissue water causing a temperature rise and subsequent denaturation of protein in the corneal surface. Wavelengths from 1400 to 3000 nm penetrate deeper and may lead to the development of cataracts resulting from the heating of proteins in the lens. The critical temperature for damage is not much above normal body temperature (~ 37°C).

12.4.2.2 Laser Radiation Effects on Skin

Skin effects are generally considered of secondary importance except for high-power infrared lasers. However, with the increased use of lasers emitting in the ultraviolet spectral region, skin effects have assumed greater importance. Erythema (sunburn), skin cancer, and accelerated skin aging are produced by emissions in the 200–280 nm range. Increased pigmentation results from exposure to light with wavelengths of 280–400 nm. Photosensitization has resulted from the skin being exposed to light from 310 to 700 nm. Lasers emitting radiation in the visible and infrared regions produce effects that vary from mild reddening to blisters and charring. These conditions are usually repairable or reversible; however, depigmentation, ulceration, and scarring of the skin and damage to underlying organs may occur from extremely high-power lasers.

A summary of the interaction of optical radiation and various tissues is shown in Figure 12.2. The wavelengths are divided into bands, as defined by the International Commission on Illumination (CIE).

12.4.2.3 Collateral Radiation

Radiation other than that associated with the primary laser beam is called collateral radiation. For example, X-rays, UV, plasma, and radiofrequency emissions are collateral radiation.

Ionizing Radiation
X-rays could be produced from two main sources in the laser laboratories. One is high-voltage vacuum tubes of laser power supplies, such as rectifiers, thyratrons, and crowbars, and the other is electric-discharge lasers. Any power supplies that require more than 15 kilovolts (kV) may produce enough X-rays to cause a health hazard. Interaction between X-rays and human tissue may cause serious diseases such as leukemia or other cancers, or permanent genetic effects that may show up in future generations.

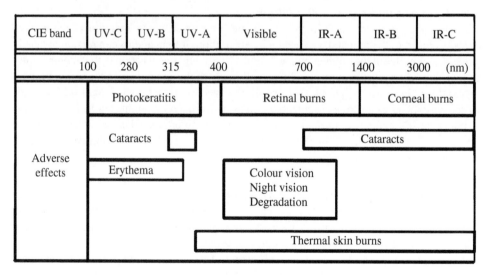

Figure 12.2 Interaction of optical radiation and various tissues [14]. Reprinted by permission of the publisher Taylor & Francis Ltd.

UV and Visible

UV and visible radiations may be generated by laser discharge tubes and pump lamps. The levels produced may exceed the maximum permissible exposure (MPE) and thus cause skin and eye damage.

Plasma Emissions

Interactions between very high-power laser beams and target materials may, in some instances, produce plasmas. The plasma generated may contain hazardous UV emissions.

12.4.2.4 Electrical Hazards

The most lethal hazard associated with lasers is the high-voltage electrical systems required to power lasers. Several deaths have occurred when commonly accepted safety practices were not followed by operators working with high-voltage sections of laser systems.

Safety Guidelines

1. Do not wear rings, watches, or other metallic apparel when working with electrical equipment.
2. Do not handle electrical equipment when hands or feet are wet or when standing on a wet floor.
3. When working with high voltages, regard all floors as conductive and grounded.
4. Be familiar with electrocution rescue procedures and emergency first aid.
5. Prior to working on electrical equipment, de-energize the power source. Lock and tag the disconnect switch.
6. Check that each capacitor is discharged, shorted, and grounded before working in the area of the capacitors.
7. Use shock-preventing shields, power supply enclosures, and shielded leads in all experimental or temporary high-voltage circuits.

12.4.2.5 Chemical Hazards

Many dyes used as lasing media of some types of lasers are toxic, carcinogenic, corrosive, or cause of a fire hazard. All chemicals must be accompanied by a material safety data sheet (MSDS). The MSDS will supply appropriate information about the toxicity, personal protective equipment (PPE), and storage of chemicals.

Various gases are exhausted by lasers and produced by targets. Proper ventilation is required to reduce the exposure levels of the products or exhausts below standard exposure limits.

Cryogenic fluids are used in the cooling systems of certain lasers. As these materials evaporate, they replace the oxygen in the air. Adequate ventilation must be ensured. Cryogenic fluids are potentially explosive when ice collects in valves or connectors that are not specifically designed for use with cryogenic hazards if the liquid oxygen comes in contact with any organic material. Although the quantities of liquid nitrogen that are used are small, protective clothing and face shields must be used to prevent freeze burns to the skin and eyes.

Compressed gases used in some types of lasers present serious health and safety hazards. Problems may arise when working with unsecured cylinders, cylinders of hazardous materials not maintained in ventilated enclosures, and gases of different categories (toxins, corrosives, flammable, and oxidizers) stored together.

12.4.2.6 Fire Hazards

Class 4 lasers represent a fire hazard. Depending on construction material, beam enclosures, barriers stop, and wiring is all potentially flammable if exposed to high-beam irradiance for more than a few seconds.

12.4.2.7 Explosion Hazards

High-pressure arc lamps, filament lamps, and capacitors may explode violently if they fail during operation. These components are to be enclosed in a housing that will withstand the maximum explosive force that may be produced. Laser targets and some optical components also may shatter if heat cannot be dissipated quickly enough. Consequently, care must be used to provide adequate mechanical shielding when exposing brittle materials to high-intensity lasers.

12.4.3 Eye Protection

The following is an account written by a researcher who sustained permanent eye damage viewing the reflected light of a Class 4 neodymium YAG laser emitting a 10-ns pulse of 6 mJ radiation at 1064 nm.

> When the beam struck my eye, I heard a distinct popping sound caused by a laser-induced explosion at the back of my eyeball. My vision was obscured almost immediately by streams of blood floating in the vitreous humor. It was like viewing the world through a round fishbowl full of glycerol into which a quart of blood and a handful of black pepper has been partially mixed. Dr. C.D. Decker.

The researcher had eye protection available but failed to wear it. Eye protection is required, and its use is enforced by the supervisor when engineering controls may fail to eliminate potential exposure above the applicable MPE. Laser radiation is generated both by systems producing discrete wavelengths and by tunable laser systems producing a variety of wavelengths. For this reason, it is impractical to select a single eye protection filter that will provide sufficient protection from all hazardous laser radiation. Therefore, it is important to choose eye protection specific to the wavelength and power of a particular laser.

12.4.4 Laser Protective Eyewear Requirements

1. Laser protective eyewear is to be available and worn by all personnel within the nominal hazard zone (NHZ) of Class 3B and Class 4 lasers where the exposures above the MPE can occur.

2. The laser protective eyewear's attenuation factor optical density at each laser wavelength shall be specified by a laser safety officer.
3. All laser protective eyewear shall be clearly labeled with the optical density and the wavelength for which protection is afforded.
4. Laser protective eyewear shall be inspected for damage before use. The use of beam attenuators to align visible lasers will reduce laser beam intensities to a level that will allow the operator to align the beam without PPE. Laser alignment cards for ultraviolet and infrared radiation allow operators to locate the beam during alignment procedures.

12.5   Electron Beam Safety

Generally, an electron beam machine comprises an electron source, an accelerator tube, and a scanning device. The electron beam supplier, also known as the "hot cathode," is a heated filament or lanthanum hexaboride (LaB6). Electrons driven by the highly vacuumed accelerating tube through the influence of high-voltage gradient show higher speed before flowing into the scanning device. The boosted electrons leave the scanner through a narrow window and go into the environment identified as beta radiation. Because of the beta radiation, the electron poses lower penetration. Once an electron beam hits a target like polymer tubing, the majority of the energy goes for ionizing atoms and results in cross-linking inside the polymer. However, if the electron is blocked by a dense medium like metal, energy may excite the metal atoms to discharge X-rays. These X-rays show characteristics of higher penetration than the original electron. Usually, X-ray ranges over a wider angle than the electron beam. Therefore, in the case of an electron beam machine, safety concerns are associated with the protection against X-rays.

The general safety matters are to keep the workforce safe from radiation and additional hazards. In the accelerator, the common hazards include:

- Radiation (X-ray)
- Ozone, noxious, and other toxic gas
- Incidental hazards (noise, etc.)
- Noise
- Fire

Ozone is developed when oxygen is oxidized with the influence of energized electron. After a while, this type of ozone will recombine and revert to oxygen, giving a short life period.

The reaction within the electron beam and the combustible material may generate fire. Therefore, any materials for irradiation need to be examined and confirmed to be nonflammable. A CO_2-based fire fighting equipment should be installed as a protective effort.

12.6   Powder Hazards

Generally, AM's metal powders are microscopic in size, which may be as small as 40 μm or less. Even under a protective environment, small powder size raises physical contact hazards through direct exposure to skins, inhalation, or digestion. Short-term direct exposure may lead to skin irritation or rashes, whereas long-term exposure to health effects through inhalation or digestion are mostly unknown.

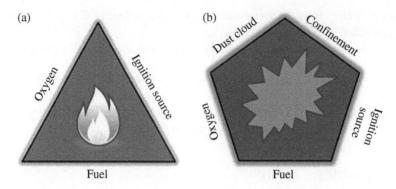

Figure 12.3 (a) Fire triangle and (b) explosion pentagon [3]. (Reprinted by permission of the publisher Taylor & Francis Ltd.

Moreover, a wide variety of AM materials are combustible and hence are explosive as dust. Also, the AM method produces dust and nano-sized particles. The dust hazard level is strongly related to the number of materials and its action in that quantity. Dust can cause fire or deflagration depending on the working environment, as described in Figure 12.3. Simple combustion and incipient fire can be fought with class D extinguishers. However, dispersion of powder in the air may result in violent ignition and afterward, massive explosion. Also, a small quantity of powder explosions usually displaces powder accumulations directing to massive blasts.

According to NFPA-484, flammable metals are all alkali metals like lithium, sodium, potassium, rubidium, cesium, and francium. Also, aluminum, magnesium, titanium, tantalum, niobium, and zirconium are identified as combustible metals. Moreover, alloys of these metals are also combustible. Powders from other common metals and alloys can also be combustible, e.g. zinc, babbitt, manganese, hafnium, and silicon.

12.6.1 Combustibility

According to NFPA 484-2012 A.3.3.6.1, particles with a minimum dimension of 500 μm or less may act as combustible when suspended in the air. The combustibility of powders can be estimated through analytical testing.

Therefore, the deflagration index (K_{st}) is calculated to show the harshness of deflagrations initiated by various combustible powders. Based on the value of K_{st}, materials are categorized to show the violence of deflagration, as presented in Table 12.2:

$$K_{st} = \left(dP/dt\right)_{max} \times \sqrt[3]{V_{vessel}}$$ (12.1)

where $\left(dP/dt\right)_{max}$ is the maximum rate of change in pressure through a deflagration. Hence, the deflagration index is the maximum rate of pressure rise during powder explosion in an equidimensional vessel, multiplied by the vessel volume's cube root.

Table 12.2 Different K_{st} values for different materials and their severity.

K_{st}(bar.m/s)	Class	Example
K_{st} <200	St. 1 (weak)	Sugar
200 < K_{st} < 300	St. 2 (moderate)	Wood flour
K_{st} > 300	St. 3 (strong)	Aluminum powder

12.7 Human Health Hazards

Because of the limited knowledge about the hazards of contact to AM basic resources, it is helpful to start the reports recording the impacts of the personal contact to identical hazardous materials manufacturing atmospheres (for example, mechanical refineries, welding stations, and so on). For example, some surveys conducted and issued by the US Department of Health and Human Services [4] present important modifications in performance test evaluating response time, eye and hand management, memory, and physical movement of aluminum foundry workers [5]. Though it is not clear yet, some data related to exposure are in use to prove the connection between aluminum powder and Alzheimer disease [6, 7].

In some research [8] about carcinogenesis, two metallic compounds are recognized as dangerous species for human health. Among them, cobalt is a material that can easily ionize and lessen oxygen, thereby producing hasty species. On the other hand, though tungsten carbide is an inert compound, it is a good electron carrier. Therefore, when these two compounds come closer, cobalt starts to transfer the electron on the surface of tungsten carbide, which reacts with oxygen. This situation creates a greater quantity of reactive oxygen species (ROS), responsible for oxidative and pro-inflammatory outcomes directing the immune system's deterioration. Cobalt is also known as neurotoxic [9] and creates cancer [10] as well as lung problems.

12.8 Comprehensive Steps to AM Safety Management

12.8.1 Engineering Controls

There already exist the proper and commonly accepted methods and controls for most of the possible hazards associated with AM. For example, the constraint of a particle as well as ultra-fine powder emission is now certified by applying local exhaust ventilation and High Efficiency Particulate Air (HEPA) filtration [11]. Also, consensus guideline on laser safety like ANSI Z136 is appropriate in AM sectors [12]. The proper use of current existing guidance, practices, and standards is contemplated in the perspective of primarily focusing on potential hazards.

But there must arise challenges in the area of unique and different types of hazards. Therefore, to focus on those matters, a holistic approach is necessary to treat hazards separately and by cycle. Further restrictions may come from the manufacturing process itself and associated economic matters. One example of rigorous economic concern is the unused and higher value materials, like superalloy powders.

Therefore, control practices maximizing recovery with limiting worker exposure is of greater importance to any management, unlike those that compromise recovery for protection.

Although some effective solutions are already existing in some techniques, further advancement in safety practices and tools is required.

12.8.2   Personal Protective Equipment

It is already known that the metal powder based on the size range can be inhaled during the AM process and cause health hazards like lung complications. Therefore, regulations to wear protective clothing for workers, specially designed for nanoparticle exposure, are considering in AM safety guidelines. For further verifications of PPE mask (fan breathing mask Sundstr"om SR 500, TH3, protection factor 250, with integrated P3 filter and prefilter), whether it is safe enough in the working environment of AM, LH (lighthouse for quantities of airborne ultrafine particles) measurements are important in and the outer surface of the mask. The SR 500 is a fan unit, which is designed to protect hazardous particles, vapors, and gases. It is supplied with two SR 510 P3 particle filters and a 221 prefilter, which makes it more protective against particles.

12.8.3   AM Guidelines and Standards

Control of hazards caused by combustible materials has been measured in the last decades. For example, the well-known ATEX directions and ISO standards (mainly ISO 60079) are applied in the European Union and some other countries. In the United States, National Fire Protection Association (NFPA) standards (mostly NFPA 652, 654 and particularly for metallic dust, NFPA 484) goal is to diminish combustible dust hazards.

NFPA standards are listed in Table 12.3. These standards and guidelines need methods to be used to catch and contain escapee dust produced through machining activities. The standards define regulatory obligations for dust collection techniques that direct the design and

Table 12.3   National Fire Protection Association (NFPA) standards [13].

Standard	Title
NFPA 484	Standard for combustible metals
NFPA 68	Standard on explosion protection by deflagrating venting
NFPA 69	Standard on explosion prevention systems
NFPA 654	Standard for the prevention of fires and dust explosions from the manufacturing, processing, and handling of combustible particulate solids
NFPA 61	Standard for the prevention of fires and dust explosions in agricultural and food processing facilities
NFPA 655	Standard for the prevention of sulfur fires and explosions
NFPA 664	Standard for the prevention of fires and explosions in wood processing and woodworking facilities
NFPA 70	The national electric code
NFPA 77	Recommended practice on static electricity
NFPA 2113	Standard on selection, care, use of maintenance of flame-resistant garments for protection of industrial personnel against short-duration thermal exposures from fire.

manufacture of the inhibition equipment or require explosion venting or explosion control structures. In regulatory obligations, performance-based styles can also be applied to control these hazards. Generally, these styles are inspected by Authorities Having Jurisdiction.

Though several standards are available for metal dust overall, some specific issues may occur in the area of AM. For example, demanding the electrical apparatus inside the printers to be categorized for usage in hazardous areas, explosion or deflagration venting, and restrictions on the managing and storing of materials, all of which can drastically affect the cost of mounting a 3D metal printer. Therefore, it is necessary to prepare a combustible dust fire and explosion managing program to comprise the following steps prior to the implementation of a 3D printer:

Creating a dust sampling and analyzing program – to detect the test materials exist in the method to verify the explosion and ignition characteristics required for the entire analysis.

Dust hazard analysis (DHA) – to explore the possible hazardous working environment and the problem associated with the design and provide recommendations for suitable process safety actions, with engineering controls, equipment, and processes.

Development of written safety procedures – to describe the methods and operating guidelines required for the safe function of the resource.

Basics of the combustible dust training program – to train the workers on-site.

Therefore, successful risk management demand a complete understanding of particular hazards related to powders or dust and others associated with individual capability based on its prospect and effective parameters. The general test applied through the preparation of the printing process includes explosibility and ignitibility of dust, electro statistic, and thermal stability. It is beneficial to develop a strategy to certify safety, including standard housekeeping, as well as dust management.

12.9   Summary

AM is expanding in many workplaces as businesses readily grab the prospects it presents in prototype making and actual production. On the other hand, AM experiences challenges in possible workplace health and safety because of the variation of processes, the growing demand for new materials and techniques, and specific characteristics of their application. Different classes of AM techniques can perform as a structure to help in hazard detection, with further dangers occurring from the workstation, workers, and associated machinery. For the separation of hazards, proper characterization statistics, contact measurement techniques, and practices are already present, while others may need improvement. However, there exist considerable information gaps which influence the capability to evaluate hazards and highlight resources. Also, there are very limited published field reports on AM emissions and worker exposure to hazards.

Moreover, very limited knowledge of how the normal AM consumer collects and applies prospective safety data is lower than the use of rules. Rules also require to be deliberated to verify their usefulness about AM hazards. The growth of nanotechnology professional safety and health area may become a standard for such advancement in AM. Although the particle sizes are much larger in most AM processes, the established safety guidelines for nanotechnology may help the AM stakeholders develop safety standards for AM more effectively. Like nanotechnology, the capability to mature general and essential knowledge will be vital to a method to track the running goal and become a state-of-the-art AM. As AM is the prominent control of a new manufacturing modernization, the prospect of professional safety and health matters is to make them safer for employees than before. Therefore, the progression of safe industrial modernization requires academic scholars, manufacturing engineers, and professional safety and health professionals to become educated about AM and highlight the improvement of awareness and skills, which will improve health and safety in this dynamic area.

References

1. P.D. Bates, "Safety Considerations for Additive Manufacturing and 3-D Printing," *UL*, 2020, 20pp. Available:https://www.ul.com/news/safety-considerations-additive-manufacturing-and-3-d-printing.
2. ISO/ASTM, "INTERNATIONAL STANDARD ISO / ASTM 52900 Additive manufacturing – General principles – Terminology," International Organization for Standardization, 2015.
3. G. A. Roth, C. L. Geraci, A. Stefaniak, V. Murashov, and J. Howard, "Potential soccupational hazards of additive manufacturing," *Journal of Occupational and Environmental Hygiene*, vol. 16, pp. 321–328, 2019.
4. Agency for Toxic Substances and Disease Registry ATSDR, A Toxicological Profile for Aluminum, 2008.
5. S. Polizzi, E. Pira, M. Ferrara, M. Bugiani, A. Papaleo, R. Albera, and S. Palmi, "Neurotoxic effects of aluminium among foundry workers and Alzheimer's disease," *NeuroToxicology*, vol. 23, pp. 761–774, 2002.
6. T. P. Flaten, "Aluminium as a risk factor in Alzheimer's disease, with emphasis on drinking water," *Brain Research Bulletin*, vol. 55, pp. 187–196, 2001.
7. D. G. Munoz, "Is exposure to aluminum a risk factor for the development of Alzheimer disease? – No," *Archives of Neurology*, vol. 55, pp. 737–739, 1998.
8. K. Czarnek, S. Terpilowska, and A. K. Siwicki, "Selected aspects of the action of cobalt ions in the human body," *Central European Journal of Immunology*, vol. 40, pp. 236–242, 2015.
9. S. Catalani, M. C. Rizzetti, A. Padovani, and P. Apostoli, "Neurotoxicity of cobalt," *Human and Experimental Toxicology*, vol. 31, pp. 421–437, 2012.
10. P. Wild, E. Bourgkard, and C. Paris, "Lung cancer and exposure to metals: The epidemiological evidence," *Methods in Molecular Biology*, vol. 472, pp. 139–167, 2009.
11. Human Services, "Protecting the Nanotechnology Workforce," 2016.
12. Laser Institute of America, American National Standard for Safe Use of Lasers, 2007.
13. National Fire Protection Association, "NFPA 921 Guide for Fire and Explosion Investigations," NFPA 921: Guide for Fire and Explosion Investigations, 2004.
14. E. Toyserkani, A. Khajepour, and S. Corbin, Laser cladding, 1st edition, CRC Press, 2004. https://doi.org/10.1201/9781420039177.

Index

Metal Additive Manufacturing, First Edition. Ehsan Toyserkani, Dyuti Sarker, Osezua Obehi Ibhadode, Farzad Liravi, Paola Russo, and Katayoon Taherkhani.
© 2022 John Wiley & Sons Ltd. Published 2022 by John Wiley & Sons Ltd.